Handbook of Scholarly Publications from the Air Force Institute of Technology (AFIT), Volume 1, 2000–2020

This handbook represents a collection of previously published technical journal articles of the highest caliber originating from the Air Force Institute of Technology (AFIT). The collection will help promote and affirm the leading-edge technical publications that have emanated from AFIT and are for the first time presented as a cohesive collection.

In its over 100 years of existence, AFIT has produced the best technical minds for national defense and has contributed to the advancement of science and technology through technology transfer throughout the nation. This handbook fills the need to share the outputs of AFIT that can guide further advancement of technical areas that include cutting-edge technologies such as blockchain, machine learning, additive manufacturing, 5G technology, navigational tools, advanced materials, energy efficiency, predictive maintenance, the internet of things, data analytics, systems of systems, modeling & simulation, aerospace product development, virtual reality, resource optimization, and operations management. There is a limitless vector to how AFIT's technical contributions can impact society.

Handbook of Scholarly Publications from the Air Force Institute of Technology (AFIT), Volume 1, 2000–2020, is a great reference for students, teachers, researchers, consultants, and practitioners in broad spheres of engineering, business, industry, academia, the military, and government.

Systems Innovation Book Series

Series Editor:
Adedeji B. Badiru

Systems Innovation refers to all aspects of developing and deploying new technology, methodology, techniques, and best practices in advancing industrial production and economic development. This entails such topics as product design and development, entrepreneurship, global trade, environmental consciousness, operations and logistics, introduction and management of technology, collaborative system design, and product commercialization. Industrial innovation suggests breaking away from the traditional approaches to industrial production. It encourages the marriage of systems science, management principles, and technology implementation. Particular focus will be the impact of modern technology on industrial development and industrialization approaches, particularly for developing economics. The series will also cover how emerging technologies and entrepreneurship are essential for economic development and society advancement.

Handbook of Scholarly Publications from the Air Force Institute of Technology (AFIT), Volume 1, 2000–2020

Edited by
Adedeji B. Badiru, Frank W. Ciarallo, and
Eric G. Mbonimpa

CRC Press
Taylor & Francis Group
Boca Raton London New York

CRC Press is an imprint of the
Taylor & Francis Group, an **informa** business

First edition published 2023
by CRC Press
6000 Broken Sound Parkway NW, Suite 300, Boca Raton, FL 33487-2742

and by CRC Press
4 Park Square, Milton Park, Abingdon, Oxon, OX14 4RN

CRC Press is an imprint of Taylor & Francis Group, LLC

ISBN: 978-1-032-11667-9 (hbk)
ISBN: 978-1-032-11668-6 (pbk)
ISBN: 978-1-003-22097-8 (ebk)

DOI: 10.1201/9781003220978

Typeset in Times
by codeMantra

Dedicated to the spirit of academic freedom and scholarship

Contents

SECTION 1   Cyber Security

SECTION 2   Data Analytics

SECTION 3   Space Technologies

SECTION 4   Human Factors

SECTION 5   Electromagnetics

SECTION 6   Materials Technologies
in Aerospace Applications

SECTION 7 Optical Technologies

Foreword

The Air Force Institute of Technology (AFIT) traces its roots to the earliest days of military aviation, with the 1919 establishment of the Air School of Application at McCook Field in Dayton, Ohio. AFIT's current mission is to "educate our Total Force military and civilian defense professionals to innovatively accomplish the deterrence and warfighting missions of the U.S. Air and Space Forces ... today and tomorrow." The technologies underpinning U.S. military capabilities are constantly evolving, and AFIT's education and research programs must anticipate future requirements while also meeting today's challenges.

This volume is a selected subset of AFIT's thousands of technical publications during 2005–2022 and provides readers a brief introduction to some of our technical interests. Consistent with our military education mission and the integration of our research with our graduate education programs, many of the articles provided here include both graduate student and military faculty authors. All research efforts have relevance to current or future requirements of the Department of the Air Force.

Research described in the Materials Technologies in Aerospace section expands understanding of new materials and their potential behavior in high-performance applications, such as in hypersonic vehicles or in turbine engines. Fundamental understanding of autonomous agent technologies as they are incorporated into high-performance military systems is also important to successful operations; the Human Factors section describes human-agent team studies designed to address that emerging need.

The Optical Technologies section explores a range of interdisciplinary projects, including on-chip multimode switch design for various applications, the computational and experimental demonstration of Fabry–Pérot cavity fabrication using a two-photon polymerization process, and more. These projects are a continuation of AFIT's long history of remote sensing and imaging research, as advanced by the Center for Directed Energy and the Center for Technical Intelligence Studies and Research.

AFIT also has a long history of contributions related to radar development, low observable technologies, and communications. The Electromagnetics section includes examples of current research related to these areas, designed to enhance understanding of the impacts of geomagnetic disturbances on 5G networks and to develop new approaches for optimizing antenna designs.

Cyber Security is of concern in all walks of life and has been a research focus area at AFIT and its Center for Cyberspace Research since 2000. The publications included here, addressing systems security engineering design principles and suspicion theory applied to operator responses during cyber attacks on a human-machine team system, are relevant to those interested in developing secure, effective systems capable of operating in real world environments.

The growing importance of successful operations in space is reflected in the articles presented in the Space Technologies section and the operations of AFIT's Center for Space Research and Assurance. Rapid development efforts will be enhanced using AFIT's CubeSat reference architecture described here. Other papers address a space debris propagation model and space system safety and security in a world with automated satellite control systems.

As a reflection of AFIT's role in educating personnel responsible for acquisitions, the Data Analytics section includes applications to cost estimation model development and multimodal data fusion approaches. These documents are just a small sample of the research at AFIT with applicability to the digital transformation currently underway throughout the defense enterprise.

While no single volume can provide comprehensive coverage of AFIT's research activities, this book nonetheless provides intriguing examples of what is being accomplished by AFIT's students and faculty to advance the nation's military capabilities. For readers inspired by these examples to seek more information about AFIT's research contributions than is presented here, the AFIT website (www.afit. edu) contains descriptions of AFIT's research centers, academic departments, and links to additional publications on AFIT Scholar via AFIT's library.

Heidi R. Ries, Ph.D.
Provost and Chief Academic Officer
Air Force Institute of Technology

Preface

HANDBOOK OF SCHOLARLY PUBLICATIONS FROM THE AIR FORCE INSTITUTE OF TECHNOLOGY (AFIT), VOLUME 1, 2000–2020

This handbook represents a rare collection of selected journal publications that have originated from the Air Force Institute of Technology (AFIT) over the past decades and have influenced science and technology developments that support national defense. The handbook presents a critical mass of intellectual publications with a theme that benefits all spheres of technology, business, industry, government, and the military. The collection is based on previously technically-refereed and published journal articles. This unique collection of articles will be a valuable resource for readers and will, hopefully, stimulate further research in appropriate technical areas. The cohesive compilation of selected papers by AFIT-affiliated students and faculty contains work that is seminal in the respective scientific fields. Hence, this handbook presents a coalition of technical expertise that readers can embrace as one-stop access to the high-level outputs of AFIT. The covered topics represent a balanced cross-section of the scholarly realm of STEM education and research at the Air Force Institute of Technology and include

- Cyber Security,
- Space Technologies,
- Data Analytics,
- Human factors,
- Electromagnetics,
- Materials Technologies in Aerospace, respectively
- Optical Technologies.

The vision of this handbook project was formed and communicated in early 2021 by the lead editor, Adedeji B. Badiru, Dean of AFIT, in a message to the AFIT community. In that message the idea of compiling some of the most influential work from AFIT faculty, students, and researchers was formed. Although this work has individually garnered attention, praise and technical following within the disparate national defense-focused and academic communities, there was no outlet to more broadly appreciate the collective scope and impact of AFIT-authored research. When the complete set of editors was assembled, we were daunted at the size of the effort before us. How would we compile such a volume of material and organize it into a coherent whole? After beginning the process of collecting material, this sense of overwhelming continued based on the amount of information gathered by the editors.

As the editors navigated the forest of renowned research from which to choose and made an effort to organize and categorize, the trees quickly emerged. From this process, many focused areas of research supporting national defense and a broader societal mission emerged. These coalesced into the sections of this volume: Cyber Security, Data Analytics, Space Technologies, Human Factors, Electromagnetics, Materials Technologies in Aerospace Applications, and Optical Technologies.

These represent a breadth of science, technology and engineering areas, while also representing the depth of academic inquiry that exists in the AFIT community.

This volume is just a sampling of this breadth and depth. The process of navigating these areas was a journey of discovery for the editors, reinforcing that initial vision. The long-term plan of the publisher, Taylor and Francis/CRC Press is to do additional volumes of the handbook. It is planned that this first volume will be followed by subsequent volumes, reporting additional selected journal publications in the respective fields of science, technology, engineering, and mathematics (STEM).

Complementary to this printed Handbook is the exciting report on the growing worldwide downloads of AFIT's intellectual products, including journal publications, theses, and dissertations, via the AFIT Scholar portal in the AFIT-AFRL library. The exponential growth in the electronic downloads demonstrates AFIT's intellectual outreach to science, technology, engineering, and mathematics (STEM) communities.

MATLAB® is a registered trademark of The Math Works, Inc. For product information, please contact:

The Math Works, Inc.
3 Apple Hill Drive
Natick, MA 01760-2098
Tel: 508-647-7000
Fax: 508-647-7001
E-mail: info@mathworks.com
Web: http://www.mathworks.com

Acknowledgments

We express our profound thanks and gratitude to our colleagues at the Air Force Institute of Technology for supporting and providing materials for this monumental literary work. Our special commendation and appreciation go to Mr. Andreas Mertens, our German Administrative and Professional Exchange Program (APEP) visitor, whose technical contributions to the manuscript got us farther along much quicker than we had anticipated. Also, deserving mention is Dr. Nils Wagenknecht, our German Engineering and Science Exchange Program (ESEP) visitor, whose scholarly mentoring put us on the proper track for making this handbook worthy of international acclaims.

Editors

Adedeji B. Badiru is the dean and senior academic officer for the Graduate School of Engineering and Management at the Air Force Institute of Technology (AFIT). He was previously professor and head of Systems Engineering and Management at AFIT, professor and department head of Industrial and Information Engineering at the University of Tennessee in Knoxville, and professor of Industrial Engineering and dean of the University College at the University of Oklahoma, Norman. He is a registered Professional Engineer (PE), a certified Project Management Professional (PMP), a fellow of the Institute of Industrial Engineers, and a fellow of the Nigerian Academy of Engineering. He holds a BS in Industrial Engineering, an MS in Mathematics, an MS in Industrial Engineering from Tennessee Technological University, and a PhD in Industrial Engineering from the University of Central Florida. His areas of interest include mathematical modeling, project modeling and analysis, economic analysis, systems engineering, and efficiency/productivity analysis and improvement. He is the author of over 35 books, 38 book chapters, 88 technical journal articles, and 220 conference proceedings and presentations. Often venerated by his colleagues as "a writing machine," Dr. Badiru has also published 35 magazine articles and 20 editorials and periodicals. He is a member of several professional associations and scholastic honor societies and a series editor for the Taylor & Francis Group book series on Systems Innovation and the Focus series on Analytics and Control. Professor Badiru was the recipient of the Taylor & Francis Group 2020 Author Lifetime Achievement Award.

Frank W. Ciarallo is an associate professor in the Department of Operational Sciences in the Graduate School of Engineering and Management at the Air Force Institute of Technology (AFIT). He was previously an associate professor at Wright State University in the College of Engineering and Computer Science, as well as an assistant professor in the College of Engineering at the University of Arizona. He has been a visiting professor at the Haas School of Business at the University of California, as well as a visiting researcher at the Eindhoven Technical University. He holds a BS in Electrical and Computer Engineering/Engineering and Public Policy from Carnegie Mellon University, an MS in Manufacturing and Operations Systems from Carnegie Mellon and a PhD in Industrial Administration from Carnegie Mellon. His research interests include predictive analytics for logistics systems, management of supply chain systems with uncertainty, simulation, biases in multi-attribute group decision-making and situation awareness/situation understanding. He has completed sponsored research for organizations such as the Air Force Materiel Command, Cardinal Health, the National Science Foundation, the Ohio Dept of Transportation and Trimble Navigation Limited. He is the author of 29 technical journal articles and 89 conference proceedings papers and presentations.

Eric G. Mbonimpa is an assistant professor in the Department of Systems Engineering and Management at Air Force Institute of Technology (AFIT). Prior to joining AFIT he held research scientist positions at U.S. Environmental Protection Agency and South Dakota State University. He holds a PhD in Civil Engineering from Purdue University, MS in Civil Engineering from University of Missouri-Columbia and BS in Civil Engineering and Environmental Technology from Kigali Institute of Science and Technology. He is a licensed Professional Engineer (PE) from Michigan. He conducts interdisciplinary research on energy systems, environmental systems, and sustainability. Dr. Mbonimpa has authored 20 journal articles, two book chapters, one patent, and 25 conference proceedings and presentations. At AFIT, he has supervised nine master's students, served on fifteen theses committees, and mentored three postdocs and two undergraduate students.

Section 1

Cyber Security

1 Evaluating Information Assurance Strategies*

J. Todd Hamill, Richard F. Deckro, and Jack M. Kloeber Jr
US Air Force Institute of Technology

CONTENTS

* Reprinted with permission: Hamill, J. T., Deckro, R. F., & Kloeber Jr, J. M. (2005). Evaluating information assurance strategies. *Decision Support Systems*, *39*(3), 463–484.

DOI: 10.1201/9781003220978-2

1.1   INTRODUCTION

The tremendous worldwide increase in reliance upon information technologies (IT) not only reaps huge benefits for their users but also threatens significant drawbacks. These technologies afford decision makers (DMs) with the capability to quickly fuse data from multiple sources, make informed decisions, and disseminate those decisions at nearly the speed of light. IT capabilities have become essential for day-to-day operations in today's global economy.

The Advanced Research Projects Agency Network (ARPANET) evolved into today's Internet. However, the "ARPANET protocols...were originally designed for openness and flexibility, not for security" [20]. The initial approach that permitted "unrestricted insiders" to easily share information is no longer appropriate for today's commercial and government use [20]. While the Internet now effectively spans the entire globe, organizations often deal with the subsequent vulnerabilities that develop on an after-the-fact basis, or worse, not at all.

Internet and internal threats employ widely available tools and easily obtainable technology to seek out and capitalize upon IT vulnerabilities. The President's Commission on Critical Infrastructure Protection (PCCIP) addressed these vulnerabilities on a national scale by identifying five sectors of industry that share common characteristics. In particular, the Commission highlighted the interconnectedness of these key sectors and their heavy reliance upon information technology. The five sectors include:

1. Information and Communications
2. Banking and Finance
3. Energy (Including Electrical Power, Oil and Gas)
4. Physical Distribution
5. Vital Human Services [24:2]

Information systems now monitor and control many of the operations of various other infrastructures. These systems are often an ad hoc mixture of components, processes, and software, which were not often designed to interoperate in a secure fashion. The resulting interdependencies and relatively easy access for a number of threats put all sectors at risk.

The increasing need for Information Assurance (IA) of government, commercial and individual information systems stems from the growing number of threats with their increasing capabilities to inflict damage upon information systems. Those techniques that provided an advantage in the past now pose a threat to not only our national infrastructure but also to the industry's current and future capabilities.

Figure 1.1 illustrates the increasing trend in incidents handled by the Computer Emergency Response Team (CERT). Noting the fact that these incidents are only

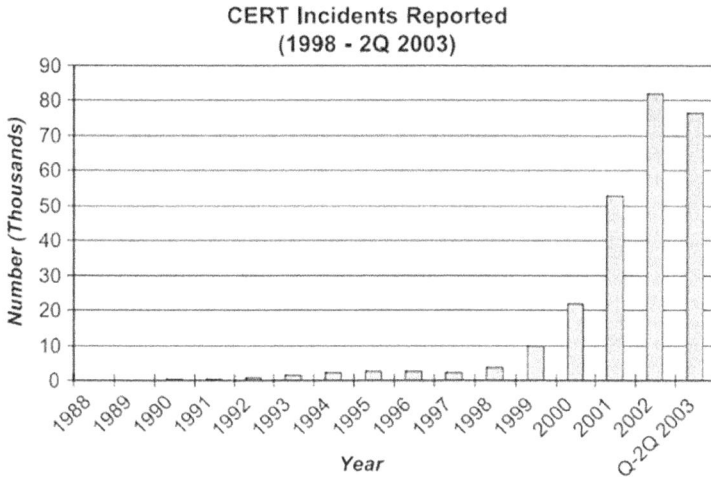

FIGURE 1.1   Security incidents [4].

those that were detected *and* reported implies that a much larger number of attacks may have actually occurred. Military exercises like "Eligible Receiver" have demonstrated, with relative ease, hackers' ability to "cripple U.S. military and civilian computer networks..." [12]. The Melissa virus and the "Love Bug" further reveal the potential threat to an organization's information infrastructure. The increasing trends in cyber assaults show that IA is, and will continue to be, a vital corporate strategy. As time and technology continue to advance, maintaining normal day-to-day operations and capability at any given moment will hinge on the continuous development, implementation and improvement of the level of IA.

To provide IA, the level of assurance attained must often be balanced with potential reductions in operational capability and the consumption of valuable resources (e.g. time, money, and people). This chapter develops a decision support tool to facilitate a three-dimensional, quantitative trade-off analysis between the level of IA gained by a collection of capabilities, the resulting effect on operational capability, and the resources required for their implementation.

1.2   BACKGROUND

There exists a large collection of documents dealing with what IA should be, methods on achieving assurance, and suggested strategies (e.g., 'defense-in-depth'). The principles of IA presented here draw heavily from DoD and government-related agencies' documents. These principles can, and have been, extended to the private sector; however, it should not be assumed that the DoD is the sole source of IA literature. A wealth of information regarding IA strategies and measures can be found through the National INFOSEC Education and Training Program webpage (http://www.nsa. gov/isso/programs/nietp/newspg1.htm). This page links to the web pages of over 20 academic institutions that have qualified as Centers of Academic Excellence for IA. Each of these pages is a trove of past and ongoing work in IA. While all of these sites are valuable, the Center for Education and Research in Information Assurance and Security, (CERIAS) at Purdue University (http://www.cerias. purdue.edu/) and the CERT http://www.cert.org/) and the Center of Academic Excellence in Information Assurance Education (http://www.heinz.cmu.edu/infosecurity/) at Carnegie Mellon University are excellent starting points. Although a wide array of academic, industrial, and government literature were reviewed for this study, we have restricted the literature reviewed in this chapter to those that were directly used in the development of the model for the sponsor, DARPA.

From the DoD's perspective, JP 3-13, entitled Joint Doctrine for Information Operations, discusses both offensive and defensive information operations (IO), stating both are equally important to ensure successful military operations. JP 3-13 offers the following, widely accepted, definition of IA.

> IA protects and defends information and information systems by ensuring their availability, integrity, identification and authentication, confidentiality, and non-repudiation. This includes providing for the restoration of information systems by incorporating protection, detection, and reaction capabilities. IA employs technologies and processes such as multilevel security, access controls, secure network servers, and intrusion detection software.

[6:III-1]

Restated, IA ensures that information and information systems are available to DMs when needed, that the information is as accurate and complete as possible, and that control over both the information and the information systems is maintained. In the event that control is lost, the capabilities to detect a loss of control, to regain control, and to restore the information systems to its original state must exist. These objectives are achieved by taking proactive measures (to protect) and allowing for detection and reaction capabilities (to defend) through the integration of secure technologies and best practices into the information system. Specific definitions of these IA requirements are shown in Table 1.1.

The objectives of IA include information environment protection, attack detection, capability restoration, and IO attack response [6:ix]. Table 1.2 describes the elements that comprise the information realm.

IA encompasses defensive IO in the context that systems are under continuous scrutiny by varying levels of threats. Although the concepts of defensive IO and

---

TABLE 1.1

Definitions of IA Objectives

Availability	Assured access by authorized users [6].
Integrity	Protection from unauthorized change [6].
Identification	Process an information system uses to recognize an entity [9].
Authentication	Verification of the originator; Security measure designed to establish the validity of a transmission, message, or originator, or a means of verifying an individual's authorization to receive specific categories of information [9].
Confidentiality	Protection from unauthorized disclosure [6].
Non-repudiation	Undeniable proof of participation [6].

TABLE 1.2

Elements of the Information Realm [6:I-9–11,10]

Information

Facts, data, or instructions in any medium or form. This includes the meaning that humans assign to
   data by means of known conventions used in their representation.

Information-Based Processes

Processes that collect, analyze, and disseminate information using any medium or form, that adds value
   to the decision-making process by performing designated functions or provide anticipated services.

Information System

The entire infrastructure, organization, personnel, and components that collect, process, store, transmit,
   display, disseminate, and act on information. The information system also includes information-based
   processes.

TABLE 1.3

IA Objective Definitions

Information and IS Protection: includes those measures taken to afford protection to information and
   IS, and ensure their availability, confidentiality, and integrity.

Detection: includes measures taken to provide detection of impending or ongoing attacks against an
   information system or the residing information.

Reaction: includes the measures taken to (i) appropriately respond to an identified attack and (ii)
   restore the information and IS capabilities to an acceptable state, their original state, or an improved
   state. (Modified from the definition of IA in Ref. [6]).

IA are similar, the definition of IA is used in this study to develop a hierarchy of
main objectives. These objectives, defined in Table 1.3, include Information and
Information System (IS) Protection, Detection, and Reaction capabilities.

Alberts clarifies areas of defensive operations within Information Warfare (IW),
an approach closely related to tackling the problem of IA. This approach formulates
the defensive IW problem as

> ...the possible environments that may be faced, one's options, and the objective that is
> being sought. This requires an identification of the variables that are relevant, that is,
> those that can significantly influence the outcome as well as the subset of these relevant
> variables that are controllable, which form the basis of designing options.
>
> *[1:19]*

Alberts noted five challenges to defensive IW capabilities that remain strong today.
These included:

- a better understanding of the nature of the threat must be achieved;
- a deterrent strategy against digital attacks must be developed;
- timely notification of indicators and warning regarding impending attacks;
- methods for successfully defending against attacks that do occur; and
- the development of "appropriate and effective responses to attacks" [1:59–62].

A number of models and methodologies have been developed to find a solution to this problem. Two models, one with a DoD focus, the other with a commercial enterprise focus, are described below.

The minimum essential information infrastructure (MEII) is defined as a process, rather than a structure [2]. A methodology to attain a feasible MEII is proposed, and the concept is described by the following four principles. The MEII...

- does not guarantee security but is instead a type of information system insurance policy by which risks are managed at some reasonable cost while pursuing information age opportunities;
- is not a central system responding to multiple threats but a set of systems defined locally to respond to local vulnerabilities;
- is not a fixed, protected entity, but a virtual functionality on top of the existing infrastructure; and
- is not a static structure, but a dynamic process—a means to protect something, instead of a thing that has to be protected. [2:xiv]

The focus is on military organizations, and it is assumed that as more organizations complete this process, an MEII will evolve, thus securing the defense information infrastructure (DII). This is in agreement with the 'weakest link' approach to security in general.

The process Anderson et al. define has six steps, as shown in Table 1.4.

The overall process is similar to other risk reduction or risk assessment processes. Categories of security techniques, shown in Table 1.5, may illustrate desirable attributes of an information system in the context of IA.

The Accreditor's Guideline, written by the National Computer Security Center (NCSC), provides guidance on the certification and accreditation process required for DoD information systems. Within this document, the risk management process is described. Risk is "something bad that might, or might not, actually come to pass" [19:136].

The notion of risk avoidance—"'the view that all risks to the information of an information system or network ought to be removed entirely before that system

---

TABLE 1.4
Six Steps of the MEII Process [2:xiv–xv]

1	Determine what information functions are essential to successful execution of the unit's missions.
2	Determine which information "systems" are essential to accomplish those functions.
3	For each essential system and its components, identify vulnerabilities to expected threats. In analyzing the system, it could (and perhaps should) be viewed in various ways: as a hierarchical set of subsystems supporting each other at different levels, or as a collection of functional elements like databases, software modules, hardware, etc.
4	Identify security techniques to mitigate vulnerabilities.
5	Implement the selected security techniques.
6	Play the solutions against a set of threat scenarios to see if the solutions are robust against likely threats. It is critical that the success of security enhancements be testable.

TABLE 1.5
MEII Security Technique Categories [2:xvii]

Heterogeneity	May be functional (multiple methods for accomplishing an end), anatomic (having a mix of component or platform types), and temporal (employing means to ensure future admixture or ongoing diversity).
Static resource allocation	The a priori assignment of resources preferentially, as a result of experience and/or perceived threats, with the goal of precluding damage.
Dynamic resource allocation	According to some assets or activities, greater importance as a threat develops; this technique calls for directed, real-time adaptation to adverse conditions.
Redundancy	Maintaining a depth of spare components or duplicated information to replace damaged or compromised assets.
Resilience and robustness	Sheer toughness; remaining serviceable while under attack, while defending, and/or when damaged.
Rapid recovery reconstitution	Quickly assessing and repairing damaged or degraded components, communications, and transportation routes.
Deception	Artifice aimed at inducing enemy behaviors that may be exploited.
Segmentation, decentralization, and quarantine	Distributing assets to facilitate independent defense and repair; containing damage locally and preventing propagation of the damaging vector.
Immunologic identification	Ability to discriminate between self and non-self; partial matching algorithms (flexible detection); memory and learning; continuous and ubiquitous function.
Self-organized and collective behavior	Valuable defensive properties emerge from a collection of autonomous agents interacting in a distributed fashion.
Personnel management	Personnel security clearances and training, design of human interfaces to reduce vulnerability of systems to human frailties.
Centralized management of information resources	(Self-explanatory)
Threat/warning response structure	Establishment of a hierarchy of increasing information attack threat levels and concomitant protective measures to be taken.

was allowed to operate"—was once supported by security professionals [22:3-1]. Eventually, it was recognized that some level of risk will always remain, and therefore, trade-offs between security and functionality must be made [22:3-2]. The current process of risk management approximates the current level of risk within a given system and relies upon rational decision-making to determine if it is at an acceptable level. An overall objective is to facilitate the cost-effective placement of countermeasures to mitigate the identified risks.

> The limited availability of resources, particularly money, is the most common problem in trying to establish IA when fiscal benefits are not readily available or obvious.
> Complex and expensive systems frequently involve lengthy approval cycles and prove to be more difficult in evaluating the benefits of such investments.
>
> *[21:2]*

Materna examined the evaluation processes of "next generation Information Technology investments..." and found that a variety of measures existed, but few ascertained the contributions that the investment made to the "business needs of the firm, however they are defined" [21:2].

Two types of benefits, "Hard" and "Soft," are discussed and differentiated. Hard benefits "refer to those benefits that can be readily quantified using standard measurement techniques," which includes dollars saved or generated, as well as time saved [21:3]. Soft benefits "refer to those benefits which are often less obvious or difficult to quantify such as worker empowerment, flexibility, or the multifarious aspects of competitive advantage" [21:3]. These are also referred to as financial and operational benefits, respectively.

Three general approaches to measuring these benefits of IT investments are discussed by Materna: Economic, Cost Reduction, and Strategic. Economic approaches include such analyses as Net Present Value, Internal Rate-of-Return, Return on Investment, and Breakeven/Payback. Unfortunately, these lend themselves to financially oriented assessments of stand-alone systems, but pose significant weaknesses when applied to interdependent systems that may involve intangible costs or benefits [21:4]. Without valid measures of IA benefits, classic financial analysis is difficult. What is "fire protection" worth to day-to-day operations when there is no apparent threat? What is that same protection worth when one is surrounded by arsonists?

The cost reduction approaches discussed include cost displacement/avoidance, work value analysis, and the cost of quality. Cost displacement (or cost avoidance) compares "the cost of the proposed system to the cost it will displace and avoid" [21:4]. Technical importance evaluates potential investments by their ability to support the achievement of long-term objectives. Although there may be no return on the investment, future operations may be impossible without it [21:7–8]. Clearly, the level of assurance must be balanced against the operational and fiscal costs of that assurance.

Considering the context of this research, employment of the latter, commercially oriented model did not suit the purposes of DoD and the ultimate interests of National Security. There are, however, valuable lessons learned with regard to measuring certain aspects of an information system's value, and the potential comparison of IA alternatives given subsequent changes in architecture. The former methodology lends itself to a potentially thorough evaluation of a given IS and its associated functions and consequential vulnerabilities. It is hypothesized, however, that in order to limit the resources spent in implementation, the MEII process will require a means to focus on the trades between IA and its resulting costs in the fiscal and operational sense.

1.3 VALUE-FOCUSED THINKING

1.3.1 INTRODUCTION

Operations research is intended to improve decision making; and values, indicating what one wants to achieve, are essential for guiding decision making [16:793]. Values are what we fundamentally care about in decision-making. Alternatives are simply means to achieve our values [16:793]. Keeney defines the typical approach used by most organizations as "alternative-focused thinking"—attacking the problem by evaluating the alternatives available and then choosing the best one

[17:4].

How to achieve an acceptable level of IA, with a minimum operational impact, at a reasonable cost is the decision opportunity addressed in this research. A focus on values aids in the evaluation of such complex decisions. Even in the alternative-focused approach, the effort of choosing an alternative involves evaluating each alternative based on the underlying values of the DMs. Because we believe a focus on values rather than alternatives is the more correct way to attack such a complex and important issue, we have based the methodology in this study on VFT.

1.3.2 OVERVIEW OF VALUE MODEL DEVELOPMENT

A value model is a hierarchical collection of a set of fundamental objectives applicable to the decision problem. These objectives are broken down until they can be measured, allowing the DM to quantitatively assess the degree to which the objectives are met. Several authors and analysts have proposed desirable properties of objectives contained within a fundamental objectives hierarchy including [3,17–19]. Keeney's desirable properties are extensive and include the following: essential, controllable, complete, measurable, operational, decomposable, non-redundant, concise, and understandable [17:82].

Although each of the authors named above submits different lists of desirable properties, they all include, in essence, the three properties we used for our study—complete, non-redundant, and operational.

The top tier of the fundamental objectives hierarchy includes only top-level objectives. The lower level tiers break each of the fundamental objectives down into sub-objectives. This process is called specification. Specification allows the organization to be more specific about what is meant by the fundamental objectives and ties the objective more closely to the achievement of the alternatives in the objectives' areas. Specification continues until the achievement of each objective is adequately measured through evaluation measures that are complete, non-redundant, and operational.

1.3.3 MEASURING THE ATTAINMENT OF OBJECTIVES

In order to assess how well an alternative does, or does not, meet a DM's objectives, an evaluation measure is developed for each lowest level objective. Each measure has characteristics that make it an appropriate and effective measure. The measure will be expressed in units that are either natural or constructed units and will be either a direct or proxy measure of the attainment of the objective being quantified. The most preferred evaluation measure is a natural-direct measure which is generally the least controversial and most transparent; whereas, a constructed-proxy measure is least preferred and must be more carefully and explicitly described to be useful in scoring the attainment of a particular objective. An in-depth discussion of types and characteristics of measures can be found in [19:24].

Once the measures are developed, the ranges of evaluation for each measure are declared and single dimension value functions are built to reflect the appropriate returns-to-scale.

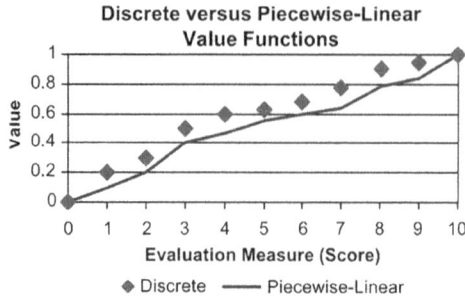

FIGURE 1.2   Discrete and piecewise-linear value functions.

1.3.4   BUILDING SINGLE DIMENSION VALUE FUNCTIONS

A single dimension value function is a monotonic (increasing or decreasing) function that captures the value a particular score represents to the DM. We will use the notation $v_i(x_i)$ to indicate the value function for the ith evaluation measure. A particular $v_i(x_i)$ may be discrete, piecewise-linear, or continuous, as shown in Figure 1.2.

The degree to which the DM prefers a higher score to a lower score level, termed a value increment, is elicited from the DM to build the value functions. However, the infinite number of scores on a continuous function may require an approximation of the functional form. A piecewise-linear function is one form of approximation we have used as well as the simple exponential function and the S-shaped sigmoid function. Methods and example interviews for constructing value functions can be found in Ref. [19].

Single dimension value functions, allowing consistent returns-to-scale of the real world units identified in the evaluation measures, are developed for all evaluation measures within the hierarchy. Having quantified the preferences of the DM within an evaluation measure, we must also elicit and quantify the preferences between evaluation measures and between objectives.

1.3.5   MULTI-OBJECTIVE SIMPLE ADDITIVE VALUE FUNCTION

A multi-objective value analysis requires a value model that "combines the multiple evaluation measures into a single measure of the overall value of each alternative" under consideration [19:53]. The overall function that combines the values resulting from the single dimension value functions will produce the overall value for each alternative—we will use the simple additive value function (see Eq. 1.1). A correctly structured hierarchy assumes the objectives in the hierarchy represent mutually exclusive and collectively exhaustive objectives. Mutual preferential independence is another requirement for using the additive value function. The objectives in our hierarchy were specifically defined in order to achieve mutual preferential independence.

$$v(x) = \sum_{i=1}^{n} w_i v_i(x_1)$$

$$(1.1)$$

where

- $\sum_i w_i = 1$ is the requirement for normalization;
- x_i is the score for the ith evaluation measure;
- η is the number of objectives (or the number of single dimension value functions);
- w_i is the *global* weight for the ith objective;
- $v_i(x_i)$ returns the *value* of the alternative with respect to the ith objective; and,
- $v(x)$ is the overall value of an alternative.

The additive function has been shown to be quite robust with respect to preferential independence, especially for large value hierarchies (number of evaluation measures >6). Bodily has shown the additive model to have no more than 4% error if the hierarchy has been built correctly. The additive model requires only single dimension values from each alternative and global weights. We used the swing weight method described in Ref. [11] for eliciting weights from the DM.

1.4 EVALUATION OF IA STRATEGY

IA, like all major decisions, is replete with trade-offs. The risk accepted by merely operating an information system within today's globally connected information infrastructure must be balanced with the needs of the organization to accomplish its intended mission and with the costs associated with the information technologies and practices that assure information systems and the information within them.

This attempt at modeling these trade-offs resulted in the construction of three distinct value models, denoted by IA, operational capability, and resource costs. All of these models are focused on a single decision context—Select the best IA strategy. For the purposes of this analysis, an IA strategy is defined as a collection of technical (hardware, software, and firmware) and non-technical (policies and procedures) means to achieve a desired or improved level of IA. An overview of these models will be presented.

The IA model was developed from a combination of 'top down' and 'bottom up' analysis of DoD doctrine and open literature. Various experts reviewed portions of this model during its development. Establishing the fundamental objectives was the purpose of the top down analysis. The IA model captures the benefits of IA.

The operational capability model was developed to provide a better understanding of the enhancements or limitations associated with implementing an IA strategy. In general, more secure environments are less capable than those with unconstrained, unmonitored access. Finally, virtually any IA strategy will incur costs in one or more aspects of funding, time, or personnel. The Resource Costs model was developed to facilitate such cost comparisons between alternatives.

A visual representation of how the three models would be integrated is shown in Figure 1.3. Once a set of IA strategies is evaluated with respect to each model, the DM must then ascertain the trade-offs between the three axes. This may be accomplished by weighting the results of the three models and comparing a single number (of overall value) from each strategy. Another approach could be to analyze

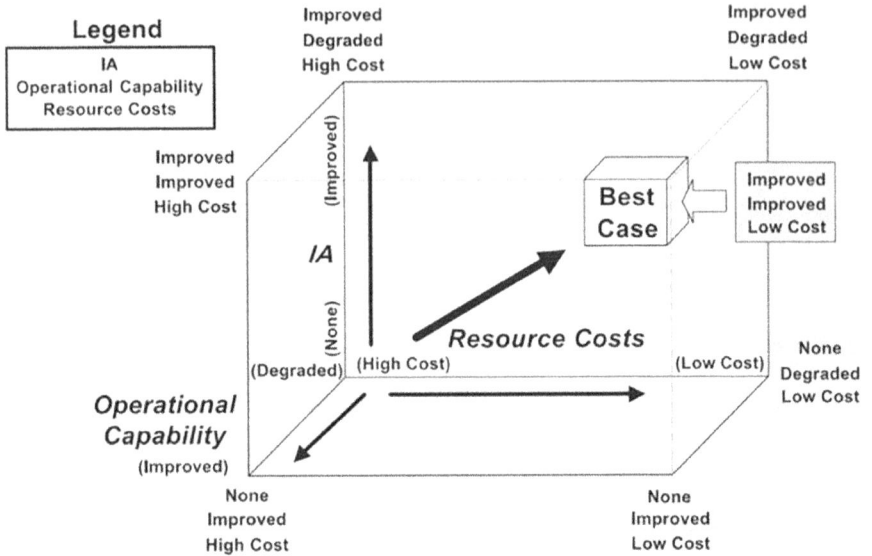

Legend

| IA |
| Operational Capability |
| Resource Costs |

Improved
Degraded
High Cost

Improved
Degraded
Low Cost

Improved
Improved
High Cost

(Improved)

IA

Best
Case

Improved
Improved
Low Cost

(None)

(Degraded) (High Cost)

Resource Costs

(Low Cost)

None
Degraded
Low Cost

**Operational
Capability**

(Improved)

None
Improved
High Cost

None
Improved
Low Cost

FIGURE 1.3   The IA balance [14].

the individual trade-offs between the hierarchies. The following sections describe the underlying rationale used for model development. The interested reader can find detailed discussions of VFT in Ref. [17]; a detailed explanation of the specifics of this value model's components and underpinnings are presented in Refs. [13,14].

1.5   MODELING IA

With the overall goal of achieving IA in mind, the fundamental objectives important to this goal must be identified. Revisiting the definition of IA, we have (Figure 1.4):

> Information Assurance "protects and defends information and information systems by ensuring their availability, integrity, identification and authentication, confidentiality, and non-repudiation. This includes providing for restoration of information systems by incorporating protection, detection, and reaction capabilities."

[6:III-1]

The stated fundamental objectives of IA are to 'protect and defend information and information systems.' Defense, however, implies that (i) forces must be aware of an impending or ongoing attack [detection], and (ii) forces have the capability to retaliate in some manner against the threat [reaction]. From this, the three main values (objectives) that support IA are derived: Information and Information System (IS) Protection, Detection, and Reaction capabilities. Each of these contributes value to the DM by taking part in assuring the intended information and information functions.

Whether a system is a stand-alone, isolated system, on a restricted access LAN, running a supervisory control and data acquisition (SCADA) system over telephone

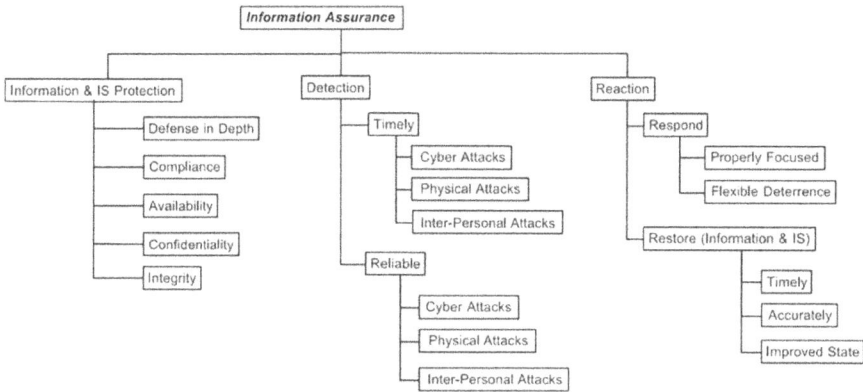

FIGURE 1.4 IA value hierarchy.

lines, or connected to the worldwide web, it will need protection and that protection will have security, operational, and financial impact. Among the top threats to each infrastructure listed in PCCIP is a cyber attack, often launched via the Internet. Just as a chain is only as strong as its weakest link, a system is only as well defend as the most vulnerable element in the system. "A risk accepted by one is a risk imposed on all." While the CERTs provide outstanding reactive support and are critical to maintaining security, it is necessary to have a level of IA that is appropriate for a system's tasks. The model presented here, coupled with the associated sensitivity analysis, provides a method to baseline a system, highlight potential system gaps and vulnerabilities, and investigate potential effects. It can be used alone to evaluate a system or in conjunction with other available techniques.

It may be argued that taking active measures to detect and react to attacks (thus mitigating their impact) also support the protection role. In order to clarify these values, and ensure mutual exclusivity in the value hierarchy, the definitions used in this analysis are those provided in Table 1.3. These specific definitions facilitate an independent assessment of each of the IA values, which are discussed further.

1.6 INFORMATION AND IS PROTECTION

The key elements from the definition (availability, confidentiality, and integrity) relate to the desired characteristics of information and information systems in order for them to support decision-making. Threats to IA, and to these key characteristics, may be defined as "any circumstance or event with the potential to harm an information system (IS) [or the information within] through unauthorized access, destruction, disclosure, modification of data, and/or denial of service" [23:45]. Note that the threats seek to adversely affect availability (through destruction and denial of service), confidentiality (through unauthorized access and disclosure), and integrity (through modification). The motivation, regardless of means, involves the reduction of the information and IS value to the DM. Therefore, measures protecting these characteristics provide value to the DM.

TABLE 1.6
IA Value Hierarchy

Identification: The process an information system uses to recognize an entity [9:13].

Authentication: A means of identifying individuals and verifying their eligibility to receive specific categories of information [7:46].

Non-repudiation: Assurance the sender of data is provided with proof of delivery and the recipient is provided with proof of the sender's identity, so neither can later deny having processed the data [23:32].

From the definition of IA, 'ensuring' identification and authentication and non-repudiation relates to the means that accomplish the protection of information and information systems. Therefore, the key elements 'identification and authentication' and 'non-repudiation' will be viewed as processes that support the confidentiality and respond objectives respectively. This is supported simply by the accepted definitions shown in Table 1.6.

One other value that may be incorporated into information and IS protection is Defense-in-Depth. Joint Publication 1-02 defines defense-in-depth as "the siting of mutually supporting defense positions designed to absorb and progressively weaken attack, prevent initial observations of the whole position by the enemy, and to allow the commander to maneuver his reserve" [7:125]. This area evaluates the cyber- and physical hardness of a system, either of which may contribute to protecting one or all of the values Availability, Confidentiality, and Integrity.

The final value contributing to Information and IS Protection objective may be termed as Compliance, which evaluates the DM's desire to minimize the potential exposure of an information system and its information system to known vulnerabilities. Learning from others' misfortunes is much better than experiencing a similar attack firsthand. Measures that permit the evaluation of this objective account for the efficiency (or lack thereof) by which known vulnerabilities, applicable to the system of interest, are reduced or eliminated altogether. Table 1A.1 in Appendix 1A lists the evaluation measures developed for the protection portion of the value hierarchy.

1.7   DETECTION

History has shown the value and need for reliable, adequate, and timely intelligence, and the harm that results from its inaccuracies and absence.

[6:III-5]

In light of the historical perspective of 'detecting' enemy actions, Joint doctrine also emphasizes, "timely attack detection and reporting are the keys to initiating capability restoration and attack response" [6:III-10]. In addition to timely detection, effective defense against IO is "...predicated on how well the intelligence processes function and on the agility of [those involved] to implement protective countermeasures" [6:III-2]. This suggests that a certain level of reliability is required to ensure that threats are indeed identified—maximizing the probability of detection and minimizing the probability of false alarms. Additionally, an effective IA strategy must also be robust in that it exhibits timeliness and reliability, regardless of the type of attack.

As stated earlier, regardless of the type of attack, the earlier an attack (or intrusion) is detected, the quicker an appropriate response can be initiated. Because of the speed at which cyber attacks may be accomplished, timeliness is a vital factor. In addition, due to the nature of available countermeasures, a distinction between internal (or "insider") attacks and external attacks must be made.

The timely detection of physical attacks is dependent upon the level of sophistication of the controls in place as well as the level of awareness of authorized personnel. More sophisticated controls rely less upon the human ability to detect an intrusion.

Social Engineering is defined as "a deception technique utilized by hackers to derive information or data about a particular system or operation" [8:F-17]. There are a number of methods to accomplish this, all of which focus on the lack of awareness or lack of training (or both) that authorized users possess. Timely detection in this context is assumed to rely upon the awareness of the users.

The reliability of intrusion detection systems (IDS) determines how often they fail to detect a valid intrusion, and how often an anomalous event is construed as an intrusion (false alarms). High false alarm rates can consume valuable resources, and could potentially be used to an adversary's advantage. However, failing to detect a valid intrusion is assumed the more serious of the two possibilities.

The detection reliability of physical attacks is assumed to be dependent upon a combination of the organization's physical controls and the level of user awareness. The scope of this model currently appraises those areas under the control of the organization—the information system of interest. However, the connectivity and interdependence of today's systems will eventually require addressing a larger scope, to include the infrastructure supporting the IS. User Training evaluates the effectiveness of training programs designed to provide authorized users with the knowledge to recognize (detect) a potential interpersonal attack. Table 1A.2 in Appendix 1A lists the evaluation measures developed for the detection portion of the value hierarchy.

1.8   REACTION

Joint doctrine addresses the importance of response and restoration capabilities [6:III-10]. In this analysis, respond and restore comprise the sub-objectives for reaction, since both are dependent upon either attack detection, attack warning, or some other, perhaps natural, event that has caused or has the potential to cause some level of disruption. The overall objective of an effective reaction capability is to provide the organization with a properly focused response mechanism and to restore the availability, confidentiality, and integrity of information and information systems to their original or an improved state.

1.8.1   RESPOND

The properly focused objective assesses the ability to correctly identify the individuals involved, the vulnerabilities exploited and the motivation for the attack in order to form the most appropriate response against the attacker (or attackers). This process may be accomplished externally or internally, measured by Indicators and Warning (I and W) Notification and ID Accuracy, respectively.

Once an attack is detected and those responsible have been identified, the organization must act to mitigate the risk posed to the organization. Flexible deterrence entails taking the appropriate action at the appropriate time. In this study, the appropriate action is either stopping the attack or collecting evidence to facilitate legal action or both. Due to scope and security considerations, more active defensive measures have not been captured. The appropriate time required to act upon threats depends upon the type of attack, the subsequent risks, and the capability of the organization.

1.8.2   RESTORE

The potentially damaging effects of today's attacks on information and information systems often require that an effective reaction capability also permit their restoration. The reliance upon these systems often requires that this process is accomplished in a timely manner, recovers the information as accurately as possible, and results in improvements to the systems, allowing their protection capability to evolve with the threat capability. Table 1A.3 in Appendix 1A lists the evaluation measures developed for the reaction portion of the value hierarchy.

1.9   CONSIDERATION OF OPERATIONAL CAPABILITIES AND IA

Increasingly complex information systems are being integrated into traditional warfighting disciplines such as mobility; logistics; and command, control, communications, computers, and intelligence (C4I). Many of these systems are designed and employed with inherent vulnerabilities that are, in many cases, the unavoidable consequences of enhanced functionality, interoperability, efficiency, and convenience to users [6:I-11].

Functionality, interoperability, efficiency, and convenience all add value to the operational capability of an information system. Just as vulnerabilities stem from trying to achieve these values, countermeasures to eliminate them often detract from the information systems' value. The operational capability hierarchy accounts for the changes that may result from IA strategy implementation. This hierarchy attempts to measure these effects, and assumes that the DM wants to minimize any adverse impact upon the existent system at a reasonable level of IA (Figure 1.5).

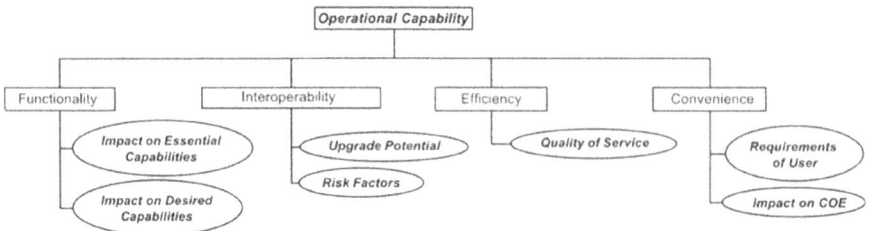

FIGURE 1.5   Value hierarchy for operational capability.

1.10   FUNCTIONALITY

Functionality is defined as the usefulness offered to system clients by providing infor-mation and information-related capabilities. Attributes that describe the value of the information system regarding functionality are desired and essential capabilities. Essential capabilities are those services that an organization currently relies heavily upon to accomplish their stated mission. If these services are no longer made avail-able, it is assumed that other means must be found to enable the organization to accom-plish mission objectives. Desired capabilities are defined as those capabilities that offer enhanced mission effectiveness, but are not required to perform their stated objectives. To ascertain the changes corresponding with an IA strategy, two constructed measures are developed: Impact on Essential Capabilities and Impact on Desired Capabilities. These assess any impact (good or bad) an IA Strategy may have upon services and information currently accessible to authorized users. This focuses only on those services (or supporting services) that are of value to the DM or the majority of authorized users.

1.11   INTEROPERABILITY

Systems that are interoperable and can be easily integrated with current and future systems provide immediate and cost-effective value to the DM. In the context of ascertaining the value of an IA strategy, Interoperability issues are measured with the two attributes Upgrade Potential and Risk Factors. These measures focus on the potential impact on future maintenance and/or the possibility for upgrades based upon the uniqueness of the components. Risk Factors evaluate the additional risk that may be associated with the implementation of certain types of countermeasures within a strategy. This risk applies to the likelihood of new vulnerabilities being introduced into the system, to include the possibility of incompatibility. It is assumed that the level of this risk is contingent upon the maturity of the technology, which serves as a constructed proxy for these types of risk.

1.12   EFFICIENCY

The efficiency of an information system is dependent upon many factors (e.g. band-width, throughput, processing capabilities, routing algorithms, and so forth). Currently, the quality of service (QoS) that an information system provides is predominantly system-specific, based upon the architecture and operating system employed, and is dependent upon the workload at any given time. Therefore, the degradation of QoS due to the addition of components (countermeasures) may not be perceived consis-tently throughout the IS, if at all. For these reasons, a categorical assessment of the impact that an IA Strategy may have upon information systems' QoS is offered.

1.13   CONVENIENCE

Convenience relates to the level of complexity involved in the human interfaces designed into the information system of interest. The trade-offs involved include buying more security at the expense of preventing users from employing the IS and its information in an operationally effective, or timely, manner. Attributes of a system that measure its convenience include the requirements a user must fulfill in order to gain authorized access, and the demands placed upon the user to employ and benefit from the IS once access is gained. These are captured by Requirements of User and Impact on Common Operating Environment, respectively.

Table 1A.4 in Appendix 1A lists the measures used to evaluate IA strategies with respect to operational capability considerations.

1.14   CONSIDERATION OF THE COST OF IA STRATEGIES

"Technology that affects an adversary's information and information systems and protects and defends friendly information and information systems will be pursued at every opportunity to ensure the greatest return on investment" [6:I-5]. This statement emphasizes the fact that, in an environment of shrinking budgets, costs associated with implementing an IA strategy must be considered. However, in addition to the acquisition costs, implementation costs must also be taken into account. Figure 1.6 illustrates the cost hierarchy addressed in this research.

For the purpose of this study, IA costs are grouped into two categories: Finite-Resource Consumption and Fiscal Resources. Finite-Resource Consumption accounts for the tangible, direct costs incurred in time and people that is required to implement an IA strategy. The Fiscal Resources account for the dollar costs associated with acquiring an IA strategy.

It is important to note that for the evaluation of costs, low-cost alternatives provide more value to the DM. Therefore, on a scale from 0 to 10, 0 is least preferred (high cost) and 10 is most preferred (low or no cost). This methodology focuses primarily on the total costs in time, people, and money required to procure and implement an IA strategy. Opportunity costs (in dollars), as well as any sunk costs of the legacy system, are not considered. Additionally, the salvage value of items being replaced is not directly addressed, but may be incorporated if the appropriate accounting procedures are available. However, the salvage value of IT items is often relatively low.

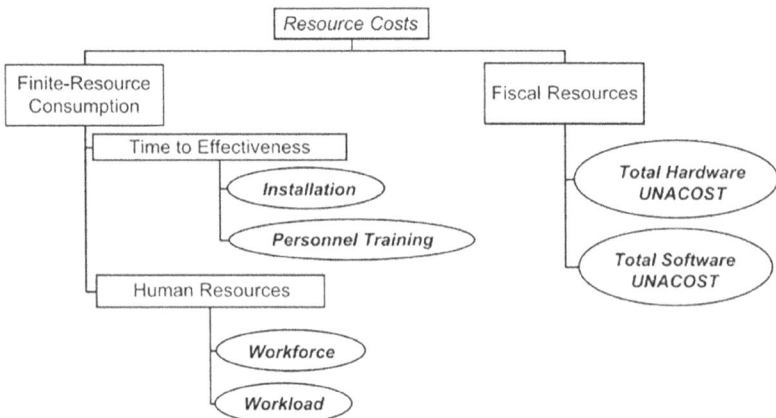

FIGURE 1.6   Resource cost hierarchy.

1.15 FINITE-RESOURCE CONSUMPTION

Finite-Resource consumption captures the amount of time and people required to implement an IA strategy.

1.15.1 TIME TO EFFECTIVENESS

The element of time is important due to the rapid evolution of technology, as well as the threats against it, suggesting that an effective IA strategy is one that can be implemented quickly. The time required in order for a particular countermeasure (CM) within an IA strategy to become effective is a function of two things—how long it takes to install the CM, and how long it takes the appropriate personnel to be trained in the CM. A CM that is easy to install and requires no training for it to be effective incurs less "cost" in time than a CM that is difficult and time-consuming to install and also requires significant training time before it becomes operationally effective. The longer a CM takes to implement, the longer the system remains vulnerable. It is assumed that the DM prefers to minimize the time that the organization's information and information system are exposed to vulnerabilities identified.

1.15.2 HUMAN RESOURCES

The personnel element is of importance due to the associated training, management, and overhead costs; however, the real concern is that of technological expertise. High training costs and turnover rates of personnel specializing in information technology and management may cause an organization to defer an IA strategy requiring more people [15]. The alternative to new workers is requiring overtime of existing personnel. Although this approach may potentially be cost-effective, it is not without consequences and therefore must be considered when evaluating IA strategies.

1.16 FISCAL RESOURCES

Recognizing the dollar costs of IA strategies is a key concern; it has been included in the hierarchy. These fiscal costs were broken down into two categories (hardware and software) to capture DM preferences for each type. Hardware Costs include the dollar costs associated with initial procurement, operations and maintenance (O and M), and supporting training dollar costs associated with hardware. Software costs are considered in an identical manner to hardware costs.

The assumptions for this evaluation consideration include:

- if salvage costs are known, they are included; otherwise, they are ignored;
- it is assumed that funds are available, and will be procured from the appropriate budget where
- applicable;
- any IA strategy under consideration is assumed to be within budgetary constraints throughout its life span; and
- an alternative that exceeds the organization's budget will not be considered.

To account for potentially varying life spans of the components within an IA strategy, the total discounted uniform annual costs (Unacost) are calculated. This provides a means to facilitate equitable comparisons between the long-term monetary impacts of IA strategies. If the DM does not require that the costs be broken down into hardware and software, then all Unacost values may be added together, while still considering only those strategies that are within the organization's budgetary constraints. Table 1A.5 in Appendix 1A lists the measures developed for the Resource Costs considerations.

1.17   ILLUSTRATIVE EXAMPLE

A notional, illustrative example is offered, demonstrating how the models discussed may be used to support the decision-making process.

Figure 1.7 shows the overall process required to implement this methodology. Using the triad of models, the organization must evaluate the levels of performance (for each model) based on current IA strategies that are already being implemented. This serves two purposes. The first establishes a 'baseline' of demonstrated performance, which can be compared to the estimated performance of potential alternatives. Second, this is, in itself, a means to find weaknesses within the current IA strategy of an organization—possibly highlighting other areas that the risk assessment may have missed. Areas that score poorly are potential candidates for improvement. New insight may be gained, offering the potential to find new and potentially better controls to construct better IA strategies.

1.17.1   ACHIEVING A BALANCED IA STRATEGY

As seen in Table 1.7. Notional results, three alternatives are offered: "Do Nothing" (which serves as a baseline), Strategy 1, and Strategy 2. The scales, in this example, are from 0 to 10, signifying the DM preference from least to most preferred, respectively. Although the values for alternatives in the illustrative example were randomly generated for proprietary and classification reasons, the underlying intentions were

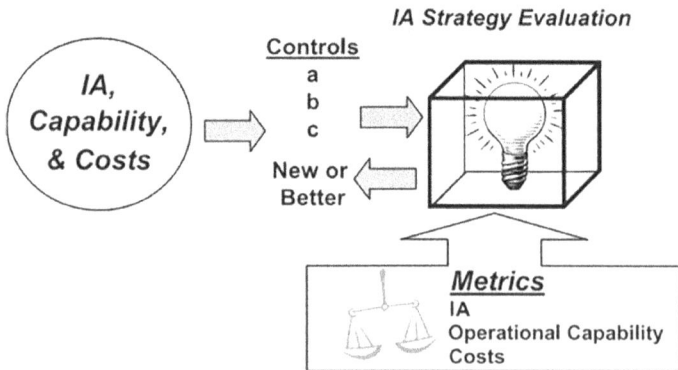

FIGURE 1.7   IA strategy evaluation process.

TABLE 1.7
Notional Results

Alternative	IA	Operational Capability	Resource Costs
"Do Nothing" (Baseline)	2.3	6.5	2.5
Strategy 1	3.2	7.2	2.6
Strategy 2	5.8	5.0	6.4

Information Assurance

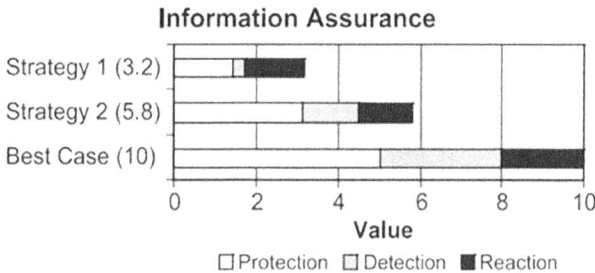

FIGURE 1.8   IA results.

to demonstrate potential differences between alternatives, and the considerations (or trade-offs) that must be made during comparisons. Application of this methodology would require (i) elicitation of weights from the DM and (ii) evaluation of each alternative within the value model. Specifics of weight elicitation techniques can be found in Refs. [18,19].

Through inspection of the table, both of the proposed strategies will yield some level of improvement in the organization's IA. Strategy 1, however, will also provide more operational capability at a slightly less resource cost (note that a score closer to 10 is preferred) than what the organization is currently incurring. Strategy 2 provides a much larger increase in IA compared to either Strategy 1 or the status quo and requires fewer resource costs. However, Strategy 2 will result in a decrease in operational capability, compared to what is currently enjoyed by the organization. It should be noted that each proposed strategy might actually be a 'basket' or portfolio of IA choices. The remainder of this section assumes that the status quo is an unacceptable option, for one reason or another.

To gain further insight from the evaluation process, the extent to which each strategy contributes to the DM preferences can be analyzed. For example, Figure 1.8 shows the portion of value from each strategy that is attributed to the three main objectives within the IA value model. The 'Best Case' at the bottom of the chart shows the weights that would be assigned to each of the objectives. In this example, these correspond to 0.5, 0.3, and 0.2 for Protection (Information and IS), Detection, and Reaction, respectively.

Again, a similar approach may be taken for the other dimensions of this problem. Figure 1.9 shows which objectives each strategy meets (or more importantly, falls short of) the DM's fundamental objectives within the operational capability value model.

Operational Capability

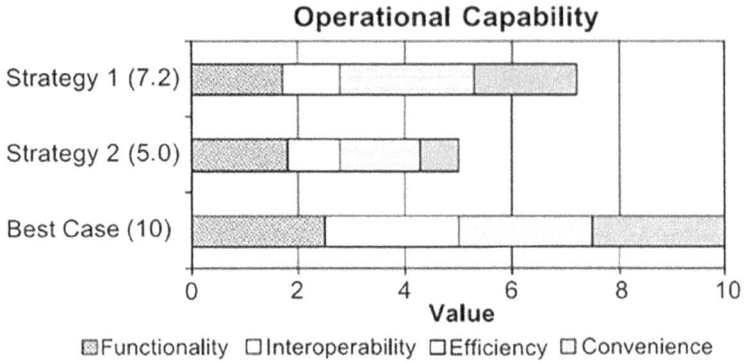

Functionality   Interoperability   Efficiency   Convenience

FIGURE 1.9 Operational capability results.

Resource Costs

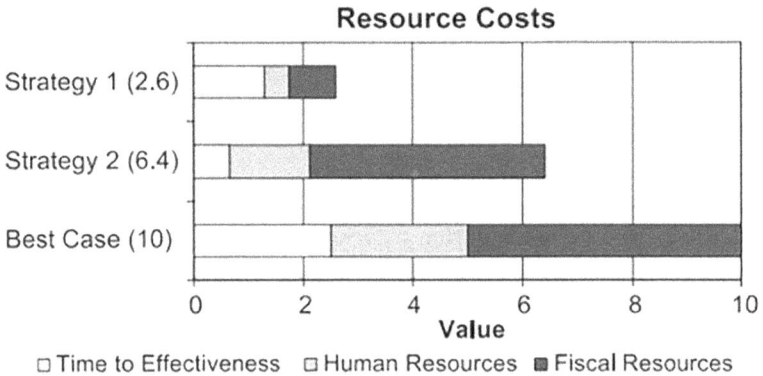

Time to Effectiveness   Human Resources   Fiscal Resources

FIGURE 1.10 Resource costs results.

Finally, the same type of graphical analysis is shown for the strategies with respect to the Resource Costs value model, shown in Figure 1.10. An added benefit to this type of modeling is the capability to assess the sensitivity of the results to the underlying assumptions, particularly the weights.

According to the 'Best Case' data in Figure 1.10, the current weights are 0.25, 0.25, and 0.5 for the Time to Effectiveness, Human Resources, and Fiscal Resources objectives, respectively. Suppose that the DM noted that Strategy 2 was much more cost-effective (Fiscal Resources) than Strategy 1; however, evaluation of Strategy 2 revealed that it would take much longer to implement (Time to Effectiveness). If uncertainty exists about the initial weight assigned to the Time to Effectiveness objective, analysis of the sensitivity of the model results to this weight is accomplished, and is shown in Figure 1.11.

At the point of the original weight for Time to Effectiveness (0.25), the values (and subsequently the rank order) of the strategies are shown. However, as the weight for Time to Effectiveness is extended beyond (approximately) 0.7, then the preference

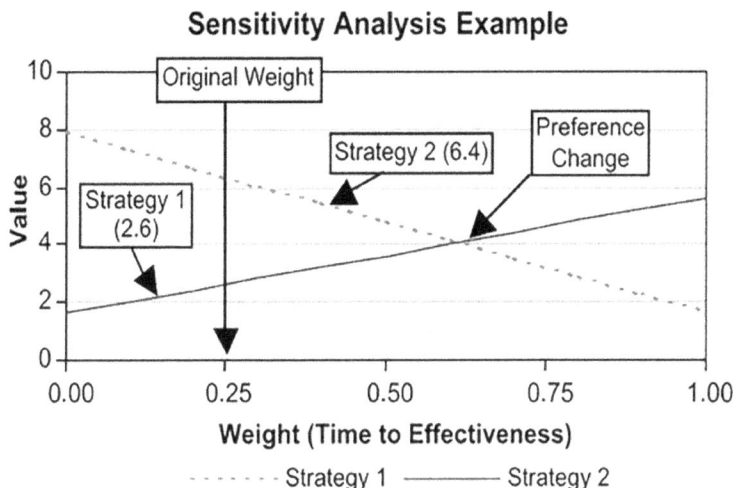

FIGURE 1.11   Sensitivity analysis results.

between the two strategies changes. Note that this only considers changing one weight (within one model) at a time, keeping the relative weighting of all other objectives constant.

Not only do these methods of analysis enable the DM to evaluate the trade-offs between and within IA strategies from a 'big picture' perspective, the results may be broken down to provide information regarding the specific areas where a strategy did (or did not) perform well and why. This further poses a potential for identifying new and improved ways to attain the organization's objectives in each of the three areas. Figure 1.12 shows a method to graphically compare the subsequent results for all three models simultaneously.

1.18   SUMMARY

The formulation of this analytical framework not only facilitates the evaluation of IA strategies, but the development of them as well. This is accomplished by focusing on what the DM values with respect to IA, operational capability, and the limited resources available. This focus quantifies the value added for each component within a strategy, providing a method to balancing the three in order to provide the most overall value to the DM.

The outputs of this effort lend themselves to the evaluation of an Information System with the triad of value models created to address IA, operational capability, and resource cost considerations.

To evaluate alternative IA strategies, measures were developed to assess the level to which these strategies meet (or do not meet) their objectives. This was accomplished mostly through bottom-up analysis, focusing on how current alternatives (countermeasures) differ and why.

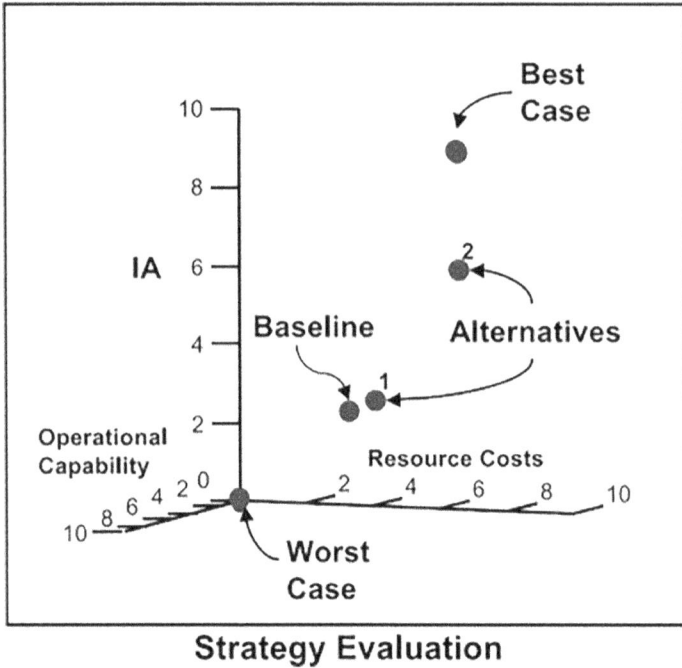

Strategy Evaluation

FIGURE 1.12 Notional comparison.

The culmination of these efforts resulted in a fully functional decision support tool (developed in Microsoft Excel©) that enables the DMs and system experts to implement the current value models. This tool is capable of accepting inputs for each evaluation measure and the weighting criteria required, as well as providing a summary of results. Minor modifications would allow sensitivity analyses and other presentation schema. Additionally, the process required to incorporate new evaluation measures is semi-automated through the use of Visual Basic© macros. All of these aspects facilitate the actual implementation of the described methodology.

1.18.1 RECOMMENDATIONS FOR FUTURE RESEARCH

This methodology assumes that the outcomes of each alternative, with respect to their appropriate evaluation measure scores, are deterministic. However, similar to the methodology developed in Ref. [10] pertaining to offensive IO evaluation, the expected scores, combined with the projected high and low scores, allow the multi-objective value model to give DM's insight into the relative uncertainty of each alternative [10].

The deterministic nature assumed for this modeling effort may not be entirely appropriate for all circumstances. For example, some controls suggested to mitigate risks are not "100%" solutions. The rapid evolution not only of technology, but of the threats against them, also lends the information technology environment to potential uncertainties. The eventual incorporation of utility may prove useful for future DMs implementing this model, enabling them to incorporate their own risk preferences in the evaluation of alternatives "in decisions where there is uncertainty about the specific consequence that will result from selecting a particular alternative" [19:245].

Another potential avenue for analytical efforts involves the application of mathematical programming techniques—linear or goal programming, in particular. This assumes that an optimal IA strategy is sought, requiring the maximization of the levels of IA and operational capability while adhering to any applicable resource constraints. One means of approaching this may be accomplished by identifying potential components (technical and non-technical) that comprise the IA strategy alternatives. Once these components are identified, the incremental changes that occur in each of the IA 'triad of models' may be determined, serving as the coefficients within the chosen mathematical model. Existing formulations of multi-dimensional knapsack problems with multiple-choice constraints and capital budgeting problems provide promising avenues in such an endeavor [5].

It is also important to note that the focus of this research was taken from an organizational perspective. The connectivity of today's organizations, and their reliance upon each other (particularly within the realm of our Nation's information infrastructures), will eventually require a broader scope. This may be accomplished by either: (1) evaluating systems of systems (an inter-organizational perspective), or (2) by building upon current policy, facilitating the creation of a common mental model of the elements that are important to everyone concerning IA.

Despite the level of the perspective, the values considered in the IA problem should remain constant. However, the level of perspective may change the underlying motivations (and therefore the shapes of the value functions, and perhaps the axis limits) concerning the benefits received from a strategy, compared to its impact on capabilities and its cost. A strategy that is regarded as least preferred from an individual organization's viewpoint might be the only acceptable alternative from a National perspective. Fortunately, this is where the strengths of VFT and the set of models will provide common ground to communicate and eliminate weaknesses in our Nation's IA posture. Nonetheless, this potential area of concern should be considered in future studies.

There is still much work to be done in this area; the need for IA will persist as long as information technologies are relied upon. The focus on DMs' values will lead to the development of alternatives that have a better chance of fulfilling their IA objectives. Through the proposed set of models, modeling IA provides a means to accomplish this goal, and a foundation to build upon—offering insight into the difficult and complex problem of IA.

ACKNOWLEDGMENTS

This work was supported by a grant from DARPA/IASET program. Additional support was provided by OSD/OT and E. We are also indebted to a vast number of individuals who shared their insights with us. We also wish to thank the editor and the reviewer for their insightful comments and suggestions.

Appendix 1A

TABLE 1A.1
Evaluation Measures Developed for Information and IS Protection

Title	Measure Unit	Measure Type[a]	Lower Bound	Upper Bound
Defense-in-Depth				
Time to penetrate essential elements	Ratio: (time required to attack)/ (time required to defend)	Ratio (S-curve)	0	4
Physical security	Probability of failure	Probability (Exponential)	0	1
Compliance				
Patches installed	Percentage of applicable patches installed	Percentage (Linear)	0	100
Latency-Implementation	Maximum age of known vulnerability	Months (Linear)	0	6
Latency-Assessment	Time since last vulnerability assessment	Years (Exponential)	0	3
Availability				
Essential Service Uptime	Percentage Availability of Essential Services	Percentage (S-Curve)	90	100
(Overall) System Uptime	Percentage Availability of the Overall System	Percentage (S-Curve)	75	100
Information Redundancy	Number of Data Sources	Quantity	1	4
Confidentiality				
Filter Technology	Filter Type	Category	Packet	Hybrid
Authentication Strength	Identification and Authentication (I&A) Method	Category	None	Combination
Supporting Policy	I and A Support	Category	No-Policy	Policy-Automated
Encryption Strength	Encryption Generation Used	Category	None	State of the Art
Integrity				
Data Integrity	Implementation of Anti-Malicious Code	Category	None	Automated-Full
System Integrity	Percentage of Validated Components	Percentage (Exponential)	0	100

[a] (Shape) of value function, if applicable.

TABLE 1A.2
Evaluation Measures Developed for Detection

Title	Measure Unit	Measure Type	Lower Bound	Upper Bound
Timely				
Internal Cyber Attacks	Detection Capability	Category	None	Real-Time (Off Duty)
External Cyber Attacks	Detection Capability	Category	None	Real-Time (Off Duty)
Physical Attacks	Time to Physical Intrusion Detection	Hours (Exponential)	0	72
Interpersonal Attacks	User Awareness	Percentage (Exponential)	0	100
Reliable				
Internal Cyber Attacks	Time Between Configuration	Days (S-Curve)	0	30
External Cyber Attacks	Time Between Configuration	Days (S-Curve)	0	30
Physical Attacks	Control Sophistication	Category	Presence	Automated
Interpersonal Attacks	Training Effectiveness	Category	Not Addressed	Trained and Evaluated

TABLE 1A.3
Evaluation Measures Developed for Reaction

Title	Measure Unit	Measure Type	Lower Bound	Upper Bound
Respond (Properly Focused)				
Indicator and Warning Sources	Number of Sources of Warning	Quantity	0	5
Identification Accuracy	Granularity of Non-repudiation	Category	None	Comprehensive
Respond (Flexible Deterrence)				
Timely Initiation of Deterrent Options	Decision Level Required	Category	Automatic	Higher Level
Stop Attack	Process to Stop Attack	Category	No Capability	Automatic
Collect Evidence	Capability to Collect Evidence	Category	No Capability	System-Benign
Restore Information and IS (Timely)				
Time to Restore Essential Elements	Time Required	Time (Linear)	0	Maximum acceptable time specified by organization

(Continued)

TABLE 1A.3 (*Continued*)
Evaluation Measures Developed for Reaction

Title	Measure Unit	Measure Type	Lower Bound	Upper Bound
Time to Restore to Fully Operational Capable Level	Time Required	Time (Linear)	0	Maximum acceptable time specified by organization
Restore Information and IS (Accurately)				
Restoration Accuracy Recoverable	Percentage of Information	Percentage (S-Curve)	0	100
Restore Information and IS (Improved State)				
Resource Inventory	Percentage of Components Inventoried	Percentage (S-Curve)	0	100
Improved State	Are procedures in place?	Yes/No	–	–

TABLE 1A.4
Measures Developed for Operational Capability Model

Title	Measure Unit	Measure Type	Lower Bound	Upper Bound
Functionality				
Impact on Essential Capabilities	Net Change in Essential Services	Quantity (Linear)	3	3
Impact on Desired Capabilities	Net Change in Desired Services	Quantity (Linear)	3	3
Interoperability				
Upgrade Potential	Component Source	Category	One-of-a-kind	COTS
Risk Factors	Technology Type	Category	Never been used	Previously used on a similar system with similar configuration
Efficiency				
Quality of Service	Impact on Network Performance	Category	Unacceptable Performance	Improved Performance
Convenience				
Requirements of User	Time to Access System	Time (S-Curve)	0 (seconds)	Maximum acceptable time designated by organization
Impact on Common Operating Environment	Impact based upon previous system	Category	Negative Impact	Positive Impact

TABLE 1A.5
Measures Developed for Resource Costs Model

Title	Measure Unit	Measure Type	Lower Bound	Upper Bound
Finite-Resource Consumption (Time to Effectiveness)				
Installation Time	Days required to install all components within the strategy	Days (Linear)	0	365
Personnel Training Time	Days required to complete required training associated with strategy	Days (Linear)	0	365
Finite-Resource Consumption (Human Resources)				
Workforce	Percentage change in workforce required	Percentage (Linear)	0	100
Workload	Overtime hours (per week, per person)	Hours (Exponential)	0	20
Fiscal Resources				
Total Hardware UNACOST	Uniform Annual Cost	Dollars (Linear)	0	Determined by applicable budget constraints
Total Software UNACOST	Uniform Annual Cost	Dollars (Linear)	0	Determined by applicable budget constraints

REFERENCES

1. D.S. Alberts, Defensive Information Warfare, National Defense University, Institute for National Strategic Studies, The Center for Advanced Concepts and Technology, Washington, 1996 (August).
2. R.H. Anderson, P.M. Feldman, S. Gerwehr, B. Houghton, R. Mesic, J.D. Pinder, J. Rothenberg, J. Chiesa, *Securing the U.S. Defense Information Infrastructure: A Proposed Approach*, RAND, Santa Monica, 1999.
3. R.T. Clemen, *Making Hard Decisions*, Brooks/Cole Publishing, Pacific Grove, CA, 1996.
4. Computer Emergency Response Team/Coordination Center (CERT/CC). CERT/CC Statistics: 1998–2001, Carnegie Mellon Software Engineering Institute (Excerpt from published report, http://www.cert.org/stats/cert_stats.html#incidents), Pittsburgh, 2003 (14 Oct).
5. C.C. Davis, R.F. Deckro, J.A. Jackson, A Methodology for Evaluating and Enhancing C4 Networks, *Military Operations Research* 4 (2) (1999) 45–60.
6. Department of Defense, Joint Chiefs of Staff, Joint Publication 3–13, Joint doctrine for Information Operations, Pentagon, Washington, 1998 (9 Oct).
7. Department of Defense, Joint Chiefs of Staff, Joint Publication 1–02, Department of Defense Dictionary of Military and Associated Terms, Pentagon, Washington, 1999 (amended through 29 June).

8. Department of Defense, Joint Chiefs of Staff, Information Assurance: Legal, Regulatory, Policy and Organizational Legal, Regulatory, Policy and Organizational Considerations, (Fourth Edition), Pentagon, Washington, 1999 (August 1999).

9. Department of the Air Force, Identification and Authentication, AFMAN 33–223, HQ USAF, Washington, 1998 (1 June).

10. M.P. Doyle, R.F. Deckro, J. Kloeber, J.A. Jackson, Measures of merit for offensive information operations courses of action, *Military Operations Research* 5 (2) (2000) 5–18.

11. W. Edwards, F.H. Barron, SMARTS and SMARTER: Improved Simple Methods for Multiattribute Utility Measurement, *Organizational Behavior and Human Decision Processes* 60 (1994) 306–325.

12. B. Gertz, *Computer Hackers Could Disable Military*, The Washington Times, (1998 April 16) A1.

13. J.T. Hamill, Modeling Information Assurance: A Value Focused Thinking Approach, Masters Thesis, AFIT/GOR/ENS/ 00M-15, Air Force Institute of Technology, Wright-Patterson AFB, OH, 2000 (March).

14. J.T. Hamill, R.F. Deckro, J. Kloeber, "A Strategy for Information Assurance," Technical Report number 2000–02, Center for Modeling, Simulation, and Analysis, Air Force Institute of Technology, Wright-Patterson AFB, OH, 2000 (September).

15. Information Operations Symposium: Key Technologies for Information Assurance, San Diego, CA, 1999 (26–28 October).

16. R.L. Keeney, *Using Values in Operations Research, Operations Research*, (1994 (September–October)) 793–813.

17. R.L. Keeney, *Value Focused Thinking: A Path to Creative Decision-making*, Harvard Univ. Press, Cambridge, 1998.

18. R.L. Keeney, H. Raiffa, *Decisions with Multiple Objectives: Preferences and Value Tradeoffs*, Cambridge Univ. Press, Cambridge, 1976.

19. C.W. Kirkwood, *Strategic Decision Making: Multiobjective Decision Analysis with Spreadsheets*, Duxbury Press, Belmont, 1997.

20. T.A. Longstaff, J.T. Ellis, S.V. Hernan, H.F. Lipson, R.D. McMillan, L.H. Pesante, D. Simmel, Security of the Internet, Carnegie Mellon University, Software Engineering Institute (Excerpt from published article, http://www.cert.org/encyc_ article/tocencyc. html) Pittsburgh, 1997.

21. R.D. Materna, *Assessing the Value of Information Technology, Strategic Consulting Group*, NCR, Dayton, 1992 (March).

22. National Computer Security Center (NCSC), Accreditor's Guideline, NCSC-TG-032, Version 1.6, 1997 (March).

23. National Security Telecommunications and Information Systems Security Committee (NSTISSI), National Information Systems Security (INFOSEC) Glossary (Revision 1), NSTISSI No. 4009, National Security Agency, Fort Meade, 1999 (January).

24. President's Commission on Critical Infrastructure Protection (PCCIP), Critical Foundations: Thinking Differently, GPO, Washington, 1997 (October).

2 Analysis of Systems Security Engineering Design Principles for the Development of Secure and Resilient Systems[*]

Paul M. Beach, Logan O. Mailloux,
Brent T. Langhals, and Robert F. Mills[†]
US Air Force Institute of Technology

CONTENTS

[*] Reprinted with permission: Beach, P. M., Mailloux, L. O., Langhals, B. T., & Mills, R. F. (2019). Analysis of systems security engineering design principles for the development of secure and resilient systems. *IEEE Access*, *7*, 101741–101757.

[†] Supported in part by the Air Force Research Laboratory Space Vehicles Directorate (AFRL/RV).

DOI: 10.1201/9781003220978-3

2.1   INTRODUCTION

Modern systems are increasingly complex compositions of system elements, subsystems, supporting & enabling systems, and extensive infrastructures that often result in a myriad of cyber dependencies, complicated interactions, and emergent behaviors. Moreover, because of their cyber dependencies (e.g., hardware, software, communications, etc.), these expansive Systems-of-Systems are inherently susceptible to a wide range of malicious and non-malicious events which can result in unexpected disruptions and unpredictable actions. Substantiating these realities, the United States Department of Defense (U.S. DoD) – arguably the world's largest acquirer of complex systems – has conceded that they cannot confidently assure their critical mission systems will operate as intended through a full-spectrum cyberattack by well-resourced nation-state actors [1]. Thus, it is necessary to undertake deliberate systems-level engineering efforts to design, develop, and field secure and resilient systems capable of operating in highly contested operational environments fraught with uncertainty, unpredictability, and attacks from intelligent adversaries as well as abuse and misuse by humans (e.g., owners, operators, maintainers, etc.) [2, 3]. Moreover, inherent in discussions of "securing" complex cyber-physical systems is often the issue of safety as a key stakeholder need [4]. For example, the safety of humans is paramount in the development of aircraft and more recently autonomous vehicles.

To address this critical systems security problem, the National Institute of Standards and Technology (NIST), the National Security Agency (NSA), MITRE, and several industry leaders from around the world collaborated on a five-year effort to produce a comprehensive Systems Security Engineering (SSE) approach: NIST Special Publication (SP) 800-160, *Systems Security Engineering* [5]. Subtitled "Considerations for a Multidisciplinary Approach in the Engineering of Trustworthy Secure Systems," NIST SP 800-160 is focused on institutionalizing engineering-driven actions to develop more defensible systems through a rigorous analysis approach in alignment with industry standards for systems development such as ISO/IEC/IEEE 15288:2015 [6]. In 2018, NIST SP 800-160 was designated as NIST SP 800-160 Volume 1 with the draft release of NIST SP 800-160 Volume 2 [7]; however, at the time of this writing, Vol. 2 is still in draft form and does not align with standardized systems engineering approaches (i.e., Vol. 1 is based on the widely accepted ISO/IEC/IEEE 15288 systems engineering standard [8] while Vol. 2 is based on the MITRE Cyber Resiliency Engineering Framework [9]). As a unified and comprehensive SSE approach, in this work, we've chosen to focus on the effective application of the systems security design principles as presented in NIST SP 800-160 Vol. 1.

In this work, we extend our previous work [10] by performing Design Structure Matrix (DSM) analysis of the security-oriented design principles presented in NIST SP 800-160 Vol. 1 and studying their mappings to systems security strategies. The contribution of this chapter is threefold:

1. Mapping the NIST SP 800-160 Vol 1. SSE design principles to the proposed resiliency security strategy
2. Studying the design principle interactions and clustering the principles for ease of implementation
3. Providing a case study which describes how the security and resiliency design principles can be applied to a given system

Section 2.2 of this chapter introduces the broader context for this work as a standardized SSE approach and identifies seminal SSE-related works. Section 2.3 details the DSM analysis methodology used to study the application of the subject security strategies and design principles. Section 2.4 of this chapter discusses the mapping of the 18 NIST SP 800-160 Vol. 1 design principles across the four SSE strategies in the design matrix, while Section 2.5 details the results of the DSM analysis into two clusters of secure systems design: Architecture and Trust. Additionally, in Section 2.6, a notional command and control System of Interest (SoI) is used to provide a concrete example of how these principles and their groupings can be used to field secure and resilient systems. Of note, we also offer commentary on "engineering considerations" for developing mission critical systems.

Finally, this work emphasizes the "design-for" purpose of the Vol. 1 security principles, provides security requirements traceability, and points toward evidences of trustworthiness (e.g., design artifacts, analyses, test results, etc.) which can aid system developers, owners, and operators in satisfying their security and resiliency needs by mapping conceptual strategies to security principles that can be implemented and tested.

2.2   A STANDARDIZED SYSTEMS SECURITY APPROACH

This chapter is part of an ongoing research activity to understand and promote the application of a standardized approach for developing secure systems. More specifically, the "systems security engineering" approach adopted in this work is considered a specialty area of systems engineering and consistent with ISO/IEC/IEEE 15288, NIST SP 800-160 Vol. 1, and the International Counsel on Systems Engineering (INCOSE) handbook [5]. For related publications which seek to promote a systems-oriented view of SSE please see [11–14], and [15]; excellent comprehensive works are available here: [16, 17]. Note, while few formally adopt the name "systems security engineering" for their work the discipline of SSE continues to rapidly mature, especially with the ever increasing, worldwide interest in fielding secure and resilient cyber-physical systems such as autonomous vehicles.

2.2.1   Defining Systems Security Engineering

While the NIST SP 800-160 Vol. 1 does not formally provide a definition for SSE, two helpful definitions are offered in Military Handbook 1785 [18]:

> System Security Engineering: An element of system engineering that applies scientific and engineering principles to identify security vulnerabilities and minimize or contain risks associated with these vulnerabilities. It uses mathematical, physical, and related scientific disciplines, and the principles and methods of engineering design and analysis to specify, predict, and evaluate the vulnerability of the system to security threats.
>
> System Security Engineering Management: An element of program management that ensures system security tasks are completed. These tasks: include developing security requirements and objectives; planning, organizing, identifying, and controlling the efforts that help achieve maximum security and survivability of the system during its life cycle; and interfacing with other program elements to make sure security functions are effectively integrated into the total system engineering effort.

With these two definitions in mind, the goal of SSE is to minimize system vulnerabilities to known and presumed security threats across the SoI's lifetime.

2.2.2   The NIST Sp 800-160 Vol. 1 Development
Strategies and Design Principles

Throughout the past several decades a number of security best practices, principles, and patterns have been proposed for system development. For example, the NSA specifies nine security "first principles" in their educational criteria [19].

In another example, dozens of security patterns are captured in [20]. In a third example, the U.S. DoD's System Survivability Key Performance Parameter suggests three pillars and ten attributes to achieve cybersecurity and survivability [21].

While none of these approaches are inherently deficient, the NIST SP 800-160 Vol. 1 uniquely captures the essence of these works and has been thoroughly refined in multiple rounds of public review [5]. Moreover, these strategies and principles are part of a compressive and standardized engineering approach. Thus, Vol. 1's strategies and principles are used to frame our analysis approach. The Vol. 1 systems security strategies design principles are described thoroughly in Section 2.4. For a detailed history of system-level security principles please see [22].

While the authors often reference works from the United States Government and Military, the topics discussed in this chapter are broadly applicable to any organization seeking to develop secure cyber-physical systems of interest.

2.3 METHODOLOGY FOR MAPPING SYSTEMS SECURITY STRATEGIES TO DESIGN PRINCIPLES

As is consistent with our previous work and Vol. 1, we aim to maintain a unified, holistic, and standardized systems-based approach for systems security and resiliency. This "top-down" approach is in contrast to a "bottom-up" compliance-based cybersecurity perspective [6]. For example, while a highly skilled cybersecurity professional might know how to best secure a piece of hardware, the systems security engineer should be asking "Why do we need to secure that piece of hardware?" – there is a very important difference which can be subtle at times. The subtle difference between the two perspectives is important especially as modern systems become increasingly autonomous with high levels of trustworthiness and safety. In a proper SSE approach, the goal is to first understand the security and resiliency requirements for the CPS of interest and then design a suitable system. Unfortunately, the necessity of defining the security problem within the SoI's operational context can be missed [23].

2.3.1 STANDARDIZED APPROACH FOR SYSTEMS SECURITY AND RESILIENCY

Shown in Figure 2.1, a simplified system development model (known as the "Vee model") is useful for illustrating the basic premise of SSE – an initial investment in "engineering-out" vulnerabilities and "designing-in" security and resiliency countermeasures is a long-term cost-saving measure [18, 24]. The Vee model depicted in Figure 2.1 is overlaid with four critical SSE stages: Determining what to secure (blue), Designing a secure system (gray), Demonstrating the system is secure (green), and Analysis to support system security decisions (yellow). Most importantly, this approach is consistent with systems engineering best practices where security and resiliency stakeholder needs are identified early in the development life cycle (i.e., the blue processes) and their solution trade space is explored at reduced cost (i.e., the gray processes) [24]. Shown in green, testing and analytical evidence are used to support claims of trustworthiness, security, and resiliency.

Across the entire developmental life cycle, the systems analysis process is used to eliminate ad hoc approaches, utilize scientific reasoning, and infuse engineering rigor to systematically identify and reduce vulnerabilities (e.g., mathematical analysis, model and simulation, defined approaches, etc.). For a recent detailed

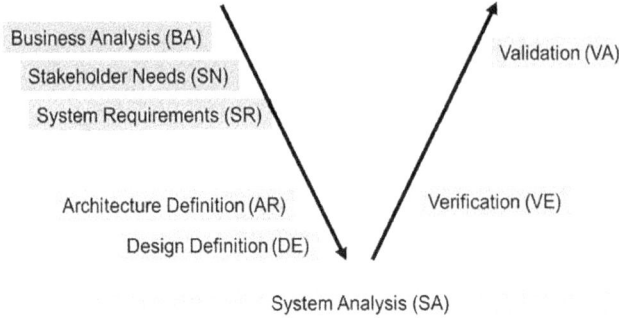

Business Analysis (BA)
Stakeholder Needs (SN)
System Requirements (SR)

Validation (VA)

Architecture Definition (AR)   Verification (VE)
Design Definition (DE)

System Analysis (SA)

FIGURE 2.1   A simplified depiction of a standardized systems security engineering approach based on the systems engineering processes of ISO/IEC/IEEE 15288 [8].

example, please see [25]. While our previous works focused on understanding the blue SSE processes [5], [26–28], this work focuses on understanding the gray SSE processes through the analysis and application of the NIST SP 800-160 Vol. 1 principles.

2.3.2   INTRODUCTION TO DESIGN STRUCTURE MATRICES

Shown in Figure 2.2, our strategy-to-principle mapping illustrates the relationships between the various security and resiliency strategies, and their associated principles. This mapping allows users to more easily understand the interdependencies associated with implementing these principles. Moreover, detailing these positive and negative relationships facilitates reasoning about important security design tradeoffs. This mapping is also useful for producing evidence of sound engineering practices [29]. For example, stakeholders, auditors, and security specialists can more easily consider which security principles should be implemented given a security or resiliency strategy. Thus, these mappings can be used to support requirements traceability, decisions of trustworthiness, and justification of limited resources. Note, detailed discussion of the mapping is provided in Section 24.

More formally, Figure 2.2 is a type of adjacency matrix known as a "Design Structure Matrix" or simply a DSM [30]. The intention of the DSM is to provide the reader with a visual depiction of the interactions between matrix elements (i.e., the strategies and principles list across the rows and columns). The DSM also takes advantage of directionality which allows additional information to be encoded in the matrix. For example, a developer can read down a particular column to ascertain which design principles inform or support the desired security strategy or principle (i.e., each marked row). Likewise, reading across each row, the developer can see how.

Examining the DSM brings insight into the tradeoffs associated with the application of each security principle as there are inherent conflicts and contradictions that must be considered when applying each security principle. Most importantly, the DSM enables a stochastic analysis of these relationships to provide an optimized clustering of the most interrelated elements.

Legend
"•" indicates a strong positive relationship
"o" indicates a weak positive relationship
"–" indicates a conflicting relationship
"X" indicates a relationship that could be either positive or negative

Structural Security Principles	Access Control	Defense in Depth	Isolation	System Resiliency	1. Clear Abstractions	2. Least Common Mechanism	3. Modularity and Layering	4. Ordered Dependencies (Partially)	5. Efficiently Mediated Access	6. Minimized Sharing	7. Reduced Complexity	8. Secure Evolvability	12. Hierarchical Protection	13. Minimize Trusted Components	16. Self-Reliance	9. Trusted Components	10. Hierarchical Trust	11. Commensurate Protection	14. Least Privilege	15. Multi-Factor Permissions	17. Secure Composition	18. Trusted Communication
	Security Strategies				Structural Security Principles																	
1. Clear Abstractions	•	o	•			•	o	o	•	•	•	•	o	•	o	o	o	o	•	o	o	o
2. Least Common Mechanism	•		•	X			o	X	o	–	•	•					o		o			
3. Modularity and Layering	•	•	•	•	o		•	•	o	•	•	•			o							o
4. Ordered Dependencies (Partially)			o			•		•		•		•					•					
5. Efficiently Mediated Access	•		•			•	o			•	o											
6. Minimized Sharing	•	o	•	o	–		•		o		•	o		o								
7. Reduced Complexity	•	–	X	X	o	o	•	•	•	•		•	•	•	•	•	o	•	•	o	•	•
8. Secure Evolvability	–	o	o	•	o	•	•	•	•	•	o		o	•	o	o	o	o	o	o	o	o
12. Hierarchical Protection	•	o	o	•			o				•	o			o							
13. Minimize Trusted Components	o	–	o	•		•	o		•	•	•	•	o		o						o	o
16. Self-Reliance			•	X	o		•	•		•	•	•										
9. Trusted Components		•		o		o	o						o				•	•		•		•
10. Hierarchical Trust	•	o		o			•						•					•				
11. Commensurate Protection	•	•	o	•								o	–		•		•		o	o	o	•
14. Least Privilege	•		•	•					•				•	•	•							
15. Multi-Factor Permissions	•	o	–	X							–	o					•					
17. Secure Composition	o	•	–		o	–	o	•	o	•	–	X	•	•	•	•	•	•	•	•		•
18. Trusted Communication	•	o	•	o	o		o	•		o		o			•	•					•	

FIGURE 2.2 Design Structure Matrix (DSM) provides a visual depiction of the positive and negative relationships between the various NIST SP 800-160 Vol. 1 systems security strategies and principles. The DSM takes advantage of directionality where a developer can read down a column to ascertain which design principles inform or support a particular security strategy or principle (i.e., each marked row). Likewise, reading across each row, the developer can see how each security principle contributes to other security strategies and principles. These details are particularly helpful for understanding tradeoffs between the security design principles. Finally, DSM analysis results in two clusters with architectural principles shown in red and trust principles shown in blue.

2.3.3 DESIGN STRUCTURE MATRIX ANALYSIS

A DSM provides a means of modeling complex interactions in a set of elements and utilizes a systematic approach to analyze, order, and organize that information [30]. Thus, a DSM is a useful mechanism for studying the inter-relationships between the security design principles and strategies of Vol.1.

Our DSM analysis builds upon the software developed by Thebeau and leverages his algorithm to perform the clustering of elements [31]. As shown in Figure 2.2, the security principles are highly interrelated and suggest a densely populated DSM; thus, to provide the most value to the practitioner, only the strong positive relationships are fully examined (i.e., those with a solid circle "•"). Analyzing only strong relationships enables the optimization algorithm to categorize the security principles into useful groupings. Additionally, the relationships between the security strategies and the design principles are not used as inputs to the clustering algorithm because their inclusion within a single homogenous DSM would not be appropriate; thus, the four security and resiliency strategies are merely included to understand the applicability of the principles, and their clustering, as they pertain to achieving stakeholder needs.

The DSM optimization uses a stochastic bidding algorithm shown in Eq. (2.1) to determine which cluster a given element should be assigned to. This process is based on factors such as the size of the overall cluster and the number of interactions within a cluster where cost penalties are assigned to larger clusters to avoid returning a single cluster and promote the creation of highly inter-related clusters. To prevent localized optimum solutions, Thebeau's algorithm also incorporates simulated annealing which periodically accepts the second highest bid rather than the highest bid to increase the probability of finding a global optimum solution. The bid for a cluster j is given by:

$$\text{ClusterBid}_j = \frac{\text{inout}^{\text{powdep}}}{\text{ClusterSize}_j^{\text{powbid}}} \tag{2.1}$$

where *inout* is the sum of DSM interactions of the chosen element with each of the elements in cluster j; powdep = 4 is a user-specified exponential to emphasize interactions; *ClusterSize* is the number of elements currently in cluster j; and *powbid* = 1 is a user-defined exponential to penalize the size of the cluster. Additionally, a *stable_ limit* parameter in the control logic of the algorithm, which specifies how long the algorithm runs before reporting results, was adjusted. Specifically, the process was modified to perform longer and broader searches in order to return higher overall *likeliness* values which is largely due to the fact that the availability of computing capacity has dramatically increased since the original algorithm was coded in 2,000. While the clustering algorithm permits clusters as small as a single unit or as large as the entire matrix, the *likeliness* value is introduced to measure the consistency of clustering results across multiple runs and increases the confidence in the repeatability of the outcomes. Our algorithm is available on the IEEE Access website as part of this publication. It is also worth noting that Borjesson and Höltt\u00e4-Otto improved the speed of the algorithm; however, given the relatively small size of this problem matrix, the original algorithm was more than sufficient when executed on modern hardware [32].

2.3.4   ADDITIONAL DEFINITIONS

Before starting a detailed discussion of the security principle analysis results, it is helpful to first introduce a few essential SSE terms from Vol. 1. Note that these definitions are not presented as such in Vol. 1 but are constructed as "working definitions" to assist the reader in understanding the DSM analysis, and particularly, detailed discussion of protection and trust principles below.

> Protection: A capability (or statement of the stakeholder's security need/requirement) with the objective of controlling the events, conditions, and consequences that contribute to unacceptable losses.

> Trust: The degree to which the security behavior of a component is demonstrably compliant with its stated functionality.

Given the frequency of the concept in this discussion we also include a definition of trustworthiness to aid the reader [33]:

> Trustworthiness: Trustworthiness implies simply that something is worthy of being trusted to satisfy its expected requirements.

For the purposes of this study, we also adopt a definition of resilience from [1] to help frame the discussion and form a basis for analyzing the security principles in the broader context of SSE:

> Resilience: Resilience is the ability to continue or return to normal operations in the event of some disruption (natural or man-made, inadvertent or deliberate).

This concise definition is specific enough to provide a grounding for the following analysis yet broad enough to adequately cover a myriad of expected and unexpected events (e.g., malicious action, unintended misuse, system failures, etc.). For a more formal discussion of "resiliency" used in the study of complex networks please see [34].

2.4   SECURITY PRINCIPLES MAPPING DISCUSSION

In this section, we discuss the 18 NIST SP 800-160 Vol. 1 security principles and their mappings as presented in Figure 2.2. Specifically, we expand upon our previous work [10] to provide detailed explanations of each principle, rationale for mappings between the principles, introduce a fourth systems security strategy (Resiliency), and provide justification for the principle to strategy mappings. Ultimately, our goal with this work is to facilitate widespread adoption of these system-level security strategies and their associated principles such that defensible and resilient SoIs can be more easily designed, built, and tested to meet stakeholders' security needs regardless of the developer background or SoI's application domain. To facilitate this knowledge transfer, our DSM software and mappings are available on the IEEE Access website for the reader to study, modify, and improve upon.

2.4.1   SECURITY STRATEGIES

In this section, we formally introduce the systems security strategies studied in this work (1. Access Control, 2. Defense in Depth, 3. Isolation, and 4. System Resiliency). These four strategies have been modified and expanded upon those presented in NIST SP 800-160 Vol. 1, Appendix F [6] to improve understandability and broaden their applicability.

Shown in the green portion of Figure 2.2, each of the four SSE strategies is supported by numerous design principles. The systems security strategies are presented in a one-way relationship within the DSM because the developer will primarily need to know which principles contribute to the desired security strategy (i.e., each upper-level strategy is supported by one or more principles). For example, if the stakeholders are developing a SoI that processes highly classified information, an isolation-focused security strategy may be desired. With the DSM of Figure 2.2, the developer can easily see which security principles directly contribute (or in some cases take away) from the intended isolation-focused outcome.

2.4.1.1   Access Control

Access control (formally known as "the Reference Monitor Concept") provides a conceptual model of the necessary access controls (i.e., requirements) that must be achieved to enforce security policies. Ideally, realizations of the access mediation concept possess three properties: (1) tamper-proof; (2) always invoked; and (3) can be subjected to analysis and testing to assure correctness. This means that any mechanism claiming to perform access mediation only does what it is supposed to do and can never be bypassed, coerced, manipulated, or deceived.

2.4.1.2   Defense in Depth

Defense in depth describes security approaches that create a series of barriers to prevent, delay, or deter an attack by an adversary. Typically, defense in depth is achieved through the application of multiple security mechanisms (i.e., conceptual and physical obstructions). It is important to note that implementing a defense-in-depth strategy is not a substitute or equivalent to choosing a sound security architecture or system design that leverages a balanced application of security concepts and design principles [17].

2.4.1.3   Isolation (Physical and Logical)

Isolation pertains to the creation of separated processing environments; they can be logical, physical, or a combination thereof. While physical isolation is somewhat more straightforward (i.e., physically distinct components and systems), the sharing of data often necessitates logical isolation mechanisms which require the use of underlying trustworthy mechanisms to minimize resource sharing.

2.4.1.4   Resiliency

In addition to the three security strategies of Vol. 1, we also considered emerging stakeholder needs such as "cyber resiliency" and "system survivability" which have generated congressional mandates, service-level policy changes, and significant organizational changes within the U.S. DoD [35]. Introduced in 2005, the System Survivability Key Performance Parameter (SS KPP) ensures the SoI can maintain critical functionality while under attack [21]. Formally, the SSKPP includes three pillars (Prevent, Mitigate, Recover) intended to reduce system susceptibility to adversaries, limit damage from exploited vulnerabilities, and increase system resiliency in order to execute the SoI's mission. Likewise, the U.S. DoD and similar industries are increasingly focusing on resiliency – the ability of a system to perform essential functions in spite of hostile actions [36].

Thus, we consider "system resiliency" or "resiliency" as an important system-level security strategy in addition to Vol. 1's three strategies. While the topic of "resiliency" is being studied in multiple domains from complex networks to biological [37], we thought it pertinent to include resiliency for advanced cyber-physical systems in our analysis as it is timely and germane to developing more secure and defensible systems. For additional background on the topic of systems-based resiliency, please see: [38–40]. Of note, during the course of our research, the draft version of NIST SP 800-160 Vol. 2, Cyber Resiliency, was released [7]. Vol. 2 is largely focused on countering the cyber-specific "Advanced Persistent Threat" as a manifestation of the MITRE Cyber Resiliency Engineering Framework (CREF) [38]. While Vol. 2, and the CREF, are very helpful in many regards, in this work we strive to maintain a standardized approach consistent with industry standards (i.e., ISO/IEC/IEEE 15288, INCOSE handbook, and specifically the security design principles of Vol. 1). More generally, we are interested in engineering resilient systems which are able to operate in and through situations where the SoI is degraded, whether due to hostile actors (insiders or outsiders) or internal failures (hardware or software).

2.4.2 Design Principles

In this section, we formally introduce the 18 SSE principles from NIST SP 800-160 Vol. 1, Appendix F [6]. In particular, we've modified and extended these principle descriptions to make them less "computer science centric" and broaden their applicability to any engineering discipline at the "systems level" of thinking. More pragmatically, we've attempted to make these principles more readily understandable and applicable to system architects, designers, developers, and especially those who may not have formal security education or training.

   Furthermore, it is important to note that these 18 security-oriented design principles are conceptual in nature and thus are used to inform a secure system design. They do not immediately provide a more secure system such as one might find with "adding encryption" or "performing security penetration testing". These principles contribute to designing and developing a complex system architecture that is fundamentally more secure and defensible from adverse cyber threats (malicious or non-malicious). In addition to being more secure, the resultant architecture should have higher levels of evidence-based trustworthy protection capabilities. Finally, these principles are important for thinking and planning for the fluid nature of modern software-based systems such that the SoI can be more easily maintained and securely modified at less cost over the total system life cycle. This last point is particularly important as security improvements often become cost prohibitive in all but the most extreme situations.

2.4.2.1 Clear Abstractions

Historically, the principle of clear abstractions has been applied to software functions and system interfaces to provide a consistent and intuitive view of the SoI's data – its data elements, data utilization, and data management. However, in a more general sense, clear abstractions mean that the SoI should have a design that is obvious to others, functions that are distinct and well-defined, and information exchanges that are labeled and fully characterized (*within reason*). Ideally, the system design should be easily understandable and seemingly obvious to interpret by an engineer (for example).

   Among the several security strategies and principles that clear abstractions supports, its contribution to the access control and isolation strategies is particularly evident given that it emphasizes the unambiguous definition of boundaries, behaviors, and relationships between subsystem elements, functions, users, dependencies, and data exchanges (i.e., all the SoI's significant objects and their associations). For example, information dependencies must be identified before access control rules can be defined.

While it is advantageous to initially define many abstractions, complete enumeration across the SoI's many users, subsystems, components, supporting/enabling systems, and various forms of data is difficult *(and arguably fleeting)*; thus, this principle must be constantly applied as the system's architecture, design, and implementation become further refined until the SoI is fully realized. Note that these abstractions also provide a detailed baseline for standardizing software and testing cybersecurity issues.

2.4.2.2   Least Common Mechanism

The principle of least common mechanism seeks to reduce unnecessary duplication within the SoI. More specifically, if multiple components in a system require the same functionality, the desired functionality should be built into a single mechanism (physical or logical). For example, utilizing a single access control mechanism to implement a security policy can significantly reduce complexity across the SoI, as well as, reduce errors of omission and commission which may introduce vulnerabilities.

The least common mechanism security principle directly contributes to both access control and isolation strategies. Additionally, this principle supports reduced life cycle costs since there are no multiple instantiations of the same functionality to document and modify. On the other hand, taken to the extreme, the least common mechanisms can result in a monoculture which contributes to reduced system resiliency where diversity helps to minimize the rapid propagation of an attack. Moreover, diversity provides alternative means of delivering required functionality so this principle has the potential to conflict with a system's resiliency goals [40, 41].

2.4.2.3   Modularity and Layering

Modularity organizes and isolates functionality and related data into well-defined conceptual groupings, while layering orders the relationships between these groupings and their associated data flows. Note that we use the term "groupings" to refer to systems, system elements, components, or software objects with a systems engineering perspective rather than a narrower software engineering perspective as would typically be the case when discussing modularity and layering. While modularity is very closely related to clear abstractions, they are certainly different (e.g., a software object can be clearly defined but not modular; likewise, it can be modular but poorly defined). Confusion often occurs with these terms because a "good software object" is clearly defined, modular, and inherently layered. Perhaps, a helpful way to think about these principles is to consider that layering implies an ordered hierarchy of groupings (systems elements or objects), modularity implies "plug-and-play" of the groupings within the hierarchy, and clear abstractions imply that the hierarchy and its groupings are well-defined.

This principle is widely used in system development and especially modern software engineering practices, and as such, it contributes to each of the four security strategies and several principles. Of note, modularity and layering are particularly critical for supporting isolation and the resiliency strategy because the extent and impact of undesired behaviors can be more easily constrained (whether they be malicious or non-malicious). Finally, it is worth mentioning that "layering" is not synonymous with "defense in depth"; the former is focused on the efficient design of a system while the latter is typically focused on providing redundant and heterogeneous forms of protection.

2.4.2.4   Ordered Dependencies (Partially)

In general, ordered dependencies refer to the logical (or hierarchical) arrangement of system elements (e.g., layers, modules, objects, etc.), and specifically, an ordering which functional calls, synchronization, and other dependencies are achieved through linearization or hierarchical design. Partially ordered dependencies simply refer to when the ordered dependency is mostly hierarchical (for example) but may require an additional non-hierarchical linkage. Ordered dependencies also seek to minimize (or ideally, remove) circular dependencies.

Logically structuring dependencies (and minimizing them where possible) contributes to isolation between layers, increases understandability of the design, reduces system complexity, and facilitates testing and analysis. A system with partially ordered dependencies is also less likely to adversely impact associated functions and elements, thus contributing to survivability. Finally, ordering dependencies helps preserve trustworthiness by avoiding linkages between components with lower and higher trust levels.

2.4.2.5   Efficiently Mediated Access

The efficiently mediated access principle ensures security mechanisms do not adversely hinder system performance to an unacceptable level. More generally, this principle seeks to "optimize" security and performance tradeoffs such that stakeholder systems security requirements (or goals) are met while the SoI performs its mission critical functionality. For example, security policies should be enforced at the lowest level possible with the most effective mechanism (physical or logical) within expressed constraints. In specific cases, the efficiently mediated access principle will overlap and result in the same outcome as the least common mechanism principle (but not as a general rule).

This principle directly contributes to the realization of the access control and isolation strategies, and it is related to several other security principles such as least common mechanism, minimized sharing and reduced complexity. Of note, whereas verification and validation should be accounted for in the development of sound security principles these requirements often are not sufficiently addressed. As such, the deliberate application of this principle is important because analytically based evidence are used to substantiate claims of trustworthiness, fulfillment of security objectives, and determination of risk.

2.4.2.6   Minimized Sharing

The minimized sharing principle states that no resources should be shared between system components unless it is absolutely necessary to do so. Historically, this principle has been used to describe single-purpose hardware such as separate computer hard drives; however, the principle is generally applicable to system elements, software processes, interfaces, and humans. This principle directly supports the access control and isolation strategies by reducing the number of interactions between various system elements.

Restricting the sharing of resources has several benefits and effectively creates boundaries within the SoI to protect critical functions, simplify design, and streamline implementation. Separation can also be used to facilitate defense in-depth solutions. However, while limiting shared resources is generally advantageous for security solutions, this principle may limit the resiliency solution trade space. Thus, the benefits of its application should be carefully considered when designing for security and resiliency.

2.4.2.7   Reduced Complexity

The principle of reduced complexity implies that the system design should be as simple and small as possible. Clarity of design simplifies the understandability of the SoI, reduces total life cycle costs, and has several benefits especially when considering the development of secure systems. For example, reduced complexity directly contributes to the proper design and correct implementation of security and resiliency mechanisms. Moreover, the said mechanisms can be thoroughly tested and provide convincing evidence that the desired protection capability is achieved (i.e., verification and validation). It also contributes to the identification of potential vulnerabilities which would have been obfuscated otherwise. Additionally, simpler designs can reduce the number of unnecessary dependencies which further reduces the system's attack surface.

Because of the additional insight gained through simplicity, this principle strongly contributes to nearly every design principle. However, if the application of this principle is taken too far, it may preclude the employment of isolation strategies, as well as, system redundancies which can negatively impact resiliency. Because of these competing dynamics, this principle merits careful attention.

2.4.2.8   Secure Evolvability

The principle of secure evolvability implies that a system be developed to facilitate upgrades and maintenance activities in a secure fashion. This means that the SoI should be designed to easily and securely adapt to changes in its configuration, functionality, architecture, structure, interfaces, and interconnections. Given the fact that complex systems typically have life cycles spanning many years and face dynamic, constantly evolving threats, this principle is key to improving system security, resiliency, and survivability. Thus, this principle needs to account for the secure development and implementation of hardware and software upgrades, modifications, and patches during operations and sustainment.

Thus, this principle supports nearly all security strategies and principles throughout the SoI's life cycle. Focusing on secure evolvability early in the life cycle also has the potential to drive down engineering costs and facilitate fewer complex mitigations to future threats once the system has been deployed. On the other hand, it can adversely impact access control strategies if subsequent upgrades to the system are not carefully designed (and as a result, inadvertently introduce vulnerabilities into the SoI).

2.4.2.9   Trusted Components (Elements)

The trusted components principle implies that a component must be at least as trustworthy as the components it supports. In more colloquial terms, this security principle simply implies that a chain is only as "trustworthy" as the "weakest link" in the chain. More formally, the trusted components principle implies that a component (or system element) must be trustworthy to at least a level commensurate with the security dependencies it supports. While originally conceived of for hardware components and devices, this principle is valid for the entire SoI and is applicable to component, elements, networks, functions, and other more subtle dependencies such as data and personnel which are often assumed to have high level of trustworthiness. This principle is foundational for the development (and operation) of assured systems because it enforces the requirement for valid evidences that support decisions of trustworthiness. Moreover, it ensures trust is not misplaced or inadvertently diminished throughout the development process.

Trusted components, and specifically the resulting trustworthy behaviors, enable the development of trustworthy secure systems such that desirable levels of trustworthiness can be achieved through a systematic approach. For example, how much testing is necessary to determine when an autonomous vehicle should be trusted to drive without a human operator? Returning to the security chain analogy, systems-level analysis of this principle enables the reasoning about and justification of the SoI's "trust chain" and highlights where trust is being hindered by less trustworthy "chain links". This principle is highly related to other trust and structurally-oriented principles including hierarchical trust and commensurate protection.

2.4.2.10   Hierarchical Trust

Building upon the principle of trusted components, hierarchical trust provides a logical basis for reasoning about levels of trustworthiness when developing a system from several components. Hierarchical trust is especially pertinent with modern SoI, as developers consider the composition of numerous components, networks, and data at multiple levels of hierarchy each with differing levels of trustworthiness. Thus, hierarchical trust is an essential principle for reasoning about, and ultimately achieving, trustworthy secure systems.

Within the context of NIST SP 800-160, hierarchical trust supports the strategy of access control and to a lesser extent resiliency. It is worthwhile to note that this principle also depends on the correct implementation of several other design principles (more so than many other principles) because complex systems are composed of various components (and their dependencies) each of which implies differing levels of trustworthiness. In particular, hierarchical trust supports the principles of trusted components, hierarchical protection, and partially ordered dependencies which collectively provide evidences for reasoning about and justifying trustworthiness decisions.

2.4.2.11   Commensurate Protection

The principle of commensurate protection implies that the degree of protection provided for a component must be appropriate to its desired level trustworthiness. For example, as the trust placed in a component increases, the protection against unauthorized modification of said component should increase to the same degree. This principle is consistent with historic security defense approaches in that the most valuable items should be very well protected. Moreover, the term "commensurate" implies that the protection expenditures should correspond to the value of the items being protected. For example, the cost to defend the SoI should not exceed the total cost of the SoI. More practically, the cost to protect the SoI will be a percentage of the system lifecycle cost and must be justifiable to stakeholders.

This principle specifically contributes to the access mediation and defense in-depth strategies, and builds on the principles of trusted components and hierarchical trust, as well as, several others to ensure that adequate trustworthiness is designed-in to the SoI. It also indirectly supports several other principles such as hierarchical trust, self-reliance and trusted communication. By deliberately focusing on the SoI's most critical components and functions, this principle ensures that supporting evidence are available for claims of trustworthiness.

2.4.2.12   Hierarchical Protection

The principle of hierarchical protection means that a component need not be protected from more trustworthy components. In addition to conventional physical components, this principle should be applied to software, information dependencies, and other system elements. It is akin to the principle of trusted components and hierarchical trust, but specifically addresses the aspect of unnecessarily over-protecting a component. In a similar way, this principle is akin to commensurate protection but applied where the subject component establishes the minimum required protection threshold for a complex system composed of multiple elements and/or components.

This principle supports all of the security strategies (to varying degrees) by protecting the SoI from components, functions, data, and even users with lower trustworthiness and reserving higher level privileges for more trusted entities. Regarding systems security design decisions, this principle is very important for guiding the application of architectural and component-level principles to ensure security resources are not wasted on protections against higher trust level components.

2.4.2.13   Minimized Trusted Components

The principle of minimized trusted components implies that a system should not have extraneous trusted elements, components, data, or functions. Similar to the foundational "keep it simple" engineering principle (described as "reduced complexity" in the NIST SP 800-160), the minimized trusted components principle suggests that the SoI should contain as few trustworthy components as possible. Thus, this principle directly impacts multiple facets of system development and especially total lifecycle costs.

This principle supports several security strategies by reducing the system's attack surface and improves the system's resiliency by reducing the SoI's likelihood of component-level failure. In particular, resiliency thinking often considers the possibility that systems and/or their components will occasionally be compromised, for example when facing a persistent intelligent adversary. Thus, minimizing the number of trusted components helps to mitigate the impact of those compromises, lowers assurance costs, and facilitates easier monitoring of a system's security posture [40]. However, minimizing the number of components may negatively restrict the SoI's design flexibility and collective resiliency capability.

2.4.2.14   Least Privilege

The principle of least privilege implies that a component should be allocated sufficient privileges to accomplish its specified function, and no more. While often described with respect to system users, when employed at the systems level, the principle is much more widely applicable to each system element, component, network, user, and more. Additionally, although the principle of least privilege is pervasive in computer security literature, it is often difficult to realize. For example, thorough testing can be accomplished to demonstrate the desired outcome occurs, however, it is much more difficult to demonstrate that undesired outcomes do not occur.

The least privilege principle directly supports access control, often with the outcome of logical or physical isolation to help minimize the impact of potential failures, corruption, misuse, and malicious activities. It also serves to reduce interdependencies, which simplifies component design, implementation, and analysis. Judicial application supports resiliency and can improve survivability as an attacker's movements are limited when they are denied privilege escalation, a key tactic employed by advanced persistent threats [40].

2.4.2.15 Multi-Factor Permissions

Formally presented as "Predicate Permission" in NIST SP 800-160, the principle of multi-factor permission requires that multiple authorizing entities (or operators) provide consent before a critical operation occurs. While traditionally thought of as two operators turning a key or offering a password, this principle should include authorization through multiple means to include human and system. Most commonly, multi-factor permission is granted through successive approval actions (i.e., a sequence of events occurring without error). Although slightly different, this concept can be loosely achieved through independent validation where the independent party can act as a second factor before granting permission or passing information.

This principle directly contributes to the access control strategy and defense in depth, while also enhancing survivability by protecting mission critical information, components, and processes. However, it is also important to consider that requiring multiple authentications also has the potential to negatively impact availability, survivability and/or resiliency. Finally, the multi-factor permissions principle adds complexity into the design, so its application needs to be considered carefully within the systems security trade space.

2.4.2.16 Self-Reliance

The principle of self-reliance means that systems and all their elements to include information and software should minimize their reliance on other systems, elements, or components. While this principle is specifically targeted at building trustworthiness, it is generally applicable for protecting the SoI as well. Applying the self-reliance principle can significantly reduce design and implementation complexity, increase testability, and improve system survivability. However, it is worth nothing that an SoI's ability to maintain situational awareness is often dependent upon the system's capability to monitor subsystems, data feeds, and personnel interactions which may be negatively impacted with strict adherence to self-reliance.

This principle directly impacts the isolation strategy by minimizing both external and internal dependencies. When applied outside the SoI, this principle serves to reduce the system's attack surface which can reduce susceptibility to vulnerabilities inherent in external systems and network links. When applied within the SoI's architecture, it serves to protect critical functions and components through isolation.

2.4.2.17   Secure Composition

The principle of secure composition means that the combination of system elements designed to enforce a security policy should result in a system that enforces said policy at least as well as the individual components do. Like many of the design principles, this principle is rather intuitive but problematic in implementation when considering modern systems with growing complexity and emergent behaviors. Note, when considering the point of multiple elements enforcing the same security policy, several principles toward reduced complexity should be considered.

This principle supports the access control and defense in-depth security strategies by ensuring the commensurate implementation of security policy across isolated systems (and their respective permission levels). Because this principle includes the composition of distributed features, components, and system elements, it supports nearly all design principles. This principle should especially be considered when evaluating interactions between various components where the developer must work to identify and assess emergent behaviors and properties.

2.4.2.18   Trusted Communication

The principle of trusted communications means that each communication channel (i.e., an interface, link, or network) must be trustworthy to a level commensurate with the security dependencies it supports. While seemingly redundant, this principle specifically considers the necessary protections for the SoI's communication links and ensures weaknesses are not introduced via communication channels. This is important because communication channels are inherently more susceptible than the SoI's internal elements. Moreover, system stakeholders often don't own the communication channels, thus additional scrutiny is required to ensure they have the same level of protection as the components they support.

This principle directly contributes to the access control and isolation strategies. Because of the preponderance of communication requirements in modern systems, the application of this principle impacts several other principles such as efficiently mediated access, trusted components, and hierarchical trust principles.

2.4.5   ANALYSIS RESULTS

Shown in Figure 2.2, our analysis generated two clusters (outlined in red and blue, respectively) with a mean likeness value of 0.752 (on a scale from 0 to 1) during a 100-iteration run of the algorithm. By limiting the DSM optimization to elements with strong positive relationships (and controlling the abovementioned cluster size and interaction strength parameters), the DSM optimization algorithm is able to provide concise, meaningful groupings.

The first cluster (shown in red) represents architectural considerations of a system, while the second cluster (shown in blue) represents considerations that collectively contribute to the trustworthiness of a system. While these results are dependent upon the mappings of Figure 2.2, and ultimately the expertise of those interviewed, the DSM clustering software is provided online for users to investigate and update as

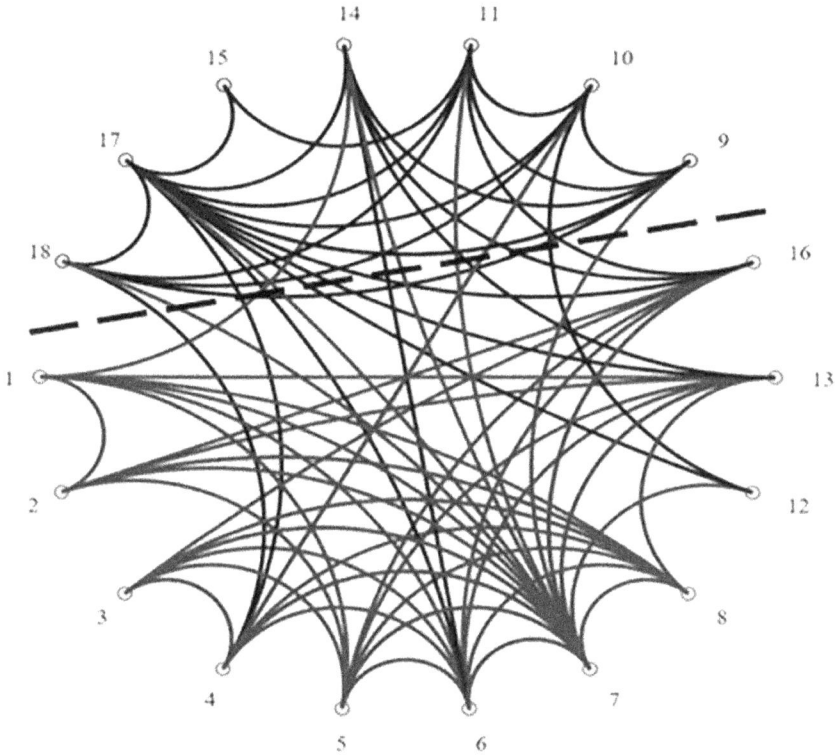

FIGURE 2.3 Relationship graph of the Design Structure Matrix (DSM) clustering optimization of interrelated systems security design principles with the names of design principles specified in Table 2.1. The trust-related principles are shown in blue and the architectural related principles are shown in red. Note, directionality of the relationships is tabulated in the last two rows of Table 2.1 and detailed in the rows and columns of Figure 2.2.

desired. With a majority of the principles being design-oriented, roughly two-thirds of the strongly positive relationships occur in the first cluster with 11 of the 18 principles.

One notable exception is that three of the four strong positive elements supporting defense in depth are from the second cluster. The fact that the second cluster held a majority of the principles strongly related to defense in depth (Trusted Components, Commensurate Protection, and Secure Composition) is noteworthy in light of our earlier remark that "there is no formalized theoretical basis to assume that defense in depth alone achieves a level of trustworthiness greater than that of the individual security components" [10]. More specifically, poor implementation of these three principles detrimentally impacts the trustworthiness of a given system.

Figure 2.3 provides an informative visual representation of the division and interconnectedness of the security principles among the two groupings with details provided in Table 2.1. In the relationship graph of Figure 2.3, the color of each node represents which cluster it belongs to and the color of each edge identifies which cluster the relationship belongs to based on its originating node. Table 2.1 enumerates the total number of relationships (either supporting, supported, or two-way) and the

TABLE 2.1

Relationship Counts Per Principle

Principle	1. Clear Abstractions	2. Least Common Mechanism	3. Modularity and Layering	4. Ordered Dependencies	5. Efficiently Mediated Access	6. Minimized Sharing	7. Reduced Complexity	8. Secure Evolvability	9. Trusted Components	10. Hierarchical Trust	11. Commensurate Protection	12. Hierarchical Protection	13. Minimize Trusted Components	14. Least Privilege	15. Multi-Factor Permissions	16. Self-Reliance	17. Secure Composition	18. Trusted Communication
Total # of Relationships	7	6	6	7	8	10	15	8	6	6	7	3	8	7	2	8	12	6
# to Architectural Cluster	6	6	6	5	8	8	10	8	1	4	5	1	6	3	2	5	11	1
# to Trust Cluster	1	0	0	2	0	2	5	0	5	2	2	2	2	4	0	3	1	5
# of Supported Principles (Across each Row)	7	2	4	3	3	3	13	6	3	3	3	0	4	4	1	5	10	4
# of Contributing Principles (Down each Column)	0	5	5	7	6	8	8	4	4	5	5	3	5	3	1	4	3	2

number of relationships to the opposing cluster for each principle. This helps highlight several properties of our mapping and clustering. First, it provides a means of identifying the relational density of a given principle. For instance, principle 7 shows the strongest inter-relationship mapping (with a value of 15), while principle 15 has the weakest (being linked to only 2 others). While all 18 design principles may be important to the security of a system, this ranking can serve as an indicator of the relative criticality of each principle. Second, links crossing the clustering divide (shown as a dashed line) and particularly the principles with higher "# to Other Cluster" counts contribute to both clusters. In particular, clusters 12 and 17 exhibit a relatively high number of links to the opposing cluster, so they may warrant further consideration in both clusters (architectural considerations and trustworthiness).

2.5.1 INTERPRETATION

Given the two groupings, we now evaluate them for actionable insights to assist those charged with executing SSE roles and responsibilities. Beginning with a cursory inspection of the first cluster (shown in the red outline of Figure 2.2), security principles associated with interfaces, boundaries, data flows, architecture, and other design-based concerns are identified. For instance, the clarity and simplicity of interfaces, data elements, and trusted components are promoted through the application of three design principles: clear abstractions, reduced complexity, and minimize trusted components. Likewise, least common mechanism, minimized sharing, and efficiently mediated access place emphasis on designs involving the fewest number of common mechanisms and shared resources to accomplish a specified function.

Longstanding engineering tenets such as modularity, layering, ordered dependencies, and hierarchical protection are concerned with essential system design decisions such as logical organization, structure, data flows, and the orderliness of function calls. Self-reliance helps to bound the design and engineering effort to minimize security, resiliency, and protection requirements which directly impact the cost and feasibility of proposed security solutions. Finally, secure evolvability considers the application of these principles over the SoI's life cycle as it is transformed.

With respect to the system life cycle, the first cluster of principles is particularly beneficial for consideration during the earliest phases of a SoI's development. Specifically, these architectural-based principles should be considered during the inception and concept phases through requirements definition and design when the solution trade space is widest [24]. For example, the effective application of these principles can improve system security and resiliency in a cost-effective manner. Conversely, rigorous application of these architectural principles to a pre-determined or fixed physical architecture is of significantly less value.

Continuing our examination to the second cluster (shown in the blue outline of Figure 2.2), this group of principles is strongly associated with the notion of trust and how the security engineer defines and ensures appropriate levels of trustworthiness are achieved. Since trust is used to form the foundation of how security analysts reason about the correctness of the system's design to its stated specification, it follows that these principles should be considered during the design phases (i.e., concept and development). Moreover, these principles need to be reconsidered throughout the system life cycle to ensure the SoI is properly operated and maintained in accordance with its intended purpose and emerging stakeholder's security needs. It is important to highlight that the SoI's required level of trustworthiness is not static, as its relative criticality to an organization (or mission) often evolves over time. Thus, the system's specified degree of trustworthiness and associated protection mechanisms should be periodically re-evaluated, especially as major system modifications occur.

Regarding the specific trust-oriented principles, trusted components assure that the SoI is developed, fielded, and maintained with components that are determined to be trustworthy. Likewise, hierarchical trust and trusted communication ensure dependencies between trusted components and communication mechanisms are correctly accounted for. These principles take on increasing importance as changes to one component often impact the trustworthiness of other components which rely upon it (including the SoI itself). These trust principles are typically employed in conjunction with the principle of commensurate protection which collectively assure that the components (and combinations thereof) are protected to a level appropriate for the security dependencies they support. As a common means for securing critical data and operations, multi-factor permissions need to be considered during design activities, and not merely costly add-on modifications.

Similarly, secure composition seeks to maintain the trustworthiness of a given systems-of-system's collective security policy implementation as modifications can cause unexpected or emergent results as the SoI changes over time and/or is used differently than expected. While an important consideration during initial design, the principle of least privilege is critical during operations and should be re-validated from time to time, lest excessive permissions begin to aggregate in a system.

Finally, the relative importance of these trust-oriented design principles cannot be overstated as it is common to place unmerited trust in components that are without evidence to support their claims of trustworthiness. Likewise, while the importance of these security principles is often emphasized during early design efforts, their importance across the entire system life cycle is less so, where planned and unplanned repairs, modifications, replacements, and upgrades can undermine critical trust relationships and protection measures.

2.5.2 NOTIONAL CASE STUDY EXAMPLE

To demonstrate how a systems security engineer can make use of the proposed principle groupings, we briefly consider the development of a notional intra-theater ballistic missile defense system shown in Figure 2.4 [42]. Early consideration of the

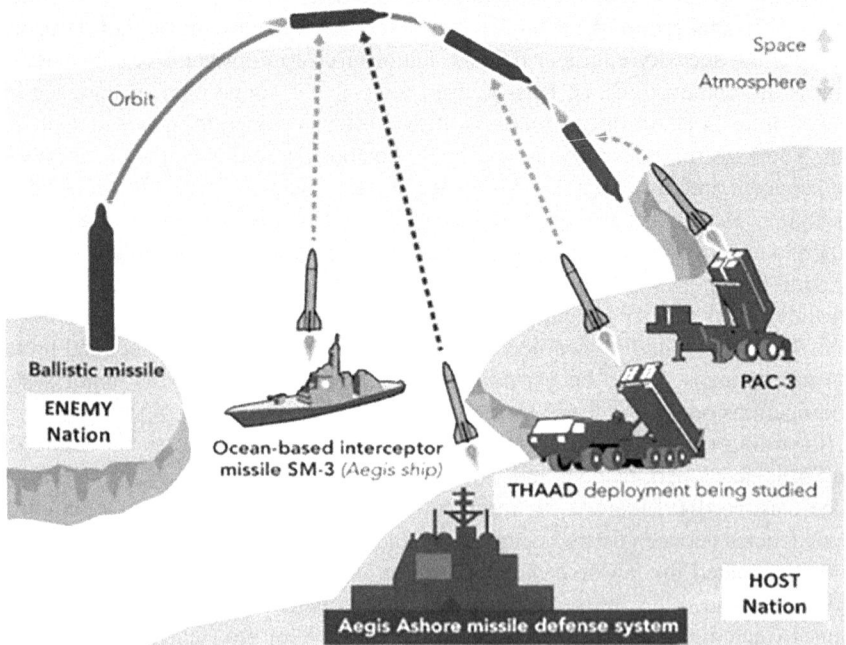

FIGURE 2.4 Notional intra-theater ballistic missile defense system for discussing the application of security design principle groupings throughout the system life cycle. (Adapted from [42].)

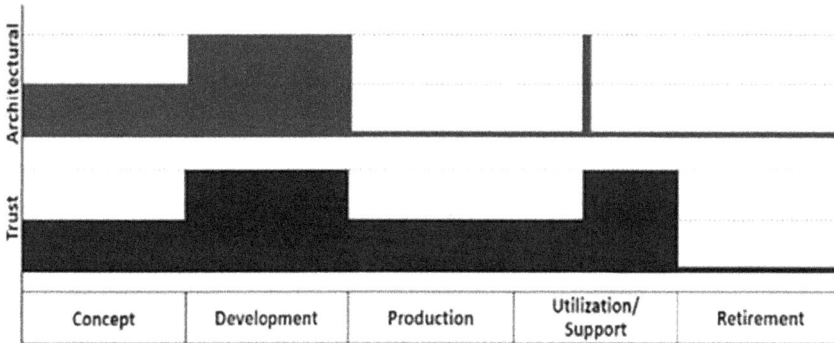

FIGURE 2.5 Notional application of the security design principle groupings throughout the system life cycle.

trust and architectural principle groupings is helpful for thinking about the entire system life cycle for a non-trivial SoI in order to meet both security and resiliency stakeholder needs. For example, Figure 2.5 provides a general overview of how the developer's level of attention to these two principle groupings might increase and decrease throughout the SoI's life cycle. Detailed in the next five Sections, we describe tradeoffs between these groupings for developing secure and resilient systems. Additionally, we provide some commentary as to how individual principles may be applied during each phase of the life cycle.

2.5.2.1 Concept Phase

Beginning with the concept phase, the architectural design principles are considered as a means for identifying, defining, considering, analyzing, and ultimately realizing a system that meets security and resiliency requirements for the desired capability. In this phase, it is prudent to consider principles such as clear abstractions to properly identify, define, decompose, and allocate the SoI's desired functionally (i.e., mission-essential behaviors). Likewise, the principle of modularity and layering affects the high-level design of the system and directly contributes to security and resiliency decisions such as required data flows and what classification of data each element of the SoI will need to process and store. In this example, this system is to be highly trustworthy; thus, the trust-related principles should also be initially considered. Ideally, each principle should be systematically assessed for applicability to provide evidence for trustworthiness of mission essential or safety-critical systems.

We would also like to highlight that we believe the key to developing secure and resilient systems is to fully understand their intended purpose (i.e., its desired capability) and proposed operational environment. In our experience, we've found that the System-Theoretic Process Analysis for Security (STPA-Sec) is a simple and straightforward approach for understanding and framing the SoI's purpose at a conceptual level which helps to elicit meaningful security and resiliency requirements within a mission-informed context [28].

2.5.2.2   Development Phase

During the development phase, scrutiny of the system's detailed design places increased emphasis on both groups of principles. This is because the SSE principles are fundamental in designing secure and resilient systems. Moreover, unintentional security design omissions often create critical problems later in the life cycle. For example, fixing vulnerabilities in a fielded system is often very costly, if not entirely prohibitive. Thus, systems security engineers should work diligently to avoid foreseeable weaknesses, diagnosable faults, and potential failure modes. Moreover, these engineers should also seek to conceptualize and develop systems that are easier for users to operate and sustain which can be achieved by applying the design principles not only for security but more generally to all aspects of the SoI.

In exercising the application of the security-oriented design principles during this case study (and a second one to be described in a future paper), we realized it is easy to become "stuck" in the process. This is because tradeoffs are inherent in complex system development as even DARPA's most secure systems to date are forced to make key security concessions [25]. With respect to the design principles, we believe this is because it is generally desirable to maximally employ each design principle, yet several conflicts and tradeoffs exist (as detailed in Figure 2.2 and Figure 2.3). For example, additional communication links may increase resiliency yet hinder security. Consequently, in order to achieve the SoI's desired protection capability, as well as, control long-term cost and maintainability, additional assistance is needed in efficiently applying these principles.

In Table 2.2, we provide a helpful listing of several engineering considerations to aid system developers and security practitioners in their SSE developmental effort. Specifically, we want to aid in the application of the NIST SP 800-160 Vol. 1 security design principles, so we've captured various practitioner-oriented questions and insights that have come to light during our case study. For example, consider the two "most connected" principles from Figure 2.2–#7. Reduced Complexity and #17. Secure Composition from the architectural and trust groupings, respectively. The principle of reduced complexity is architecturally significant and contributes strongly toward designing defensible and resilient systems where eliminating excess interfaces and dependencies reduces the SoI's attack surface. Likewise, the secure composition principle is important for building systems with a justifiable trust level as each element of the SoI contributes to or diminishes from the SoI's resultant trustworthiness level. Following the questions and considerations presented in Table 2.2 provides a systematic way to address the application of the NIST SP 800-160 Vol. 1 principles that is relatively easy to follow and not too confusing.

While Table 2.2 is particularly focused on the development phase of the system life cycle, these principles are generally applicable to several SSE processes, activities, and tasks across the entire life cycle; however, please note these questions must not be blindly followed without understanding the SoI's mission and/or business processes.

2.5.2.3   Production Phase

Once the production phase is reached, the architectural-based principles are deemphasized. Presumably by this time the principles have served their purpose to help define a system that, when operational, will be more secure and resilient (this is especially true

TABLE 2.2

Systems Security Design Principle Engineering Considerations

Principle Name	Simplified Descriptions	Considerations and Guidance
Clear Abstractions	A system should have simple, well-defined interfaces and functions	Is the system design readily understandable to a third party? Are the SoI's functions and subsystems obvious to an independent third party? Are the data elements precisely names and well-defined? Are there defined interfaces for every information input, output, and control?
Least Common Mechanism	If multiple components in a system require the same functionality, the desired functionality should be built into a single mechanism (logical or physical)	Does the SoI have security features which are repeated across the SoI's functions and subsystems? Can the security functionality be consolidated (where appropriate)?
Modularity and Layering	Organize and order functionality and related date flows between entities	Is the system design readily understandable to a third party? Are the SoI's functions and subsystems obvious to an independent third party? Separate and eliminate dependencies where possible. Is there a defined interface for every information input/output? Are the date elements precisely named and well-defined?
Ordered Dependencies	Logically arrange modules and layers such that linear (or hierarchical) function calls, synchronization, and other dependencies are achieved	Are security dependencies identified? Are security dependencies one way – to prevent circular dependencies?
Efficiently Mediated Access	Security enforcement mechanisms (logical or physical) should not unnecessarily impact the SoI's performance/purpose.	What is the applicable security policy? Are the SoI's security requirements and constraints specified? What means are available for meeting these policies, requirements, and constraints?
Minimized Sharing	Only share between components when absolutely necessary.	Reduce and/or eliminate resource sharing as much as possible.
Reduced Complexity	Develop a simple and small system.	Is the system design readily understandable to a third party? Have unnecessary requirements been eliminated? Is it possible to reduce any functions, subsystems, dependencies, or interfaces?
Secure Evolvability	Plan for secure maintenance and sustainment activities over the SoI's life cycle.	Has maintainability been designed into the system? Have user-level security monitoring capabilities been designed into the system? Have consideration been made for human learning and process changes? Have dependencies on legacy systems been minimized?

(*Continued*)

TABLE 2.2 (Continued)
Systems Security Design Principle Engineering Considerations

Principle Name	Simplified Descriptions	Considerations and Guidance
Trusted Components	A component must be trustworthy to at least a level commensurate with the security dependencies it supports.	Identify the components that transmit, receive, and store critical information. Ensure said components operate at the highest information classification level. Are the SOI's functions and subsystems obvious to an independent third party? Are the data elements precisely named and well-defined? Is there a defined interface for every information input/output?
Hierarchical Trust	An ordered hierarch should be used when composing a system from multiple components.	Is there a defined interface for every information input/output? Are the data elements precisely named and well-defined? Does the security hierarchy build appropriately?
Commensurate Protection	The degree of security provided should be commensurate with an asset's value.	Is the security requirement appropriate for the asset being protected? Are the high trust assets and information protected appropriately? Is too much protection being applied to a low value asset? Is too little protection being applied to high value assets?
Hierarchical Protection	It is not necessary to protect a less secure component from a more secure component.	Does a system-level security hierarchy exist? Are there any circular relationships?
Minimize Trusted Components	A system should have a limited set of trusted elements, components, data, and functions.	Eliminate high security functions, data, and components whenever possible.
Least Privilege	Each component should only have sufficient privilege to accomplish its specified function.	Are the SoI's privileges identified, defined, and captured in a single location? For each function, subsystem, or component is the principle of minimum access employed?
Multi-Factor Permissions	Highly critical operations and data should require multiple authorization steps.	Do mission critical activities and data require multi-factor authorization? Do these multi-factor permissions inhibit mission execution?
Self-Reliance	Minimize dependencies on other systems, components, infrastructure, and data.	Are the internal dependencies reduced? Are external dependencies reduced?
Secure Composition	Combining secure components or functions should not reduce their collective level of trustworthiness.	Has the SoI's assembly reduced the trustworthiness of the security capability? Has the SoI's assembly reduced the trustworthiness of the resiliency capability?
Trusted Communication	Communication links and networks must be secured to the same level as the components and information that they carry.	Are the communication requirements known and well understood? Are secure communication channels properly identified in an architectural document (conceptual and/or physical)? Are the secure communication channels adequately secured? Have unnecessary high security information networks and communication links been eliminated or minimized?

when engineering rigor and systems analysis is applied early in the life cycle). At the production phase, the engineering effort is typically focused on ensuring that the SoI is built in accordance with the design specifications with additional consideration for the trustworthiness of the system such that appropriate levels of trustworthy security and resiliency are maintained throughout implementation. For example, nation-state adversaries have demonstrated the ability to inject vulnerabilities during each phase of the development life cycle [1], so it is important that the trust-related principles (and the trustworthiness of the components produced) are continually monitored and assessed. Likewise, suitable training must be created (and successfully delivered) for operational personnel to be able to detect and respond when the SoI is under cyberattack. Unfortunately, this important human-system integration aspect of cyber systems is often neglected for more tangible hardware and software issues.

2.5.2.4   Utilization/Support Phase

The initial transition to the utilization phase is unlikely to merit any changes to the developer's level of effort as there is little extra to be gained from the design and architectural principles once the SoI is fielded; however, the trust principles will always require a degree of constant attention to ensure the system is being operated as designed. During the support phase (i.e., whenever modifications occur whether small or large), it's incumbent on the user and developers to pay careful attention to changes in the SoI, its operational environment, and its performance to include any supporting personnel, processes, and technology to properly understand the SoI's trustworthiness.

For example, during operations, those responsible for assuring the security of the SoI should remain vigilant in monitoring how the system is being used to understand if changes to its usage and/or the threats it faces require modifications to the SoI. To illustrate this point, suppose that during the utilization/support phase there are increasing tensions in a given geopolitical region with a nuclear-armed adversary who possess inter-continental ballistic missiles. Accordingly, the military elects to modify the system by installing a space-based early warning detection capability and utilize the existing SoI to relay indications of enemy launches. Moreover, these intra-theater notifications are to be integrated into a national missile defense system via the satellite ground station. What was once merely a theater-specific ballistic missile warning system is now going to be part of a larger system-of-systems which supports a critical, strategic-level military capability. As indicated by the utilization/support spike in Figure 2.5, the system developer should once again revisit the architectural and trust principles to ensure the necessary modifications and resulting system composition still meets the SoI's trust worthiness requirements (which should be commensurate with its new role).

For instance, the developer may need to determine if the communications channels are sufficiently protected for this new mission (i.e., Trusted Communication) or address the vulnerability of commercial off-the-shelf elements in the supply chain (i.e., Trusted Components). At this point, the system developer is presumably thankful that consideration was given for the Secure Evolution of the SoI. Once the system is modified, the architectural-based principles are de-emphasized once again, but the trust-related principles should be continually monitored to assure the trustworthiness of the system (which is now at a higher level due to the SoI's relative increase in strategic importance to its key stakeholder–the military).

2.5.2.5 Retirement Phase

Once the system has reached the end of its useful life, it should be properly decommissioned with respect to both systems security and trust. For example, ensuring no sensitive configuration information remains in the SoI prior to its disposal where an adversary may gain operational insight. In another example, proprietary interfaces should be destroyed or removed to prevent ease of access.

2.6 CONCLUSION

This article is part of a series of works that aims to assist developers, owners, and operators in understanding and achieving a rigorous, yet cost-effective SSE approach. This work uniquely provides an analysis of the NIST SP 800-160 Vol. 1 security design principles with a detailed mapping and analysis of conceptual security strategies to design principles that can be more effectively designed-for, built-in, and tested to meet security and resiliency objectives. Insights regarding the interrelationships among the 18 design principles are described along with practical guidance on when and how they can be applied to more effectively perform SSE activities.

While attempting to systematically apply the design principles to both the case study described within and an additional autonomous vehicle example (not described), we found it difficult at times to differentiate between the various principles. We also found redundancy between the design principles, especially as we abstracted up the principles to the systems-level from their original lower-level, computer security roots. This is not a criticism, per se, but merely a finding that leads us to believe that a smaller set of systems security design principles is probably likely. Additionally, in considering the application of these security design principles to an autonomous system, we found the principles had limited applicability to this growing area of interest. Again, we state this not as a criticism – as the original design principles are not specifically for autonomous systems – but merely to suggest that design principles for trustworthy autonomous cyber-physical systems would be an interesting area to study.

Our future work includes: 1. Identifying associated technical performance measurements for these design principles; 2. Analyzing the application of these principles through modeling and simulation; and 3. Further understanding the applicability of these principles for autonomous systems.

ACKNOWLEDGMENT

The authors would like to thank Air Force Research Laboratory, Space Vehicles Directorate, Kirtland AFB, NM and Michael A. McEvilley, co-author of NIST SP 800-160 Systems Security Engineering for their support in this research effort. They would also like to thank the participants of NATO panel IST-164 for attempting to apply these design principles to an autonomous SoI. The lessons learned from the NATO working group are very helpful in refining the key ideas presented in this chapter.

DISCLAIMER

The views expressed in this paper are those of the authors and do not reflect the official policy or position of the U. S. Air Force, the Department of Defense, or the U.S. Government.

REFERENCES

1. J. R. Gosler and L. Von Thaer, *Task Force Report: Resilient Military Systems and the Advanced Cyber Threat*. Washington, DC: Dept. Defense Bus. Board, 2013.
2. K. Baldwin, J. F. Miller, P. R. Popick, and J. Goodnight, "The United States department of defense revitalization of system security engineering through program protection," in *Proc. IEEE Int. Syst. Conf. SysCon*, Mar. 2012, pp. 1–7.
3. D. Snyder, J. D. Powers, E. Bodine-Baron, B. Fox, L. Kendrick, and M. H. Powell, *Improving the Cybersecurity of U.S. Air Force Military Systems Throughout Their Life Cycles*. Santa Monica, CA: RAND Corporation, 2015.
4. *A Path Towards Cyber Resilient and Secure Systems*, NDIA Syst. Eng. Division, Virginia, VA, USA, 2016.
5. L. O. Mailloux, M. A. McEvilley, S. Khou, and J. M. Pecarina, "Putting the 'Systems' in security engineering: An examination of NIST special publication 800-160," *IEEE Secur. Privacy*, vol. 14, no. 4, pp. 76–80, Jul./Aug. 2016.
6. R. Ross, M. McEvilley, and J. C. Oren, *Systems Security Engineering: Considerations for a Multidisciplinary Approach in the Engineering of Trustworthy Secure Systems*. Gaithersburg, MD: NIST, 2018.
7. R. Ross, R. Graubart, D. Bodeau, and R. McQuaid, *Systems Security Engineering: Cyber Resiliency Considerations for the Engineering of Trustworthy Secure Systems (DRAFT*. Gaithersburg, MD: NIST, 2018.
8. *IEC/IEEE International Standard-Systems and Software Engineering– System Life Cycle Processes*, 1st ed., ISO/IEC/IEEE International Standard, ISO/IEC/IEEE, 15288-2015, 2015.
9. D. J. Bodeau, R. D. Graubart, J. Picciotto, and R. McQuaid, *Cyber Resiliency Engineering Framework*. Bedford, MA: MITRE Corporation, 2011.
10. P. M. Beach, L. O. Mailloux, and M. T. Span, "Examination of security design principles from NIST SP 800-160," in *Proc. Annu. IEEE Int. Syst. Conf. (SysCon)*, Vancouver, BC, Canada, Apr. 2018, pp. 1–8.
11. C. Irvine and T. D. Nguyen, "Educating the systems security engineer's apprentice," *IEEE Secur. Privacy*, vol. 8, no. 4, pp. 58–61, Jul. 2010.
12. J. Bayuk, "Systems security engineering," *IEEE Secur. Privacy*, vol. 9, no. 2, pp. 72–74, Mar./Apr. 2011.
13. K. Baldwin, "System security engineering: A critical discipline of systems engineering," *Insight*, vol. 12, no. 2, pp. 11–13, 2009.
14. L. Masso and B. Wilson, "Management initiatives to integrate systems and security engineering," *Insight*, vol. 16, no. 2, pp. 10–12, 2013.
15. J. C. Oren, "What does a systems security engineer do and why do systems engineers care?" *Insight*, vol. 16, no. 2, pp. 16–18, 2013.
16. R. Anderson, *Security Engineering*, 2nd ed. Indianapolis, IN: Wiley, 2008.
17. J. Bayukm, D. Barnabe, J. Goodnight, D. Hamilton, and B. Horowitz, *Systems Security Engineering Final Technical Report*. Hoboken, NJ: Stevens Institute, 2010.
18. *Military Handbook 1785. System Security Engineering Program Management Requirements*. Washington, DC: Dept. Defense, 1989.
19. N. S. Agency. (Nov. 2016). *National Centers of Academic Excellence*. Accessed: May 30, 2017]. [Online]. Available: https://www.nsa.gov/ resources/educators/centers-academic-excellence/cyberoperations/requirements.shtml.

20. M. Schumacher, E. Fernandez-Buglioni, D. Hybertson, F. Buschmann and P. Sommerlad, *Security Patterns: Integrating Security and Systems Engineering.* Hoboken, NJ: Wiley, 2013.

21. Defense Acquisition University. (May 2017). *System Survivability Key Performance Parameter.* Accessed: Jun. 1, 2017. [Online]. Available: https://dap.dau.mil/acquipedia/ Pages/ArticleDetails.aspx?aid=626ace5ffc1f-4638-9321-fe4345451558.

22. T. V. Benzel, C. E. Irvine, T. E. Levin, G. Bhaskara, T. D. Nguyen, and P. C. Clark, "Design principles for security," Dept. Comput. Sci., Nav. Postgraduate School, Monterey, CA, Tech. Rep. NPS-CS-05-010, 2005.

23. X. Sun, P. Liu, and A. Singhal, "Toward cyberresiliency in the context of cloud computing [resilient security]," *IEEE Secur. Privacy*, vol. 16, no. 6, pp. 71–75, Nov./Dec. 2018.

24. *INCOSE Systems Engineering Handbook: A Guide for System Life Cycle Processes and Activities, Version 4*, Int. Council Syst. Eng., San Diego, CA, 2014.

25. D. Cofer, A. Gacek, J. Backes, M. W. Whalen, L. Pike, A. Foltzer, M. Podhradsky, G. Klein, I. Kuz, J. Andronick, G. Heiser, and D. Stuart, "A formal approach to constructing secure air vehicle software," *Computer*, vol. 51, no. 11, pp. 14–23, Nov. 2018.

26. S. Khou, L. O. Mailloux, and J. M. Pecarina, "System-agnostic security domains for understanding and prioritizing systems security engineering efforts," *IEEE Access*, vol. 5, pp. 3465–3474, 2017.

27. S. Khou, L. O. Mailloux, J. M. Pecarina, and M. Mcevilley, "A customizable framework for prioritizing systems security engineering processes, activities, and tasks," *IEEE Access*, vol. 5, pp. 12878–12894, 2017.

28. M. Span, L. O. Mailloux, R. F. Mills, and W. Young, "Conceptual systems security requirements analysis: Aerial refueling case study," *IEEE Access*, vol. 6, pp. 46668–46682, 2018.

29. E. Crawley, B. Cameron, and D. Selva, *System Architecture: Strategy and Product Development for Complex Systems.* Upper Saddle River, NJ: Prentice-Hall Press, 2015.

30. D. V. Steward, "The design structure system: A method for managing the design of complex systems," *IEEE Trans. Eng. Manage.*, vol. EM–28, no. 3, pp. 71–74, Aug. 1981.

31. R. E. Thebeau, *Knowledge Management of System Interfaces and Interactions From Product Development Processes.* Cambridge, MA: Massachusetts Institute of Technology, 2001.

32. F. Borjesson and K. Höltta-Otto, "Improved clustering algorithm for design structure matrix," in *Proc. ASME Int. Design Eng. Tech. Conf. Comput. Inf. Eng. Conf.*, Aug. 2012, pp. 921–930.

33. P. G. Neumann, "Reflections on system trustworthiness," *Adv. Comput.*, vol. 70, pp. 269–310, Jan. 2007.

34. A. Avižienis, "A visit to the jungle of terminology," in *Proc. 47th Annu. IEEE/IFIP Int. Conf. Dependable Syst. Netw. Workshops*, Jun. 2017, pp. 149–152.

35. National Defense Industrial Association. (2017). *NDIA Systems Engineering Cyber Resilient & Secure Weapon System Summit 2017.* Accessed: May 5, 2017. [Online]. Available: http://www.ndia.org/events/2017/4/18/ndia-systems-engineering-cyber-resilient -and-secure-weapon-systemsummit-2017.

36. O. Sokolsky and N. Bezzo, "Resiliency in cyber-physical systems [Guest Editors' Introduction]," *Computer*, vol. 51, no. 11, pp. 10–12, Nov. 2018.

37. J. Gao, B. Barzel, and A. L. Barabási, "Universal resilience patterns in complex networks," *Nature*, vol. 530, no. 7590, pp. 307–312, Feb. 2016.

38. D. J. Bodeau, R. D. Graubart, J. Picciotto, and R. McQuaid, *Cyber Resiliency Engineering Framework*. Bedford, MA: MITRE Corporation, 2011.

39. J. Hughes and G. Cybenko, "Three tenets for secure cyber-physical system design and assessment," *Proc. SPIE*, vol. 9097, Jun. 2014, Art. no. 90970A-1.

40. D. Bodeau and R. Graubart, *Cyber Resiliency Design Principles*. Bedford, MA: MITRE, 2017.

41. J. Knight, J. Davidson, A. Nguyen-Tuong, J. Hiser, and M. Co, "Diversity in Cybersecurity," *Computer*, vol. 49, no. 4, pp. 94–98, Apr. 2016.

42. K. Sakai. (Aug. 17, 2017). Japan to Deploy New Land-Based Missile Defense System. Nikkei Asian Review. Accessed: Mar. 29, 2019. [Online]. Available: https://asia.nikkei.com/Politics/Japan-to-deploy-newland-based-missile-defense-system.

3 Operator Suspicion and Human–Machine Team Performance under Mission Scenarios of Unmanned Ground Vehicle Operation[*]

Chris Gay and John J. Elshaw
US Air Force Institute of Technology,

Barry Horowitz and Inki Kim
University of Virginia

Philip Bobko
Management and Psychology

CONTENTS

[*] Reprinted with permission: Gay, C., Horowitz, B., Elshaw, J. J., Bobko, P., & Kim, I. (2019). Operator suspicion and human-machine team performance under mission scenarios of unmanned ground vehicle operation. IEEE Access, 7, 36371–36379.

DOI: 10.1201/9781003220978-4

3.1   INTRODUCTION

A considerable effort is ongoing to prevent, detect and mitigate cyberattacks on the Department of Defense networks and information technology (IT) systems; in contrast, the effort to address these concerns in cyber-physical system (CPS), such as unmanned vehicle systems, pales in comparison. These systems represent an intrinsic vulnerability and allow adversaries to attempt cyberattacks with the malicious intention of undermining military assets. As an example, Iranian cyber capabilities were believed to have forced down the Central Intelligence Agency-operated RQ-170 Sentinel drone while operating near the Iranian border in 2011 [1], causing concern over the potential compromise of highly sensitive surveillance capabilities. This incident sparked much research directed toward the hardware and software security of unmanned vehicle systems [2, 3]. However, research addressing the human dimensions of cyberattack detection and response in the mission operation context remains sparse and represents an emergent area of research needed to fully address cyberattacks against CPS.

Our research took an operator-centric approach toward exploring the human dimensions of cyberattack detection and responses through a scenario-based, human-in-the-loop experiment with Air Force personnel as operators of an unmanned vehicle system in a military context. In prior work, we took a systems-oriented approach to the problem by considering the interaction of a Human–Machine Team (HMT) [4, 5] responding to cyberattacks and defining a framework of performance measurement [6].

In this work, HMT is defined as a team of an operator and Sentinel, an automated cyberattack detection aid. For machine design, the operators' biases associated with suspicion in their responses to cyberattacks shed light on the development of an adaptive sentinel. For human operators, the findings on the relationship between HMT performance and level of suspicion have implications for the selection, evaluation and training of appropriate personnel.

3.2   BACKGROUND

A challenge in designing for high-performance HMT is a lack of theory to help understand how humans interact with machines in work contexts. A recent paradigm in human–machine automation considers autonomy as a variable, rather than a fixed parameter, which can be distributed between human and artificial agents to achieve optimal performance at work [7]. An ultimate vision for human–machine teamwork is to "race with machines" [8], not against ones, by continuously redefining human roles under new work processes. The promise of complementary engagement of human and machine abilities for enhanced performance has seen some positive examples [9, 10]. Yet, it is difficult to fully accomplish this vision without knowing the constraints of the human, the machine, and the environment [11].

In military operations, mission complexity is outpacing the ability to manage disruptions, which calls for systemic approaches that span technology, human, and mission space [12]. At a minimum, any framework that addresses this complexity should enable the evaluation of human–machine interactions with regard to the nature of problem and solution sets [13, 14], under the situational constraints of mission context. The traditional framework of Level of Automation (LOA) and its alternatives [15–17] are confined to the concept of function allocation, not reflecting situational constraints.

So far, many unmanned systems [18] have attained assurance by counting on human supervision as the last resort. Some systems attempt to augment human cognitive abilities in particular tasks, such as spatial detection [19] and path planning [20]. The cognitive support in HMT [4, 21, 22], focuses on team cognition and mental workload. In particular, human–machine collaboration for emergency management has gained attention, with a focus on risk management and resiliency [23, 24]. Under emergency situations, HMTs are forced to make decisions within tight time schedules often with incomplete information, while the new situational complexity is likely to overload team cognitive resources [25]. In military unmanned systems, a failure to first-respond to the emergency situations can result in catastrophic damages, and there are growing concerns over the potential of cyber threats to impede the timely responses [26].

There are methods proposed to help analyze and guide cognitive responses of human supervisor under cyberattacks (see [27], for instance), but they do not fully consider the dynamic interdependence of human, machine, and situational context. The Instance-based Learning (IBL) model for cyber situation awareness [28] predicted security analysts' recognition of cyberattacks based on the situational attributes and on their similarity to past instances (to be retrieved from memory). Another example of analyzing cyber situation awareness in [29], proposed a distribution-based simulation model to identify cyber-behaviors and their cognitive aspects based on browser log data. In a hybrid approach, the work in [30] proposed a decision-support scheme to assist in response selection against cyber threats by combining qualitative expert assessment, event history, and multi-criteria decision analysis [31]. Although these works presented formal models and methods to represent performance in cognitive aspects, they focused exclusively on humans, rather than on the dynamics of the HMT.

The dynamics associated with the analysis of HMT performance can be internal (i.e., between human and machine), or external (i.e., situation-specific relations between the team and work-related factors). Regarding internal dynamics, the concept of "trust" is key to successful emergency responses – i.e., how trust is formed, developed and confirmed with the automated agents [32]. The literature on operator trust abounds [33–36], including when the autonomous systems are under potential cyberattacks [37]. A wide array of factors has been identified that influence the level of trust in human-automation interaction [38]. Not only formation, but the confirmation of trust becomes critical particularly when an unmanned system is under cyberattack. On the contrary, relatively little attention has been paid to understand the external dynamics of the HMT. Such investigations are not straightforward because it is not always feasible to keep the situational factors transparent to the supervisor or the machine [39]. For example, in which task-related conditions can the HMT performance be weakened (or strengthened)? Are there particular cognitive states of the human supervisor that can help improve HMT performance? What are the effective ways for the machine to support the supervisor under cyberattacks?

This paper determined the construct of suspicion to be particularly useful for investigating HMT performance in response to cyberattacks. In recent work, the theory of suspicion [40] defines state-suspicion as "a person's simultaneous state of cognitive activity, uncertainty, and perceived malintent about underlying information that is being electronically generated, collated, sent, analyzed, or implemented by an external agent". This work also describes the sequential structure of state-suspicion development across three stages.

Stage 1 refers to perceptual cues and indications from the task environment that can trigger suspicious states in the mind of the operators. For example, missing information, patterns of negative discrepancy, or other system and interface characteristics can serve to provoke different levels of suspicion. In UGV control, an operator and Sentinel collaborate in a team for detection and response to cyberattacks, and the Sentinel alert messages, or their lack, on a control interface can serve as stage-1 cues to initiate operator suspicion. Therefore, this research manipulated the sentinel alert messages to stimulate state-suspicion. For example, the Sentinel alert message popped up in the mission video window that read "Cyberattack: Throttle Control," and it remained visible for 30 s, see Figure 3.1. Not only the display of alert, but the lack of alert when the vehicle was maneuvering abnormally could also trigger suspicion from the interface.

Stage 2 of the suspicion model identifies individual levels of trust, distrust, training and other personal traits that can affect state-level suspicion [41, 42]. Especially, an operator's trait-level attributes, including creativity, cognitive demand and capacity, and propensity to trust [43], can form an internal condition to the arousal of state-suspicion [40]. This research incorporated a set of pre-test surveys prior to the experiment about operator self-ratings of intelligence scores, creativity, general attitude toward complex problems, and propensity to trust.

Finally, stage 3 refers to behavioral, cognitive and emotional outcomes of becoming suspicious. In particular, the State-Suspicion Index (SSI) [40] has been developed to quantify the level of suspicion through a 20-item questionnaire that assesses the suspicion components of uncertainty, malintent, and cognitive load, as well as overall suspicion. To reflect the operational context of UGV missions, the original SSI was adapted to a 13-item questionnaire in collaboration with one of those authors.

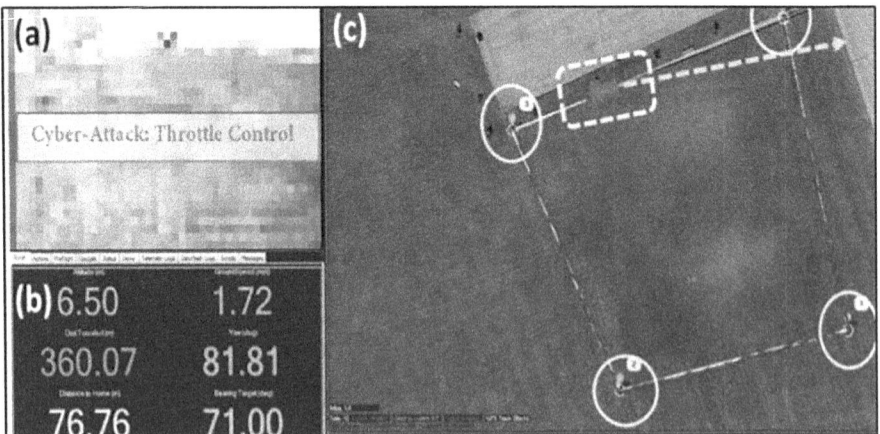

FIGURE 3.1   A Mission Scenario on a UGV Control Interface. (a) Ego-Centric view of the UGV with a sentinel alert, (b) numerical indicators of the UGV control parameters, (c) Bird's-Eye view with way-points (Circled) and the UGV's location and direction (dotted lines).

3.3 METHODS

This research primarily revolves around the relationship between the level of operator suspicion and HMT performance in the mission operation context of an unmanned ground vehicle (UGV). The definition of key variables and their measurements, and experimental process are described in this section.

3.3.1 RESEARCH HYPOTHESES, VARIABLES, AND MEASUREMENTS

This research primarily revolves around the relationship between the level of operator suspicion and HMT performance in the mission operation context of an UGV. To answer how suspicion affects HMT performance in a human-in-the-loop simulation, this research paired a UGV operator with a sentinel for automated cyberattack detection. Guided by suspicion theory, a set of visual cues in the sentinel alarm and control environment was simulated for anomalous system events under different mission scenarios. On completing each mission scenario, HMT performance, as well as suspicion level, was quantified. HMT performance was evaluated on the two general criteria of speed and accuracy [44], for the detection and selection of responses to suspected cyberattacks. To elaborate on the research question, the following hypotheses were set.

H_1: Sentinel alert has significant effects on operator suspicion.
H_2: Operator suspicion is positively related to HMT performance.
H_3: Cyberattack (yes/no)/Sentinel alert (alert/no alert) combinations have significant effects on operator suspicion.
H_4: Operator suspicion is positively related to operator response time (i.e. response times are delayed when operators are more suspicious).

Figure 3.2 depicts how these four hypotheses associate the operator's suspicion with the responses to cyberattacks on unmanned systems. Based on suspicion theory,

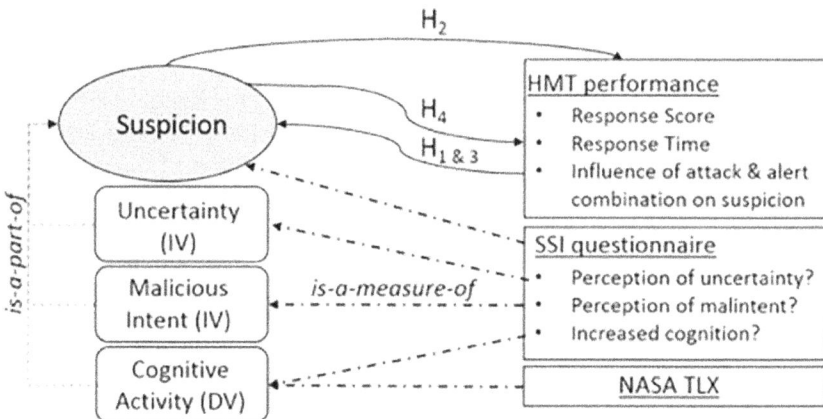

FIGURE 3.2 Overview of experimental variables, relationships among them, and methods of measurement.

operator suspicion is presumed to be a latent variable that has three components, "Uncertainty", "Malicious Intent", and "Cognitive Activity." The experimental levels, either high or low, of both uncertainty and malicious intent were manipulated as independent variables (IV) through each mission scenario, while cognitive activity was measured as a dependent variable (DV) at the end of each mission run. For the estimation of cognitive activity, NASA-TLX [45, 46] and the related items in the SSI questionnaire were used. The two levels for each one of the IVs were verified based on the responses to the corresponding items in the SSI questionnaire, which helped confirm if different mission scenarios effectively set different levels of perceptions as intended. The 13-item SSI questionnaire determined the overall level of suspicion (by linear combination of its components).

The performance measures of response time and score were recorded while the mission videos dynamically played back anomalous system events, including cyber-attacks and sentinel alert messages. The time to respond to such events was recorded using an interactive polling software (TurningPoint, TurningTechnologies, Ltd.) during the experiment, and the performance score was determined post-experiment based on rubrics. Each mission scenario has its own unique score rubric defined by subject matter experts. The operator's response selections from a given set of decision trees were logged in the software and were then evaluated against the rubric. Furthermore, the four-way combinations of cyberattacks (attacks vs. no attacks) and sentinel alert messages (alert vs. no alert) enabled us to analyze operator suspicion under different circumstances.

3.3.2 EXPERIMENTAL PROCEDURES, DESIGN, AND SETTING

The human-in-the-loop experiments were designed and conducted in three phases. In phase 1, we obtained consent from thirty-two military operators (IRB: FWR20160115H) and collected personal information, including demographic and personality-related questionnaires. Phase 2 familiarized participants with the experimental tasks through instruction and demonstrations, so that an acceptable level of fluency was ensured in the operational context. In phase 3, participants were presented, in a random order, a series of eight mission scenarios, each with a pair of mission briefing and mission videos. Once the mission briefing was done, the participant responded to events on the mission videos that occurred during each mission scenario while response selection and response times were recorded simultaneously. On completion of each mission scenario, participants' perceptions of uncertainty, malicious intent, and cognitive workload during the mission were obtained via the NASA TLX and SSI questionnaires.

To the operator, a mission scenario was characterized by the combination of mission briefings, illustrated in Figure 3.3, and mission videos. The mission briefings described mission type, mission context, and descriptive profiles for the operation of the unmanned ground vehicle system (UGVS). The mission type was either training or operational missions for transport and re-supply. The mission context was set in the U.S. or Middle Eastern locations, with the corresponding estimated frequencies of cyberattacks in the past.

For the machine-side, a mission profile configured the UGV behaviors: when a profile was deployed, the UGVS autonomously ran it and generated mission views

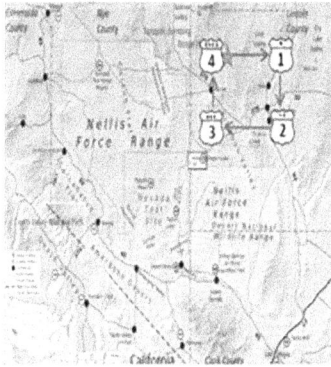

<Mission Brief>
[Mission Location] United States, Nellis Air Force Range, Nevada
[Description] Your Unit is participating in a Flag Exercise at Nellis Air Force Range, Nevada, and its mission is the autonomous delivery UGV with supplies to a remote operating unit.
[Risk] Missions in this area have been protested due to its proximity to the Wildhorse Management Area..., but there is no intel suggesting activist actions during the Flag Exercise...Failure of your UGV mission will result in a missed opportunity to exercise integrated support of a remote unit assault and waste valuable / limited ISR and CAS assets.

FIGURE 3.3   Illustration of a mission briefing (part).

TABLE 3.1

Descriptive Statistics for Cyberattack/Sentinel Alert Combinations (mean ± SD)

Dependence Variable	(a) TP Attack Yes /Alert Yes	(b) TN Attack No/ Alert No	(c) FP Attack No/ Alert Yes	(d) FN Attack Yes /Alert No
Suspicion; (7-Likert)	4.38 ± 0.97	3.97 ± 0.96	3.90 ± 0.96	4.52 ± 1.00
Score; (0–100)	90.8 ± 18.1	94.8 ± 12.1	92.7 ± 15.7	81.2 ± 27.7
Time (s)	14.91 ± 16.45	1.00 ± 2.56	4.61 ± 3.69	13.48 ± 12.79
NASA-TLX; (Rating 0–25)	22.1 ± 14.1	16.6 ± 11.4	16.2 ± 9.8	23.5 ± 13.8

for playback to use in the simulation experiments. Overall, both verbal and visual elements of mission scenarios were constructed to indirectly manipulate the operator's state-suspicion by forming the two independent variables (IVs), "uncertainty" and "malicious intent", into a two-level full-factorial design.

After being oriented to the mission briefing, the participants were tasked to record the UGV speed every thirty seconds while monitoring the mission video as well as instrument readouts for anomalous events from the UGV mission. On detecting anomalous events, the participants were instructed to select the most appropriate response from a decision tree that was provided as a guide to standardize the range of possible operator responses (see Table 3.1). These tasks were closely aligned with typical unmanned vehicle system operator tasks. At the conclusion of each mission scenario, the operator completed two questionnaires: (i) the NASA TLX questionnaire which quantifies the operator's self-assessment of cognitive workload on six dimensions, with each dimension rated on a 0–100 scale, and (ii) the 13-item SSI questionnaire which was developed specifically for this research and evaluated (on a 7-point Likert scale) the operator's perception of uncertainty, malicious intent, cognitive activation, and overall suspicion. The 13 items measuring suspicion were then aggregated to form an overall quantitative measure of operator suspicion. Cronbach's alpha [47] for the 13 items was.88, indicating acceptable internal consistency of the measure. Participants were thirty-two Air Force officers from the Air Force Institute of Technology (AFIT), with each experiment taking 2–2 1/2 h to complete. Since many current operations associated with unmanned vehicle missions occur in an office environment, the experiment took place in such a space.

3.4   RESULTS

The experiment yielded significant outcomes on the relationship between operator suspicion and HMT performance. The overall level of suspicion derived from the 13-item SSI questionnaire had a significant ($p < 0.001$) Pearson correlation with the response score ($\rho = -0.251$), as well as with the response time ($\rho = 0.379$). As was expected with the suspicion theory, the subgroups of the SSI questionnaire items that correspond to perception of uncertainty, malicious intent, and cognitive activation also showed a strong, significant correlation with overall suspicion ($p < 0.001$). Correlations were estimated to be $\rho = 0.803$, $\rho = 0.905$, and $\rho = 0.828$, respectively. In addition, there were predictable relationships among the HMT performance metrics. The response score was negatively correlated with response time ($\rho = -0.225$), as well as with the standard deviation of its own score ($\rho = -0.354$), implying that less-desirable decision responses tend to accompany slow and inconsistent responses.

Contrary to H_1, which proposed Sentinel alerts are related to operator suspicion, the result of the one-way ANOVA was not significant (F_1, $254 = 0.688$, $p = 0.408$); hence Sentinel alerts alone did not create operator suspicion. The variability within each group of the sentinel alert being activated versus not activated, outweighed that of the between-group ($MSE_{Within-group} = 1.009$, $MSE_{Between-group} = 0.694$). It contrasts, for the factor of cyberattack (or not), a second one-way ANOVA showed a significantly different level of suspicion (F_1, $254 = 18.393$, $p < 0.001$); i.e., a higher level of suspicion was observed when attacks occurred. These results that Sentinel alerts are not an independent factor of suspicion despite their visual saliency on a display, while cyberattacks did significantly arouse suspicion, imply a complicated cognitive structure of judgment based on the uncertainty of perceptual information [48, 49]. Of particular statistical concern with this uncertainty, is a shared-variance structure of an individual operator's suspicion, which might be determined by the combined effects of the Sentinel alert and cyberattack scenarios.

To resolve these combined effects, a Hierarchical Linear Model (HLM) was applied. The HLM is capable of accounting for the shared-variance structure in a nested data with hierarchical levels of variables, by using a complex form of Ordinary Least Squares (OLS) regressions [50]. This HLM method can effectively compensate for the known risks [51]; a risk of ignoring the between-scenario effects on suspicion (as was seen in the first one-way ANOVA of the previous paragraph), as well as ignoring the individual propensity to trust the Sentinel alert (as seen in the second one-way ANOVA of the previous paragraph).

In general, the final outcome of HLM takes on a form of simple regression, where a dependent variable Y_{ij} is predicted by using an ith-level variable X_{ij} that is nested within a higher-level variable, j.

$$Y_{ij} = \beta_{0j} + \beta_{1j}X_{ij} + \varepsilon_{ij} \tag{3.1}$$

In HLM, this low-level model (3.1) further incorporates the higher-level models of equations (3.2) and (3.3) below, for each of the coefficients, β_{0j} and β_{1j}, in terms of the interim variable Q_j of the jth-level variable, and the random effects, U_{0j} and U_{1j}, that are adjusted for Q_j. The statistical significance of β_{1j} can be tested to determine if the combined levels i and j influence the dependent variable Y_{ij}.

$$\beta_{0j} = \gamma_{00} + \gamma_{01}Q_j + U_{0j} \tag{3.2}$$

$$\beta_{1j} = \gamma^{10} + \gamma_{11}Q_j + U_{1j} \tag{3.3}$$

Finally, the overall model in (3.4) incorporates both the ith-level and the jth-level predictors X_{ij} and Q_j, respectively, by combining (3.2) and (3.3) into (3.1).

$$Y_{ij} = \gamma_{00} + \gamma_{10}X_{ij} + \gamma_{01}Q_j + \gamma_{11}Q_jX_{ij} + U_{1j}X_{ij} + U_{0j} + \varepsilon_{ij} \tag{3.4}$$

In order to apply HLM, the dependent variables as summarized in Figure 3.2 for state-suspicion, HMT performance, and cognitive workload, respectively, were structured for each combination of the level-i of Sentinel alert ($i=0$ if no alert; $i=1$ if an alert message was shown on a display) and the level j of cyberattacks ($j=0$ if no attacks; $j=1$ if any attacks occurred in an experimental scenario). Table 3.1 summarizes the mean and the standard deviation for each combination of the nested levels. Such orthogonal dichotomies of True/False (of cyberattacks) and Positive/Negative (of sentinel alarm) on a 2-by-2 contingency table allows us to further analyze the experimental results around the classic framework of signal detection theory [52].

Since cyberattacks are by nature malicious events and require consideration of multiple solutions for the observed behavior, H_2 hypothesized that operator suspicion is positively related with HMT performance, suggesting a suspicious operator would score better on the tasks. This hypothesis was the opposite of the experimental results. The linear coefficient of the HLM analysis was significant when modeled after (1) ($\beta_{10} = -5.63$, $p < 0.001$), and the direction of the relationship was negative, meaning increased operator suspicion had a significantly negative relationship to HMT performance as depicted in Figure 3.4.

Additionally, H_4 proposed operator suspicion is positively related to operator task response time, which meant higher suspicion is associated with a longer task response time. The linear coefficient of the HLM analysis supported H_4; i.e., the relationship is

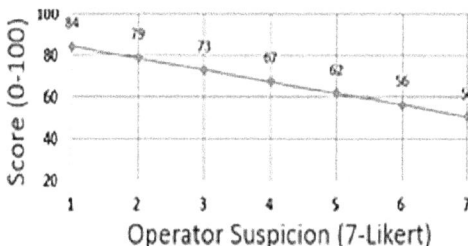

FIGURE 3.4 Performance score as a function of operator suspicion.

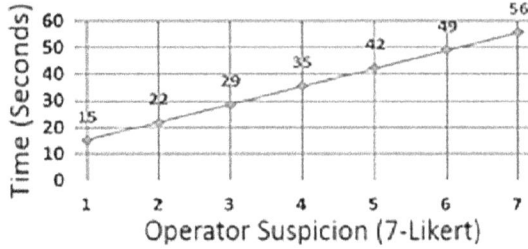

FIGURE 3.5   Time as a function of operator suspicion.

TABLE 3.2
Combined Effects on Operator Suspicion

		Cyberattacks	
		Yes	**No**
Sentinel	**Yes**	**(a) True Positive (TP)**	**(c) False Positive (FP)**
Alert		Increases suspicion ↑	Decreases suspicion ↓
		($\beta_{10} = +0.255$, $p = 0.047$)	($\beta_{10} = -0.394$, $p = 0.002$)
	No	**(d) False Negative (FN)**	**(d) True Negative (TN)**
		Increases suspicion ↑	Decreases suspicion ↓
		($\beta_{10} = +0.440$, $p = 0.001$)	($\beta_{10} = -0.301$, $p = 0.019$)

statistically significant and in a positive direction ($\beta_{10} = 6.95$, $p < 0.001$). This linear relationship is depicted in Figure 3.5.

Finally, the four cyberattack and Sentinel alert combinations were tested in the experiment and analyzed by using the HLM as summarized in Table 3.2. The two combinations without cyberattacks, both (b) True Negative (TN) and (c) False Positive (FP), had a significant ($p < 0.05$) negative impact on operator suspicion, meaning that operator suspicion was lowered in both cases.

In contrast, the two combinations containing cyberattacks, (a) True Positive (TP) and (d) False Negative (FN), had a significantly ($p < 0.05$) positive impact on operator suspicion by increasing operator suspicion. These results are consistent with the finding for H_1 that Sentinel alerts alone do not always create suspicion.

The combined effects of Table 3.2 warrant further discussion. Table 3.3 presents a frequency analysis of HMT actions for each combination (a–d) in terms of the four dependent variables: operator decision selections, suspicion, HMT performance score evaluated in terms of the desirability of the decision response to a given mission scenario, and response time. As previously noted, all operators in the experiment responded to suspicious events by referring to a pre-defined tree of decision responses, and the frequencies associated with those response options are summarized in the first section of Table 3.3. The HMT actions in the combinations of (a) True Positive (TP) and (b) True Negative (TN) are predictable based on the findings

TABLE 3.3
Frequency Analysis of Cyberattack/Sentinel Alert Combinations

	(a) True Positive	(b) True Negative	(c) False Positive	(d) False Negative
Decision Responses on a Decision Tree				
0 - No response	-	51	1	<u>1</u>
1 - Continue Mission	2	4	<u>46</u>	<u>6</u>
2 - Take action; Sentinel fixes the problem; continue	54	5	11	<u>14</u>
3 - Take action; Operator fixes the problem; continue	5	4	6	<u>38</u>
4 - Take action; Call backup; continue	-	-	-	<u>2</u>
5 - Abort; recovery; backup	2	-	-	2
6 - Abort; recovery; no backup	1	-	-	1
Subtotal (N)	64	64	64	64
Suspicion (SSI Total range of 1–7) * Higher indicates more suspicious				
Low (SSI Total: 1–3)	5	10	12	1
Medium (SSI Total: 5–10)	40	43	41	40
High (SSI Total: 10–60)	19	11	11	23
Subtotal (N)	64	64	64	64
HTF Performance (Score range 0–100) * Evaluated by desirability				
Low (Score: 0–5)	3	-	1	11
Medium (Score: 50–75)	4	5	2	3
High (Score: 75–100)	57	59	61	50
Subtotal (N)	64	64	64	64
Response Time (Time range 1–60 s)				
Fast (Time: 0–5)				
Medium (Time: 5–10)				
Slow (Time: 10–60)				
Subtotal (N)	64	64	64	64

of other hypotheses and will not require further discussion. The more interesting behaviors are from situations (c) False Positive (FP) that represent scenarios in which no cyberattacks occurred, but the Sentinel sent an alert to the operator anyway, and (d) False Negative (FN) that represent scenarios in which cyberattacks occurred, but the Sentinel failed to send an alert.

In FP scenarios, 71.8% of responses (i.e., 46 out of 64) were judged desirable for the mission context by subject matter experts: when the operators received the Sentinel alert, most of them collected information available from the system to tell if a cyberattack was in effect and decided to over-ride the Sentinel alert by continuing the mission without taking additional action. This quick search-and-override decision resulted in a relatively higher HMT performance, and faster response times compared with other combinations as summarized in Table 3.3. Furthermore, there were no "call for backup" or "abort" actions, which may have come with high cost in mission operation. Overall, these responses in False Positive (FP) scenarios are generally desirable.

In contrast, the HMT actions in False Negative (FN) scenarios were considerably less desirable to the mission context. Regardless of the fact that the operators did not receive a Sentinel alert to prompt information search, they grew more suspicious when cyberattacks occurred, and it took longer for them to respond, yielding lower HMT performance scores. Of 64 responses, 38 chose to develop their own solutions, 2 called for backup, and 14 even allowed the Sentinel to act, which can be considered instances of *over-reliance* on the Sentinel although it did not detect the attack [53]. Another issue that emerged in this operational context is the frequency of missed detections. The operators completely missed the cyberattack seven times (the responses with codes 0 or 1). Overall, the HMT behaviors around the FN scenarios were potentially more damaging to mission outcomes.

3.5   DISCUSSIONS AND CONCLUSION

3.5.1   SENTINEL ALERT, SUSPICION, AND INFORMATION-SEEKING BEHAVIOR

The analysis of recent cyberattacks on cyber-physical infrastructures reveals that adversaries promptly adapt their attack strategies to mitigation actions [54]. This makes early detection and recognition of incoming cyberattacks even more critical to effective mitigation. So far, much research has focused on engineering cyberattack detection aids [55], while not necessarily considering their cognitive effects on the human or human–machine collaboration in mission contexts.

The finding that Sentinel alerts did not necessarily arouse operator suspicion (i.e., rejection of H_1) has implications for vigilant human–machine integration. Perhaps, rather than the Sentinel alert, visual cues of unexpected system behaviors in mission environment are more likely to determine suspicion.

In fact, our finding could be related to the perceived risk that might have been triggered by a Sentinel alert. In the decision science literature [56–58], a positive correlation of perceived risk and information-seeking behavior is widely observed in decision under uncertainty. For instance, when a consumer has to choose a service that does not allow feature-by-feature comparison, information-seeking behavior is a common strategy to reduce perceived risk. In particular, information search triggered by perceived risk is more likely to be thorough if decision-makers have less knowledge about their choice and its consequences [59], leading to increased search time. The operators not knowing the true system states on cyberattacks, Sentinel alert could have triggered perceived risk, which then initiated wider information search to resolve suspicion.

In this regard, operator suspicion is a state of suspended or postponed decision-making, and it significantly lengthened mission time as observed both in (a) TP and (d) FN of Table 3.1. This strong linear relationship of suspicion and time is depicted in Figure 3.5. The negative correlation of suspicion and performance score also suggests it is wider information-seeking behavior, rather than more elaborated response selection, which actually lengthened the mission time. If the increased mission times were due to more effort investigated into response selection, the operator would have obtained a better score.

Yet, one cannot rule out the possibility that the scenarios which evoked suspicion were inherently more difficult to respond to, and thus increased mission time. Besides, causal relations among alert, state-suspicion and information-seeking behavior are not fully established. The current results do not allow us to conclude how state-suspicion is aroused, modulated, and resolved in the context of HMT collaboration.

3.6 CONCLUSION

The novel application of suspicion theory to UGV operations in a military context demonstrated the potential of that theory – particularly in relation to understanding the operation of a human–machine (sentinel) team. We suggest that operator suspicion needs to be managed in order for a HMT to achieve the best results in regard to the detection of cyberattacks, and subsequent responses when unmanned vehicle systems incur those cyberattacks. This research provides an understanding of suspicion effects on HMT performance and offers insights about moving quickly (or not) from a position of state-suspicion to making a decision.

A Sentinel alert on cyberattack symbolizes the roles that automation can play in responding to cyberattacks and sheds light on how HMT design can help exploit operator suspicion. As systems developers consider the balance of false-positive and false-negative errors in the design of cyberattack detection aids, the results of this experiment suggest erring on the side of false positives as more desirable. In addition, the Sentinel design that was used in the experiment did not provide operators with any indication of the need for a more or less immediate response to the attack that was detected. Providing such information could potentially help operators in managing the undesirable delays that were experienced during the experiments. Satisfying such a need could be difficult as it places requirements on the Sentinel to develop more detailed assessments of the attacks that it detects and may also require access to additional data sources that would serve this purpose.

ACKNOWLEDGMENT

Any opinions, findings and conclusions or recommendations expressed in this material are those of the authors and do not necessarily reflect the views of the United States Department of Defense.

REFERENCES

1. K. Hartmann and C. Steup, "The vulnerability of UAVs to cyber attacks—An approach to the risk assessment," in *Proc. 5th Int. Conf. Cyber Conflict (CYCON)*, Jun. 2013, pp. 1–23.
2. B. M. Horowitz and K. M. Pierce, "The integration of diversely redundant designs, dynamic system models, and state estimation technology to the cyber security of physical systems," *Syst. Eng.*, vol. 16, no. 4, pp. 401–412, 2013.
3. R. A. Jones and B. M. Horowitz, "A system-aware cyber security architecture," *Syst. Eng.*, vol. 15, no. 2, pp. 225–240, 2012.

4. M. A. Neerincx et al., "The mission execution crew assistant: Improving human-machine team resilience for long duration missions," in *Proc. 59th Int. Astron. Congr. (IAC)*, Sep. 2008, pp. 7910–7921.

5. J. M. Hoc, "From human-machine interaction to human-machine cooperation," *Ergonomics*, vol. 43, no. 7, pp. 833–843, Jul. 2000.

6. C. Gay, B. Horowitz, J. Elshaw, P. Bobko, and I. Kim, "Operator suspicion and decision responses to cyber-attacks on unmanned ground vehicle systems," *Proc. Hum. Factors Ergonom. Soc. Annu. Meeting*, vol. 61, no. 1, pp. 226–230, Sep. 2017.

7. S. A. Mostafa, M. S. Ahmad, and A. Mustapha, "Adjustable autonomy: A systematic literature review," *Artif. Intell. Rev.* vol. 51, no. 2, pp. 149–186, 2019.

8. K. Kelly, *The Inevitable: Understanding the 12 Technological Forces that Will Shape our Future*. New York City: Viking Press, 2017.

9. A. Birk and M. Pfingsthorn, "A hmi supporting adjustable autonomy of rescue robots," in *Robot Soccer World Cup*. Berlin, Heidelberg: Springer, 2005, pp. 255–266.

10. K. Petersen and O. V. Stryk, "Towards a general communication concept for human supervision of autonomous robot teams," in *Proc. 4th Int. Conf. Adv. Comput.-Hum. Interact. (ACHI)*, 2011, pp. 228–235.

11. N. B. Sarter, D. D. Woods, and C. E. Billings, "Automation surprises," in *Handbook of Human Factors Ergonomics*, 2nd ed. G. Salvendy (Ed.), New York: Wiley, 1997, pp. 1926–1943.

12. *A World in Motion: Systems Engineering Vision 2025*, INCOSE, San Diego, CA, USA, 2014.

13. J. M. Bradshaw et al., "Adjustable autonomy and human-agent teamwork in practice: An interim report on space applications," in *Agent Autonomy*. Boston, MA, USA: Springer, 2003, pp. 243–280.

14. A. Alzahrani, V. Callaghan, and M. Gardner, "Towards adjustable autonomy in adaptive course sequencing," in *Proc. 9th Int. Conf. Intell. Environments*, 2013, pp. 466–477.

15. T. B. Sheridan, *Telerobotics, Automation, and Human Supervisory Control*. Cambridge, MA: MIT Press, 1992.

16. J. E. Allen, C. I. Guinn, and E. Horvtz, "Mixed-initiative interaction," *IEEE Intell. Syst. their Appl.*, vol. 14, no. 5, pp. 14–23, Sep. 1999.

17. T. Fong, C. Thorpe, and C. Baur, "Collaborative control: A robot-centric model for vehicle teleoperation," *School Comput. Sci., Carnegie Mellon Univ.*, Pittsburgh, PA, Tech. Rep. CMU-RI-TR-01-34, 2001.

18. M. L. De Brun et al., "Mixed-initiative adjustable autonomy for human/unmanned system teaming," in *Proc. AUVSI Unmanned Syst. North Amer. Conf.*, 2008, pp. 732–746.

19. J. Y. C. Chen and M. J. Barnes, "Supervisory control of robots using RoboLeader," *Proc. Hum. Factors Ergonom. Soc. Annu. Meeting*, vol. 54, no. 19, pp. 1483–1487, Sep. 2010.

20. J. Wang and M. Lewis, "Assessing cooperation in human control of heterogeneous robots," in *Proc. 3rd ACM/IEEE Int. Conf. Hum. Robot Interact.*, Mar. 2008, pp. 9–16.

21. E. Salas, M. A. Rosen, C. S. Burke, D. Nicholson, and W. R. Howse, "Markers for enhancing team cognition in complex environments: The power of team performance diagnosis," *Aviation, Space, Environ. Med.*, vol. 78, no. 5, pp. B77–B85, May 2007.

22. P. Millot and M.-P. Pacaux-Lemoine, "A common work space for a mutual enrichment of human-machine cooperation and team-situation awareness," *IFAC Proc. Volumes*, vol. 46, no. 15, pp. 387–394, 2013.

23. S. Zieba, P. Polet, and F. Vanderhaegen, "Using adjustable autonomy and human–machine cooperation to make a human–machine system resilient—Application to a ground robotic system," *Inf. Sci.*, vol. 181, no. 3, pp. 379–397, Feb. 2011.

24. L. A. M. Bush, A. J. Wang, and B. C. Williams, "Risk-based sensing in support of adjustable autonomy," in *Proc. IEEE Aerosp. Conf.*, Mar. 2012, pp. 1–18.

25. L. Carver and M. Turoff, "The human and computer as a team in emergency management information systems," *Commun. ACM*, vol. 50, no. 3, pp. 33–38, Mar. 2007.

26. G. Loukas, D. Gan, and T. Vuong, "A review of cyber threats and defence approaches in emergency management," *Future Internet*, vol. 5, no. 2, pp. 205–236, Jun. 2013.

27. L. Rothrock and S. Narayanan, *Human-in-the-Loop Simulations*. London: Springer, 2011.

28. V. Dutt, Y.-S. Ahn, and C. Gonzalez, "Cyber situation awareness: Modeling the security analyst in a cyber-attack scenario through instance-based learning," in *Proc. IFIP Annu. Conf. Data Appl. Secur. Privacy*, 2011, pp. 280–292.

29. D. Robinson and G. Cybenko, "A cyber-based behavioral model," *J. Defense Model. Simul., Appl., Methodol., Technol.*, vol. 9, no. 3, pp. 195–203, Jul. 2012.

30. M. R. Grimaila and A. Badiru, "A hybrid dynamic decision making methodology for defensive information technology contingency measure selection in the presence of cyber threats," *Oper. Res.*, vol. 13, no. 1, pp. 67–88, Apr. 2013.

31. A. B. Badiru, P. S. Pulat, and M. Kang, "DDM: Decision support system for hierarchical dynamic decision making," *Decis. Support Syst.*, vol. 10, no. 1, pp. 1–18, Jul. 1993.

32. K. E. Schaefer, E. R. Straub, J. Y. C. Chen, J. Putney, and A. W. Evans, III, "Communicating intent to develop shared situation awareness and engender trust in human-agent teams," *Cogn. Syst. Res.*, vol. 46, pp. 26–39, Dec. 2017.

33. J. D. Lee and K. A. See, "Trust in automation: Designing for appropriate reliance," *Hum. Factors, J. Hum. Factors Ergonom. Soc.*, vol. 46, no. 1, pp. 50–80, Mar. 2004.

34. D. H. McKnight, V. Choudhury, and C. Kacmar, "The impact of initial consumer trust on intentions to transact with a web site: A trust building model," *J. Strategic Inf. Syst.*, vol. 11, nos. 3–4, pp. 297–323, Dec. 2002.

35. E. T. Chancey, J. P. Bliss, A. B. Proaps, and P. Madhavan, "The role of trust as a mediator between system characteristics and response behaviors," *Hum. Factors, J. Hum. Factors Ergonom. Soc.*, vol. 57, no. 6, pp. 947–958, Apr. 2015.

36. J. Xu, K. Le, A. Deitermann, and E. Montague, "How different types of users develop trust in technology: A qualitative analysis of the antecedents of active and passive user trust in a shared technology," *Appl. Ergonom.*, vol. 45, no. 6, pp. 1495–1503, Nov. 2014.

37. F. Boroomand et al., "Cyber security for smart grid: A human-automation interaction framework," in *Proc. IEEE PES Innov. Smart Grid Technol. Conf. Eur. (ISGT Eur.)*, Oct. 2010, pp. 1–6.

38. K. E. Schaefer, J. Y. Chen, J. L. Szalma, and P. A. Hancock, "A meta-analysis of factors influencing the development of trust in automation: Implications for understanding autonomy in future systems," *Hum. Factors, J. Hum. Factors Ergonom. Soc.*, vol. 58, no. 3, pp. 377–400, Mar. 2016.

39. A. R. Selkowitz, S. G. Lakhmani, and J. Y. C. Chen, "Using agent transparency to support situation awareness of the autonomous squad member," *Cogn. Syst. Res.*, vol. 46, pp. 13–25, Dec. 2017.

40. P. Bobko, A. J. Barelka, L. M. Hirshfield, and J. B. Lyons, "Invited article: The construct of suspicion and how it can benefit theories and models in organizational science," *J. Bus. Psychol.*, vol. 29, no. 3, pp. 335–342, Sep. 2014.

41. D. B. Buller and J. K. Burgoon, "Interpersonal deception theory," *Commun. Theory*, vol. 6, no. 3, pp. 203–242, Aug. 1996.

42. J. L. Hilton, S. Fein, and D. T. Miller, "Suspicion and dispositional inference," *Pers. Social Psychol. Bull.*, vol. 19, no. 5, pp. 501–512, Oct. 1993.

43. J. Mayer and T. Mussweiler, "Suspicious spirits, flexible minds: When distrust enhances creativity," *J. Pers. Social Psychol.*, vol. 101, no. 6, p. 1262, Dec. 2011.

44. I. Kim and J. H. Jo, "Performance comparisons between thumb-based and finger-based input on a small touch-screen under realistic variability," *Int. J. Hum.-Comput. Interact.*, vol. 31, no. 11, pp. 746–760, Jun. 2015.

45. S. G. Hart and L. E. Staveland, "Development of NASA-TLX (Task load index): Results of empirical and theoretical research," *Adv. Psychol.*, vol. 52, pp. 139–183, Jan. 1988.
46. S. G. Hart, "NASA-task load index (NASA-TLX); 20 years later," *Proc. Hum. Factors Ergonom. Soc. Annu. Meeting*, vol. 50, no. 9, pp. 904–908, Oct. 2006.
47. J. R. A. Santos, "Cronbach's alpha: A tool for assessing the reliability of scales," *J. Extension*, vol. 37, no. 2, pp. 1–5, 1999.
48. D. E. Bell, H. Raiffa, and A. Tversky, "Descriptive, normative, and prescriptive interactions in decision making," in *Descriptive, Normative, and Prescriptive Interactions Decision Making*, vol. 1. New York: Cambridge Univ. Press, 1988, pp. 9–32.
49. R. W. Cooksey, *Judgment Analysis: Theory, Methods, and Applications.* New York: Academic, 1996.
50. S. W. Raudenbush and A. S. Bryk, *Hierarchical Linear Models: Applications and Data Analysis Methods*, vol. 1. London: Sage, 2002.
51. H. Woltman, A. Feldstain, J. C. MacKay, and M. Rocchi, "An introduction to hierarchical linear modeling," *Tuts. Quant. Methods Psychol.*, vol. 8, no. 1, pp. 52–69, 2012.
52. T. Harlow, "Deputy praised for not sweving in deer crash," *Star Tribune*, Oct. 2017.
53. K. Drnec, A. R. Marathe, J. R. Lukos, and J. S. Metcalfe, "From trust in automation to decision neuroscience: Applying cognitive neuroscience methods to understand and improve interaction decisions involved in human automation interaction," *Frontiers Hum. Neurosci.*, vol. 10, p. 290, Jun. 2016.
54. R. Lee, M. Assante, and T. Conway, *Analysis of the Cyber Attack on the Ukrainian Power Grid*, Washington, DC: E-ISAC, 2016.
55. P. Nader, P. Honeine, and P. Beauseroy, "Detection of cyberattacks in a water distribution system using machine learning techniques," in *Proc. 6th Int. Conf. Digit. Inf. Process. Commun. (ICDIPC)*, Apr. 2016, pp. 25–30.
56. K. E. Crocker, "The influence of the amount and type of information on individuals' perception of legal services," *J. Acad. Marketing Sci.*, vol. 14, no. 4, pp. 18–27, Dec. 1986.
57. R. J. Lutz and P. J. Reilly, "An exploration of the effects of perceived social and performance risk on consumer information acquisition," in *Advances in Consumer Research*, S. Ward and P. Wright, Eds. Ann Abor, MI: Association for Consumer Research, 1974, pp. 393–405.
58. D. L. Davis, J. P. Guiltinan, and W. H. Jones, "Service characteristics, consumer search, and the classification of retail services," *J. Retailing*, vol. 55, no. 3, p. 3, 1979.
59. K. Mitra, M. C. Reiss, and L. M. Capella, "An examination of perceived risk, information search and behavioral intentions in search, experience and credence services," *J. Services Marketing*, vol. 13, no. 3, pp. 208–228, 1999.

4 Performance Evaluations of Quantum Key Distribution System Architectures*

Logan O. Mailloux, Michael R. Grimaila, and Douglas D. Hodson
US Air Force Institute of Technology

Gerald Baumgartner
Laboratory for Telecommunication Sciences

Colin McLaughlin
Naval Research Laboratory

CONTENTS

* Reprinted with copyright permission: Mailloux, L. O., Grimaila, M. R., Hodson, D. D., Baumgartner, G., & McLaughlin, C. (2015). Performance evaluations of quantum key distribution system architectures. *IEEE Security & Privacy*, *13*(1), 30–40.

DOI: 10.1201/9781003220978-5

4.1   THE EMERGENCE OF QKD-ENABLED CRYPTOGRAPHY

QKD systems are emerging in the cryptographic solution space, where many claim they function as unconditionally secure key distribution devices. (The term "key distribution" is somewhat misleading as QKD systems generate or grow shared secret keys from previously established keys and don't merely distribute them.) Figure 4.1 illustrates a QKD system configured to generate shared secret key K for use in external bulk encryptors. The architecture consists of a sender Alice, a receiver Bob, an optical fiber quantum channel, and a classical channel (that is, a conventional networked connection). Alice and Bob each consist of a CPU, network interface, and quantum channel module (QCM). The quantum channel is sensitive to physical disturbances such as ambient light and generally employs otherwise unused "dark" fiber. The quantum and classical channels are often separated due to security and performance sensitivities but might be multiplexed on a single fiber.

The architecture consists of a sender Alice, a receiver Bob, an optical fiber quantum channel, and a classical channel. Alice and Bob each consist of a CPU, network interface, and QCM. Additional administrative and control signals are omitted for clarity.

QKD systems can be paired with—and configured to increase the security posture of—traditional symmetric encryption algorithms, such as Data Encryption Standard (DES), 3DES, and Advanced Encryption Standard, through frequent rekeying. Alternatively, QKD is often discussed in conjunction with one-time pad (OTP) encryption—enabling "unconditionally secure communications"—as it provides a feasible solution to OTP's stringent requirements: a truly random key, key length equal to or greater than the length of the message to be encrypted, and a key that's never reused.[1]

However, the challenge is to provide sufficient key generation rates to meet realistic applications. For example, the ID Quantique Cerberis QKD server advertises key generation rates up to 3 Kbps over a transmission distance of 50 km.[2] Other commercial offerings from SeQureNet, Quintessence Labs, MagiQ Technologies, and Quantum Communication Technology have similar performance limitations.[3]

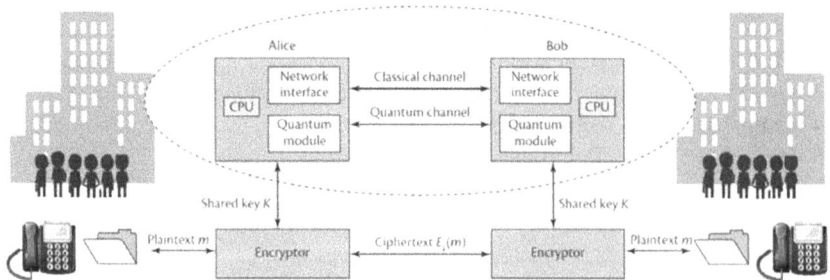

FIGURE 4.1   Quantum key distribution (QKD) system.

4.2 EXPLOITING QUANTUM PHYSICS FOR SECURITY

QKD's genesis can be traced to Stephen Wiesner, who came up with the idea of fraud-proof quantum money in the late 1960s.[4] In 1984, Charles Bennett and Gilles Brassard operationalized this concept when they proposed the first QKD protocol (BB84) in which single photons, representing quantum bits (qubits), are used to securely generate a shared cryptographic key (see the "QKD 'Prepare-and-Measure' Encoding" **sidebar**[1] for examples).[5]

QKD security is based on quantum uncertainty inherent in the BB84 encoding scheme and captured in formal proofs (see the "Demystifying the 'Quantum Physics' of QKD" sidebar for further discussion).[6,7] Modern BB84 implementations generally consist of eight operational phases—authentication, quantum exchange, raw key sifting, error estimation, error reconciliation, entropy loss estimation, privacy amplification, and final key generation (see Table 4.1). Only the second phase, quantum exchange, uses nonclassical communication techniques.[8]

TABLE 4.1
BB84 Protocol Operational Phases

Phase	Description
Authentication	Authentication occurs over the classical channel—typically a point-to-point networked connection—and might or might not involve encryption. (QKD security proofs don't require the classical channel to be encrypted.) The authenticated classical channel is used to control the QKD process, and transactional authentication is assumed to ensure secure transmission.
Quantum exchange	The sender encodes information onto quantum bits (qubits) using a randomly selected basis and bit value and sends them to the receiver over the quantum channel—typically a dedicated optical fiber. The receiver randomly selects a basis to measure each qubit. This phase generates raw key bits in which the quality of randomness can be assured through certified quantum random number generators in both the sender and receiver.[7–9]
Raw key sifting	The sender and receiver exchange basis information for each qubit. If the receiver measures the qubit on the same basis as the sender encoded it, the bit value will be obtained with a high degree of accuracy. If the receiver measures qubits on the wrong basis, a random result occurs and it needs to be sifted out from both the sender's and receiver's raw key bits. This results in a shared sifted key (in both Alice and Bob) approximately half the length of the raw key. Note that the sender and receiver expose only the bases and not the bit values.
Error estimation	The sender and receiver exchange a portion of sifted key bits over the public channel to estimate the sifted key's quantum bit error rate (QBER). If the estimated QBER is higher than a user-defined threshold, an unauthorized third party is assumed to be eavesdropping on the quantum channel, and the process is aborted and restarted.
	The estimated QBER can be used as an input parameter for error reconciliation.

(Continued)

TABLE 4.1 (*Continued*)
BB84 Protocol Operational Phases

Phase	Description
Error reconciliation	Error reconciliation is a two-way error correction of discrepancies between the distributed sifted key bits. Reconciliation occurs over the classical channel and leaks information about the potential key based on the efficiency of the algorithm selected. The actual QBER is calculated after error reconciliation and used as the primary QKD secret key check. If the QBER is higher than a predetermined security threshold, eavesdropping is assumed and the process is aborted and restarted.
Entropy loss estimation	Entropy loss attempts to quantify the amount of information exposed to an eavesdropper during the key distribution process based on the QBER and amount of information exposed during error reconciliation. While the QBER is relatively fixed for a given architecture, the amount of information lost can be thought of as exposed parity bits on a variable number of increasingly smaller block sizes based on the number and placement of errors in the sifted key.
Privacy amplification	Privacy amplification is an information theory technique that ensures the eavesdropper has negligible information regarding the final key based on the entropy loss estimation. This results in a smaller, securer key.
Final key generation	A hash of the privacy-amplified key is produced and shared with the sender and receiver to ensure the final key is the same. If the hashes match, the QKD system has successfully generated shared secret keying material.

4.3 IMPLEMENTATION NONIDEALITIES

In this article, we focus on a "prepare-and-measure" BB84 QKD system with polarization-based encoding, where qubits are encoded and decoded using four polarization states $\leftrightarrow$, $\updownarrow$, $\diagdown$, and $\diagup$. The states are encoded by randomly selecting a bit value (0 or 1) and a basis (horizontal/vertical "$\oplus$" or diagonal/antidiagonal "$\otimes$") and decoded by randomly selecting a measuring basis ($\oplus$ or $\otimes$). (See the "QKD 'Prepare-and-Measure' Encoding" sidebar.) The BB84 protocol assumes several idealities, including[6–10]

- on-demand single photon sources in Alice,
- perfect single photon detection in Bob,
- a lossless quantum channel, and
- perfect basis alignment across the quantum channel.

However, these critical assumptions aren't valid in real-world systems. Reliable on-demand single photon sources aren't currently available, single photon detectors (SPDs) have low detection efficiencies, and quantum channels (optical fiber and direct line-of-sight optical links) have transmission losses, with the compensation mechanisms' accuracy limiting basis alignment.

SIDEBAR 1: QKD "PREPARE-AND-MEASURE" ENCODING

Practical quantum key distribution (QKD) implementations often use "prepare-and-measure" protocols wherein a sender Alice prepares a quantum bit (qubit) by encoding a randomly selected bit and basis, and a receiver Bob measures the qubit according to a randomly selected basis. If Alice's and Bob's bases agree, the encoded qubit is read correctly with a very high probability; otherwise, a random result occurs. Prepare-and-measure protocols primarily use two types of modulation techniques, as **Table 4A.1** illustrates. Polarization-based protocols encode and decode qubits using horizontal ($H=0°$), vertical ($V=90°$), diagonal ($D=45°$), and antidiagonal ($A=-45°$) polarization states in two mutually unbiased, conjugate basis sets: "rectilinear" representing the horizontal/vertical orthogonal basis and "diagonal" representing the diagonal/antidiagonal orthogonal basis. Polarization-based implementations are typically used for free-space line-of-sight laser applications because they're less susceptible to disturbances in the atmosphere. Phase-based protocols are generally used for transmission through optical fibers, where wave interference is used to encode and decode information at phase shifts of 0 and π.

4.4 EAVESDROPPING ON QKD SYSTEMS

QKD security relies on the fact that eavesdropping on the quantum communication channel necessarily introduces detectable errors evident in the quantum bit error rate (QBER). We calculate the QBER from the error-reconciled key by dividing the number of errors in the sifted key by the total number of sifted bits. Typical values are approximately 3% to 5% due to device imperfections and transmission errors.[7,8]

If eavesdropper Eve attempts to listen to the quantum exchange, she introduces additional errors and increases the QBER. Therefore, the QBER must be monitored closely with allowable rates of 11%–12% and practical rates as low as 8%[7]. If the QBER exceeds the threshold, the key generation process generally aborts, as it's assumed an adversary is interfering with the key exchange. Because QKD systems attribute all errors to eavesdroppers (even if they're caused by the environment), it's important to characterize the QBER well to prevent false positives and provide reliable system performance.

4.5 QUANTUM HACKING

With increased interest in and availability of QKD technologies, "quantum hacking" has become a specialty area.[11] QKD systems are vulnerable to attacks over the quantum channel, including man in the middle (authentication failures), intercept and resend (measuring and replacing photons), photon splitting (stealing photons), and blinding optical receivers (overpowering photon detectors to control their detections). QKD systems are also subject to threats common to networked devices such as vulnerabilities in operating systems, applications, and system interfaces.

TABLE 4A.1

Example Prepare-and-Measure Encoding Types

| Polarization-Based BB84 | | | | |
| Sender (Alice) | | | Receiver (Bob) | |
Encoded Bit Value	Encoding Basis	Encoded Polarization State (α) in Degrees	Measuring Basis	Measured Bit Value
0	H/V	H = 0	H/V	0
0	H/V	H = 0	D/A	0 or 1
1	H/V	V = 90	H/V	1
1	H/V	V = 90	D/A	0 or 1
0	D/A	A = −45	H/V	0 or 1
0	D/A	A = −45	D/A	0
1	D/A	D = 45	H/V	0 or 1
1	D/A	D = 45	D/A	1

| Phase-Based BB84 | | | | |
| Sender (Alice) | | Receiver (Bob) | | |
Encoded Bit Value	Encoding Phase Shift (φ_A in Radians)	Measuring Phase Shift (φ_B in Radians)	Phase Interference ($\varphi_B - \varphi_A$ in Radians)	Measured Bit Value
0	0	0	0	0
0	0	$\pi/2$	$3\pi/2$	0 or 1
1	π	0	π	1
1	π	$\pi/2$	$\pi/2$	0 or 1
0	$\pi/2$	0	$\pi/2$	0 or 1
0	$\pi/2$	$\pi/2$	0	0
1	$3\pi/2$	0	$3\pi/2$	0 or 1
1	$3\pi/2$	$\pi/2$	π	1

Note: "H," "V," "D," and "A" represent the horizontal, vertical, diagonal, and antidiagonal polarization states; these polarization states are sometimes represented with bidirectional arrows ↔, ↕, ↘, and ↗. Furthermore, while "*H/V*" and "*D/A*" represent the rectilinear and diagonal orthogonal basis sets, they're often represented as ⊕ and ⊗.

(See "The Black Paper of Quantum Cryptography: Real Implementation Problems" for further discussion on practical QKD security considerations.[10])

As security devices, QKD systems should be protected against physical attacks and manipulation in controlled telecom environments. In addition, practical security techniques, such as ensuring properly calibrated systems, characterizing a secure baseline, and continuous monitoring, should be enforced; little discussion of this type of practical system security is available in the published QKD literature. These issues should further motivate a proper understanding of practical engineering considerations for secure system design, implementation, and operation.

4.6 ADVANCEMENT OF QKD TECHNOLOGY

Realizing QKD's potential, researchers from academic, commercial, and government sectors in Canada, the US, Europe, Australia, Japan, Singapore, China, and other countries continue to advance the technology. Most notably, researchers in Tokyo recently achieved a 1-Mbps key distribution rate to support real-time video encryption,[12] and the Battelle Memorial Institute seeks to daisy-chain 10 QKD links from Columbus, Ohio, to Washington, DC, in a 1,000-km trusted node configuration.[3] Such technological advancements have resulted in a diverse trade space of competing design and implementation choices, including several encoding schemes; quantum exchange protocols; photon sources; transmission mediums; error reconciliation methods; privacy amplification solutions; photon detector technologies; and various combinations of optical, electro-optical, and electronic devices.

SIDEBAR 2: DEMYSTIFYING THE "QUANTUM PHYSICS" OF QKD

"It is safe to say that nobody understands quantum mechanics." – Richard Feynman

QKD security is based on the laws of quantum mechanics, where photons representing 0s and 1s (known as quantum bits or qubits) are exchanged to generate a shared secret key. To appreciate QKD's security aspects, we first need to examine the differences between classical and quantum physics. We attempt to explain these quantum behaviors in a brief and accessible way, specific to QKD. More comprehensive discussions can be found in Protecting Information,[1] Quantum Computation and Quantum Information,[2] and Introductory Quantum Optics.[3]

4.7 QUBITS EXIST IN A STATE OF SUPERPOSITION

Classical bits exist in a deterministic state of either 0 or 1, whereas qubits exist in a simultaneous combination of states, often represented in a two-dimensional complex vector space. For example, Figure 4A.1 (part 1) depicts a polarization-based quantum superposition described as $|\psi\rangle = \alpha|H\rangle + \beta|V\rangle$, where α and β represent complex numbers, and the orthogonal basis is represented by horizontal $|H\rangle$ and vertical $|V\rangle$ states. This means QKD systems can encode qubits in a continuum of states between $|H\rangle$ and $|V\rangle$, which directly impacts their measurement.

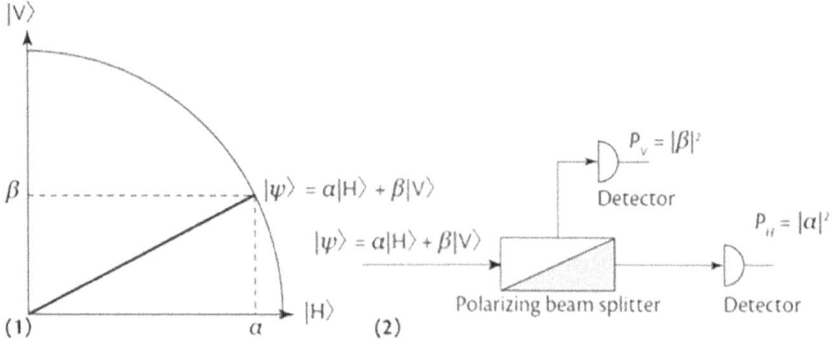

FIGURE 4A.1   Examples of (1) quantum superposition and (2) quantum measurement.

4.8 QUBITS CAN'T BE MEASURED DIRECTLY

A classical bit can be measured—and remeasured if necessary—to determine its state 0 or 1. In quantum mechanics, directly measuring a qubit in a state of superposition, specifically α or β, isn't possible. Instead, when a qubit is measured, a 0 occurs with probability $|\alpha|^2$ and a 1 occurs with probability $|\beta|^2$, where $|\alpha|^2 + |\beta|^2 = 1$. This means that whenever a qubit is measured, it's forced to collapse into a state corresponding to the measurement basis. For example, as Figure 4A.1 (part 2) depicts, when an arbitrary quantum state $|\psi\rangle = \alpha|H\rangle + \beta|V\rangle$ encounters a polarizing beam splitter (a common component used to differentiate encoded states) and is subsequently measured in the rectilinear (horizontal/vertical) basis, it probabilistically collapses into the state $|H\rangle$ or $|V\rangle$. Likewise, when an arbitrary quantum state is measured on the diagonal (diagonal/antidiagonal) basis, it collapses into the state $|D\rangle$ or $|A\rangle$. Consequently, QKD systems can't perfectly measure unknown quantum states encoded in mutually unbiased, nonorthogonal states (that is, the rectilinear and diagonal bases), as is the case in BB84.

Note on Figure 4A.1: Quantum superposition allows a single quantum bit (qubit) to simultaneously exist in more than one state; however, when the qubit is measured, it collapses into one of two states based on the measurement apparatus. Thus, measurement of a quantum state disturbs the state, preventing perfect copies from being made.

4.9 QUBITS CAN'T BE PERFECTLY COPIED

Classical bits can be perfectly copied, whereas qubits can't, as the no-cloning theorem proves.[2,4] In classical physics, we can measure an object without inducing a change and can therefore make perfect copies. In contrast, the act of observing a quantum state imparts a change in the system, introducing noise and preventing exact copies. For example, if a qubit is diagonally encoded

$$|D\rangle = \frac{1}{\sqrt{2}}|H\rangle + \frac{1}{\sqrt{2}}|V\rangle$$

and measured in the rectilinear basis, as Figure 4A.1 demonstrates, it has an equal probability $|\alpha|^2 = |\beta|^2 = 1/2$ of being detected in the $|H\rangle$ or $|V\rangle$ state and can't create a copy of the $|D\rangle$ state. Furthermore, because the quantum state collapses, all previously encoded information is lost during the act of measurement (that is, the $|D\rangle$ state is gone and can't be recovered). This quantum phenomenon is unavoidable due to interactions between the subject photons and the measurement device, which prevent the possibility of making perfect copies.

The principles of quantum mechanics provide the necessary foundation for provably secure QKD systems, where adversaries measuring or manipulating photons in flight necessarily introduce detectable errors. However, quantum mechanics doesn't protect against physical or remote attacks against QKD hardware or software.

SIDEBAR 2 REFERENCES

1. S. Loepp and W.K. Wooters, *Protecting Information*, Cambridge Univ. Press, 2006.
2. M.A. Nielsen and I.L. Chuang, *Quantum Computation and Quantum Information*, Cambridge Univ. Press, 2010.
3. C.C. Gerry and P.L. Knight, *Introductory Quantum Optics*, Cambridge Univ. Press, 2005.
4. V. Scarani et al., "Quantum Cloning," *Rev. Modern Physics*, vol. 77, no. 4, 2005, pp. 1225–1254.

4.10   THE QKD SECURITY–PERFORMANCE TRADE SPACE

QKD systems are designed to securely distribute cryptographic keys, and security–performance tradeoffs should factor in the system's purpose and operational environment. System performance is generally defined by the desired capability (that is, a secret key rate) and the intended application (that is, a transmission distance), while security is described by theoretical proofs tempered by nonideal implementations assuming a quality random key certified by the National Institute of Standards and Technology.[7-9] The creation of an ideal single photon source is an excellent example of a security–performance tradeoff. Because on-demand single photon sources aren't currently feasible, Alice attenuates a classical laser pulse down from millions of photons to a mean photon number (MPN) of 0.1, according to QKD security proofs.[7] A Poisson distribution probabilistically represents this low energy level

$$P(n \mid \mu) = \frac{\mu^n e^{-\mu}}{n!},$$

where $\mu = $ MPN and n is the number of photons in the pulse. When MPN $= 0.1$, 90.48% of the pulses have no photons, 9.05% have one photon, and 0.47% have two or more photons.[8] The low MPN mitigates vulnerabilities associated with multiphoton pulses, which let Eve siphon photons for use in cryptanalysis. Although system designers would like to increase the MPN to improve throughput, the MPN must be relatively low to maintain a reasonable security posture.

Other works have addressed QKD's theoretical and practical security–performance assumptions[6–8]; we attempt to understand their component- and system-level impact in realized systems. It's clear that QKD system performance is a function of competing requirements (e.g., key throughput, cost, security, usability, maintainability, and resilience) and implementation design decisions (e.g., quantum exchange protocols, modulation types, encoding schemes, and photon sources). However, system architects often make these design choices without fully understanding the impact of their decisions.

This complex and dynamic trade space is precisely where modeling and simulation provide great benefit to system designers, implementers, and customers. Simulation studies allow system architects to more efficiently compare designs and make objective decisions with increased confidence. To this end, we've enabled the efficient study of QKD system architectures through a modularized and parameterized simulation framework that lets system architects and analysts design, build, and evaluate QKD systems.

4.11 QKD TRADE SPACE EVALUATIONS

Here, we present our QKD reference architecture and three example studies to show the design tradeoffs between critical QKD components and system performance. We ordered these studies according to practical issues in the implementation space related to the sender, receiver, and transmission channel; they also correspond to the implementation nonidealities we mentioned. These evaluations are intended to demonstrate the ability to model and study practical security–performance tradeoffs in QKD systems and aren't intended to be individually comprehensive.

4.12 MODELED QKD SYSTEM ARCHITECTURE

Figure 4.2 provides an end-to-end depiction of the QKD system's quantum communication path from Alice's laser source to Bob's photon detectors. Each QCM comprises multiple subsystems with one or more optical components and associated controllers; we present only the most important. In total, this architecture consists of 35 optical components modeled in an event-driven paradigm and 16 controllers modeled in a process-based paradigm. The model provides a hardware-focused architectural representation in which each component is configured with adjustable performance parameters.

Note in Figure 4.2: The quantum communication path is used to prepare and measure polarization-based qubits. Each subsystem includes one or more optical components and controllers modeled in a modular fashion with configurable operational parameters.

Designers built the model in a modular fashion from a library of components with user-defined levels of detail and verified each component against commercial specifications. System and subsystem verification of critical behaviors such as expected signal strength, transmission losses, final key rate, and QBER was also conducted to

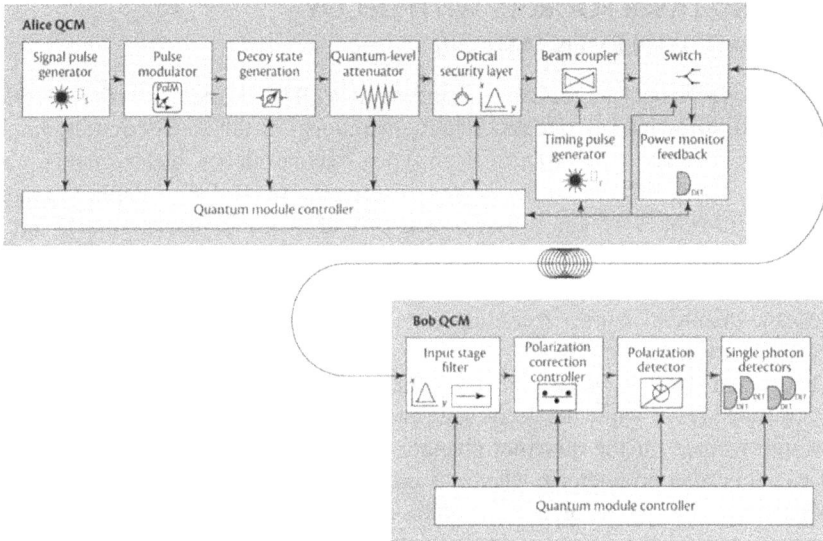

FIGURE 4.2 Decomposed Alice and Bob quantum channel modules (QCMs).

ensure valid operation. In total, designers made dozens of decisions and assumptions to model these devices and supporting processes based on product specifications; reference literature; available publications; and input from senior quantum communication physicists, optical research physicists, computer scientists, and electrical, computer, and systems engineers.

Alice's signal pulse generator creates optical pulses with wavelength $\lambda_S = 1{,}550\,\text{nm}$ while the pulse modulator polarization-encodes the qubits. The decoy state generator employs additional security states, the quantum-level attenuator reduces the classical-level optical pulses (that is, millions of photons) to quantum-level pulses (that is, MPN = 0.1), and the optical security layer detects adversaries probing Alice's pulse modulator to read the encoded pulse value.

A timing pulse generator creates synchronization pulses with wavelength $\lambda_T = 1{,}540\,\text{nm}$, which are combined with the qubit pulses λ_S in the coupler. A power monitor measures the optical pulses to ensure they have the desired output MPN. Bob's input stage filters out unwanted wavelengths and reduces the threat of optical probing attacks. The polarization correction controller compensates for transmission-induced polarization errors using the timing pulse λ_T as a frame of reference. The polarization detector is configured to passively separate the four BB84 encoded polarization states (horizontal, vertical, diagonal, and antidiagonal), while SPDs are configured to detect arriving qubits. The SPDs are synchronized with the timing pulse λ_T to reduce noise by precisely gating the detectors—temporarily placing the SPD in a more sensitive state to detect single photons.

4.13 QKD LASER SOURCES WITH DECOY STATE IMPLEMENTATIONS

Decoy states represent a significant advancement in QKD technology as they increase the systems' effective key rate and secure operational distance.[7] Decoy state implementations comprise three transmission types—signal, decoy, and vacuum—each configured with a different MPN and occurrence percentage. For example, the signal state has an MPN of 0.6 and is transmitted 70% of the time, the decoy state has an MPN of 0.2 and is transmitted 20% of the time, and the vacuum state has an MPN near 0 and is transmitted 10% of the time. Each of these states is randomly sent across the quantum channel according to its occurrence percentage and is otherwise indistinguishable to Eve.

The signal state facilitates higher key rates and greater operational distances due to increased MPNs, while the decoy state enables the system to detect photon-dependent interference on the quantum channel. The vacuum state is intended to have pulses containing no photons, where it's used to determine the dark count (a spontaneous error rate) of the receiver's photon detectors. The system's security posture is ensured by monitoring and conducting statistical comparisons between the signal and decoy states to determine if Eve is interfering with the quantum channel, thereby preventing her from gaining information on the secret key.

Table 4.2 demonstrates simulated results of two decoy state configurations (A and B) over two operation distances—5 and 50 km. Each pulse type is transmitted according to the stated MPN and occurrence percentage, with the resulting QBER and final key size reported. The decoy state implementation has the net effect of increasing the system key rate; however, the security–performance tradeoff is unclear (that is, how the decoy state MPNs and occurrence percentages affect the desired system security capability). For example, in "Field Test of a Practical Secure Communication Network with Decoy-State Quantum Cryptography," two similar 20-km transmission links were established with comparable signal MPNs of 0.65 and 0.60, while the decoy MPNs were significantly different at 0.08 and 0.20, yet there was no discussion of the system's ability to detect an eavesdropper.[13]

In general, decoy state research has focused on increasing theoretical secure key rates with little discussion of their overall security posture. A comprehensive characterization and sensitivity analysis is needed for a fuller understanding of security–performance design tradeoffs. This type of analysis helps system designers and security specialists determine appropriate performance parameters to meet user requirements and, ultimately, system certification.

4.14 COMPETING SPD TECHNOLOGIES

In QKD systems, SPDs are critical components, configured to detect very small energy levels (approximately $1.28 \times 10{-}19$ J per photon at 1,550 nm). Commercially available SPDs severely constrain system throughput owing to poor detection efficiencies and relatively low maximum detection rates (due to long "dead" times necessary to prevent erroneous "after pulse" detections).[14] SPD performance is further limited by dark counts and jitter time (that is, variance in detector response

TABLE 4.2
Comparisons of Decay State Performance

A	Mean Photon Number	%	Propagation over 15 km (More Than 50% Loss)				Propagation over 50 km (More Than 90% Loss)			
			Pulses Sent	Detection Count	QBER	Final Key Size	Pulses Sent	Detection Count	QBER	Final Key Size
Signal	0.5	60	251,501	41,140	0.0119	15,439	926,269	41,529	0.0033	15,562
Decoy	0.2	30	124,837	8,854			463,235	8,459		
Vacuum	Approx. 0+	10	41,662	6			154,496	12		
Total	—	100	418,000	50,000			1,544,000	50,000		

B	Mean Photon Number	%	Propagation over 15 km (More Than 50% Loss)				Propagation over 50 km (More Than 90% Loss)			
			Pulses Sent	Detection Count	QBER	Final Key Size	Pulses Sent	Detection Count	QBER	Final Key Size
Signal	0.8	70	217,825	47,775	0.0164	17,975	781,727	47,930	0.0051	18,020
Decoy	0.1	20	62,179	2,223			223,123	2,063		
Vacuum	Approx. 0+	10	30,996	2			112,150	7		
Total	—	100	311,000	50,000			1,117,000	50,000		

once a photon is received). Although these limitations can be partially mitigated by advanced control circuitry, they're inherent to the device's material makeup and operational environment. In addition, SPDs can play an important role in security as the capability to precisely determine the number of photons received—and therefore multiphoton pulses sent—helps to mitigate photon-splitting attacks on the quantum channel. In this study, we consider the efficiency, maximum count rate, dead time, jitter time, dark count probability, and temperature sensitivity of avalanche photodiodes, superconducting nanowire SPDs, and transition edge sensors (see Table 4.3). We determine system performance from a simulated transmission of 1,000,000 qubits through the architecture, reporting the scenario detection count and operational time. As the relatively low detection counts indicate, QKD systems generally have poor throughput due to the source's Poissonian nature (1/10 pulses contains a photon) and high transmission losses through the fiber channel (approximately 50% over 15 km).

Avalanche photodiodes are classical optical detectors reverse-biased with higher-than-normal voltage, causing them to become sensitive to single photons. Despite their seemingly poor performance, avalanche photodiodes are implemented with inexpensive thermal-electric coolers owing to their relatively low production costs and ability to operate in close proximity to room temperature.

Offering increased efficiency and high throughput potential with little noise, superconducting nanowire SPDs detect single photons through current fluctuations caused by arriving photons. These devices have been used in experimental QKD systems and show promise as the next-generation SPD of choice. Simulating this device had the added advantage of modeling significantly increased pulse rates, which revealed temporal dependencies and limitations in the architecture, such as the relationship between pulse rate and encoding mechanism performance.

Superconducting transition edge sensor arrays offer very high efficiencies, triggered by minute temperature changes caused when absorbing photons. Their ability to accurately resolve photons is very appealing for security, but they require elaborate multistage cooling devices to achieve near-absolute-zero operating temperatures, which preclude them from all but the costliest QKD implementations.

This example demonstrates the ability to quickly compare competing technologies, including multiple key performance parameters, in an end-to-end system configuration with less overall expense than building or modifying existing systems. In addition, this study highlights the flexibility of modeling and evaluating desired capabilities, such as photon resolving detectors to meet user performance and security requirements.

4.15   IMPACT OF POLARIZATION STATE ERRORS

Here, we examine the relationship between polarization error compensation and system-level performance. Accurate timing and polarization alignment are necessary for successful quantum communications and are particularly important for polarization-based QKD systems, such as the original BB84 protocol, terrestrial line-of-sight lasers, and satellite-based QKD. Our study is loosely based on results from the Tokyo QKD network experiment in which environmentally induced vibrations over

TABLE 4.3

Single Photon Detector (SPD) Evaluation for Secret Key Generation

Competing Detector Technologies	Detector Efficiency at 1,550 nm	Maximum Detection Rate (Hz)	Dead Time (μs)	Jitter Time (ps)	Dark Count (1/s)	Operating Temperature (K)	Relative Cost	Photon Resolving	Scenario Detection Count	Scenario Operation Time (s)
Ideal SPD	100%	∞	Approx. 0+	0.0	0.0	295	N/A	Y	39,016	—
Avalanche photodiode	10%[a]	10 kHz[a]	100[a]	370[a]	Approx. 91[a]	200[a]	Low	N	4,059	100
Superconducting nanowire SPD	57%[a]	1 GHz[a]	0.01[a]	30[a]	Approx. 0+[a]	1.5–4[a]	High	N	22,457	0.001
Transition edge sensor	95%[a]	100 kHz[a]	10[a]	100,000[a]	Approx. 0+[a]	0.1[a]	High	Y	37,323	10

[a] Values are from "Single-Photon Detectors for Optical Quantum Information Applications."[14] The ideal SPD is a baseline to compare performance parameters.

a 45-km transmission link strung from telephone poles caused continuous recalibrations and temporary system outages.[15] In this study, the modeled QKD system architecture is configured to transmit frames of qubits, where each timing pulse λ_T begins a frame of 1,000 individually modulated signal pulses λ_S. These frames propagate through 45 km of aerial fiber subject to simulated environmental disturbances such as temperature change, vibration, sway, and inclement weather. These disturbances induce changes to the polarization state of the signal pulses λ_S, and when left uncorrected, they can cause channel misalignment errors proportional to the angle of drift from the reference angle. The receiver's polarization controller is designed to correct this error but has a fixed slew rate, potentially limiting the QKD secret key generation rate.

Figure 4.3 depicts our examination of this scenario over a simulated 30-second quantum communication period, where Alice sends Bob 15 million reference pulses λ^T subject to physical disturbances. A general random approximation (that is, Brownian motion) is used to model these disturbances resulting from induced stresses on the quantum channel. Under normal operating conditions, including minor temperature changes, winds, and vibrations, or even moderate stresses, including light winds and nearby traffic vibration, the polarization error can be corrected. However, with significant environmental or physical stresses, such as high winds or close-proximity subway trains, adverse effects on key throughput will likely occur,

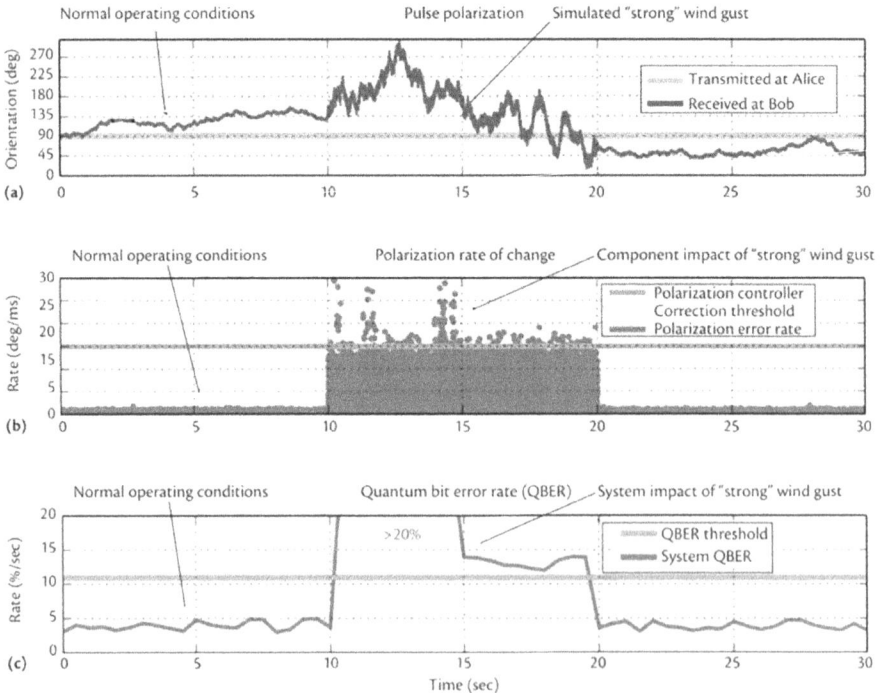

FIGURE 4.3   Polarization correction for aerial fiber disturbances.

as Figure 4.3 demonstrates with a simulated strong wind gust occurring between 10 and 20 s.

Note on Figure 4.3: During this simulated 30-s quantum communication period, Alice sends Bob 15 million reference pulses λ_T subject to physical disturbances. Under normal operating conditions or moderate stresses, the polarization error can be corrected. However, when significant environmental or physical stresses exist, adverse effects on key throughput are likely.

We simulated a 10-s strong wind gust with increased randomness, exceeding the controller's ability to compensate for rapid changes to the polarization state of the transmitted optical pulses (15°/ms). Severe disturbances on the quantum channel result in system-level outages in which the QBER exceeds the established security threshold of 11%[15] The results of this simulation demonstrate the careful consideration designers must take regarding the operation and performance of critical components. For instance, polarization controller performance ranges from high-end controllers capable of correcting errors to within 1° in less than 1 ms for tens of thousands of dollars to low-end devices costing hundreds of dollars with far less precision. Holistically considering the architectural trade space, QKD system designers might adjust the pulse frame size, lower the pulse rate, use alternative synchronization methods, or take advantage of more robust QKD encoding protocols. Designers should also factor in tradeoffs between buried (protected) and aerial (unprotected) fiber. A similar argument can be made for line-of-sight laser QKD applications, which might have lower installation costs but are subject to additional environmental constraints such as rain, fog, dust, and smog.

Our future efforts will include studying alternate QKD architectures, exploring emerging applications such as satellite-based QKD, and modeling notional capabilities. We also want to model and conduct multicriteria performance analyses to mitigate the risk of unwanted emergent behaviors, discover unknown dependencies, and confidently meet strict performance and security requirements.

ACKNOWLEDGMENTS

The Laboratory for Telecommunication Sciences grant 5743400-304-6448 supported this work. The views expressed in this article are those of the authors and do not reflect the official policy or position of the US Air Force, the US Department of Defense, or the US government.

REFERENCES

1. C. Elliott, "Quantum Cryptography," *IEEE Security & Privacy*, vol. 2, no. 4, 2004, pp. 57–61.
2. *"Layer 2 Link Encryption with Quantum Key Distribution,"* Cerberis, IDQ, Jan. 2012; www.idquantique.com/images/ stories/PDF/cerberis-encryptor/cerberis-specs.pdf.
3. L. Oesterling, D. Hayford, and G. Friend, "Comparison of Commercial and Next Generation Quantum Key Distribution: Technologies for Secure Communication of Information," *Proc. IEEE Conf. Technologies for Homeland Security (HST 12)*, 2012, pp. 156–161.
4. S. Wiesner, "Conjugate Coding," *ACM Sigact News*, vol. 15, no. 1, 1983, pp. 78–88.

5. C.H. Benne_ and G. Brassard, "Quantum Cryptography: Public Key Distribution and Coin Tossing," *Proc. IEEE Int'l Conf. Computers, Systems, and Signal Processing*, 1984, pp. 175–179.

6. R. Renner, N. Gisin, and B. Kraus, "An Information-theoretic Security Proof for QKD Protocols," *Physical Rev. A*, vol. 72, no. 1, 2005, p. 012332.

7. V. Scarani et al., "_ e Security of Practical Quantum Key Distribution," *Rev. Modern Physics*, vol. 81, no. 3, 2009, pp. 1301–1350.

8. N. Gisin et al., "Quantum Cryptography," *Rev. Modern Physics*, vol. 74, no. 1, 2002, pp. 145–195.

9. "Random Number Generation," Nat'l Inst. Standards and Technology, 16 July 2014; h_ p://csrc.nist.gov/groups/ ST/toolkit/rng/index.html.

10. V. Scarani and C. Kurtsiefer, "_ e Black Paper of Quantum Cryptography: Real Implementation Problems," arXiv:0906.4547v2, 2009.

11. "Quantum Hacking Lab," Inst. Quantum Computing, Univ. Waterloo, 2014; www.vad1. com/lab.

12. A. Tanaka et al., "High-Speed Quantum Key Distribution System for 1-Mbps Real-Time Key Generation," *IEEE J. Quantum Electronics*, vol. 48, no. 4, 2012, pp. 542–550.

13. T.-Y. Chen et al., "Field Test of a Practical Secure Communication Network with Decoy-State Quantum Cryptography," *Optics Express*, vol. 17, no. 8, 2009, pp. 6540–6549.

14. R.H. Had_ eld, "Single-Photon Detectors for Optical Quantum Information Applications," *Nature Photonics*, vol. 3, no. 12, 2009, pp. 696–705.

15. K. Shimizu et al., "Performance of Long-Distance Quantum Key Distribution over 90-km Optical Links Installed in a Field Environment of Tokyo Metropolitan Area," *J. Lightwave Technology*, vol. 32, no. 1, 2014, pp. 141–151.

Section 2

Data Analytics

5 Multimodal Data Fusion for Systems Improvement
A Review*

Nathan Gaw, Safoora Yousefi, and
Mostafa Reisi Gahrooei
US Air Force Institute of Technology

CONTENTS

5.1 INTRODUCTION

Evaluating, analyzing, and modeling complex systems require a multi-perspective data acquisition framework. Such a framework gathers data through different instruments, sensors, and experiments to be translated into information and knowledge

* Reprinted with permission: Gaw, N., Yousefi, S., & Gahrooei, M. R. (2021). Multimodal data fusion for systems improvement: A review. *IISE Transactions*, 1–19.

DOI: 10.1201/9781003220978-7

about the system. Like Rumi's elephant in the dark room, each instrument or sensor collects partial information about a few characteristics of the system. However, when the partial information is combined, one may be able to discern a more complete picture. We refer to information collected by each instrument as a data mode and the full dataset obtained from the multi-perspective acquisition framework as a multimodal dataset. A multivariate dataset may also contain features collected from multiple instruments; however, there is no distinction of particular perspectives represented by one or more of the features. In contrast, a multimodal dataset is organized in such a way that data can be grouped into multiple perspectives, for which each perspective consists of at least one feature. Improvements in sensing technology have created opportunities for the acquisition of multimodal data, which can be used for analysis that goes beyond separate evaluation of each mode of the data. In this article, data fusion is defined as the process of integrating different modes of multimodal data to achieve a more comprehensive or accurate understanding of the system that could not be achieved otherwise by analysis of each mode.

The relationship between multiple data sets was first analyzed in the breakthrough work by Hotelling (1936). Since then, the number of analysis methods for multimodal datasets has increased rapidly. Techniques such as multiset CCA, parallel factor analysis (PARAFAC), and tensor decomposition were introduced in the 1960s and 1970s (Tucker, 1964; Ilarshman, 1970; Kettenring, 1971). Nevertheless, only a limited number of domains, including chemometrics, benefited from these developments. With the recent insurgence of multimodal datasets, an increasing number of domains, including manufacturing, healthcare, and renewable energy, demonstrate interest in exploiting the potential benefits of these datasets.

A systematic analysis and fusion of multimodality datasets result in increased understanding that can facilitate decision-making and result in systems improvement. We define systems improvement as the practice of using data-driven actions that increase the efficiency of the system's processes. This increase of efficiency can be due to more effective descriptive and predictive models, more reliable abnormality detection methods, or more accurate and interpretable features extracted from data that facilitate decision-making. Examples of such systems improvement include the following applications: in systems prognostics, where the remaining useful lifetime of a system component is predicted for better maintenance scheduling; in healthcare, where medical imaging and other patient data are used for more accurate identification of a disease; in renewable energy, where consumption data is used to improve energy management; and in agriculture, where imaging and weather data can be used for crop yield predictions and crop health management. Even though the benefits of analyzing multimodal data sets are evident, the knowledge of how to exploit the similarities and differences of modalities is still limited.

Problems such as heterogeneity of data (i.e., modes are in different dimensions or structure), differences in scale, resolution, accuracy, conflicting modes, and redundant modes create significant challenges that hinder the advancements of multimodal data analysis.

We group the multimodal data fusion algorithms into two classes: (i) algorithms that do not use neural networks and mainly focus on decomposition techniques; (ii) methods that use neural networks to perform data fusion. Tensor data analysis, factor analysis, and generalized principal complement analysis belong to the first group of algorithms and are suitable when the sample size is small compared with the number of variables. Deep neural networks with architectures that contain several layers of fusion belong to the second class of algorithms. In this article, we refer to the first class of algorithms as decomposition-based fusion algorithms and to the second class of algorithms as neural network-based algorithms. Within both of these categories, three ways of fusion are available: (i) early fusion (low-level fusion), (ii) late fusion (high-level fusion), and (iii) intermediate fusion. Some application areas tend to take one way over others. For example, in healthcare applications, late fusion is more common (Zhang and Ma, 2012; Suk et al., 2017; Khasha et al., 2019; Liu, Chen, Wu, Weidman, Lure, Li and Alzheimer's Disease Neuroimaging Initiative, 2020c), in systems monitoring and prognostics early fusion is more prevalent (Liu et al., 2013; Liu and Huang, 2014; Liu et al., 2015; Fang et al., 2017; Chehade et al., 2018; Song and Liu, 2018).

Early fusion (or low-level fusion) is the process of fusing modalities by only using information from the predictors (i.e., independent variables). Early fusion can either occur as a preprocessing task before incorporation into the main model or as a purely unsupervised task to generate features that best describe the underlying patterns across modalities. In feature preprocessing, the main goal is to combine raw features from different modalities to generate new features that combine complementary information of the raw features from different modalities. These new features are then inputted into a supervised model for a training task. Early fusion as a purely unsupervised task has the goal of combining features across modalities to discern underlying patterns present across different modalities or generate visualization that aptly describes information from the different modalities (i.e., combining different types of medical imaging to generate another image that displays complementary information) (He et al., 2010; Moin et al., 2016; Rajalingam and Priya, 2017). Principal component regression is an example of early fusion, in which Principal Component Analysis (PCA) is performed to extract input features that are then employed to predict an output value.

Late fusion (or high-level fusion) is fusion at the decision level. When modalities have been processed and modeled separately, the individual predictions from each modality can be combined in a number of ways depending on the importance of each modality for the prediction task, the appropriateness of the modality combination (whether the fusion should be modeled as an element-wise summation (Brentan et al., 2017), weighted average (Kahou et al., 2016), bi-linear product (Chen and Irwin, 2017), etc.), the noise level present in each modality, and/or other considerations deemed appropriate by the practitioner. Some popular examples of late fusion include ensemble learning (Yokoya et al., 2017; Sagi and Rokach, 2018; Samareh et al., 2018) and deep late fusion (Simonyan and Zisserman, 2014; Kahou et al., 2016; Wu et al., 2016; Ramachandram and Taylor, 2017).

Intermediate fusion is the process of incorporating features from different modalities in the model training process, using both predictors (i.e., independent variables) and response (i.e., dependent variables). These methods incorporate fusion directly in the model training process and make decisions on fusion in a way to optimize the objective (e.g., accuracy, detection rate, etc.). Partial Least Square (PLS) is an example of intermediate fusion, in which the fused features are extracted in a supervised fashion to best explain the output (Zhao et al., 2012). Another example of intermediate fusion can be found in tensor regression, which can extract features from tensors that contain multiple sources to estimate the output (Gahrooei et al., 2020). Deep learning architectures can also be designed to perform intermediate fusion (Karpathy et al., 2014).

To better understand the intuition behind early, late, and intermediate fusion, let us consider an example in prognostics where the remaining useful lifetime (or Time To Failure (TTF)) of an asset is predicted based on the available sensors data (Song and Liu, 2018; Song et al., 2019; Li et al., 2021). Consider a rotary machine whose condition is monitored by three different sensors that produce vibration signals (A), noise signals (B), and infrared thermal images (C). Early fusion (see Figure 5.1a) focuses on generating latent variables by combining these signals and then using the latent variables for model training and prediction. For example, early fusion first performs

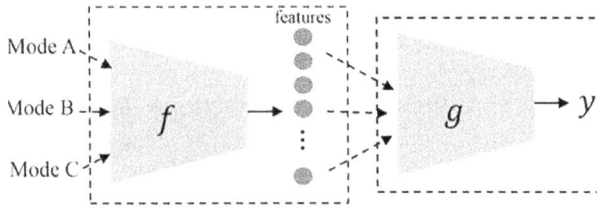

(a) Early Fusion

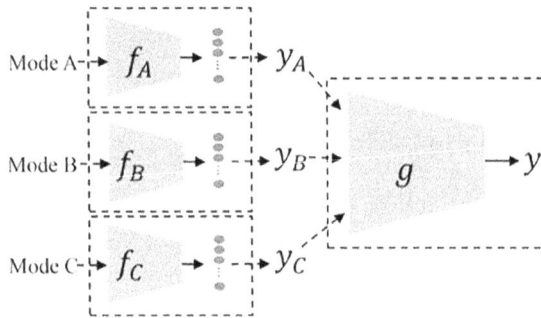

(b) Late Fusion

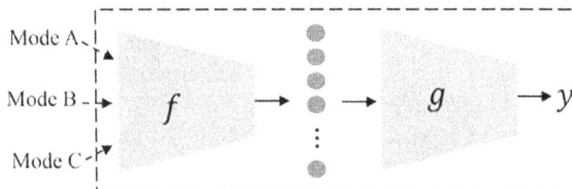

(c) Intermediate Fusion

FIGURE 5.1   Illustration of different levels of fusion.

PCA on the merged dataset and then uses the PC scores to create a model that predicts TTF. In this scenario, f and g are trained separately. Late fusion (see Figure 5.1b) focuses on developing models fA, fB, and fC separately from sensor types A, B, and C to estimate separate TTF values, yA, yB, and yC, then fuses the separate predictions into one overall prediction, for example by taking the average or median of the TTFs. Intermediate fusion (see Figure 5.1c) performs fusion during the model training process and simultaneously finds latent information shared between the different sensor modalities, while also determining the best settings to accurately predict the TTF.

Let us also consider another data fusion example in healthcare. In a research hospital, there is a cohort of patients with a brain disease along with healthy controls for which three different types of neuroimaging were collected: (i) structural magnetic resonance images (sMRI), (ii) functional magnetic resonance images (fMRI), and (iii) magnetoencephalography (MEG). sMRI conveys brain structure and provides the highest spatial resolution, but has no temporal resolution. fMRI indicates blood oxygen level and provides an acceptable spatial resolution along with a lower temporal resolution. MEG records magnetic fields generated by brain electrical activity and provides a higher temporal resolution at the cost of a lower spatial resolution. The task is to build a statistical model capable of optimally combining the information available in images A, B, and C to accurately quantify patient disease severity and highlight particular brain locations or functions that may be causing impairment. Early fusion (see Figure 5.1a) can use an algorithm, such as Independent Component Analysis (ICA), to identify spatially independent signals that convey underlying brain networks/ structures (f), from which features can be extracted and used to predict disease severity (g). Late fusion (see Figure 5.1b) can develop separate machine learning models trained on each image type, fA, fB, and fC, to make predictions of disease severity, yA, yB, and yC, that can then be fused into an overall prediction score by g (ex., through averaging the scores). Intermediate fusion (see Figure 5.1c) performs neuroimaging fusion during the model training process (e.g., via fused group lasso) and can simultaneously fuse information from images A, B, and C while building an accurate model that can predict the patient disease severity.

5.1.1 Contribution and Article Organization

This article reviews data fusion algorithms as categorized earlier into decomposition-based algorithms and neural network-based algorithms. Within each category, this article covers a broad range of algorithms in multimodality fusion that are mainly developed in recent years with a focus on Industrial Engineering (IE) applications, such as healthcare and prognostics. Other review papers are available in the area of multimodality data analysis (Atrey et al., 2010; Khaleghi et al., 2013; Lahat et al., 2015) with different points of focus. Atrey et al. (2010) concentrated on multimodal fusion techniques for multimedia analysis. They group the available techniques into rule-based, classification-based, and estimation-based algorithms and provide several traditional techniques such as linear modeling, Support Vector Machines (SVMs), entropy maximization, and Kalman and particle filtering. This review article is distinct from the one by Atrey et al. (2010), both in terms of domain application and the algorithms covered. Khaleghi et al. (2013) provided a review of multi-sensor fusion with an information theoretic perspective. The main focus of that review is on low-level fusion and discusses available frameworks in addressing multi-sensor fusion challenges such as imperfections, correlation, inconsistency, and disparateness. The frameworks introduced in that paper are related to probability theory, fuzzy set theory, possibility theory, rough set theory, and Dempster–Shafer evidence theory. Although that paper is a great source for understanding formal definitions of data fusion challenges, it does not provide a practical understanding of data fusion and available algorithms, particularly to the IE audience.

One of the more recent review papers on multimodal data analysis is authored by Lahat et al. (2015). The main focus of that paper is on early fusion using decomposition-based techniques such as ICA, canonical analysis, and tensor analysis. Our manuscript extends Lahat et al. (2015) in several ways. First, we will introduce methods beyond matrix/tensor decomposition and canonical analysis, as well as discuss techniques that perform the intermediate fusion. Second, we introduce recent developments that were published after 2015. For example, tensor and factor analysis gained significant attention in the IE community in the past few years, and are covered in this article. Finally, we tailor the methods to IE applications through examples as our main audience is this community.

Section 5.2 discusses recent developments in decomposition-based algorithms with a specific focus on IE applications; Section 5.3 presents neural network-based fusion; Section 5.4 covers a brief discussion on data and domain knowledge integration; Section 5.5 discusses current challenges and future research directions, and Section 5.6 concludes the article.

5.2   DECOMPOSITION-BASED FUSION ALGORITHMS

In this article, we define decomposition-based fusion as fusion techniques that decompose data matrices or tensors to extract patterns and features. These methods do not employ neural networks. Factor analysis, tensor analysis, and dimensionality reduction methods are examples of decomposition-based fusion that are covered in the following discussion. The focus of this section is not particularly on supervised or unsupervised methods, and we assume that a knowledgeable reader can distinguish them from the provided context. This section does not also explicitly mention whether a technique is suitable for early, late, or intermediate fusion as it should be clear from the context. As an example, tensor regression techniques are supervised and intermediate fusion methods, whereas coupled tensor decomposition are unsupervised methods that can be employed in both early or late fusion. Among the described methods, tensor analysis mainly focuses on the fusion of heterogeneous datasets that are different in their form (e.g., images and profiles). The other approaches mainly take each observation as a vector. Several regularization techniques are also discussed in this section. These techniques are complementary to decomposition-based fusion and are often used to improve data fusion by pinpointing and eliminating noninformative modes or features. Therefore, regularization methods are subtractive. That is, while most of the methods combine different modes to obtain a feature, regularization techniques omit the uninformative modes/features. Table 5.1 reports a brief summary of decomposition-based fusion literature.

TABLE 5.1

Summary of the Decomposition-Based Methods and Their Corresponding Capabilities (C) and Limitations (L)

Framework	Descriptions	Capabilities (C) and Limitations (L)
Tensor Analysis (Bro,1996; Zhao et al., 2012; Zhou et al., 2013; Acar et al., 2014; Lock,2018; Mou et al., 2019; Fang et al., 2019; Yan et al., 2019; Yue et al., 2020; Gahrooei et al., 2020)	Extract common and uncommon structures among modalities via tensor decomposition and tensor calculus	(C1) Suitable for heterogeneous datasets with different dimensions (C2) Suitable for extracting interpretable patterns between and within the modes (L1) Lack of identifiability and uniqueness (L2) Tensor rank selection
Factor Analysis (Bro et al., 1997; Li et al., 2003; Wang et al., 2012; Virtanen et al., 2012; Klami et al., 2013; Acar et al., 2015; Argelaguet et al., 2018; Li and Li, 2019)	Describes covariance among observed modes and variables in terms of a potentially smaller set of latent variables (factors)	(C1) Interpretable model for understanding the underlying between-mode relationships (C2) Generates a lower-dimension representation of the original data (L1) Limited inference approaches are available (L2) Most estimation algorithms are based on Expectation-Maximization (EM) algorithms that are computationally expensive (L3) Unknown number of latent variables (L4) Models need to be better understood for particular applications for accurate factor recovery (Acar et al., 2015)
Generalized PCA and beyond (Maaten and Hinton, 2008; Lampertand and Kr€omer, 2010; Candes et al., 2011; Zhang et al., 2011; White et al., 2012; White and Schuurmans, 2012; Li et al., 2017; Xiao et al., 2018)	Finds a low-dimensional representation across modalities	(C1) Suitable for visualization (C2) Dimension reduction of both paired and unpaired modalities (Lampertand and Kr€omer, 2010) (C3) Linear and nonlinear dimensionality reduction (Maatenand and Hinton, 2008) (L1) Lack of scalability to incorporate into parallel and distributed computing structures (Candes et al., 2011) (L2) Incorporation of decision-level fusion is lacking (Xiao et al., 2018) (L3) t-SNE not guaranteed to converge to the global optimum of its cost function (Maatenand Hinton, 2008) (L4) t-SNE cannot differentiate modalities.
Regularization (Xiang et al., 2014; Paynabar et al., 2015; Zhang et al., 2018a; Si et al., 2020)	Selects informative modes and informative features within a mode	(C1) Suitable for automatic extraction of relevant modes and features (C2) Outputs interpretable, parsimonious models (L1) Increases parameters estimation complexity (L2) Highly correlated modes and features may cause unstable estimations

5.2.1 FACTOR AND CANONICAL ANALYSIS

Factor Analysis (FA) is a method that describes covariance among observed variables in terms of a potentially smaller set of latent variables (called factors). FA takes the following basic form

$$X = ZW^T + E$$

where X is an $n \times p$ matrix, where n represents the number of instances and p represents the number of observed variables; Z, the latent factor matrix, is an $n \times p'$ matrix, where $p' \leq p$; W, the factor loading matrix, is an $p \times p'$ matrix and performs the transformation between latent and observed variables; and E is an error matrix. FA can be solved by traditional Maximum Likelihood Estimation (MLE) (Gaskin and Happell, 2014).

One of the first known instances of using FA in the context of multimodality fusion is CANDECOMP/PARAFAC (CP) decomposition (Bro et al., 1997; Li, Choi, Perros, Sun, and Vudue, 2017). CP decomposition is often used for three datasets, but has the capability to be expanded to more. For simplicity, this chapter describes the three-way formulation, which is summarized as follows:

$$x_{ijk} = \sum_{r=1}^{R} a_{ir} b_{jr} c_{kr} + \epsilon_{ijk}; \quad i = 1, \ldots I; \; j = 1, \ldots, J, \; k = 1, \ldots, K$$

With the associated sum of squares loss:

$$\min_{ijk} = \sum_{ijk} \left\| x_{ijk} - \sum_{r=1}^{R} a_{ir} b_{jr} c_{kr} \right\|^2 ,$$

where $A = (a_1, \ldots, a_R)$, $B = (b_1, \ldots, b_R)$, and $C = (c_1, \ldots, c_R)$ denote the $I \times R$, $J \times R$, and $K \times R$ matrices containing the R different factor loadings in the three datasets. The model may also be written as

$$\sum_{r=1}^{R} a_r \otimes b_r \otimes c_r ,$$

where a_r, b_r, and c_r are the rth columns of A, B, and C, respectively. Under assumptions of Gaussian noise, CP decomposition can be solved via MLE. CP decomposition has previously been applied to raw, high-dimensional electronic health records to identify useful phenotypes or medical concepts that can be utilized for patient diagnosis, prognosis and treatment (Ho et al., 2014; Wang et al., 2015; Li, Cerise, Yang and Han, 2017). CP has also been used for change detection and monitoring systems with image or multichannel data (Yan et al., 2014; Li et al., 2015).

Bayesian Group FA (Virtanen et al., 2012) is a method that is capable of finding factors of different types–namely, those specific to all sources, a combination of some data sources, a single data source, or "noise" factors. The main task of Bayesian

Group FA is to find a set of factors that explains dependencies between all possible subsets of the data sources. Bayesian Group FA takes the following form:

$$[\mathbf{X}_1|\mathbf{X}_2|\ ...\ |\mathbf{X}_M] \approx \mathbf{ZW}^T,$$

where $\mathbf{X}_m$, $m = 1, ..., M$, are $n \times p_m$ matrices represent the different sets of possible data sources; pm is the number of features in modality m; $\mathbf{Z}$ is a $n \times p'$ matrix that represents the latent components; $\mathbf{W}$, the factor loading matrix, is an $p \times p'$ matrix that is group-wise sparse (i.e., a sparsity constraint is applied with respect to the variables in each modality), so each factor is active only in some subset of data sources, all of them, or only one. If the factor is active in only one modality, it is associated with noise or independent variation of that particular modality. $\mathbf{W}$ is made sparse by a group-wise Automatic Relevance Determination (ARD) prior. Figure 5.2 demonstrates an illustration of this method. Bayesian Group FA has been applied in the area of drug discovery by identifying systems-level drug-response phenotypes from genome-wide transcriptomic profiles (Khan et al., 2014; Yadav et al., 2015; Kibble et al., 2016).

Bayesian Interbattery FA (BIBFA) is a method motivated by Canonical Correlation Analysis (CCA) (Klami et al., 2013; Acar et al., 2015). However, instead of merely extracting correlated components between datasets like CCA, BIBFA takes into consideration both shared and unshared components (originally based on the non-probabilistic form, InterBattery FA (IBFA) (Tucker, 1958). Given two data sources, $\mathbf{X}_1$ and $\mathbf{X}_2$, BIBFA relies on the following probabilistic form of IBFA:

$$\mathbf{Z}_0 \sim N(\mathbf{0}, \mathbf{I}); \mathbf{Z}_m \sim N(\mathbf{0}, \mathbf{I}); \mathbf{X}_m \sim N(\mathbf{A}_m\mathbf{Z}_0 + \mathbf{B}_m\mathbf{Z}_m, \mathbf{\Sigma}_m),$$

where $N(\mathbf{\mu}, \mathbf{\Sigma})$, $R\mathbb{P}$ is the normal distribution with mean $\mathbf{\mu}$ and covariance matrix $\mathbf{\Sigma}$; $\mathbf{\Sigma}_m$ is a diagonal matrix that describes the covariance of modality m; $\mathbf{I}$ and $\mathbf{0}$ are the identity and zero matrices; $\mathbf{Z}_0$ denotes the factors shared between the data sources; $\mathbf{Z}_m$ denotes the unshared factors in each data source $\mathbf{X}_m$, $m = 1, 2$.

To derive an effective method to solve the model, it can be reformulated as follows:

$$\mathbf{Z} \sim N(\mathbf{0},\mathbf{I}); \mathbf{X} \sim N(\mathbf{WZ},\mathbf{\Sigma})$$

where

$$\mathbf{W} = \begin{pmatrix} \mathbf{A}_1 & \mathbf{B}_1 & \mathbf{0} \\ \mathbf{A}_2 & \mathbf{0} & \mathbf{B}_2 \end{pmatrix}, \mathbf{Z} = \begin{pmatrix} \mathbf{Z}_0 \\ \mathbf{Z}_1 \\ \mathbf{Z}_2 \end{pmatrix}, \mathbf{\Sigma} = \begin{pmatrix} \mathbf{\Sigma}_1 & \mathbf{0} \\ \mathbf{0} & \mathbf{\Sigma}_2 \end{pmatrix}.$$

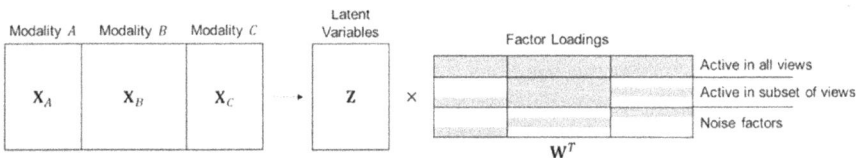

FIGURE 5.2 Group factor analysis. (Adapted from Virtanen et al. 2012.)

An appropriate structure in $\mathbf{W}$ is accomplished by imposing group-wise sparsity via an ARD prior. For inference, variational approximation is used (based on priors that assume maximally orthogonal latent factors). See Klami et al. (2013) for more information on solving BIBFA. BIBFA has been applied to cancer gene prioritization via DNA copy number data and integrative analysis of mRNA expression (Lahti et al., 2013).

Cross-modal Factor Analysis (CFA) combines two data sources, $\mathbf{X}_1$ and $\mathbf{X}_2$ by finding two linear transformations, $\mathbf{W}_1$ and $\mathbf{W}_2$ for each data source (Li et al., 2003). $\mathbf{X}_m$ is a $n \times p_m$ matrix, whereas $\mathbf{W}_m$ is a $p_m \times p'_m$ matrix, where $p'_m \leq p_m$ and $m = 1, 2$. CFA is formulated as follows:

$$\min_{\mathbf{W}_1, \mathbf{W}_2} \left\| \mathbf{X}_1 \mathbf{W}_1^T - \mathbf{X}_2 \mathbf{W}_2^T \right\|_F^2$$

$$\text{s.t. } \mathbf{W}_m^T \mathbf{W}_m = \mathbf{I}, m = 1, 2 ,$$

where $\| \ \|_F^2$ is the Frobenius norm and $\mathbf{I}$ is an identity matrix of appropriate dimensions. Li et al. (2003) showed that the formulation can be reduced to

$$\max_{\mathbf{W}_1, \mathbf{W}_2} \text{Tr} \mathbf{X}_1 \mathbf{W}_1 \mathbf{W}_2^T \mathbf{X}_2^T$$

$$s.t. \ \mathbf{W}_m^T \mathbf{W}_m = \mathbf{I}, m = 1, 2 ,$$

which can be solved via Singular Value Decomposition (SVD) to obtain the final transformed latent factors, $\mathbf{Z}_m$ as,

$$\hat{\mathbf{Z}}_m = \mathbf{X}_m \mathbf{W}_m^T, m = 1, 2$$

Unlike CCA, CFA does not need to calculate the inverse of the covariance matrices, provides orthogonal transformations, and does not require independence of vectors, $\mathbf{X}_m$. However, CFA cannot provide correct information associations if the modalities are not linearly related. An extension of CFA, called Kernel CFA (KCFA) handles this issue by mapping the $\mathbf{X}_m$ vectors in the original space to a high-dimensional space via the kernel trick (Wang et al., 2012). One application for CFA is in the integration of multimedia sources (i.e., audio and video); CFA can compensate for missing or noisy media sources and effectively integrate multiple streams of information together (Li et al., 2003).

Multi-Omics Factor Analysis (MOFA) (Argelaguet et al., 2018) is a recently developed method made specifically for integrating multiple omics data modalities, but is capable to be applied to other applications as well. MOFA builds upon Bayesian Group FA (Virtanen et al., 2012), by (i) enabling fast inference via variational approximation, (ii) inducing sparse solutions to help interpretation, (iii) handling missing values in an efficient manner, and (iv) allowing for flexibility in combining different likelihood models for each data modality. MOFA takes on the following form:

$$\mathbf{X}_m = \mathbf{Z} \mathbf{W}_m^T + \mathbf{E}_m, m = 1, ..., M$$

where $\mathbf{X}_m$ denotes the original feature matrix for modality m, $\mathbf{Z}$ is the factor matrix (common for all data modalities), $\mathbf{W}_m$ is the weight matrix corresponding to modality m, and $\mathbf{E}_m$ is the error term for the particular modality m. The MOFA model is formulated in a probabilistic Bayesian framework, in which a prior distribution is applied to all unobserved variables. MOFA utilizes two levels of regularization: (i) view- and factor-wise sparsity via an ARD prior, which helps identify which factor is active in which view, and (ii) feature-wise sparsity via a spike-and-slab prior that usually results in a smaller number of features with active weights.

Statistical inference for FA of multimodal data has mainly remained unexplored; however, one recent work examines (i) how to infer the significance of one data source given other sources in the model, (ii) how to infer the significance of a combination of variables across different modalities, or from a single modality, and (iii) how to quantify the contribution of one data source given the other data sources, using a goodness-of-fit measure (Li and Li, 2019).

Multimodal datasets have also been employed for more effective clustering of the data. One key challenge in designing clustering algorithms is to identify which modes and which features within each mode are informative for distinguishing clusters. Regularization techniques combined with factor analysis have been introduced for this purpose. For example, Si et al. (2020) introduced a hierarchical clustering approach that uses an L_{12} penalty to effectively identify informative modes and features. Specifically, let $\mathbf{x}_{m,i}$ denote the vector of features of mth mode in the ith sample. Then, the following factor analysis is considered,

$$\mathbf{x}_{m,i} = \mathbf{H}_m \mathbf{f}_{m,i} + \mathbf{B}_m \mathbf{z}_i = \mathbf{e}_{m,i},$$

where $\mathbf{f}_{m,\,i}$ is a latent factor, $\mathbf{z}_i$ is a vector of known covariates, $\mathbf{H}_m$ and $\mathbf{B}_m$ are the loading matrices, and $\mathbf{e}_{m,\,i}$ represents the model errors that follow a multivariate normal distribution with mean zero and covariance matrix $\mathbf{\Sigma}_m$. The latent factors are then linked to clusters $\mathbf{s}_i$:

$$\mathbf{f}_{m,i} = \mathbf{A}_m \mathbf{s}_i + \mathbf{v}_{m,i},$$

where $\mathbf{A}_m$ is a loading matrix and $\mathbf{v}_{m,\,i}$ is an error term that follows a multivariate normal distribution with mean zero and covariance matrix $\mathbf{\Psi}_m$. The goal is to estimate the model parameters, $\mathbf{\Theta} = \{\mathbf{H}_m, \mathbf{B}_m, \mathbf{A}_m, \mathbf{\Sigma}_m, \mathbf{\Psi}_m\}$ while imposing sparsity on loading matrices $\mathbf{H}_m$ and $\mathbf{A}_m$. Imposing sparsity facilities the selection of informative modes and features. The following objective function has been minimized through Expectation-Maximization (EM) technique to achieve this goal:

$$-l(\Theta) + \sum_j \sum_m (\| \mathbf{h}_m^j \|_2) + \sum_m (\| \mathbf{A}_m \|_2),$$

where $\mathbf{h}_m^j$ denotes the jth column of $\mathbf{H}_m$ and $\| \ \|_2$ represents the L_{21} norm. The model was applied to a multimodal MRI dataset of brain cortical area, thickness, and volume to identify subgroups of migraine patients.

5.2.2 Tensor and Functional Data Analysis

One of the main challenges in integrating multimodal datasets is identifying a unified and flexible representation of each mode of a dataset without losing information. Defining such representation allows the use of mathematical tools that apply to all modes of data. A straightforward approach is to model all the data as vectors. However, this approach breaks down and loses the structural information of data instances such as images. Recent advances in multilinear algebra (Kolda, 2006) create an unprecedented opportunity for the use of tensors (i.e., higher-order arrays) and tensor calculus for multimodal data fusion. In this section, we describe different existing approaches used for integrating data using tensor analysis.

5.2.2.1 Tensor Regression Approaches

Tensor regression is an early or intermediate data fusion approach that uses tensor decomposition techniques, including CP and Tucker decomposition (Kolda, 2006), to extract features out of an input tensor that contains data from multiple sources to estimate the output. Most approaches focus on single tensor input (Bro, 1996; Zhao et al., 2012; Zhou et al., 2013; Lock, 2018; Fang et al., 2019; Yan et al., 2019; Yue et al., 2020) but multiple tensor-input regression frameworks are also available (Gahrooei et al., 2020). Scalar-on-tensor, tensor-on-scalar, and tensor-on-tensor models are different forms of tensor regression modeling that have been introduced in the literature. Let us denote a tensor by calligraphy font. For example, $\mathcal{A} \in \mathbb{R}^{I_1 \times I_2 \times \ldots \times I_d}$ is a tensor with d modes, whose kth mode is of size Ik. A linear scalar-on-tensor model (Yan et al., 2019) is usually formulated as

$$\mathcal{Y} = \mathcal{B} \times_{d+1} \mathbf{X} + \mathcal{E},$$

where $\mathcal{Y} \in \mathbb{R}^{I_1 \times \ldots \times I_d}$; $\mathbf{X} \in \mathbb{R}^{I_1 \times p}$; $\mathcal{B} \in \mathbb{R}^{I_1 \times \ldots \times I_d \times p}$; ε is the tensor of errors; and $\times_i$ denotes the ith-mode multiplication of a tensor and a matrix. The linear tensor-on-scalar (Zhou et al., 2013; Fang et al., 2019) models are those with scalar output and tensor inputs as follows:

$$y = <\mathcal{B}, \mathcal{X}> + e,$$

where y is a scalar; $\mathcal{X} \in \mathbb{R}^{Q_1 \times \ldots \times Q_l}$; $\mathcal{B} \in \mathbb{R}^{Q_1 \times \ldots \times Q_l}$; e is the error term; and $< ., . >$ denotes the tensor inner product, which is equivalent to the inner product of the vectorized version of the tensors. Finally, the tensor-on-tensor regression model (Lock, 2018; Gahrooei et al., 2020) is defined as

$$\mathcal{Y} = <\mathcal{B}, \mathcal{X}>_l + \mathcal{E},$$

where $\mathcal{Y} \in \mathbb{R}^{I_1 \times \ldots \times I_d}$; $\mathcal{X} \in \mathbb{R}^{Q_1 \times \ldots \times Q_l}$; $\mathcal{B} \in \mathbb{R}^{Q_1 \times \ldots \times Q_l \times I_1 \times \ldots \times I_d}$; and ε is the tensor of errors. The operation $< ., . >_l$ is called the tensor contraction operation over l modes. In all the cases, the tensor of parameters $\mathcal{B}$ is assumed to be low-rank and has a decomposition form that results in a lower number of parameters to be estimated. Depending on the learning algorithm, these models can be viewed as either early or intermediate fusion. In particular, if the decomposition of the parameter tensor is learned during the training, the model is of intermediate fusion type. Tensor regression models are suitable for the fusion of heterogeneous sources of data when multichannel and multi-dimensional data sets are available (Gahrooei et al., 2020). Tensor regression models have been applied in modeling and optimization of the lathe-turning process (Yan et al., 2019), in semiconductor manufacturing to predict overlay error of the lithographic process based on wafer shape (Gahrooei et al., 2020) as well as prediction of the aged state of Ni-based superalloys (Gorgannejad et al., 2019). It has also been applied in prognostics applications for the prediction of useful remaining lifetime based on thermal images (Fang et al., 2019)

5.2.2.2 Coupled Decomposition

Modes of the multimodality data may contain common and uncommon features to be discovered. The coupled decomposition technique decomposes the tensors assuming they share a common subspace spanned by a set of bases.

Simultaneous matrix and tensor factorization is an approach for discovering the common bases. Let $\mathbf{X} \in \mathbb{R}^{P \times Q}$ and $\mathcal{Y} \in \mathbb{R}^{P \times I_1 \times I_2}$. By minimizing the following objective function, Acar et al. (2013) identify the common space, spanned by basis matrix $\mathbf{U}$, between the tensor and matrix modes:

$$L = \left\| \mathcal{Y} - [\![\lambda; \mathbf{U}, \mathbf{V}_1, \mathbf{V}_2]\!] \right\|^2 + \left\| \mathbf{X} - \mathbf{U}\mathbf{W}^\mathsf{T} \right\|^2,$$

where $[\![\lambda; \mathbf{U}, \mathbf{V}_1, \mathbf{V}_2]\!]$ denotes CP decomposition of the tensor with basis matrices $\mathbf{U}$, $\mathbf{V}_1$, and $\mathbf{V}_2$:

Zhao et al. (2012) proposed Higher-order PLS (HOPLS) as an approach for identifying the common features between two tensors that represent different modalities. Let $\mathcal{X} \in \mathbb{R}^{P \times Q_2 \times \ldots \times Q_l}$ and $\mathcal{Y} \in \mathbb{R}^{P \times I_2 \times \ldots \times I_d}$. Then, the coupled Tucker decomposition of these two tensors is written as:

$$\mathcal{X} = \mathcal{C}_1 \times_1 \mathbf{U} \times_2 \mathbf{U}_2 \ldots \times_l \mathbf{U}_l$$

and

$$\mathcal{Y} = \mathcal{C}_2 \times_1 \mathbf{U} \times_2 \mathbf{V}_2 \ldots \times_d \mathbf{V}_d.$$

Here, $\mathcal{C}_1$ and $\mathcal{C}_2$ are decomposition core tensors and $\mathbf{U}_2, \ldots, \mathbf{U}_l, \mathbf{V}_2, \ldots, \mathbf{V}_d$ are bases that do not coincide between the two tensors. The matrix $\mathbf{U}$ denotes the common basis that spans the common subspace between the two modes. Zhao et al. (2012) proposed an approach based on the SVD of the covariance matrices to estimate the basis matrices. The approach was applied to decoding electrocorticography (ECoG) signals in relation to 3D hand trajectories of monkeys performing movement tasks. Potentially, this approach can be applied to before and after medical intervention data, including the MRI and Electroencephalography (EEG) signals of patients before and after an intervention.

5.2.2.3 Structural Revealing Decomposition

The main limitation of coupled decomposition is that it only finds the common bases for the shared mode. However, the shared mode may contain common and uncommon bases. The structural revealing technique is designed to resolve this issue. Nevertheless, it is only designed for situations where the data modes are a 3-D tensor and a matrix. More specifically, let $\mathbf{X} \in \mathbb{R}^{P \times Q}$ and $\mathcal{Y} \in \mathbb{R}^{P \times I_2 \times I_3}$. Then the structural revealing technique will identify the common and uncommon features related to the shared dimension of the tensor and matrix by minimizing the following objective function:

$$L = \left\| \mathcal{Y} - [\![\lambda; \mathbf{U}, \mathbf{V}_1, \mathbf{V}_2]\!] \right\|^2 + \left\| \mathbf{X} - \mathbf{U\Sigma W} \right\|^2 + \left\| \lambda \right\|_1 + \left\| \sigma \right\|_1,$$

where $\lambda; \mathbf{U}, \mathbf{V}_1, \mathbf{V}_2$ denotes CP decomposition of the tensor with basis matrices $\mathbf{U}, \mathbf{V}_1$, and $\mathbf{V}_2$; $\mathbf{U\Sigma W}$ denotes the SVD decomposition of the matrix with singular values $\sigma = diag(\mathbf{\Sigma})$. In this formulation, $\mathbf{U}$ contains common and uncommon bases, which are identified by the lasso penalization over λ and σ. The utility of this method was tested on a dataset of mixtures with known chemical composition (inferred from nuclear magnetic resonance and mass spectrometry) and it was found that the method can successfully determine the chemicals in the mixtures, as well as their relative concentrations. For more details, please refer to Acar et al. (2014).

5.2.2.4 Self-Expressive Models

Subspace clustering has been used in many applications, including systems monitoring and diagnosis based on profile data (Zhang et al., 2020). One main approach for subspace clustering is the self-expressive model, which clusters a set of signals into subspaces (Elhamifar, 2016). Let $\mathbf{Y} = [\mathbf{y}_1, \mathbf{y}_2,..., \mathbf{y}_n]$ to be a matrix containing n signals of length D. Then, the self-expressive model assumes that each signal can be explained by other signals within the matrix:

$$Y = YC + E,$$

where $diag(C) = 0$ and E is a matrix of errors. To find clusters of signals, i.e., a set of similar signals, one may impose sparsity or low-rankness on C. Therefore computing the set of model parameters C requires solving the following optimization algorithm:

$$\frac{1}{2} \| \mathbf{Y} - \mathbf{YC} \|_2^2 + \lambda \| \mathbf{C} \|_1 + \| \mathbf{C} \|_*,$$

where $\| \ \|_1$ and $\| \ \|_*$ refer to L_1 and nuclear norms, respectively. The nuclear norm of a matrix is the sum of its singular values and is used for imposing low-rankness.

The extension of self-expressive models has been introduced for multimodal datasets (Abavisani and Patel, 2018). Let $\{\mathbf{Y}_1,..., \mathbf{Y}_m\}$ be an m-mode dataset where each mode contains a set of signals similar to the case of the uni-mode self-expressive model. Then the goal of multimodal self-expressive models is to simultaneously cluster the signals in distinct modalities according to their subspaces. For this purpose, one can minimize the following objective function:

$$\frac{1}{2}\sum_{i=1}^{m}\|\mathbf{Y}_i - \mathbf{Y}_i\mathbf{C}\|_2^2 + \lambda\|\mathbf{C}\|_1 + \|\mathbf{C}\|_*$$

The model has been applied to cluster face images using various facial components (i.e., eyes, nose, and mouth) (Elhamifar, 2016).

5.2.2.5 Regularization for Functional and Multimodal Data

Regularization techniques have been used for the identification of the informative modes of data, as well as informative features within each mode. The most common regularization techniques used for this purpose are based on non-negative Garrote and L_{21} norm. Paynabar et al. (2015) integrated profiles data to predict a scalar output. The informative profiles and features within the selected profiles were identified using a hierarchical non-negative Garrote technique. Let $\mathbf{y} \in \mathbb{R}^n$ denote the vector of outputs, $\mathbf{C}_k$ be the design matrix of the kth ($k = 1, 2,..., K$) group (profile), and β_k^{ols} be the ordinary least square parameter. Then by minimizing

$$L = \left\|\mathbf{y} - \sum \mathbf{C}_k\beta_k^{ols}d_k\right\|_2^2 + \lambda\sum_k^K d_k,$$

the shrinkage term $d_k \geq 0$ is estimated. If $d_k = 0$, then the group is considered as uninformative. By only considering selected groups (profiles) and imposing a lasso penalty on the feature shrinkage terms, the informative features can be identified. This model was applied to develop a predictive model that can estimate vehicle design comfort from the three-dimensional motion signals of test drivers.

Zhang et al., (2018b) employed sparsity penalty to monitor a weakly correlated set of profiles (multichannel data). Specifically, they expanded the multivariate functional eigen-decomposition technique to the situation where not all eigen-functions are informative for spanning a given profile.

Let $\mathbf{Y}_i \in \mathbb{R}^{n \times p}$ denote the ith ($i = 1,..., N$) sample that contains p profiles (channels), each observed at n points, $V = \{\mathbf{v}_1, \mathbf{v}_1, ... \mathbf{v}_K\}$ whose columns are of eigen-functions; and $\mathbf{\Psi}_i = [\psi_{i1},..., \psi_{iK}]$ be the spanning coefficients. Then by minimizing

$$\sum_{i=1}^{N}\|\mathbf{Y}_i - V\mathbf{\Psi}_i\|_2^2 + \sum_i^K\sum_{j=1}\||\psi_{ij}\||$$

subject to

$$V^TV_i = \mathbf{I},$$

the sparse multichannel decomposition is achieved. This technique is used for monitoring semiconductor manufacturing, where a large number of weekly correlated sensors are available. Gahrooei, Payanbar, Pacella and Shi (2019) also combined a large number of profile inputs to estimate a profile output using a functional group lasso penalty and applied the model to retrieve joint motion trajectory based on other joint trajectories from sensors located on the human body (e.g., hip, neck, elbows, knees, etc.).

A generalized sparse model for multimodality data (assuming every instance has all modalities) was formulated by Xiang et al. (2014) using a multi-task framework. The two stages of this model are (i) to learn different models for each data modality and (ii) combine the learned models appropriately. The formulation is summarized below:

$$\min_{\alpha,\gamma} \frac{1}{2} \left\| \mathbf{y} - \sum_{m=1}^{M} \gamma_m \cdot \mathbf{X}_m \alpha_m \right\|_2^2 + \sum_{m=1}^{M} \frac{\lambda_m}{p} \|\alpha_m\|_p^p + \sum_{m=1}^{M} \frac{\eta_m}{q} |\gamma_m|^q ,$$

where $\mathbf{y}$ is the response variable; $\mathbf{X}_m$ are the features for modality m; α_m are the weights of the linear model learned for the mth modality; γ are the weights that combine the learned models together; p and q are adjustable integers that can induce the desired sparsity on the feature (p) and instance level (q); and finally, λ_m and η_m are tuning parameters. This model can be reduced to common regularization methods and solved by standard multi-task learning algorithms.

A more realistic (and challenging) scenario occurs when not every modality is available for every instance. Xiang et al. (2014) developed a bi-level multi-source learning algorithm for heterogeneous block-wise missing data to handle missing medical modalities in an Alzheimer's Disease dataset. That paper developed methods of handling situations in multimodality datasets when not all instances have all data sources. First, the instances are divided into different groups according to which data modalities are available. Then, using a similar strategy as the generalized sparse model for multimodality data (the model presented previously) the two stages of this model were (i) to learn different models for each data group and (ii) combine the learned models appropriately. The formulation is as follows:

$$\min_{\alpha,\beta} \frac{1}{|\mathbf{pf}|} \sum_{m \in \mathbf{pf}} \frac{1}{n} \mathcal{L} \left(\sum_{m=1}^{M} \alpha_m X_m \beta, \mathbf{y}_m \right) + \lambda \mathbf{R}_\beta(\beta)$$

$$\text{s.t.} \quad \mathbf{R}_\alpha(\alpha_m) \leq 1, \forall m \in \mathbf{pf}$$

where β corresponds to the coefficients of features across all modalities; α is the weighted combination across the different modalities; $\mathbf{pf}$ is an n-dimensional vector that encodes binary indicators for which modalities are present for each of the n instances; X_m are the features for modality m ($m = 1,..., M$); $\mathbf{y}_m$ is the response; $\mathcal{L}$ is any convex loss function (e.g., least squares, logistic loss, etc.); and $\mathbf{R}_\alpha$, $\mathbf{R}_\beta$ are regularizations on a and b, respectively. The solution to this model can be approximated via alternating optimization between α and β: The benefit of this method is that out-of-sample test instances with different modality combinations can still be predicted, as the model is designed to have a generalized β across all modality combinations.

5.2.3 Generalized PCA and Beyond

The goal of dimension reduction methods such as PCA is to find a low-dimensional representation across multiple heterogeneous data modalities that can be utilized for a variety of tasks (e.g., data compression, clustering, and model training). Dimension reduction falls under one of two categories (Lampert and Kr€omer, 2010): (i) inductive and (ii) noninductive. Inductive methods inherently include a function that can be used for future data. Noninductive methods do not include such a function and must be re-applied any time new data is obtained. There is not much work performed in noninductive multimodality dimension reduction (Fernandez-Beltran et al., 2018a, 2018b), so the focus of the discussion will be on inductive methods.

One of the first known examples of multimodality reduction is Maximum Covariance Analysis (MCA), a generalization of PCA (Tucker, 1958). Given two centered data sources, $\mathbf{X}_1$ and $\mathbf{X}_2$, with dimensions $n \times p_m$ ($m = 1, 2$), MCA performs multimodal dimension reduction by solving for $\mathbf{W}_1$ and $\mathbf{W}_2$ in the objective function below:

$$\max_{\mathbf{W}_1, \mathbf{W}_2} \operatorname{Tr}\mathbf{W}_1^T \mathbf{X}_1^T \mathbf{X}_2, \mathbf{W}_2.$$

where $\mathbf{W}_1$ and $\mathbf{W}_2$ are orthogonal matrices with dimensions $p_m \times p_m'$, where $p_m' < p_m$, $m = 1, 2$. One of the main disadvantages of MCA is that it requires instances to be completely paired, i.e., both data sources have identical instances. This scenario is often unrealistic in larger datasets.

Weakly Paired MCA (WMCA) allows for data sources to have differing instances, which provides greater flexibility with using all the available data (Lampert and Kromer, 2010). Given two centered data sources, $\mathbf{X}_1$ and $\mathbf{X}_2$, with dimensions $n_m \times p_m$ ($m = 1, 2$), MCA performs multimodal dimension reduction by solving for $\mathbf{W}_1$ and $\mathbf{W}_2$ in the objective function below:

$$\max_{\mathbf{W}_1, \mathbf{W}_2} \operatorname{Tr}\mathbf{W}_1^T \mathbf{X}_1^T \Pi \mathbf{X}_2, \mathbf{W}_2.$$

where $\mathbf{W}_1$ and $\mathbf{W}_2$ are orthogonal matrices with dimensions $p_m \times p_m'$, where $p_m' < p_m$, $m = 1, 2$; Π is an $n_1 \times n_2$ binary matrix that encodes the different grouping structures to take into account the weakly paired data. There is no closed form solution for this objective function, but it can be estimated via an alternating maximization of $\mathbf{W}_1$, $\mathbf{W}_2$, and P: WMCA has been used to discriminate between different textures of materials (e.g., styrofoam, bricks, wallpaper, etc.) based on images and audio signals recorded over the surfaces (Lampert and Krömer, 2010). Additional description for the mechanics of Π, an extension to the kernelized space, and expansion to handling more than two modalities can be found in Lampert and Kromer (€ 2010).

Comparing traditional CCA (from Section 2.1) to these dimension reduction methods, CCA is limited in that it is restricted to a square loss under a particular normalization. In the single data modality case, there have been several papers that have been able to relax these constraints and allow for more generalizable cases that can incorporate several convex losses while maintaining a reasonable computational complexity (Candes et al., 2011; Zhang et al., 2011). One proposal for the multimodality setting is convex multi-view subspace learning (White et al., 2012). Assuming that two data sources, $\mathbf{X}_1$ ($n \times p_1$) and $\mathbf{X}_2$ ($n \times p_2$) are conditionally independent given their shared latent representation $\mathbf{H}$ ($n \times p'$), where n is the number of instances and $p' < p_1 + p_2$ is the reduced dimension, an optimal data reconstruction can be found via utilizing an implicit convex regularizer that recovers $\mathbf{H}$ jointly. One formulation of this method is summarized below:

$$\min_{\mathbf{W}_1, \mathbf{W}_2, \mathbf{H}} \mathcal{L}(\mathbf{HW};\mathbf{X}) + \alpha \left\| \mathbf{H}^T \right\|_{2,1}$$

$$\text{s.t.} \left[\mathbf{W}_{1(:,i)} \mathbf{W}_{2(:,i)} \right] \in \mathcal{C}, \forall i$$

$$\text{where } \mathcal{C} := \left\{ \|\mathbf{w}_1\|_2 \leq \beta_1, \|\mathbf{w}_2\|_2 \leq \beta_2, \right\}, \mathbf{W} = \left[\mathbf{W}_1 \mathbf{W}_2 \right], \mathbf{X} = \left[\mathbf{X}_1 \mathbf{X}_2 \right]$$

where $\mathcal{L}$ is the convex loss function between $\mathbf{HW}$ and $\mathbf{Z}$ (some examples can be found in White and Schuurmans (2012)); $\mathbf{W}_m$ represents the loading matrices such that $\mathbf{HW}_m \approx \mathbf{X}_m$, $m = 1, 2$; α, β_1 and β_2 are tuning parameters of the objective function. This method was applied to classify a face image dataset consisting of various poses and lighting conditions (Georghiades et al., 2001). t-distributed Stochastic Neighborhood Embedding (t-SNE) is a relatively recent dimension reduction and visualization technique that can be used to combine data features from single or multiple modalities and reduce to a two- or three-dimensional dataset (Maaten and Hinton, 2008; Li, Cerise, Yang and Han, 2017; Xiao et al., 2018). Using conditional probability theory, t-SNE assumes that coordinates in the low dimension follow a t-distribution, which has the effect of increasing the distance between formed clusters allowing for greater distinction between different instances. One instance of t-SNE for Multimodalities performs misalignment fault diagnosis of wind turbines by fusing time and frequency features from the vibration, temperature, and stator current signals, and generates two information-dense features (Xiao et al., 2018). The two features were then used as input for a least square SVM that was optimized by the artificial bee colony algorithm. This application of t-SNE to multiple modalities does not take into account differences between modalities (i.e., covariance), but rather inputs all features from the modalities as a single unit. More work will need to be performed in this area to better incorporate individual modality differences.

5.3   NEURAL NETWORK-BASED FUSION

Neural networks, in particular deep learning models, have demonstrated great promise in several applications including medical, manufacturing, internet of things, remote sensing, and urban big data. There have also been several recent implementations proposed for multimodality fusion in each of these application areas (Calhoun and Sui, 2016; Schmitt and Zhu, 2016; Li et al., 2018; Wang et al., 2018; Liu, Li, Xie, Du, Teng and Yang, 2020b; Qi et al., 2020). A handful of review papers have also been written on this topic (Ramachandram and Taylor, 2017; Gao et al., 2020). Multimodal neural network approaches have the goal of training an end-to-end architecture that achieves both high accuracy and informative modality fusion (Table 5.2).

TABLE 5.2

Summary of the Neural Network-Based Fusion Methods and Their Corresponding Capabilities (C) and Limitations (L)

Framework	Descriptions	Capabilities (C) and Limitations (L)
Early Fusion (Ng, 2011; Srivastava and Salakhutdinov, 2012; Liu et al., 2017; Zhang et al., 2018a)	Extracts and fuses information from each modality before model training (commonly through the use of autoencoders)	(C1) Generates a lower-dimensional representation of the original data (C2) Flexibility to use generated features for any statistical learning algorithm (L1) Incorporation of decision-level fusion is lacking (L2) Interpretability of generated features is not straightforward (L3) Most models assume conditional independence between modalities; but in practice, modalities tend to be highly correlated, e.g., multimodal medical images, video/audio, etc. (Ramachandran and Taylor, 2017).
Late Fusion (Simonyan and Zisserman, 2014; Kahou et al., 2016; Wu et al., 2016)	Fuses predictions between multiple neural networks trained on different modalities using averaging, maximum value, Bayes decision rule, metaclassifiers, etc.	(C1) Easy to implement since fusion is performed at a high-level (C2) Errors from multiple neural networks tend to be uncorrelated, making the late fusion feature independent (L1) Less flexibility in regards to when multimodal representations are learned and where multimodality fusion occurs (L2) No conclusive evidence that late fusion is better than early fusion (Ramachandran and Taylor, 2017).
Intermediate Fusion (Karpathy et al., 2014; Neverova et al., 2015; Gao et al., 2018)	Fuses different modalities at various levels of the neural network during a model training task	(C1) Offers additional flexibility beyond early or late fusion as to where multimodality fusion occurs in the network (C2) Allows for fusion of modalities at different levels and can generate many multimodal representations that can be used at the decision level (L1) Requires careful design for when and where to apply modality fusion in the network

There are several advantages that multimodal neural networks have over conventional multimodal learning (Ramachandram and Taylor, 2017). Neural networks can learn both inter- and intramodality representations with minimal preprocessing of input data, whereas decomposition-based methods often require manual design and are more sensitive to preprocessed data. Neural networks also provide implicit dimensionality reduction within the architecture, whereas with other techniques conventional feature selection methods (i.e., filter, wrapper, or embedded) must be incorporated. Additionally, neural networks allow for the intermodality fusion architecture to be learned during training, whereas other approaches usually must resort to hand-crafted fusion methods. Some challenges involved with using neural network methods include handling the high number of hyperparameters that must be tuned, which can require high computation and need powerful computer processing units (CPUs) or graphics processing units (GPUs) to train the model in a reasonable amount of time (this problem is especially found in deep learning models). On the other hand, other learning models typically do not suffer from as many hyperparameters and having a CPU or GPU cluster is typically not necessary.

Multimodal neural network-based fusion can be divided into early, late, and intermediate fusion. Figure 5.2 summarizes the capabilities and limitations of the different deep fusion methods.

5.3.1 EARLY FUSION

Oftentimes it can be difficult to fuse multiple modalities, due to disparities between the modalities. For instance, two sensors being fused for a prediction model may be in different forms (e.g., have different sampling rates, one is analog and another is digital, etc.). To ameliorate some of these disparities, early fusion can be used to extract information from each modality and fuse before model training.

One of the common forms of early fusion comes in the form of autoencoders. An autoencoder is an unsupervised neural network that sets the target values to be equal to the input values (Ng, 2011). One of the hidden layers in the neural network (which has fewer elements than the input/output) serves as an information bottleneck, from which a compressed representation of the input features can be derived. This concept can easily be implemented to find underlying shared representations in a multimodality setting. Zhang et al., (2018a) obtained eigenvectors from video and audio sources, which were then transformed via autoencoding into reconstructed eigenvectors that have a shared representation. Srivastava and Salakhutdinov (2012) developed two Boltzmann machines to combine text and image feature vectors into a new feature vector that was used as input for a SVM classifier. Liu et al. (2017) integrated eigenvectors from different views of a face to improve face recognition by unifying the eigenvectors of the different views into a more descriptive and integrative eigenspace.

5.3.2 LATE FUSION

Late fusion involves the integration of predictions from multiple neural networks trained separately on different modalities. This method is attractive to many practitioners, as the combination of predictions from different modalities is more straightforward, especially when there are very different dimensionalities or sampling rates between the modalities (Ramachandram and Taylor, 2017). Various fusion rules are available in neural networks including averaging, maximum value, Bayes decision rule, and metaclassifiers (Ramachandram and Taylor, 2017). Simonyan and Zisserman (2014) fuse Convolutional Neural Networks (CNNs) trained on image and optic flow data for the purpose of action recognition. Kahou et al. (2016) combine audio and video data using CNNs, recursive neural networks, SVM and autoencoders.

Wu et al. (2016) fuse output from a deep belief network and CNN trained on skeletal and image features to provide a posterior estimate of gesture recognition.

5.3.3 INTERMEDIATE FUSION

Due to their hierarchical nature, neural networks allow for the fusion of features at all intermediate levels and offer flexibility beyond early or late fusion. For example in Vielzeuf et al. (2018), uni-modal features, as well as a central joint representation at every layer, are trained toward a multi-task objective. In Joze et al. (2020), multimodal Squeeze-and-Excitation (SE) modules are used to perform fusion at any intermediate level. Guided by the end learning objective, the SE modules adaptively adjust the contribution of the features of each modality, explicitly encouraging modalities to collaborate.

Deciding which features of each modality to fuse in neural networks can be a combinatoric search problem. Perez-Rua et al. (2019) propose a sequential, model-based architecture search approach to find the optimal fusion architecture, instead of empirically deciding what layer to fuse intermediate features. A recurrent surrogate model takes candidate model descriptions as input and predicts their performance on the end task, guiding the sampling process from the architecture search space. Ramachandram et al. (2018) employ Bayesian optimization using a graph-induced kernel for the same purpose.

The goal of intermediate fusion is to combine early and late fusion into a single framework. The typical workflow of an intermediate fusion network involves (i) transforming features from modalities into latent representations, (ii) fusing the representations from each modality into a single hidden layer, and (iii) learning a joint representation across the modalities to make a single prediction. Additionally, there is great flexibility with this framework since one can develop a neural network architecture that fuses various representations of the multimodal data at varying depths. Neverova et al. (2015) implemented a progressive fusion approach with visual, audio and motion capture data by fusing highly correlated modalities, then moving to fusion of less correlated modalities later in the architecture. Gao et al. (2018) used a combination of shallow and deep CNN architectures to perform image reconstruction of different views of breast cancer images and generate two different types of feature representation sets that were combined for classification using a gradient boosting tree. Karpathy et al. (2014) introduced a model that fuses video stream representations in a gradual manner, using multiple fusion layers, and were able to show the superiority of intermediate fusion to early and late fusion approaches.

5.4 DISCUSSION ON DOMAIN KNOWLEDGE AND DATA FUSION

One increasingly popular multimodal fusion research area we would like to highlight is data and domain knowledge integration. These hybrid models combine both objective information collected from a real application as well as theoretical knowledge of the underlying process via a mathematical/physical model. This process known as hybridization can occur in a variety of manners including arithmetic combination, mathematical model parameter estimation, Bayesian estimation, and feature input. These areas are summarized in Table 5.3 and briefly described below:

- *Arithmetic combination:* Arithmetic combination refers to the fusion of mathematical and machine learning model outputs in an arithmetic fashion. There is no change to the inner workings of either model, but instead both are treated as black boxes and their output is combined in a posthoc fashion. Brentan et al. (2017) added the outputs of Support Vector Regression (SVR) and a Fourier time series model for forecasting urban water demand. Chen and Irwin (2017) multiplied the outputs of a machine learning and physical model to improve solar forecasting.
- *Mathematical model parameter estimation:* Another class of hybridization methods uses machine learning to provide an estimate of some parameters for a mathematical/ physical model to better inform the prediction capabilities. To make the model more patient-specific, Clifton et al. (2017) used statistical linear estimation to fine-tune parameters for a mechanistic model of mobile health intervention in chronic pain. Meng et al. (2019) used machine learning to estimate the parameters of a mechanistic model of cane sugar crystallization. Dong et al. (2016) utilized a statistical model to estimate parameters for a mechanistic model to improve the forecasting of residential electricity. Mak et al. (2018) utilized Gaussian processes to estimate parameters in a mechanistic model to quantify turbulent flows in swirl injectors with varying geometries.
- *Bayesian framework:* A Bayesian framework can also be used to incorporate prior information from the mathematical model to the machine learning model and vice versa. Mascheroni et al. (2020) used a mechanistic model of tumor growth to inform the prior of a Bayesian model that incorporates additional empirical information to better inform the prediction. Albers et al. (2018) leveraged a Bayesian methodology to integrate physiologic knowledge into phenotype. Li and Shi (2007) used Bayesian networks along with manufacturing domain knowledge to discover the causal relationships between the process quality and process variables.
- *Feature input:* Another way to implement a hybrid model is to include output from the mathematical model as input to the training of a machine learning model. Liu, Clemente, Poirier, Ding, Chinazzi, Davis, Vespignani and Santillana (2020a) utilized a mechanistic model to inform a machine learning model's prediction of future Covid19 cases in China. Gaw et al. (2019) used the output of a mechanistic model of brain tumor growth and integrated it with a machine learning model as an input feature. They also encoded differences in the mechanistic model in the form of a graph to regularize tumor cell density predictions. Liu and Guo (2018) utilized output from a mechanistic model as a feature in a tree-based gradient boosting method (along with process conditions, such as cutting speed, feed per tooth, etc.) to predict the specific cutting energy of steel.

TABLE 5.3

Summary of the Data and Domain Knowledge Integration as well as Their Corresponding Capabilities (C) and Limitations (L)

Framework	Descriptions	Capabilities (C) and Limitations (L)
Arithmetic Combination (Brentan et al., 2017; Chen and Irwin, 2017)	Fuses mathematical and machine learning model outputs in an arithmetic fashion (e.g., addition, subtraction, multiplication, division, etc.)	(C1) Easy to implement since fusion is performed at a high-level (L1) Fusion is performed superficially and does not fully consider the intricacies of either model (L2) Less flexibility in regards to what aspects of each model to fuse and where multimodality fusion occurs
Mathematical Model Parameter Estimation (Dong et al., 2016; Clifton et al., 2017; Mak et al., 2018; Meng et al., 2019)	Uses machine learning to provide an estimate of some parameters for a mathematical/ physical model	(C1) Can make mathematical model estimation more efficient (Mak et al., 2018) (C2) Can introduce direct influence from empirical data directly into the estimation of the mathematical model (L1) Can introduce unnecessary noise in mathematical model estimation (if empirical data is not relevant to the prediction/classification task) and make mathematical model estimation more imprecise (L2) Methods are generally limited to specific mathematical models and cannot be easily translated to other problems
Bayesian Framework (Li and Shi, 2007; Albers et al., 2018; Mascheroni et al., 2020)	Incorporates prior information from the mathematical model to the machine learning model and vice versa	(C1) Allows more flexibility as to where fusion occurs between the models (C2) Enables fusion of modalities in multiple ways (ex., hierarchical, graphical, etc.) (L1) Most estimation algorithms are based on Markov Chain Monte Carlo methods that are computationally expensive (L2) Requires careful design for when and where to apply modality fusion in the network
Feature Input (Liu and Guo, 2018; Gaw et al., 2019; Liu, Clemente, Poirier, Ding, Chinazzi, Davis, Vespignani and Santillana, 2020c)	Includes output from the mathematical model as input to training of a machine learning model	(C1) Can significantly improve machine learning model accuracy (Gaw et al., 2019) (C2) Flexibility to use features for any statistical learning algorithm of interest (L1) Model fusion is indirect/imprecise because it does not fully utilize the inner mechanics of the mathematical model (L2) Communication is only one way from the mathematical model to the machine learning algorithm (no influence of machine learning on mathematical model estimation)

5.5   CHALLENGES AND FUTURE RESEARCH DIRECTIONS

There are several challenges associated with multimodality fusion. Atrey et al. (2010), Khaleghi et al. (2013), Lahat et al. (2014) and Ramachandram and Taylor (2017) have performed comprehensive reviews of these challenges and we provide these as additional references for the reader. Below, we highlight the main multimodality data fusion challenges that demand further research to be fully addressed:

- *Missing modalities:* In many instances, data from all modalities are not available across all instances or phases of model training (i.e., training, validation, and test sets). For example, in prognostics missing sensor data is prevalent (Fang et al., 2015). There have been a handful of methods developed in recent years to address these issues (Xiang et al., 2014; Liu et al., 2016; Galan et al., 2017; He et al., 2017; Adhikari et al., 2019; Liu, Chen, Wu, Weidman, Lure, Li and Alzheimer's Disease Neuroimaging Initiative, 2020c), but methodological progress is limited. Galan et al. (2017) and He et al. (2017) handle incomplete modality datasets via imputation algorithms. There are also deep learning analogs to imputation, for example, in the medical field, to transform images from one modality to another (e.g., transforming MRI to Positron Emission Tomography (PET) images (Li et al., 2014). However, the current imputation approaches are limited when there are too many missing values to impute. Separate modeling is another solution, in which different models are trained for different cohorts (determined by the available data modalities). Nevertheless, this approach is also limited because the sample size in each cohort may be small and prevent the construction of a generalizable model. One may improve this problem by including data modalities from some cohorts to append the instances in other cohorts (i.e., if one cohort has data modalities 1 and 2, and another cohort only has data modality 2, the instances in the first cohort for modality 2 can be appended to the instances in the second cohort). However, these methodologies still do not use all available information for model training. Another method addresses this issue by developing a transfer learning model that has the flexibility to train on instances with differing missing modalities and predict out-of-sample instances with a different combination of modalities (Liu, Chen, Wu, Weidman, Lure, Li and Alzheimer's Disease Neuroimaging Initiative, 2020c). The model accomplishes this task via EM using the assumption that input features follow a normal distribution. More work needs to be performed to consider features and response variables that follow different distributions.
- *Noncommensurability:* Additionally, modalities may be difficult to combine when they are at different resolutions or aggregation levels (i.e., not commensurate with each other). However, there can be a large advantage to combining multiple modalities to make use of their strengths. For example, in medical imaging, functional imaging techniques (e.g., fMRI, EEG) are capable of collecting information about a patient's brain function over

time, which allows an additional dimension of temporal resolution (Lahat et al., 2015). Unfortunately, having temporal resolution comes at the cost of a reduced spatial resolution. However, information from these images can be complemented by higher resolution medical images that do not have a time component, such as structural MRI and diffusion tensor imaging (Lahat et al., 2015). Gaw et al. (2018) created a machine learning framework that can combine features from structural and functional imaging at different aggregation levels. Additionally, in the area of meteorology, radar and satellite images provide large spatial coverage, but at the cost of not being able to measure precipitation at the ground level (Seyyedi, 2010). However, this can be overcome by utilizing information from rain gauges and microwave links to enhance the resolution of the actual amount of ground precipitation (Seyyedi, 2010; Liberman et al., 2014). Many of the challenges found in noncommensurability problems lie in the specific applications themselves. More work will need to be performed by practitioners to best understand ways to overcome specific issues that arise in practice.

- *Noise:* There may also be issues with different sources of noise across data modes. Each mode often has a different type of measurement tool or device, which can subsequently produce different magnitudes and kinds of error (Van Mechelen and Smilde, 2010; Lahat et al., 2014). There has been some work performed to address discrepancies between noise in these different modes and how to properly weigh them (Khaleghi et al., 2013; Sﹰimﹰsekli et al., 2013). One work developed a wavelet transform-based fusion method that can combine multiple medical images (i.e., computed tomography, MRI, and PET) that is resilient to Gaussian or speckle noise (Prakash et al., 2019). Additional work also needs to be performed in examining the correlation between the noise of different modalities to improve predictive performance (Chlaily et al., 2016).

- *Discordance:* It is possible that there is conflicting information in the modalities that are being fused, due to different views conveyed by the data in each modality. This may cause discrepancy in modality fusion, resulting in less confident fusion and/or prediction. Methods need to be developed that can be robust to such a phenomenon, and focus on features in each modality that are complementary to each other. In early or intermediate fusion, this may result from inconsistency in multimodal sensors in regards to their mutual information (Tmazirte et al., 2013) or random events that may be a result of the nature of data collection or type of the sensor (Kumar et al., 2007). In late fusion, this situation can be improved by a voting rule (Van Mechelen and Smilde, 2010).

- *Correlation considerations:* Correlation between modalities can be seen between different individual features and also between modalities as a whole. Constructing a model to take advantage of these connections still remains a challenging task. Highly correlated modalities may lead to models with high collinearity, requiring additional consideration with how to handle high correlation (Gaw et al., 2018). Additionally, some sensors in multimodal problems may also be subject to the same external noise, causing bias in their measurements that may lead to over or under-confident predictions (Khaleghi et al., 2013). In contrast, independent modalities (correlation ¼ 0) may also present challenges with modality fusion. In these cases, one cannot rely on the correlation between modalities to help with fusion, and must rely on other ways (e.g., confidence of individual modalities (Castellano et al., 2008), etc.).

- *Intermodality correlation:* Often the incorporation of features from multiple modalities only indirectly considers intermodality correlation to inform model building. There is a greater need to build models that consider intermodality correlation explicitly and incorporate it directly into the model building process. The authors have only found a limited number of works that harness intermodality correlation. In early fusion, Guo et al. (2018) proposed a CCA algorithm that performs joint intermodal and intramodal fusion for semi-paired scenarios. The advantage of this method is that it considers scenarios for which not all modalities have a strong pairing with each other ("semi-paired" scenarios), while also performing fusion that considers intermodality correlations that preserve intramodal correlations. Additionally, there are a couple of methods that develop neural network architectures that more explicitly consider intermodality correlations (Peng et al., 2017; Said et al., 2017). Said et al. (2017) demonstrated a deep learning approach that utilizes intermodality correlation to improve the classification of EEG and EMG signals. Peng et al. (2017) utilized both intramodality and intermodality correlation in a hierarchical classification network that takes as input texts and images to inform image and text retrieval tasks.

- *Varying confidence levels:* Each modality will often have a different level of confidence (Siegel and Wu, 2004), e.g., due to noise level, nature of the data collection task, correlation with response variable, etc. As an example, if given an audio and video data of a person crying, one may have higher confidence in predicting this event using audio instead of video (Atrey et al., 2010). In another example, Rankawat and Dubey (2017) fuse noisy ECG and atrial blood pressure signals by defining a beat Signal Quality Index (SQI) that indicates the level of noise in each modality and incorporating the SQI into a majority voting scheme. A model that can take these aforementioned factors into account will be able to emphasize the "higher confidence" modalities, while still utilizing useful information from the less confident modalities that can improve the fusion quality. There is a need to more optimally quantify confidence in specific modalities for a particular task (e.g., in regard to measure of information content, relevance to prediction task, etc.) for the purpose of improving flexibility and adaptivity in modality fusion. Having this information would enhance the capability to choose the degree of fusion and which modalities to fuse for each instance in the dataset (e.g., whether to place a higher emphasis on some modalities over others, not select particular modalities entirely).

- *Negative transfer reduction:* There is limited work in reducing negative transfer between different modalities. Developing models that can successfully integrate modality covariances (i.e., positive transfer), while also preventing the fusion of conflicting information or correlations that are irrelevant to model training (i.e., negative transfer), will provide valuable insights into how to best integrate interactions between modalities while also producing more accurate models. In one example, Yoon and Li (2018) developed a Positive Transfer Learning (PTL) model on telemonitoring data of Parkinson's Disease patients that is robust to negative transfer between patients' individual sub-models. The PTL model is built on the premise that not all information from patient data is useful for building accurate models of other patients, and it is necessary to identify the conditions for which negative transfer can happen and negatively affect the model. However, this work does not handle multiple modalities, highlighting a need for more methods such as this one to be incorporated into multimodal fusion.

- *Computation:* Because of the complexity of some multimodality fusion algorithms, they often cannot be solved analytically and approximation algorithms are needed. There have been studies that have successfully made approximation techniques that will eventually converge (Virtanen et al., 2012; Xiang et al., 2014; Zhang et al., 2018b). However, there is still a need to make more efficient approximation methods and improve efficiency to bring the convergence rate to a more acceptable level.

- *Theoretical criteria and verification:* Even though extra modes of data provide additional information in most circumstances, integration of modes is not always beneficial and may cause deterioration in the performance of a model. For example, when integrating multi-accuracy data in applications such as geometric inspection and metrology (Gahrooei, Payanbar, Pacela and Colosimo, 2019) or building simulation (Safarzadegan Gilan et al., 2016), a data mode with high non-stationary bias or variance may harm the development of a surrogate model. Establishing a set of criteria for identifying suitable modes is a challenging task and requires theoretical analysis. Other theoretical questions includes: what is the best quantitative measure of success? What is the measure that quantifies the gain of integrating several modalities? Can we obtain a theoretical lower bound of gain? What is the best error that is achievable by fusion of data?

- *Multimodal data collection:* Although several techniques are available for the integration of multimodal data, the literature is very limited on how to design the collection of multimodal data to minimize the data collection cost while obtaining an adequate level of information. In telemedicine applications, for example, a visit from a doctor (accurate mode of health data) is an expensive means of data collection in comparison with wearable devices (low-accuracy mode). How often a patient should visit a doctor given the low-accuracy data is the question to be addressed by the multimodal data collection policies. How to select the right modes, collect data, and integrate the modes is a major challenge. Gahrooei, Payanbar, Pacella and Colosimo (2019) proposed an adaptive sampling of high-accuracy data when low-accuracy data is available, and applied it to vehicle engine calibration application. However, this approach is limited to a static case and assumes the high-accuracy mode is already known.

- *Dynamic data fusion:* Limited work is available for dynamic data fusion of multimodal data. Such techniques can be applied in smart cities and tele-medicine applications, where several multi-accuracy sensors are collecting information over time. Appropriate synchronization (i.e., when and how much data should be processed from each modality) should also be taken into consideration (Atrey et al., 2010).
- *High-dimensional data:* Due to the nature of multimodal data, there is often a high number of features from which to choose for fusion. Sparse methods (Xiang et al., 2014; Argelaguet et al., 2018) can be used to select individual features or latent components can also be derived to infer the underlying patterns expressed in the data (Kibble et al., 2016; Li Choi, Perres, Sun and Vudue, 2017). Additional consideration should be made for determining when to reduce the dimension of the data (whether it should be as a pre-processing step within individual modalities before fusion, as a part of the fusion process across all modalities, or a combination of the two).
- *Deep learning:* The literature on IE-related applications has been limited in regard to incorporating deep learning models (Said et al., 2017; Peng et al., 2017; Gao et al., 2018; Ramachandram et al., 2018; Perez-Rua et al., 2019; Joze et al., 2020). The applications of deep learning are limited in this area due to the lack of instances available for model training. However, there is a possibility to use pre-trained neural networks that have a reasonable degree of discriminative ability which can be transferred to new datasets (DenseNet121, 2018; InceptionV3, 2019). Additionally, generative shallow learning models that have the capability to produce new artificial instances can augment the training set and overcome the issues of small sample size.
- *Privacy:* Due to regulatory constraints from a particular application (e.g., medical, military, industry, etc.), one may be limited in ability to fuse different modalities because of concerns of compromising confidentiality. There are only a limited number of works in this area (Kefayati et al., 2007a, 2007b; Gao, 2020; Kumar and Diwakar, 2021). One such work utilizes the nonsubsampled shearlet transform combined with noise reduction to transform, share, and fuse imaging data through a secure environment (Kumar and Diwakar, 2021). Another work focuses on the privacy and preservation of information fused between multiagent systems in regards to synchronization, information fusion, decentralized control and load balancing (Gao, 2020).
- *Objective measures of modality fusion:* More work should be performed in finding objective metrics that evaluate the degree to which information from modalities are fused (Zhu et al., 2018), along with measures of positive or negative transfer (for example, Yoon and Li (2018)). This will better assist researchers in this area to compare methods they are developing by providing objective benchmarks of "fusion performance".

5.6   CONCLUSION

As systems have become more advanced, there has been an increasing abundance of available data from different modality types. To advance the performance of statistical learning algorithms, it is crucial to understand how to best incorporate the relationships of these different modalities, while also avoiding a negative transfer of knowledge (i.e., transfer of discordant information between modalities). The key understanding that is necessary to implement and advance such algorithms is discerning the connections between the different data modes and how to best exploit them. Because multimodal datasets can be incredibly diverse, there is no one-size-fits-all model, which requires the practitioner to understand the particular application while developing a multimodal approach. As this work has demonstrated, there is potential for huge performance improvement from training a model on a single data mode if relationships between different available data modes are appropriately considered. It will be important to focus efforts on both decomposition-based models and neural networks, as some applications will require decomposition-based models when fewer training instances are at hand, while others with being able to employ the advantages of neural networks (especially deep learning) when many training instances are available. Progress in this area will span a broad number of applications, including systems monitoring/prognostics, healthcare, renewable energy, and many others since multimodal measurement technologies are emerging everywhere.

DATA AVAILABILITY STATEMENT

There is no data set associated with this article.

REFERENCES

Abavisani, M. and Patel, V.M. (2018) Multimodal sparse and low-rank subspace clustering. *Information Fusion*, 39, 168–177.

Acar, E., Bro, R. and Smilde, A.K. (2015) Data fusion in metabolomics using coupled matrix and tensor factorizations. *Proceedings of the IEEE*, 103(9), 1602–1620.

Acar, E. Papalexakis, E.E., Gurdeniz, G., Rasmussen, M.A., Lawaetz, € A.J., Nilsson, M. and Bro, R. (2014) Structure-revealing data fusion. *BMC Bioinformatics*, 15(1), 1–17.

Acar, E., Rasmussen, M.A. Savorani, F., Naes, T. and Bro, R. (2013) Understanding data fusion within the framework of coupled matrix and tensor factorizations. *Chemometrics and Intelligent Laboratory Systems*, 129, 53–63.

Adhikari, S. Lecci, F., Becker, J.T., Junker, B.W., Kuller, L.H., Lopez, O.L. and Tibshirani, R.J. (2019) High-dimensional longitudinal classification with the multinomial fused lasso. *Statistics in Medicine*, 38(12), 2184–2205.

Albers, D.J., Levine, M.E., Stuart, A., Mamykina, L., Gluckman, B. and Hripcsak, G. (2018) Mechanistic machine learning: How data assimilation leverages physiologic knowledge using Bayesian inference to forecast the future, infer the present, and phenotype. *Journal of the American Medical Informatics Association*, 25(10), 1392–1401.

Argelaguet, R., Velten, B., Arnol, D., Dietrich, S., Zenz, T., Marioni, J.C., Buettner, F., Huber, W. and Stegle, O. (2018) Multi-Omics Factor Analysis–A framework for unsupervised integration of multiomics data sets. *Molecular Systems Biology*, 14(6), e8124.

Atrey, P.K., Hossain, M.A., El Saddik, A. and Kankanhalli, M.S. (2010) Multimodal fusion for multimedia analysis: A survey. *Multimedia Systems*, 16(6), 345–379.

Brentan, B.M., Luvizotto Jr, E., Herrera, M., Izquierdo, J. and PerezGarcıa, R. (2017) Hybrid regression model for near real-time urban water demand forecasting. *Journal of Computational and Applied Mathematics*, 309, 532–541.

Bro, R. (1996) Multiway calibration. multilinear PLS. *Journal of Chemometrics*, 10(1), 47–61.

Bro, R. et al. (1997) Parafac. tutorial and applications. *Chemometrics and Intelligent Laboratory Systems*, 38(2), 149–172.

Calhoun, V.D. and Sui, J. (2016) Multimodal fusion of brain imaging data: A key to finding the missing link(s) in complex mental illness. *Biological Psychiatry: Cognitive Neuroscience and Neuroimaging*, 1(3), 230–244.

Candes, E.J., Li, S., Ma, Y. and Wright, J. (2011) Robust principal component analysis? *Journal of the ACM (JACM)*, 58(3), 1–37, 2011.

Castellano, G., Kessous, L. and Caridakis, G. (2008) *Emotion Recognition through Multiple Modalities: Face, Body Gesture, Speech, in Affect and Emotion in Human-Computer Interaction*, Springer, Milwaukee, WI, pp. 92–103.

Chehade, A., Song, C., Liu, K., Saxena, A. and Zhang, X. (2018) A data-level fusion approach for degradation modeling and prognostic analysis under multiple failure modes. *Journal of Quality Technology*, 50(2), 150–165.

Chen, D. and Irwin, D. (2017) Black-box solar performance modeling: Comparing physical, machine learning, and hybrid approaches. *ACM SIGMETRICS Performance Evaluation Review*, 45(2), 79–84.

Chlaily, S., Amblard, P.-O., Michel, O. and Jutten, C. (2016) *Impact of Noise Correlation on Multimodality, in 2016 24th European Signal Processing Conference (EUSIPCO)*, IEEE Press, Piscataway, NJ, pp. 195–199.

Clifton, S.M., Kang, C., Li, J.J., Long, Q., Shah, N. and Abrams, D.M. (2017) Hybrid statistical and mechanistic mathematical model guides mobile health intervention for chronic pain. *Journal of Computational Biology*, 24(7), 675–688.

DenseNet121. (2018) Densenet121. https://github.com/keras-team/ keras-applications/blob/ master/keras_applications/densenet..py (accessed 27 December 2020).

Dong, D., Li, Z., Rahman, S.M.M. and Vega, R. (2016) A hybrid model approach for forecasting future residential electricity consumption. *Energy and Buildings*, 117, 341–351.

Elhamifar, E. (2016) High-rank matrix completion and clustering under self-expressive models, in *Advances in Neural Information Processing Systems*, Barcelona, Spain, pp. 73–81.

Fang, X., Gebraeel, N.Z. and Paynabar, N. (2017) Scalable prognostic models for large-scale condition monitoring applications. *IISE Transactions*, 49(7), 698–710.

Fang, X., Paynabar, K. and Gebraeel, N. (2019) Image-based prognostics using penalized tensor regression. *Technometrics*, 61(3), 369–384.

Fang, X., Zhou, R. and Gebraeel, N. (2015) An adaptive functional regression-based prognostic model for applications with missing data. *Reliability Engineering & System Safety*, 133, 266–274.

Fernandez-Beltran, R., Haut, J.M., Paoletti, M.E., Plaza, J., Plaza, A. and Pla, F. (2018a) Multimodal probabilistic latent semantic analysis for sentinel-1 and sentinel-2 image fusion. *IEEE Geoscience and Remote Sensing Letters*, 15(9), 1347–1351.

Fernandez-Beltran, R., Haut, J.M., Paoletti, M.E., Plaza, J., Plaza, A. and Pla, F. (2018b) Remote sensing image fusion using hierarchical multimodal probabilistic latent semantic analysis. *IEEE Journal of Selected Topics in Applied Earth Observations and Remote Sensing*, 11(12), 4982–4993.

Gahrooei, M.R., Paynabar, K., Pacella, M. and Colosimo, M.B. (2019) An adaptive fused sampling approach of high-accuracy data in the presence of low-accuracy data. *IISE Transactions*, 51(11), 1251–1264.

Gahrooei, M.R., Paynabar, K., Pacella, M. and Shi, J. (2019) Process modeling and prediction with large number of high-dimensional variables using functional regression. *IEEE Transactions on Automation Science and Engineering*, 17(2), 684–696.

Gahrooei, M.R., Yan, H., Paynabar, K. and Shi, J. (2020) Multiple tensor-on-tensor regression: An approach for modeling processes with heterogeneous sources of data. *Technometrics*, 63(2), 147–159.

Galan, C.O., Lasheras, F.S., de Cos Juez, F.J. and Sanchez, A.B. (2017) Missing data imputation of questionnaires by means of genetic algorithms with different fitness functions. *Journal of Computational and Applied Mathematics*, 311, 704–717.

Gao, F., Wu, T., Li, J., Zheng, B., Ruan, L., Shang, D. and Patel, B. (2018) SD:CNN: A shallow-deep CNN for improved breast cancer diagnosis. *Computerized Medical Imaging and Graphics*, 70, 53–62.

Gao, H. (2020) Coordination and Privacy Preservation in Multi-Agent Systems (Doctoral dissertation, Clemson University).

Gao, J., Li, P., Chen, Z. and Zhang, J. (2020) A survey on deep learning for multimodal data fusion. *Neural Computation*, 32(5), 829–864.

Gaskin, C.J. and Happell, B. (2014) On exploratory factor analysis: A review of recent evidence, an assessment of current practice, and recommendations for future use. *International Journal of Nursing Studies*, 51(3), 511–521.

Gaw, N., Hawkins-Daarud, A., Hu, L.S., Yoon, H., Wang, L., Xu, Y., Jackson, P.R., Singleton, K.W., Baxter, L.C., Eschbacher, J. et al. (2019) Integration of machine learning and mechanistic models accurately predicts variation in cell density of glioblastoma using multiparametric MRI. *Scientific Reports*, 9(1), 1–9.

Gaw, N., Schwedt, T.J., Chong, C.D., Wu, T. and Li, J. (2018) A clinical decision support system using multi-modality imaging data for disease diagnosis. *IISE Transactions on Healthcare Systems Engineering*, 8(1), 36–46.

Georghiades, A.S., Belhumeur, P.N. and Kriegman, D.J. (2001) From few to many: Illumination cone models for face recognition under variable lighting and pose. *IEEE Transactions on Pattern Analysis and Machine Intelligence*, 23(6), 643–660.

Gorgannejad, S., Gahrooei, M.R., Paynabar, K. and Neu, R.W. (2019) Quantitative prediction of the aged state of ni-base superalloys using pca and tensor regression. *Acta Materialia*, 165, 259–269.

Guo, H., Wang, S., Tie, Y., Qi, K. and Guan, L. (2018) Joint intermodal and intramodal correlation preservation for semi-paired learning. *Pattern Recognition*, 81, 36–49.

He, C., Liu, Q., Li, H. and Wang, H. (2010) Multimodal medical image fusion based on IHS and PCA. *Procedia Engineering*, 7, 280–285.

He, D., Wang, Z., Yang, L. and Dai, W. (2017) Study on missing data imputation and modeling for the leaching process. *Chemical Engineering Research and Design*, 124, 1–19.

Ho, J.C., Ghosh, J. and Sun, J. (2014) Marble: high-throughput phenotyping from electronic health records via sparse nonnegative tensor factorization, *in Proceedings of the 20th ACM SIGKDD International Conference on Knowledge Discovery and Data Mining*, pp. 115–124. Hotelling, H. (1936) Relations between two sets of variates. *Biometrika*, 28(3/4), 321–377.

Ilarshman, RA. (1970) Foundations of the PARAFAC procedure: Models and methods for an "explanatory" multi-mode factor analysis. *UCLA Working Papers in Phonetics*, 16, 1–84.

InceptionV3.Inceptionv3.https://github.com/keras-team/keras-applications/blob/master/keras_applications/inception_v3.py (accessed 27 December 2020).

Joze, H.R.V., Shaban, A., Iuzzolino, M.L. and Koishida, K. (2020) MMTM: Multimodal transfer module for CNN fusion, in Proceedings of the IEEE/CVF Conference on Computer Vision and Pattern Recognition, IEEE/CVF, Seattle, WA, pp. 13289–13299.

Kahou, S.E., Bouthillier, X., Lamblin, P., Gulcehre, C., Michalski, V., Konda, K., Jean, S., Froumenty, P., Dauphin, Y., Boulanger-Lewandowski, N., et al. (2016) EmoNets: Multimodal deep learning approaches for emotion recognition in video. *Journal on Multimodal User Interfaces*, 10(2), 99–111.

Karpathy, A., Toderici, G., Shetty, S., Leung, T., Sukthankar, R. and Fei-Fei, L. (2014) Large-scale video classification with convolutional neural networks, *in Proceedings of the IEEE Conference on Computer Vision and Pattern Recognition,* IEEE/CVF, Columbus, OH, pp. 1725–1732.

Kefayati, M., Talebi, M.S., Khalaj, B.H. and Rabiee, H.R. (2007a) Secure consensus averaging in sensor networks using random offsets, in *2007 IEEE International Conference on Telecommunications and Malaysia International Conference on Communications*, IEEE, Penang, Malaysia, pp. 556–560.

Kefayati, M., Talebi, M.S., Rabiee, H.R. and Khalaj, B.H. (2007b) On secure consensus information fusion over sensor networks, in *2007 IEEE/ACS International Conference on Computer Systems and Applications*, IEEE/ACS, Amman, Jordan, pp. 108–115.

Kettenring, J.R. (1971) Canonical analysis of several sets of variables. *Biometrika*, 58(3), 433–451.

Khaleghi, B., Khamis, A., Karray, F.O. and Razavi, S.N. (2013) Multisensor data fusion: A review of the state-of-the-art. *Information Fusion*, 14(1), 28–44.

Khan, S.A., Virtanen, S., Kallioniemi, O.P., Wennerberg, K., Poso, A. and Kaski, S. (2014) Identification of structural features in chemicals associated with cancer drug response: A systematic data-driven analysis. *Bioinformatics*, 30(17), i497–i504.

Khasha, R., Sepehri, M.M. and Mahdaviani, S.A. (2019) An ensemble learning method for asthma control level detection with leveraging medical knowledge-based classifier and supervised learning. *Journal of Medical Systems*, 43(6), 1–15.

Kibble, M., Khan, S.A., Saarinen, N., Iorio, F., Saez-Rodriguez, J., M€akel€a, S. and Aittokallio, T. (2016) Transcriptional response networks for elucidating mechanisms of action of multitargeted agents. *Drug Discovery Today*, 21(7), 1063–1075.

Klami, A., Virtanen, S. and Kaski. S. (2013) Bayesian canonical correlation analysis. *Journal of Machine Learning Research*, 14(Apr), 965–1003.

Kolda, T.G. (2006) Multilinear operators for higher-order decompositions. Technical report, Sandia National Laboratories.

Kumar, M., Garg, D.P. and Zachery, R.A. (2007) A method for judicious fusion of inconsistent multiple sensor data. *IEEE Sensors Journal*, 7(5), 723–733.

Kumar, P. and Diwakar, M. (2021) A novel approach for multimodality medical image fusion over secure environment. *Transactions on Emerging Telecommunications Technologies*, 32(2), e3985.

Lahat, D., Adaly, T. and Jutten, C. (2014) Challenges in multimodal data fusion, in *2014 22nd European Signal Processing Conference (EUSIPCO)*, IEEE Aerospace and Electronic Systems Society (AESS), Lisbon, Portugal, pp. 101–105.

Lahat, D., Adali, T. and Jutten, C. (2015) Multimodal data fusion: An overview of methods, challenges, and prospects. *Proceedings of the IEEE*, 103(9), 1449–1477.

Lahti, L., Sch€afer, M., Klein, H.-U., Bicciato, S. and Dugas, M. (2013) Cancer gene prioritization by integrative analysis of MRNA expression and DNA copy number data: A comparative review. *Briefings in Bioinformatics*, 14(1), 27–35.

Lampert, C.H. and Kromer, O. (2010) Weakly-paired maximum covari-€ ance analysis for multimodal dimensionality reduction and transfer learning, in *European Conference on Computer Vision*, Springer, Heraklion, Crete, Greece, pp. 566–579.

Li, D., Dimitrova, N., Li, M. and Sethi, I.K. (2003) Multimedia content processing through cross-modal association, in *Proceedings of the Eleventh ACM International Conference on Multimedia*, ACM, Berkeley, CA, pp. 604–611.

Li, J. and Shi, J. (2007) Knowledge discovery from observational data for process control using causal Bayesian networks. *IIE Transactions*, 39(6), 681–690.

Li, N., Gebraeel, N., Lei, Y., Fang, X., Cai, X. and Yan, T. (2021) Remaining useful life prediction based on a multi-sensor data fusion model. *Reliability Engineering & System Safety*, 208, 107249.

Li, Q. and Li, L. (2019) Integrative factor regression and its inference for multimodal data analysis. *arXiv preprint arXiv*:1911.04056.

Li, R., Zhang, W., Suk, H.-Il, Wang, L., Li, J., Shen, D. and Ji, S. (2014) Deep learning based imaging data completion for improved brain disease diagnosis, in *International Conference on Medical Image Computing and Computer-Assisted Intervention*, MICCAI, Boston, MA, pp. 305–312.

Li, S., Wang, W., Qi, H., Ayhan, B., Kwan, C. and Vance, S. (2015) Low-rank tensor decomposition based anomaly detection for hyperspectral imagery, in *2015 IEEE International Conference on Image Processing (ICIP)*, IEEE Press, Piscataway, NJ, pp. 4525–4529.

Li, W., Cerise, J.E., Yang, Y. and Han, H. (2017) Application of t-sne to human genetic data. *Journal of Bioinformatics and Computational Biology*, 15(04), 1750017.

Li, Y., Wu, F.-X. and Ngom, A. (2018) A review on machine learning principles for multi-view biological data integration. *Briefings in Bioinformatics*, 19(2), 325–340.

Liberman, Y., Samuels, R., Alpert, P. and Messer, H. (2014) New algorithm for integration between wireless microwave sensor network and radar for improved rainfall measurement and mapping. *Atmospheric Measurement Techniques*, 7(10), 3549–3563.

Liu, D., Clemente, L., Poirier, C., Ding, X., Chinazzi, M., Davis, J.T., Vespignani, A. and Santillana, M. (2020a) A machine learning methodology for real-time forecasting of the 2019–2020 covid-19 outbreak using internet searches, news alerts, and estimates from mechanistic models. *arXiv preprint arXiv*:2004.04019.

Liu, K., Chehade, A., and Song, C. (2015) Optimize the signal quality of the composite health index via data fusion for degradation modeling and prognostic analysis. *IEEE Transactions on Automation Science and Engineering*, 14(3), 1504–1514.

Liu, J., Li, T., Xie, P., Du, S., Teng, F. and Yang, X. (2020b) Urban big data fusion based on deep learning: An overview. *Information Fusion*, 53, 123–133.

Liu, K., Gebraeel, N.Z. and Shi, J. (2013) A data-level fusion model for developing composite health indices for degradation modeling and prognostic analysis. *IEEE Transactions on Automation Science and Engineering*, 10(3), 652–664.

Liu, K. and Huang, S. (2014) Integration of data fusion methodology and degradation modeling process to improve prognostics. *IEEE Transactions on Automation Science and Engineering*, 13(1), 344–354.

Liu, M., Zhang, J., Yap, P.-T. and Shen, D. (2016) Diagnosis of Alzheimer's disease using view-aligned hypergraph learning with incomplete multi-modality data, in *International Conference on Medical Image Computing and Computer-Assisted Intervention*, MICCAI, Athens, Greece, pp. 308–316.

Liu, X., Chen, K., Wu, T., Weidman, D., Lure, F. YM, Li, J., Alzheimer's Disease Neuroimaging Initiative (ADNI), et al. (2020c) A novel transfer learning model for predictive analytics using incomplete multimodality data. medRxiv.

Liu, Z. and Guo, Y. (2018) A hybrid approach to integrate machine learning and process mechanics for the prediction of specific cutting energy. *CIRP Annals*, 67(1), 57–60.

Liu, Z., Zhang, W., Quek, T.Q.S. and Lin, S. (2017) Deep fusion of heterogeneous sensor data, in *2017 IEEE International Conference on Acoustics, Speech and Signal Processing (ICASSP)*, IEEE Press, Piscataway, NJ, pp. 5965–5969.

Lock, E.F. (2018) Tensor-on-tensor regression. *Journal of Computational and Graphical Statistics*, 27(3), 638–647.

Mak, S., Sung, C.-L., Wang, X., Yeh, S.-T., Chang, Y.-H., Joseph, V.R., Yang, V. and Wu, C.F.J. (2018) An efficient surrogate model for emulation and physics extraction of large eddy simulations. *Journal of the American Statistical Association*, 113(524), 1443–1456.

Mascheroni, P., Alfonso, J.C.L., Meyer-Hermann, M. and Hatzikirou, H. (2020) Bayesian combination of mechanistic modeling and machine learning (BAM3): Improving clinical tumor growth predictions. bioRxiv.

Meng, Y., Yu, S., Zhang, J., Qin, J., Dong, Z., Lu, G. and Pang, H. (2019) Hybrid modeling based on mechanistic and data-driven approaches for cane sugar crystallization. *Journal of Food Engineering*, 257, 44–55.

Moin, A., Bhateja, V. and Srivastava, A. (2016) Weighted-PCA based multimodal medical image fusion in contourlet domain, in *Proceedings of the International Congress on Information and Communication Technology*, Springer, IEEE, Udaipur, India pp. 597–605.

Neverova, N., Wolf, C., Taylor, G. and Nebout, F. (2015) Moddrop: Adaptive multi-modal gesture recognition. *IEEE Transactions on Pattern Analysis and Machine Intelligence*, 38(8), 1692–1706.

Ng, A. (2011) Sparse autoencoder. *CS294A Lecture Notes,* 72(2011), 1–19.

Paynabar, K., Jin, J. and Reed, M.P. (2015) Informative sensor and feature selection via hierarchical nonnegative garrote. *Technometrics*, 57(4), 514–523.

Peng, Y., Qi, J., Huang, X. and Yuan, Y. (2017) CCL: Cross-modal correlation learning with multigrained fusion by hierarchical network. *IEEE Transactions on Multimedia*, 20(2), 405–420.

Perez-Rua, J.-M., Vielzeuf, V., Pateux, S., Baccouche, M. and Jurie, F. (2019) MAFS: Multimodal fusion architecture search, in *Proceedings of the IEEE Conference on Computer Vision and Pattern Recognition*, IEEE Press, Piscataway, NJ, pp. 6966–6975.

Prakash, O., Park, C.M., Khare, A., Jeon, M. and Gwak, J. (2019) Multiscale fusion of multimodal medical images using lifting scheme based biorthogonal wavelet transform. *Optik*, 182, 995–1014.

Qi, J., Yang, P., Newcombe, L., Peng, X., Yang, Y. and Zhao, Z. (2020) An overview of data fusion techniques for internet of things enabled physical activity recognition and measure. *Information Fusion*, 55, 269–280.

Rajalingam, B. and Priya, R. (2017) Multimodality medical image fusion based on hybrid fusion techniques. *International Journal of Engineering and Manufacturing Science*, 7(1), 22–29.

Ramachandram, D., Lisicki, M., Shields, T.J., Amer, M.R. and Taylor, G.W. (2018) Bayesian optimization on graph-structured search spaces: Optimizing deep multimodal fusion architectures. *Neurocomputing*, 298, 80–89.

Ramachandram, D. and Taylor, G.W. (2017) Deep multimodal learning: A survey on recent advances and trends. *IEEE Signal Processing Magazine*, 34(6), 96–108.

Rankawat, S.A. and Dubey, R. (2017) Robust heart rate estimation from multimodal physiological signals using beat signal quality index based majority voting fusion method. *Biomedical Signal Processing and Control*, 33, 201–212.

Safarzadegan Gilan, S.S., Goyal, N. and Dilkina, B. (2016) Active learning in multiobjective evolutionary algorithms for sustainable building design, in *Proceedings of the Genetic and Evolutionary Computation Conference 2016*, ACM, Denver, CO, pp. 589–596.

Sagi, O. and Rokach, L. (2018) Ensemble learning: A survey. *Wiley Interdisciplinary Reviews: Data Mining and Knowledge Discovery*, 8(4), e1249.

Said, A.B., Mohamed, A., Elfouly, T., Harras, K. and Wang, Z.J. (2017) Multimodal deep learning approach for joint EEG-EMG data compression and classification, in *2017 IEEE Wireless Communications and Networking Conference (WCNC)*, IEEE Press, Piscataway, NJ, pp.1–6.

Samareh, A., Jin, Y., Wang, Z., Chang, X. and Huang, S. (2018) Detect depression from communication: How computer vision, signal processing, and sentiment analysis join forces. *IISE Transactions on Healthcare Systems Engineering*, 8(3), 196–208.

Schmitt, M. and Zhu, X.X. (2016) Data fusion and remote sensing: An ever-growing relationship. *IEEE Geoscience and Remote Sensing Magazine*, 4(4), 6–23.

Seyyedi, H. (2010) Comparing satellite derived rainfall with ground based radar for Northwestern Europe. University of Twente Faculty of Geo-Information and Earth Observation (ITC).

Si, B., Schwedt, T.J., Chong, C.D., Wu, T. and Li, J. (2020) A novel hierarchically-structured factor mixture model for cluster discovery from multi-modality data. *IISE Transactions*, 53(7), 1–13.

Siegel, M. and Wu, H. (2004) Confidence fusion, in *IEEE International Workshop on Robot Sensing*, IEEE Press, Piscataway, NJ, pp. 96–99.

Simonyan, K. and Zisserman, A. (2014) Two-stream convolutional networks for action recognition in videos. *Advances in Neural Information Processing Systems*, 27, 568–576.

S¸im¸sekli, U., Ermis¸, B., Cemgil, A.T. and Acar, E. (2013) Optimal weight learning for coupled tensor factorization with mixed divergences, in *21st European Signal Processing Conference (EUSIPCO 2013)*, IEEE Press, Piscataway, NJ, pp. 1–5.

Song, C. and Liu, K. (2018) Statistical degradation modeling and prognostics of multiple sensor signals via data fusion: A composite health index approach. *IISE Transactions*, 50(10), 853–867.

Song, C., Liu, K. and Zhang, X. (2019) A generic framework for multisensor degradation modeling based on supervised classification and failure surface. *IISE Transactions*, 51(11), 1288–1302.

Srivastava, N. and Salakhutdinov, R. (2012) ldquo, multimodal learning with deep Boltzmann machines, in *Proceedings of Neural Information and Processing Systems*. Neural Information Processing Systems, Lake Tahoe, Nevada, pp. 2222–2230.

Suk, H.-Il, Lee, S.-W., Shen, D., Alzheimer's Disease Neuroimaging Initiative, et al. (2017) Deep ensemble learning of sparse regression models for brain disease diagnosis. *Medical Image Analysis*, 37, 101–113.

Tmazirte, N.A., El Najjar, M.E., Smaili, C. and Pomorski, D. (2013) Dynamical reconfiguration strategy of a multi sensor data fusion algorithm based on information theory, in *2013 IEEE Intelligent Vehicles Symposium (IV)*, IIE Press, Piscataway, NJ, pp. 896–901.

Tucker, L.R. (1958) An inter-battery method of factor analysis. *Psychometrika*, 23(2), 111–136.

Tucker, L.R. (1964) The extension of factor analysis to three-dimensional matrices. *Contributions to Mathematical Psychology*, 110119.

Van der Maaten, L. and Hinton, G. (2008) Visualizing data using t-SNE. *Journal of Machine Learning Research*, 9, 2579–2605.

Van Mechelen, I. and Smilde, A.K. (2010) A generic linked-mode decomposition model for data fusion. *Chemometrics and Intelligent Laboratory Systems*, 104(1), 83–94.

Vielzeuf, V., Lechervy, A., Pateux, S. and Jurie, F. (2018) Centralnet: A multilayer approach for multimodal fusion, in *Proceedings of the European Conference on Computer Vision (ECCV)*, ECCV, Munich, Germany, pp. 1–15.

Virtanen, V., Klami, A., Khan, S. and Kaski, S. (2012) Bayesian group factor analysis, in *Artificial Intelligence and Statistics, PMLR, La Palma*, Canary Islands, pp. 1269–1277.

Wang, J., Ma, Y., Zhang, L., Gao, R.X. and Wu, D. (2018) Deep learning for smart manufacturing: Methods and applications. *Journal of Manufacturing Systems*, 48, 144–156.

Wang, Y., Chen, R., Ghosh, J., Denny, J.C., Kho, A., Chen, Y., Malin, B.A. and Sun, J. (2015) Rubik: Knowledge guided tensor factorization and completion for health data analytics, in *Proceedings of the 21th ACM SIGKDD International Conference on Knowledge Discovery and Data Mining*, pp. 1265–1274.

Wang, Y., Guan, L. and Venetsanopoulos, A.N. (2012) Kernel crossmodal factor analysis for information fusion with application to bimodal emotion recognition. *IEEE Transactions on Multimedia*, 14(3), 597–607.

White, M. and Schuurmans, D. (2012) Generalized optimal reverse prediction, in *Artificial Intelligence and Statistics, PMLR, La Palma*, Canary Islands, pp. 1305–1313.

White, M., Zhang, X., Schuurmans, D. and Yu, Y.-l. (2012) Convex multi-view subspace learning, in *Advances in Neural Information Processing Systems, Neural Information Processing Systems*, Lake Tahoe, Nevada, pp. 1673–1681.

Wu, D., Pigou, L., Kindermans, P.-J., Le, N.D.-H., Shao, L., Dambre, J. and Odobez, J.-M. (2016) Deep dynamic neural networks for multimodal gesture segmentation and recognition. *IEEE Transactions on Pattern Analysis and Machine Intelligence*, 38(8), 1583–1597.

Xiang, S., Yuan, L., Fan, W., Wang, Y., Thompson, P.M., Ye, J., Alzheimer's Disease Neuroimaging Initiative, et al. (2014) Bi-level multi-source learning for heterogeneous block-wise missing data. *NeuroImage*, 102, 192–206.

Xiao, Y., Wang, Y. and Ding, Z. (2018) The application of heterogeneous information fusion in misalignment fault diagnosis of wind turbines. *Energies*, 11(7), 1–15.

Yadav, B., Gopalacharyulu, P., Pemovska, T., Khan, S.A., Szwajda, A., Tang, J., Wennerberg, K. and Aittokallio, T. (2015) From drug response profiling to target addiction scoring in cancer cell models. *Disease Models & Mechanisms*, 8(10), 1255–1264.

Yan, H., Paynabar, K. and Pacella, M. (2019) Structured point cloud data analysis via regularized tensor regression for process modeling and optimization. *Technometrics*, 61(3), 385–395.

Yan, H., Paynabar, K. and Shi, J. (2014) Image-based process monitoring using low-rank tensor decomposition. *IEEE Transactions on Automation Science and Engineering*, 12(1), 216–227.

Yokoya, N., Ghamisi, P. and Xia, J. (2017) Multimodal, multitemporal, and multisource global data fusion for local climate zones classification based on ensemble learning, in *2017 IEEE International Geoscience and Remote Sensing Symposium (IGARSS)*, IEEE Press, Piscataway, NJ, pp. 1197–1200.

Yoon, H. and Li, J. (2018) A novel positive transfer learning approach for telemonitoring of Parkinson's disease. *IEEE Transactions on Automation Science and Engineering*, 16(1), 180–191.

Yue, S., Park, J.G., Liang, Z. and Shi, J. (2020) Tensor mixed effects model with application to nanomanufacturing inspection. *Technometrics*, 62(1), 116–129.

Zhang, C. and Ma, Y. (2012) *Ensemble Machine Learning: Methods and Applications*. Berlin/Heidelberg, Germany: Springer.

Zhang, C., Yan, H., Lee, S. and Shi, J. (2018) Weakly correlated profile monitoring based on sparse multi-channel functional principal component analysis. *IISE Transactions*, 50(10), 878–891.

Zhang, C., Yan, H., Lee, S. and Shi, J. (2020) Dynamic multivariate functional data modeling via sparse subspace learning. *Technometrics*, 1–33.

Zhang, L., Xie, Y., Xidao, L. and Zhang, X. (2018a) Multi-source heterogeneous data fusion, in *2018 International Conference on Artificial Intelligence and Big Data (ICAIBD)*, IEEE Press, Piscataway, NJ, pp. 47–51.

Zhang, C., Yan, H., Lee, S. and Shi, J. (2018b) Weakly correlated profile monitoring based on sparse multi-channel functional principal component analysis. *IISE Transactions*, 50(10), 878–891.

Zhang, X., Yu, Y., White, M., Huang, R. and Schuurmans, D. (2011) Convex sparse coding, subspace learning, and semi-supervised extensions, in *Twenty-Fifth AAAI Conference on Artificial Intelligence*, AAAI, San Francisco, CA.

Zhao, Q., Caiafa, C.F., Mandic, D.P., Chao, Z.C., Nagasaka, Y., Fujii, N., Zhang, L. and Cichocki, A. (2012) Higher order partial least squares (hopls): A generalized multilinear regression method. *IEEE Transactions on Pattern Analysis and Machine Intelligence*, 35(7), 1660–1673.

Zhou, H., Li, L. and Zhu, H. (2013) Tensor regression with applications in neuroimaging data analysis. *Journal of the American Statistical Association*, 108(502), 540–552.

Zhu, Z., Yin, H., Chai, Y., Li, Y. and Qi, G. (2018) A novel multimodality image fusion method based on image decomposition and sparse representation. *Information Sciences*, 432, 516–529.

6 Cost Estimating Using a New Learning Curve Theory for Non-Constant Production Rates*

Dakotah Hogan
US Air Force Cost Analysis Agency

John Elshaw, Clay Koschnick, Jonathan Ritschel, and Adedeji B. Badiru
US Air Force Institute of Technology

Shawn Valentine
US Air Force Lifecycle Management Center

CONTENTS

6.1   INTRODUCTION

The U.S. Government Accountability Office (GAO) critiqued the cost and schedule performance of the Department of Defense (DoD)'s $1.7 trillion portfolio of 86 major weapons systems in their 2018 "Weapons System Annual Assessment." The GAO cited realistic cost estimates as a reason for the relatively low-cost growth of the portfolio in comparison to earlier portfolios [1]. Congress and its oversight committees maintain a watchful eye on the DoD's complex and expensive weapons system portfolio. Inefficient programs are scrutinized and may be terminated if inefficiencies persist.

* Reprinted with permission: Hogan, D., Elshaw, J., Koschnick, C., Ritschel, J., Badiru, A., & Valentine, S. (2020). Cost estimating using a new learning curve theory for non-constant production rates. *Forecasting*, 2(4), 429–451.

Funding of inefficient programs will also lead to the underfunding of other programs. In the public sector, these terminated and underfunded programs may result in capability gaps that negatively impact our nation's defense. In the private sector, the inefficient use of resources often spells failure for a company.

A key to the efficient use of resources is accurately estimating the resources required to produce an end item. Learning curves are a popular method of forecasting required resources as they predict end-item costs using the item's sequential unit number in the production line. Learning curves are especially useful when estimating the required resources for complex products. The most popular learning curve models used in the government sector are over 80 years old and may be outdated in today's technology-rich production environment. Additionally, researchers have demonstrated both theoretically and empirically that the effects of learning slow or cease over time [2–4].

A new model, named Boone's learning curve, has been recently proposed to account for diminishing rates of learning as more units are produced [5]. The purpose of this research is to survey the need for alternative learning curve models and further examine how Boone's learning curve performs in comparison to the traditional learning curve theories in predicting required resources. This research uses a large number of diverse production items to compare Boone's model to the traditional theories of Wright and Crawford. While many different learning curve models exist (i.e., DeJong, Stanford B, Sigmoid, etc.), some of these others may not be as accurate in cases where the learning rate decreases over time. The next section is a review of the learning curve literature relevant to diminishing learning rates, followed by a description of our methodology and analysis to compare Boone's learning curve to traditional models. We conclude the paper by discussing managerial implications and limitations followed by recommendations for the way forward.

6.2   LITERATURE REVIEW AND BACKGROUND

The two learning curve models cited by the GAO Cost Estimating and Assessment Guide (2009) are Wright's cumulative average learning curve theory developed in 1936 and Crawford's unit learning curve theory developed in 1947. Although both learning curve theories use the same general equation, the theories have contrasting variable definitions. Wright's learning curve is shown in Eq. (6.1):

$$\overline{Y} = Ax^b \tag{6.1}$$

where $\overline{Y}$ is the cumulative average cost of the first x units, A is the theoretical cost to produce the first unit, x is the cumulative number of units produced, and b is the natural logarithm of the learning curve slope (LCS) divided by the natural logarithm of two. Note that the LCS is the complement of the percent decrease in cost as the number of units produced doubles. For example, with a learning curve slope of 80% and a first unit cost of 100 labor hours, the average cost of the first two units would be 80 labor hours or 60 labor hours for the second unit. Regardless of the number of units produced, there is a constant decrease in labor costs with each doubling of units due to the constant learning rate. Several years following the creation of Wright's cumulative average learning curve theory, J.R. Crawford formulated the unit learning curve theory. Crawford's theory deviates from Wright's by assuming that the individual unit cost (as opposed to the cumulative average unit cost) decreases by a constant percentage as the number of units produced doubles. Crawford's model is shown in Eq. (6.2):

$$Y = Ax^b \tag{6.2}$$

where Y is the individual cost of unit x, A is the theoretical cost of the first unit, x is the unit number of the unit cost being forecasted, and b is the natural logarithm of the LCS divided by the natural logarithm of two. For example, with a learning curve slope of 80% and a first unit cost of 100 labor hours, the cost of the second unit would be 80 labor hours. Note that Crawford's unit theory is similar to Wright's in function form; but the difference that arises in the variable interpretation leads to a different forecast.

Figure 6.1 shows a comparison between Wright's and Crawford's theories using the two numerical examples provided. Cumulative average theory and unit theory will produce different predicted costs provided the same set of data despite all predicted costs being normalized to unit costs. Figure 6.1 demonstrates this point where unit theory was used to generate data using a first unit cost of 100 and a learning curve slope of 90%. The original unit theory data was converted to cumulative averages in order to estimate cumulative average theory learning curve parameters.

Cumulative average theory learning curve parameters. Cumulative average theory estimated a learning curve slope of 93% and a first unit cost of 101.24. These cumulative average theory parameters were then used to predict cumulative average costs. These predicted costs were then converted to unit costs. This conversion allows for the cumulative average predictions to be directly compared to the original Unit Theory generated data. As shown in Figure 6.1, the cumulative average learning curve predictions first overestimate, then underestimate, and ultimately overestimate the generated unit theory data for all remaining units. Together, Wright's and Crawford's theories form the basis of the traditional learning curve theory.

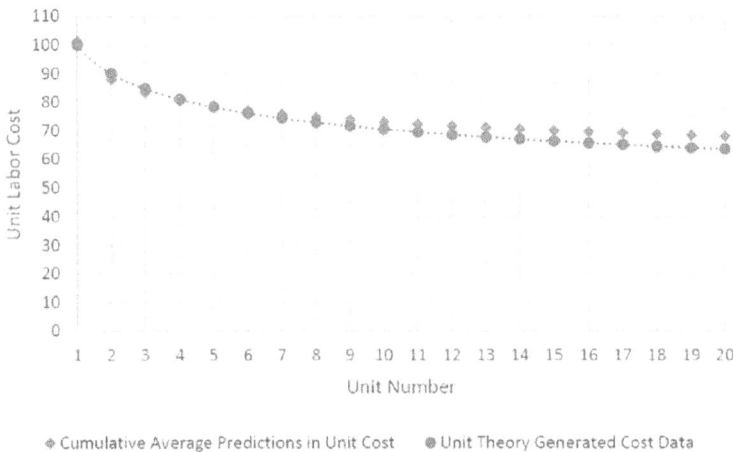

FIGURE 6.1 Wright's cumulative average theory vs. Crawford's unit theory.

One assumption of these traditional learning curve theories is that they only apply to processes that may benefit from learning. Typically, these costs are only a subset of total program costs; hence appropriate costs must be considered when applying learning curve theory to yield viable parameter estimates. In a complex program, costs can be viewed in a variety of ways to include recurring and non-recurring costs, direct and indirect costs, and costs for various activities and combinations of end items that can be stated in units of hours or dollars. Learning curve analysis focuses solely on recurring costs in estimating parameters because these costs are incurred repeatedly for each unit produced [6]. Researchers have also focused solely on direct labor costs due to the theoretical underpinnings of learning occurring at the laborer level [2,3]. Additionally, researchers have historically studied end-items that include only the manufactured or assembled hardware and software elements of the end-item [2,3]. Finally, labor hours in lieu of labor dollars are generally used in analysis so that data can be compared across fiscal years without the need to adjust for inflation. Therefore, the literature indicates using direct, recurring, labor costs in units of labor hours. These costs should be considered only for certain elements that include the manufacturing or assembly of hardware and software of an end item.

An implicit assumption in the traditional learning curve theories is that knowledge obtained through learning does not depreciate. However, empirical evidence demonstrates that knowledge depreciates in organizations [7,8]. Argote [7] showed that knowledge depreciation occurs at both the individual and the organizational levels. Many variations of the traditional models make use of the concept of performance decay (commonly called forgetting) to model non-constant rates of learning. Forgetting and its relationship to learning can take many forms and is essential to consider in contemporary learning curve analysis.

Forgetting is the concept that an individual or organization will experience a decline in performance over time resulting in non-constant rates of learning. Badiru [4] theorizes that forgetting and resulting performance decay is a result of factors "including lack of training, reduced retention of skills, lapse in performance, extended breaks in practice, and natural forgetting" (p. 287). According to Badiru [4], these factors may be caused by internal processes or external factors. Badiru [4] lists three cases in which forgetting arises. First, forgetting may occur continuously as a worker or organization progresses down the learning curve due in part to natural forgetting [4]. The impact of forgetting may not wholly eclipse the impact of learning but will hamper the learning rate while performance continues to increase at a slower rate. Second, forgetting may occur at distinct and bounded intervals, such as during a scheduled production break [4] or towards the end of production as workers are transferred to other duties. Finally, forgetting may intermittently occur at random times and for stochastic intervals such as during times of employee turnover [4]. Others have expanded on the causes of forgetting and have drawn similar conclusions to Badiru [4,9–11]. This decline in performance decays the learning rate and causes longer manufacturing times and higher costs than would be forecasted using traditional learning curve theory.

The concept of forgetting and its impact on non-constant rates of learning has proven relevant in contemporary learning curve research. Several forgetting models have been developed to include the learn-forget curve model (LFCM) [11], the recency model (RCM) [12], the power integration and diffusion (PID) model [13], and the Depletion-Power-Integration-Latency (DPIL) model [13] among others [10]. However, these forgetting models focus solely on the phenomenon of forgetting due to interruptions in the production process [9,10,14]. Jaber [9] states that "there has been no model developed for industrial settings that consider forgetting as a result of factors other than production breaks" (pp. 30–31) and mentions this as a potential area of future research. Although forgetting models have emerged after Jaber's [9] article, a review of the popular forgetting models cited confirms Jaber's statement.

A related concept to the forgetting phenomenon is the plateauing phenomenon. According to Jaber [9] (2006), plateauing occurs when the learning process ceases and manufacturing enters a production steady state. This ceasing of learning results in a flattening or partial flattening of the learning curve corresponding to rates of learning at or near zero. There remains debate as to when plateauing occurs in the production process or if learning ever ceases completely [3,9,15–17]. Jaber [9] provides several explanations to describe the plateauing phenomenon that include concepts related to forgetting. Baloff [18,19] recognized that plateauing is more likely to occur when capital is used in the production process as opposed to labor. According to some researchers, plateauing can be explained by either having to process the efficiencies learned before making additional improvements along the learning curve or to forgetting altogether [20]. According to other researchers, plateauing can be caused by labor ceasing to learn or management's unwillingness to invest in capital to foster induced learning [21]. Related to this underinvestment to foster induced learning, management's doubt as to whether learning efficiencies related to learning can occur is cited as another hindrance to constant rates of learning [22]. Li and Rajagopalan [23] investigated these explanations and concluded that no empirical evidence supports or contradicts them while ascribing plateauing to depreciation in knowledge or forgetting. Jaber [9] concludes that "there is no tangible consensus among researchers as to what causes learning curves to plateau" and alludes that this is a topic for future research (pp. 30–39).

Despite the controversy in the research surrounding forgetting and plateauing effects, empirical studies have shown learning curves to exhibit diminishing rates of learning. For instance, the plateauing phenomenon at the tail end of production was investigated by Harold Asher in a 1956 RAND study. The U.S. Air Force contracted RAND after the service noticed traditional learning curves were underestimating labor costs at the tail end of production [3]. Asher intended to study if the logarithmically transformed traditional learning curves were approximately linear. This linearity would indicate constant rates of learning throughout the production cycle. The alternative hypothesis for these learning curves was a convexity of the logarithmically-transformed traditional learning curves that would indicate diminishing rates of learning as the number of units increased [3]. An example of a learning curve with a diminishing learning rate is shown in Figure 6.2 on logarithmic scale. The first unit cost is 100 with an initial learning curve slope of 80% decaying at a rate of 0.25% with each additional unit. For example, the second unit's learning curve slope is 80.25%.

Asher investigated this hypothesis of convex logarithmically transformed learning curves by analyzing the learning curves of the various shops within a manufacturing department producing aircraft. Asher used airframe cost data with the appropriate

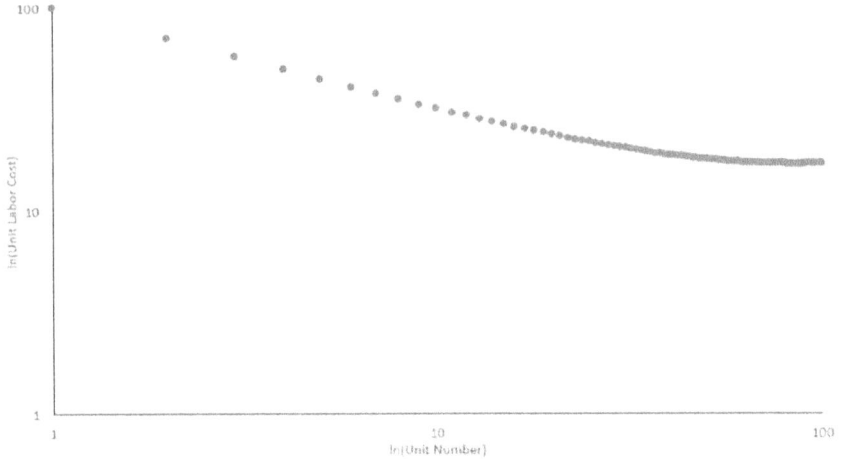

FIGURE 6.2   Unit theory learning curve with a decaying learning curve slope.

amount of detail to perform a learning curve analysis on the lower-level job shops within the manufacturing department. He divided the eleven major kinds of aircraft manufacturing operations into four shop groups each with a set of direct labor cost data [3]. If non-constant rates of learning were present, the shop group curves would differ in their rates of learning and may themselves be convex on a logarithmic scale. This would indicate their aggregate learning curve would also be convex on a logarithmic scale.

Asher's results showed that the learning curves of the manufacturing shop group had different learning slopes and were convex on the logarithmic scale [3]. Asher claims the convexity within the manufacturing shop group learning curves is due to the disparate operations within the job shops and stated that each had their own unique learning curve [3]. He asserts that a linear approximation is reasonable for a relatively small quantity of airframes produced but becomes increasingly unwarranted for larger quantities. This is due in part because larger quantities of produced end-items are likely to experience diminishing rates of learning. Moreover, highly aggregated learning curves are also likely to experience diminishing rates of learning. Because the aggregated manufacturing cost curve is usually the lowest level of detail on which learning curve analysis is performed, the manufacturing cost curve will have diminishing rates of learning as cumulative output increases. These results further justify a learning curve model with diminishing rates of learning.

Wright's and Crawford's learning curve theories provided the basis of the traditional approach that learning occurs at a constant rate as the number of units produced increases. Since this initial discovery, several log-linear learning curve models were founded in attempts to more accurately model data from manufacturing processes. These contemporary models diverge from constant rates of learning by including adjustments in various forms. The six most popular models (including the traditional model) are shown in Figure 6.3 on a logarithmic scale and include log-log graphing lines to more clearly illustrate the differences between models. These illustrated models include the traditional log-linear model or Wright/Crawford

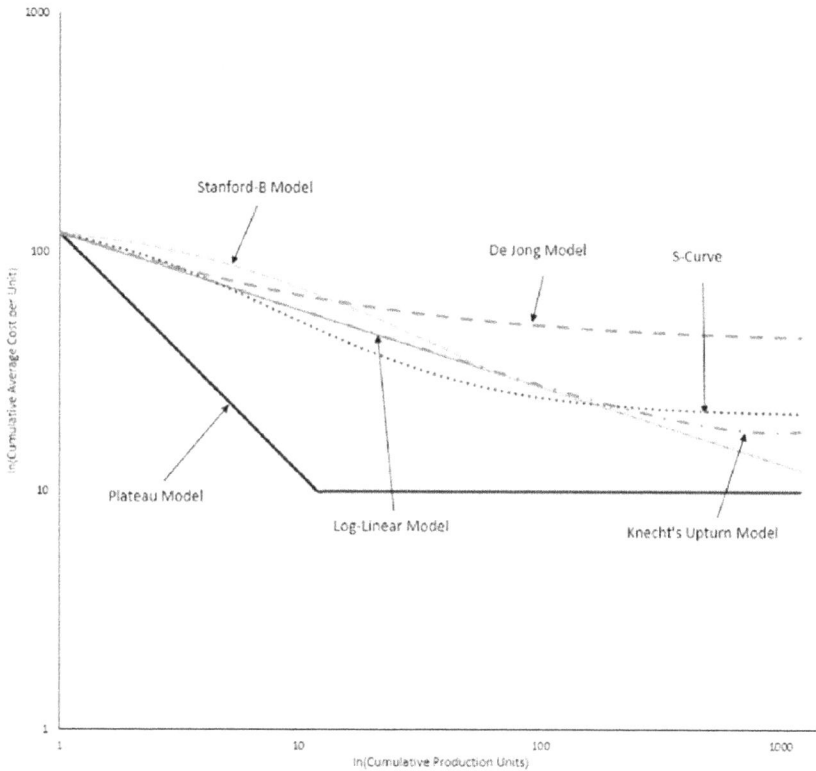

FIGURE 6.3   Comparison of learning curve *Models*. (Adapted from Badiru [27].)

curves, the plateau model [19], the Stanford-B model [24], the De Jong model [25], the S-curve model [21], and Knecht's upturn model [26].

Recent studies have investigated whether the Stanford-B, De Jong, and S-Curve models more accurately predict program costs in comparison to the traditional theories. Moore [16] and Honious [17] studied how prior experience in the manufacturing of an end-item along with the proportion of touch labor in the manufacturing process affected the accuracy of the Stanford-B, De Jong, and S-curve models in comparison to the traditional models. The authors concluded that these models improved upon the traditional curves for only a narrow range of parameter values. Their research provided insight that the traditional learning curve models become less accurate at the tail-end of production when the proportion of human labor is high in the manufacturing process. Moreover, Honious [17] explicitly references a plateauing effect at the end of production. These findings provide further justification for investigating non-constant rates of learning.

The Stanford-B, De Jong, and S-Curve univariate models illustrated in Figure 6.3 alter the resulting learning curve slope based on alterations to the theoretical first unit cost parameter *A*. However, the learning curve slopes of these models are not directly a function of the number of cumulative units produced. The plateau model and Knecht's upturn model also illustrated in Figure 6.3; each produces a learning curve whose slope is directly affected by the number of cumulative units produced. The plateau model uses a step function to reduce the learning rate to 0% (i.e., the learning curve slope is 100%) past a certain number of cumulative units produced. In contrast, Knecht's Upturn Model amends the learning curve exponent term b by multiplying b by Euler's number e raised to the term of a constant multiplied by the number of cumulative units produced. Mathematically, this is expressed $\overline{Y} = Ax^{b-e^{xc}}$, where $\overline{Y}$ is the cumulative average unit cost, A is the theoretical first unit cost, x is the number of cumulative units produced, b is the natural logarithm of the learning curve slope divided by the natural logarithm of 2, and c is a constant. The forgetting models stated within the manuscript also amend the learning curve slope based indirectly on the number of cumulative units but only apply when interruptions to the production process occur.

In response to these researchers' findings, Boone [5] developed a learning curve model with a learning rate that diminishes as more units are produced. Conversely, the traditional learning curve theories diminish the rate of cost reductions as the number of units produced doubles. However, the existing literature provides evidence that the cost reductions with each doubling of units may not be constant as the number of units produced increases. Therefore, Boone [5] sought to attenuate the cost reductions that occur with each doubling of units produced by decreasing the learning rate as the number of units increases.

Boone [5] devised a model that decreases the learning curve exponent b as the number of units produced x increases. He first considered a model without an additional parameter to reduce the learning curve exponent b directly by the unit number. However, he decided to temper the effect each additional unit has on parameter b by adding an additional parameter c. The resulting learning curve is shown in Eq. (6.3):

$$\overline{Y} = Ax^{\frac{b}{1 + \frac{x}{c}}} \tag{6.3}$$

where Y is the cumulative average cost of the first x units, A is the theoretical cost to produce the first unit, x is the cumulative number of units produced, b is the natural logarithm of the learning curve slope (LCS) divided by the natural logarithm of two, and c is a positive decay value. For example, a learning curve slope of 80%, first unit cost of 100 labor hours, and decay value of 100, Boone's model yields a cumulative average cost at the second unit of 80.35 labor hours—or 60.70 labor hours for the second unit. What began as an 80% learning curve model has decayed to an 80.35% learning curve for the second unit. In comparison to Wright's learning curve using the same parameters, the effect of learning has decreased slightly in the production of unit two. The inclusion of the decay value increases the learning curve slope and hence decreases the learning rate as more units are produced. Note that Boone's model can also be modified to incorporate Crawford's unit theory–refer to Eq. (6.3) for the necessary modifications.

Boone's learning curve diverges from the constant learning assumptions in both Wright's and Crawford's learning curve models by incorporating the unit number in the denominator of the exponent—thus decreasing the effect of b as the number of units produced increases. Furthermore, the decay value moderates this diminishing effect, so the amount of learning decreases more slowly. In general, Boone's model is flatter near the end of production and steeper in the early stages compared to the traditional theories. Note, as the decay value approaches zero (holding other factors constant), the exponent term approaches zero representing a learning curve slope approaching 100%. As the decay value approaches infinity, the parameter b remains constant, and Boone's learning curve simplifies to the traditional learning curve [5].

Boone [5] tested his learning curve using unit theory to provide a consistent comparison to Crawford's learning curve. Based on the scope of his research and lack of comparison using cumulative average theory, a more robust examination and analysis of Boone's learning curve should be accomplished.

6.3 METHODOLOGY

One goal of this research is to examine the accuracy of Boone's learning curve in comparison to the popular Wright and Crawford learning curve theories. In order to perform this analysis, production cost and quantity data from a diverse set of DoD systems were collected from government Functional Cost-Hour Reports, Progress Curve Reports, and the Air Force Life Cycle Management Center Cost Research Library. The dataset consisted of recurring costs (either in dollars or labor hours) by production lot for 169 unique end-items. Our data included end-items from a variety of systems (i.e., bomber, cargo, and fighter aircraft, missiles, and munitions), contractors, and time periods (1957–2018). Additionally, only production runs with at least four lots were included. The dataset for the Cumulative Average Theory analysis only includes 140 of the 169 end-items. This theory relies on continuous data because each lot's cumulative average cost and cumulative quantity is a function of all previous lots' costs and quantities. In order to compare Boone's model to the traditional theories, each model will be fitted to data: (i) Boone's and Wright's models using cumulative average theory, and (ii) Boone's and Crawford's models using unit theory. Then, the predicted values for each model will be compared to the actual costs using root mean squared error (RMSE) and mean absolute percentage error (MAPE).

Labor costs were collected from the work breakdown structure (WBS) for the specific item being manufactured (e.g., aircraft frame) or from the documentation provided by the government. Our data included three broad functional cost categories: labor, material, and other. These costs are included in both forms of recurring and non-recurring costs. There are also four functional labor categories delineated that include manufacturing, tooling, engineering, and quality control labor. These four labor category costs, when summed with the material costs and other costs, comprise the total cost for each WBS element for recurring and non-recurring costs.

The definition for the manufacturing labor cost category most clearly aligns with the extant literature to be the focus as the pertinent labor cost category for learning curve research. According to the WBS elements, the manufacturing labor category "includes the effort and costs expended in the fabrication, assembly, integration, and functional testing of a product or end item. It involves all the processes necessary to convert raw materials into finished items [28]." This manufacturing labor category aligns with the categories examined by Wright, which he called "assembly operations [2]," along with those cost categories Crawford studied, which he called "airframe-manufacturing processes [3]." Therefore, the manufacturing labor cost category as defined by the government is associated with the types of labor costs studied by traditional learning curve theorists and succeeding research.

The learning curve parameters for each model (i.e., Eq. (6.1–6.3)) will be estimated by minimizing the sum of squares error (SSE) using Excel's generalized reduced gradient (GRG) nonlinear solver and evolutionary solver. The SSE is calculated by squaring the vertical difference of the observed data and predicted data for each lot and summing these squared differences across all lots.

With lot of data, cumulative theory models can be estimated directly. Conversely, when utilizing unit learning curve theory, Crawford's and Boone's models are estimated using an iterative process based on lot midpoints, adapted from Hu and Smith [29]. The algebraic lot midpoint is defined as "the theoretical unit whose cost is equal to the average unit cost for that lot on the learning curve" [6]. The lot midpoint supplants using sequential unit numbers when using lot cost data.

Lot midpoints and model parameters are calculated iteratively due to the lack of a closed-form solution for the lot midpoint. First, an initial lot midpoint (for each lot) is determined using a parameter-free approximation formula [6]—see Eq. (6.4):

$$\text{Lot Midpoint(LMP)} = \frac{F + L + 2\sqrt{FL}}{4} \tag{6.4}$$

where F is the first unit number in a lot and L is the last unit number in a lot. These lot midpoint estimates are then used to estimate the learning curve parameters for Crawford's model (Eq. (6.2)) using the GRG nonlinear optimization algorithm. Next, using the estimated parameter b, a new set of lot midpoints are determined using a simple and popular formula—Asher's Approximation [6]; see Eq. (6.5):

$$\text{Lot Midpoint} \approx \left[\frac{\left(L + \frac{1}{2}\right)^{b+1} - \left(F - \frac{1}{2}\right)^{b+1}}{(L - F + 1)(b + 1)} \right]^{\left(\frac{1}{b}\right)} \tag{6.5}$$

where F is the first unit number in a lot, L is the last unit number in a lot, and b is the estimated value from Eq. (6.2). Learning curve parameters will then be re-estimated using these more precise lot midpoint estimates. The iterative process is repeated until changes between successive values of the estimated lot midpoints and b are sufficiently small [29] (see Appendix 6A for a summary of this process). In order to use an iterative process for Boone's model, Asher's Approximation from Eq. (6.5) was adapted to incorporate Boone's decaying learning curve slope. This adaptation allows the lot costs of Boone's learning curve to decrease as more units are produced which affects the lot midpoint estimates; the formula is shown in Eq. (6.6):

$$\text{Lot Midpoint}_i \approx \left[\frac{\left(L + \frac{1}{2}\right)^{b'+1} - \left(F - \frac{1}{2}\right)^{b'+1}}{(L - F + 1)(b' + 1)} \right]^{\left(\frac{1}{b'}\right)} \tag{6.6}$$

where F is the first unit number in a lot, L is the last unit number in a lot, $b' = \dfrac{b}{1 + \left(\frac{LMP_i - 1}{c}\right)}$, and i is the iteration number.

This iterative process of calculating the lot mid-point and then solving a nonlinear least squares problem requires the execution of a series of nonlinear optimization algorithms. Boone's model requires the GRG algorithm which found solutions in a longer but still reasonable amount of time. While more burdensome than the traditional models due to the longer run time and the requirement to provide bounds for the parameters. For Boone's model, the bounds for A and b have a fairly straightforward basis by which to define the bounds. In practice, parameter A is often supported by a point estimate of the cost of the first theoretical unit. Thus, a bound can be built around this value with tools such as a confidence interval. Parameter b is defined by the learning curve slope, which, for all practical purposes, will be in the (0, 1) interval—most likely on the higher end. As for parameter c, the basis for the bound is more of a challenge. From a model implementation standpoint, the bound can be arbitrarily large if a long solve time is not limiting. Practically, the bound should be reasonably set; this aspect of the model is an avenue of future research which is discussed in the conclusion. This algorithm does allow the analyst to define stopping conditions such as convergence threshold, maximum number of iterations, or maximum amount of time. Additionally, there is an option called multi-start which uses multiple initial solutions to help locate a global solution verse possibly only finding a local solution. These options allow the user to mitigate the extra burden if necessary. Overall, the computing burden to calculate these models was on the order of minutes per weapon system.

The final estimated parameters for Boone's model and the traditional learning curves were used to create predicted learning curves. These predicted curves were then compared to observed data. Total model error was calculated by comparing the difference between observations and predicted values to understand how accurately the models explained variability in the data. Two measures were used to determine the overall model error. The first error measure was Root Mean Square Error (RMSE) that is calculated by taking the square root of the total SSE divided by the number of lots. RMSE is not robust to outliers—i.e., the effects of outliers may unduly influence this measure. RMSE is often interpreted as the average amount of error of the model as stated in the model's original units.

The second measure was mean absolute percentage error (MAPE). MAPE is calculated by subtracting the predicted value from the observed value, dividing this difference by the observed value, taking the absolute value, and multiplying by 100%. These absolute percent errors are then summed over all observations and divided by the total number of observations. MAPE provides a unit-less measure of accuracy and is interpreted as the average percent of model inaccuracy. Unlike RMSE, MAPE is robust to outliers.

After calculating these measures of overall model error, a series of paired difference t-tests are conducted to determine if reductions in error from Boone's learning curve are statistically significant. In order to conduct the first paired difference t-test, Boone's learning curve RMSE using cumulative average theory will be subtracted from Wright's learning curve RMSE, and the difference will be divided by Wright's learning curve RMSE. This calculation will yield a percentage difference rather than a raw difference to compare end-items of varying differences in magnitude equitably. The null hypothesis posits that Boone's learning curve results in an equal amount (or more) of error in predicting observed values compared to Wright's learning curve. The alternative hypothesis is that the percentage difference is greater than zero. Support for the alternative hypothesis signifies that Boone's learning curve results in less error predicting observed values than Wright's learning curve. This methodology will be repeated five times to examine each learning curve theory using the two error measures and the different units of production costs—see Table 6.1.

An assumption to utilize the paired difference t-test is that the data are approximately normally distributed. For hypothesis tests with large sample sizes, the central limit theorem can be invoked. Alternatively, a Shapiro–Wilk test will be used to evaluate the normality assumption for small samples. If the Shapiro–Wilk test does not support the normality assumption, the non-parametric Wilcoxon Rank Sum test will be used. A 0.05 level of significance will be used for all statistical tests.

6.4 ANALYSIS & RESULTS

The detailed results for Wright's and Boone's learning curves using cumulative average theory are provided in Appendix 6B Tables 6B.1 and 6B.2. A total of 118 end-items in units of total dollars and 22 components in units of labor hours were analyzed. Each entry lists the program number, number of production lots, number of items produced, type of end-item, and units of the production costs. Additionally, each entry lists both error measures and the respective percent difference between the models. Positive (negative) differences indicate Boone's model has less (more) error than Wright's.

Boone's curve performs better for two reasons. First, Boone's model can explain costs to at least the same degree of accuracy as the traditional learning curve theories due to the extra parameter. Second, increased accuracy could also be explained by Boone's functional form. Despite these theoretical explanations, Boone's model had more error than Wright's for some observations; these negative percentage differences occur because an upper bound was placed on Boone's decay value. An upper bound of 5,000 was used for the decay value (same as Boone's original paper).

TABLE 6.1
Paired Difference Hypothesis Tests Conducted

Learning Curve Theory	Error Measure	Units of Measure
	Root Mean Squared Error	Total Dollars(K)
Cumulative Average Theory	Percentage Difference	Labor Hours
	Mean Absolute Percent Error	Total Dollars (K) & Labor Hours Combined
	Percentage Difference	
	Root Mean Squared Error	Total Dollars (K)
Unit Theory	Percentage Difference	
	Mean Absolute Percent Error	Total Dollars (K) & Labor Hours Combined
	Percentage Difference	

TABLE 6.2

Cumulative Average Theory Descriptive and Inferential Statistics

| | | | | | Hypothesis Test: $H_0: \mu \le 0$ $H_A: \mu > 0$ | | | |
Learning Curve Theory	Error Measure	Units of Measure	Sample Mean ($\bar{x}$)	Sample Standard Deviation (s)	Number of Observations	Test Statistic	p-Value	Result
Cumulative Average Theory	Root Mean Squared Error	Total Dollars (K)	19.3%	28.90%	118	7.23	<0.001	Reject H0
	Percentage Difference	Labor Hours	15.20%	31.20%	22	18.5	0.28	Fail to reject H0
	Mean Absolute Percent	Total Dollars (K) & Labor Hours Combined	18.60%	29.50%	140	7.45	<0.001	Reject H0

The practical effect of this particular bound can be observed by the number of end-items where the traditional models significantly outperformed Boone's (i.e., a MAPE difference larger than 0.5%): 7 out of 140 for cumulative average theory and 15 out 169 for unit theory. Thus, the majority of the results were not affected by this artificial limitation, which was chosen by trial and error. In practice, the bound could be set arbitrarily large so that it is not binding. This upper bound was necessary since the GRG algorithm requires bounds on the estimated parameters.

Some percentage error differences are approximately (but not exactly) zero. Observations with percentage error differences of approximately zero were defined as those within the bounds (−0.25%, 0.25%). These bounds were used by the researchers to distinguish between observations with approximately zero and non-zero percentage error differences in order to inform the descriptive statistics.

Boone's model had less error for 41% of observations, was approximately equal to Wright's for 50% of observations, and had more error for 9% of observations. While Boone's model is an improvement on Wright's for some observations, many times the models fit the data equally well (i.e., an approximate zero difference).

The results of the paired difference t-tests for cumulative average theory are shown in Table 6.2 and a sample graph is shown in Figure 6.4. No outliers, as defined by a value that fell more than three interquartile ranges from the upper 90% and lower 10% quantiles, were present in any of the tests.

The results of these hypothesis tests were mixed. For the RMSE percentage difference (measured in total dollars) and MAPE percentage difference, the paired difference t-tests led to a rejection of the null hypothesis—indicating the increase in accuracy is statistically significant. Conversely, RMSE percentage difference (measured in hours) failed to reject the null hypothesis. Due to the small sample size, large sample theory could not be used, and the data failed a Shapiro–Wilk test

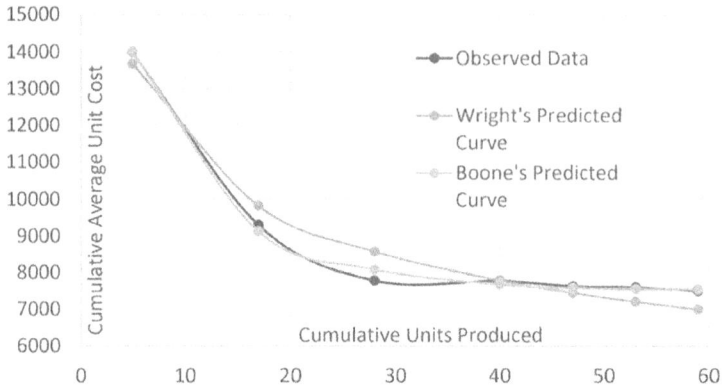

FIGURE 6.4　Comparison of program 20 PME air vehicle.

(p-value $=0.721$). Therefore, a Wilcoxon rank signed test was used. This indicates that Boone's improvement in accuracy over Wright's is not statistically significant when costs are measured in labor hours. However, small sample sizes can cause paired difference tests to have low power that may cause hypothesis tests to incorrectly fail to reject the null hypothesis [30].

Now considering unit theory, the results from Crawford's and Boone's learning curve models are presented in Appendix 6B. A total of 141 end-items (measured in total dollars) and 28 end-items (measured in labor hours) were analyzed.

Similar to cumulative average theory, observations with percent error differences of approximately zero were defined as those within the bounds $(-0.25\%, 0.25\%)$. Boone's model had less error for 43% of observations across all percent difference error measures in comparison to Crawford's learning curve.

Boone's learning curve error was approximately equal for 52% of observations and had more error for 5% of observations.

The results of the paired difference testing for unit theory are provided in Table 6.3, and a sample graph is shown in Figure 6.5. Again, no outliers were present in any of the paired difference t-tests.

The results of these paired difference tests indicate the improvement with Boone's model is statistically significant. Again, the RMSE percent difference (for labor hours) used a Wilcoxon rank sum test (due to the failure of the Shapiro–Wilk test with a p-value less than 0.001).

6.5　CONCLUSIONS

A large, diverse dataset of DoD production programs was used to test if Boone's learning curve more accurately explained error in comparison to traditional learning curve theories. The direct recurring cost data from bomber, cargo, and fighter aircraft along with missiles and munitions programs in units of total dollars and labor hours were analyzed using Cumulative Average and Unit Learning Curve theories. Various components of these programs were analyzed from wings and data link systems to

TABLE 6.3

Unit Theory Descriptive and Inferential Statistics

					Hypothesis Test: $H_0: \mu \le 0$ $H_A: \mu > 0$			
Learning Curve Theory	Error Measure	Units of Measure	Sample Mean ($\bar{x}$)	Sample Standard Deviation (s)	Number of Observations	Test Statistic	p-Value	Result
Unit Theory	Root Mean Squared Error	Total Dollars (K)	13.80%	22.70%	141	7.23	<0.001	Reject H0
	Percentage Difference	Labor Hours	6.00%	14.80%	28	74.00	0.046	Reject H0
	Mean Absolute Percent Error Percentage Difference	Total Dollars (K) & Labor Hours Combined	11.30%	23.10%	169	6.36	<0.001	Reject H0

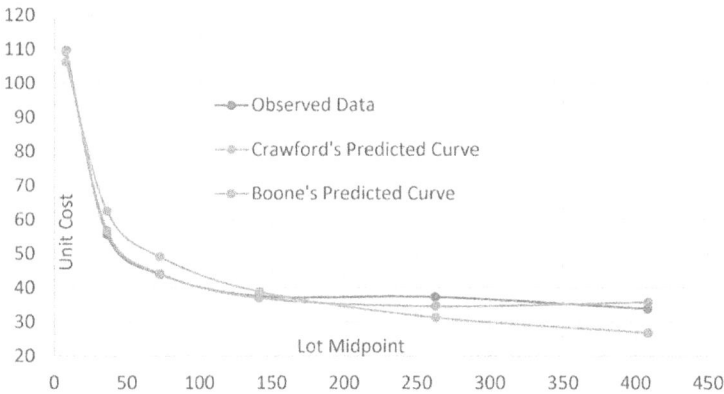

FIGURE 6.5 Comparison of program 1 PME air vehicle.

the airframes and air vehicles. Boone's learning curve was tested against both cumulative average and unit learning curve theories using two different measures of model error that resulted in six paired difference tests. This methodology resulted in 998 total observations across all measures and ensured the generalizability of Boone's learning curve was tested.

Boone's learning curve improved upon the traditional learning curve estimates for approximately 42% of the sampled program components while approximately equaling the traditional learning curve error for approximately 51% of program components. Boone's learning curve resulted in a range of mean percentage difference reductions of 6%–18.6% across all measures. The standard deviations of these improvements were high with coefficients of variation ranging from 150% to 247% across all measures. Absent additional analysis, these high amounts of variability make it challenging to conclude the degree to which Boone's learning curve will improve the accuracy of explaining program component costs in comparison to the traditional estimation methods. Specifically, more research is needed to understand the shape of the learning curve and how it behaves related to production circumstances. It remains unclear which programs are more accurately modeled using Boone's learning curve and to what degree Boone's learning curve will more accurately model program component costs.

The paired difference tests between Boone's learning curve and the traditional theories indicate that Boone's learning curve reduces the error to a significant degree across a wide range of measures. Five of the six paired difference tests resulted in rejecting the null hypothesis that Boone's learning curve had an equal amount or more error than the traditional theories at a significance level of 0.05.

Due to data availability, program lot data was used instead of unitary data. Although Boone's learning curve should perform just as well using either type of data, this research cannot conclusively state that Boone's learning curve will more accurately explain programs in unitary data. Also, the majority of data utilized were end-item components in units of total dollars. The total dollar cost includes all cost categories rather than solely labor costs. These data are not ideal when applying learning curve theory and may bias learning curves to display diminishing rates of learning. Despite these potential issues, total dollar cost data are regularly utilized by cost estimators in the field due to data availability. Therefore, the practical applications of this analysis remain valid despite the limitations of using imperfect total dollar cost data in learning curve analysis.

Boone's learning curve was tested on programs whose lot costs were already known and whose parameters can be directly estimated. In other words, Boone's learning curve was tested against the traditional theories on how well it explained rather than predicted program costs. In order to utilize Boone's learning curve to predict costs, a decay value would be selected a priori. Similar to the learning curve slope, an analyst could use the decay value from similar programs to provide a range of values to make predictions. Additionally, future research should investigate if Boone's Decay Value can be predicted using various attributes of a program. Tests could be performed on how well Boone's learning curve predicts costs for a program using analogous programs in comparison to the traditional theories. Lastly, additional labor hour data should be collected and analyzed in order to dispel the potential bias of learning curves displaying diminishing rates of learning when analyzed in units of total dollars.

Author Contributions: Conceptualization, D.H., J.E., C.K. and J.R.; methodology, D.H., J.E., C.K., J.R. and A.B.; software, D.H. and S.V.; validation, D.H., J.E., C.K. and J.R.; formal analysis, D.H., J.E., C.K. and J.R.; investigation, D.H., J.E., C.K., J.R. and S.V.; resources, D.H., A.B. and S.V.; data curation, D.H., J.E. and S.V.; writing—original draft preparation, D.H., J.E. and C.K.; writing—review and editing D.H., J.E., C.K., J.R. and A.B.; visualization, D.H. and A.B.; supervision, J.E., C.K., J.R. and A.B.; project administration, D.H., J.E., C.K. and J.R.; funding acquisition, J.E., J.R. and A.B. All authors have read and agreed to the published version of the manuscript.

Funding: This research received no external funding.

Conflicts of Interest: The authors declare no conflict of interest.

Publisher's Note: MDPI stays neutral with regard to jurisdictional claims in published maps and institutional affiliations.

Appendix 6A. Calculation Process for Lot Midpoint Estimation

The following process was implemented to estimate parameters for lot midpoint estimation.

1. Parameter-free lot midpoint approximations (Eq. (6.4)) were calculated for each production lot.
2. Crawford's learning curve parameters A and b were initially estimated using OLS regression.
   a. Average unit cost was the dependent variable while lot midpoint, calculated in Step 1, was the independent variable.
3. These initial learning curve parameter estimates were used as starting values to more precisely estimate Crawford's learning curve parameters using GRG nonlinear solver. This process generated intermediate estimates of Crawford's learning curve parameters.
4. The intermediate estimate of Crawford's learning curve b parameter was used to calculate a more precise set of lot midpoints using Asher's approximation (Eq. (6.5)).
5. Applying these more precise lot midpoint approximations, Crawford's learning curve parameters A and b were more accurately estimated using GRG nonlinear solver.

Steps 4 and 5 were repeated until the iterative process converged on a solution to produce final estimates of Crawford's learning curve parameters and lot midpoint approximations.

Appendix 6B. Learning Curve Error Comparisons Using Cumulative Average and Unit Theories

TABLE 6B.1

Error Comparison Using Cumulative Average Theory

Program	Number of Lots	Number of Units	Component Estimated	Traditional Units RMSE	RMSE Boone	Percentage Difference	Traditional MAPE	Boone MAPE	MAPE Percentage Difference
Program 1	6	483	PME–Air Vehicle	Dollars 557.9	111.7	80.0%	3.6%	0.7%	80.9%
Program 1	6	483	PME–Air Vehicle	Hours 15.5	0.3	98.0%	27.2%	0.5%	98.2%
Program 1	6	483	Airframe	Dollars 411.2	114.1	72.3%	2.8%	0.7%	74.7%
Program 1	6	483	Airframe	Hours 21.7	1.5	93.0%	31.0%	1.7%	94.6%
Program 2	5	638	PME–Air Vehicle	Dollars 129.8	6.5	95.0%	2.6%	0.1%	95.6%
Program 3	5	500	PME–Air Vehicle	Dollars 1,630.3	291.1	82.1%	20.8%	3.9%	81.5%
Program 4	19	205	PME–Air Vehicle	Dollars 581.7	581.8	0.0%	3.1%	3.1%	0.0%
Program 4	19	205	Airframe	Dollars 546.0	546.4	-0.1%	3.2%	3.2%	-0.1%
Program 5	7	459	PME–Air Vehicle	Dollars 400.8	44.7	88.8%	2.7%	0.3%	88.2%
Program 5	7	459	Electronic Warfare (1)	Dollars 4.8	3.2	32.3%	7.2%	4.8%	33.7%
Program 6	6	98	PME–Air Vehicle	Dollars 99.3	32.2	67.6%	1.1%	0.3%	69.4%
Program 6	6	98	Electronic Warfare (1)	Dollars 12.7	1.7	86.8%	3.6%	0.6%	82.4%
Program 6	6	98	Electronic Warfare (2)	Dollars 15.0	13.3	11.4%	2.3%	2.0%	12.9%
Program 6	6	98	Electronic Warfare (3)	Dollars 1.8	1.1	40.3%	1.3%	0.8%	39.6%
Program 7	7	110	PME–Air Vehicle	Dollars 145.0	98.3	32.2%	1.0%	0.7%	32.6%
Program 7	7	110	Electronic Warfare (1)	Dollars 8.4	3.6	57.2%	2.7%	1.0%	61.3%
Program 7	7	110	Electronic Warfare (2)	Dollars 140.3	107.2	23.6%	1.2%	0.8%	27.5%
Program 7	7	110	Electronic Warfare (3)	Dollars 0.9	0.9	0.0%	0.5%	0.5%	-0.1%
Program 7	7	110	Electronic Warfare (4)	Dollars 140.7	111.3	20.9%	1.3%	1.0%	24.2%
Program 7	7	110	Electronic Warfare (5)	Dollars 21.3	21.0	1.1%	2.2%	2.1%	5.2%
Program 8	8	3,529	PME–Air Vehicle	Dollars 27.7	23.6	14.8%	1.4%	1.3%	7.8%
Program 8	8	3,529	PME–Air Vehicle	Hours 0.1	0.1	-27.5%	1.1%	1.3%	-27.9%
Program 9	9	3,798	PME–Air Vehicle	Dollars 166.5	170.7	-2.5%	8.4%	8.8%	-3.7%
Program 10	10	3,803	PME–Air Vehicle	Dollars 8.0	4.8	39.6%	2.5%	1.2%	51.7%
Program 10	10	3,803	PME–Air Vehicle	Hours 24.4	14.0	42.7%	4.3%	2.0%	54.0%

(Continued)

TABLE 6B.1 (Continued)
Error Comparison Using Cumulative Average Theory

Program	Number of Lots	Number of Units	Component Estimated	Traditional Units RMSE	RMSE Boone	RMSE Percentage Difference	Traditional MAPE	Boone MAPE	MAPE Percentage Difference
Program 11	6	180	PME–Air Vehicle	Dollars 514.0	508.4	1.1%	0.9%	0.8%	4.2%
Program 12	10	20	PME–Air Vehicle	Dollars 699.2	694.1	0.7%	5.8%	5.7%	1.0%
Program 12	10	20	PME–Air Vehicle	Hours 1,042.5	906.5	13.1%	9.5%	8.4%	11.8%
Program 12	7	11	Mission Computer (1)	Dollars 44.3	44.3	0.0%	2.5%	2.5%	0.0%
Program 13	5	100	PME–Air Vehicle	Dollars 53,386.7	21,143.7	60.4%	12.8%	4.8%	62.1%
Program 13	5	100	Airframe	Dollars 6,569.7	6,578.0	-0.1%	3.7%	3.7%	0.0%
Program 14	5	275	PME–Air Vehicle	Dollars 3,114.0	145.5	95.3%	3.8%	0.2%	95.5%
Program 15	10	77	PME–Air Vehicle	Dollars 44,386.0	44,390.2	0.0%	9.5%	9.5%	0.0%
Program 15	12	83	PME–Air Vehicle	Hours 79,242.0	79,247.5	0.0%	6.5%	6.5%	0.0%
Program 15	11	83	Airframe	Dollars 39,624.4	39,628.0	0.0%	10.6%	10.6%	0.0%
Program 15	10	68	Mission Computer (1)	Dollars 1,959.3	1,959.4	0.0%	17.0%	17.0%	0.0%
Program 16	9	76	PME–Air Vehicle	Dollars 436.3	144.4	66.9%	2.6%	1.0%	62.9%
Program 17	5	50	PME–Air Vehicle	Dollars 13,023.6	13,029.8	0.0%	2.8%	2.8%	-0.1%
Program 18	9	31	PME–Air Vehicle	Dollars 2,942.5	2,941.9	0.0%	1.0%	0.9%	0.0%
Program 19	6	98	PME–Air Vehicle	Dollars 313.3	313.4	0.0%	0.5%	0.5%	-0.1%
Program 20	11	84	PME–Air Vehicle	Dollars 1,568.7	1,121.9	28.5%	1.7%	1.5%	7.8%
Program 20	7	59	Electronic Warfare (1)	Dollars 452.8	143.0	68.4%	4.6%	1.3%	71.5%
Program 20	11	84	Electronic Warfare (2)	Dollars 98.7	76.5	22.5%	3.4%	3.6%	-6.3%
Program 20	7	59	Electronic Warfare (5)	Dollars 562.5	517.4	8.0%	1.8%	1.8%	1.7%
Program 21	6	326	PME–Air Vehicle	Dollars 5,267.1	2,408.8	54.3%	8.0%	4.2%	47.4%
Program 21	7	344	Airframe	Dollars 4,819.5	2,544.3	47.2%	9.1%	5.4%	40.4%
Program 21	7	344	Avionics	Dollars 763.2	429.9	43.7%	6.6%	3.9%	40.8%

(Continued)

TABLE 6B.1 (Continued)

Error Comparison Using Cumulative Average Theory

Program	Number of Lots	Number of Units	Component Estimated	Traditional Units RMSE	Boone RMSE	RMSE Percentage Difference	Traditional MAPE	Boone MAPE	MAPE Percentage Difference
Program 21	14	453	PME–Air Vehicle	Hours 3,493.6	3,495.9	−0.1%	4.8%	4.8%	0.1%
Program 21	14	453	Airframe	Hours 4,338.4	4,339.7	0.0%	6.2%	6.2%	0.1%
Program 22	8	538	PME–Air Vehicle	Hours 856.7	857.7	−0.1%	2.5%	2.6%	−0.1%
Program 22	8	538	Airframe	Hours 5,608.5	5,609.7	0.0%	15.8%	15.9%	−0.1%
Program 23	5	469	PME–Air Vehicle	Dollars 637.5	339.3	46.8%	5.4%	2.9%	47.3%
Program 24	10	59	PME–Air Vehicle	Dollars 3,032.5	3,033.0	0.0%	2.2%	2.2%	0.0%
Program 25	9	348	PME–Air Vehicle	Dollars 117.8	118.1	−0.2%	0.9%	0.9%	−0.2%
Program 26	5	109	PME–Air Vehicle	Dollars 3,247.4	1,676.8	48.4%	11.0%	6.0%	45.7%
Program 26	5	109	PME–Air Vehicle	Hours 607.1	453.5	25.3%	5.7%	4.2%	25.9%
Program 27	18	631	PME–Air Vehicle	Dollars 1,669.6	913.3	45.3%	3.6%	1.9%	46.2%
Program 28	6	425	PME–Air Vehicle	Dollars 320.0	322.0	−0.6%	0.9%	0.9%	−0.6%
Program 28	7	522	PME–Air Vehicle	Hours 1,776.1	1,785.6	−0.5%	1.8%	1.8%	−0.1%
Program 28	7	522	Airframe	Hours 1,389.9	1,393.9	−0.3%	1.2%	1.2%	−0.2%
Program 29	9	358	PME–Air Vehicle	Hours 610.6	611.1	−0.1%	0.9%	0.9%	0.4%
Program 29	9	358	Airframe	Hours 4,804.8	2,124.2	55.8%	7.3%	2.9%	60.1%
Program 30	5	204	PME–Air Vehicle	Dollars 513.5	212.7	58.6%	1.2%	0.5%	56.1%
Program 31	5	605	PME–Air Vehicle	Dollars 1,482.6	629.1	57.6%	6.1%	2.9%	53.1%
Program 32	5	870	PME–Air Vehicle	Dollars 61.3	61.6	−0.5%	0.4%	0.4%	−0.3%
Program 33	10	178	PME–Air Vehicle	Dollars 7,093.5	7,101.1	−0.1%	3.5%	3.5%	−0.1%
Program 33	10	178	PME–Air Vehicle	Hours 8,131.1	8,144.1	−0.2%	2.9%	2.9%	−0.1%
Program 33	10	178	Airframe	Dollars 1,906.9	1,910.8	−0.2%	1.7%	1.7%	−0.2%
Program 33	10	712	Body	Dollars 232.2	234.9	−1.2%	1.5%	1.6%	−1.3%

(Continued)

TABLE 6B.1 (*Continued*)
Error Comparison Using Cumulative Average Theory

Program	Number of Lots	Number of Units	Component Estimated	Traditional Units RMSE	RMSE Boone	Percentage Difference	Traditional MAPE	Boone MAPE	MAPE Percentage Difference
Program 33	10	178	Alighting Gear	Dollars 76.6	76.6	0.0%	7.9%	7.9%	0.0%
Program 33	10	178	Auxiliary Power Plant	Dollars 90.7	90.7	-0.1%	3.9%	3.9%	-0.1%
Program 33	10	178	Electronic Warfare (1)	Dollars 775.5	776.1	-0.1%	6.5%	6.5%	-0.1%
Program 33	10	178	Electronic Warfare (2)	Dollars 360.1	273.4	24.1%	58.3%	46.0%	21.2%
Program 33	10	178	Electronic Warfare (3)	Dollars 62.5	62.4	0.2%	5.7%	5.7%	0.1%
Program 33	10	178	Empennage	Dollars 352.2	352.3	0.0%	5.1%	5.1%	-0.1%
Program 33	10	178	Hydraulic	Dollars 22.7	22.7	-0.1%	2.2%	2.2%	-0.1%
Program 33	10	178	Wing	Dollars 296.5	296.9	-0.1%	2.3%	2.3%	-0.1%
Program 34	6	67	PME–Air Vehicle	Dollars 11,059.1	11,061.2	0.0%	6.6%	6.6%	0.0%
Program 34	6	67	PME–Air Vehicle	Hours 9,058.6	9,061.7	0.0%	4.4%	4.4%	0.0%
Program 34	6	67	Airframe	Dollars 2,798.1	2,004.6	28.4%	2.8%	1.7%	37.9%
Program 34	6	201	Body	Dollars 1,924.5	828.9	56.9%	19.0%	8.7%	54.0%
Program 34	6	67	Alighting Gear	Dollars 316.5	166.9	47.3%	17.2%	8.3%	51.9%
Program 34	6	67	Electrical	Dollars 50.7	50.7	-0.1%	1.9%	1.9%	-0.1%
Program 34	6	67	Electronic Warfare (1)	Dollars 428.3	428.4	0.0%	5.3%	5.3%	0.0%
Program 34	5	49	Empennage	Dollars 202.2	202.2	0.0%	4.1%	4.1%	0.0%
Program 34	6	67	EO/IR	Dollars 45.6	36.6	19.7%	1.2%	1.1%	13.1%
Program 34	6	67	EOTS	Dollars 347.6	347.7	0.0%	6.5%	6.5%	0.0%
Program 34	6	67	Hydraulic	Dollars 122.3	101.5	17.0%	8.4%	6.2%	26.8%
Program 34	6	67	Mission Computer (1)	Dollars 484.8	484.9	0.0%	0.9%	0.9%	-0.2%
Program 34	6	67	Surface Controls	Dollars 196.0	196.0	0.0%	4.9%	4.9%	0.0%
Program 34	6	67	Wing	Dollars 998.4	998.6	0.0%	3.3%	3.3%	-0.1%

(*Continued*)

TABLE 6B.1 (*Continued*)
Error Comparison Using Cumulative Average Theory

Program	Number of Lots	Number of Units	Component Estimated	Traditional Units RMSE	RMSE Boone	Boone Percentage RMSE Difference	Traditional MAPE	Boone MAPE	MAPE Percentage Difference
Program 35	5	41	PME–Air Vehicle	Dollars 3,578.6	3,579.8	0.0%	1.5%	1.5%	0.0%
Program 35	5	41	PME–Air Vehicle	Hours 2,003.7	2,004.7	0.0%	1.1%	1.1%	0.0%
Program 35	5	50	Airframe	Dollars 609.3	610.4	−0.2%	0.6%	0.6%	−0.3%
Program 35	5	150	Body	Dollars 235.8	156.5	33.6%	1.9%	1.4%	28.0%
Program 35	5	50	Alighting Gear	Dollars 13.2	13.2	−0.1%	0.5%	0.5%	0.0%
Program 35	5	50	Electronic Warfare (1)	Dollars 259.6	259.7	0.0%	3.2%	3.2%	0.0%
Program 35	5	50	EO/IR	Dollars 121.6	121.7	0.0%	1.3%	1.3%	−0.1%
Program 35	5	50	EOTS	Dollars 177.9	177.9	0.0%	2.8%	2.8%	−0.1%
Program 35	5	50	Hydraulic	Dollars 58.2	58.2	0.0%	3.1%	3.1%	0.0%
Program 35	5	50	Radar	Dollars 256.8	256.9	0.0%	3.2%	3.2%	0.0%
Program 35	5	50	Surface Controls	Dollars 121.5	121.5	0.0%	2.6%	2.6%	0.0%
Program 35	5	50	Wing	Dollars 1,213.5	1,213.6	0.0%	3.8%	3.8%	0.0%
Program 36	13	1,285	PME–Air Vehicle	Dollars 28.8	29.4	−2.1%	0.6%	0.6%	−2.2%
Program 37	6	432	PME–Air Vehicle	Dollars 791.3	793.8	−0.3%	3.4%	3.4%	−0.4%
Program 38	6	52	PME–Air Vehicle	Dollars 253.6	154.9	38.9%	1.2%	0.7%	41.6%
Program 38	6	44	PME–Air Vehicle	Hours 831.5	614.2	26.1%	1.3%	0.8%	42.8%
Program 39	19	1,023	PME–Air Vehicle	Dollars 19.3	19.3	−0.2%	0.7%	0.7%	−0.2%
Program 40	5	1,725	PME–Air Vehicle	Dollars 19.2	0.6	96.7%	2.0%	0.1%	97.0%
Program 41	10	16	PME–Air Vehicle	Dollars 14,787.6	14,787.8	0.0%	5.2%	5.2%	0.0%

(*Continued*)

TABLE 6B.1 (*Continued*)
Error Comparison Using Cumulative Average Theory

Program	Number of Lots	Number of Units	Component Estimated	Traditional Units RMSE	RMSE Boone	RMSE Percentage Difference	Traditional MAPE	Boone MAPE	MAPE Percentage Difference
Program 41	10	16	Data Link (1)	Dollars 138.8	138.8	0.0%	3.7%	3.7%	0.0%
Program 42	11	203	PME–Air Vehicle	Dollars 1,000.0	1,000.1	0.0%	7.0%	7.0%	0.0%
Program 42	11	899	Electronic Warfare (1)	Dollars 67.5	67.7	−0.2%	13.9%	13.9%	−0.5%
Program 43	11	203	PME–Air Vehicle	Dollars 1,121.7	1,121.9	0.0%	5.5%	5.5%	0.0%
Program 43	13	251	PME–Air Vehicle	Hours 1,944.2	1,762.2	9.4%	3.4%	3.2%	6.1%
Program 44	5	136	PME–Air Vehicle	Dollars 57.1	16.3	71.4%	1.1%	0.3%	71.4%
Program 45	9	155	PME–Air Vehicle	Dollars 149.6	149.7	−0.1%	0.3%	0.3%	−0.1%
Program 46	6	68	PME–Air Vehicle	Dollars 3,435.9	3,436.0	0.0%	1.7%	1.7%	0.1%
Program 46	6	68	PME–Air Vehicle	Hours 2,286.4	2,286.6	0.0%	2.6%	2.6%	0.0%
Program 46	6	68	Airframe	Dollars 539.1	527.6	2.1%	2.3%	2.1%	10.9%
Program 46	6	68	Data Link (1)	Dollars 44.0	44.0	0.0%	3.0%	3.0%	0.0%
Program 46	6	68	Electronic Warfare (1)	Dollars 221.8	221.9	0.0%	5.4%	5.4%	0.0%
Program 46	6	68	Electronic Warfare (2)	Dollars 220.0	220.0	0.0%	6.5%	6.5%	0.0%
Program 46	6	68	Electronic Warfare (3)	Dollars 17.7	8.8	50.4%	2.2%	1.0%	54.6%
Program 46	6	68	Electronic Warfare (4)	Dollars 530.0	530.0	0.0%	5.2%	5.2%	0.0%
Program 46	6	68	EO/IR	Dollars 120.7	120.8	0.0%	15.7%	15.7%	0.0%
Program 46	6	68	Mission Computer (1)	Dollars 477.9	478.0	0.0%	4.3%	4.3%	0.0%
Program 47	9	36	PME–Air Vehicle	Dollars 1,039.4	1,039.4	0.0%	2.5%	2.5%	0.0%
Program 47	9	36	PME–Air Vehicle	Hours 8,278.7	8,278.6	0.0%	15.5%	15.5%	0.0%
Program 47	9	36	Data Link (1)	Dollars 170.2	170.2	0.0%	17.7%	17.7%	0.0%
Program 48	5	179	PME–Air Vehicle	Dollars 1,858.3	391.3	78.9%	3.1%	0.6%	79.4%
Program 49	6	180	PME–Air Vehicle	Dollars 435.3	99.8	77.1%	4.4%	1.0%	76.5%

(*Continued*)

TABLE 6B.1 (*Continued*)
Error Comparison Using Cumulative Average Theory

Program	Number of Lots	Number of Units	Component Estimated	Traditional Units RMSE	RMSE	Boone Percentage Difference	Traditional MAPE	Boone MAPE	MAPE Percentage Difference
Program 50	5	488	PME–Air Vehicle	Dollars 349.3	350.7	−0.4%	3.3%	3.4%	−0.8%
Program 51	6	663	PME–Air Vehicle	Dollars 5.6	3.6	36.6%	0.6%	0.4%	24.8%
Program 52	5	380	PME–Air Vehicle	Dollars 456.9	454.6	0.5%	9.0%	8.9%	0.3%
Program 53	6	749	PME–Air Vehicle	Dollars 37.2	36.6	1.7%	0.5%	0.5%	4.3%
Program 54	8	194	PME–Air Vehicle	Dollars 28.8	28.8	−0.1%	0.6%	0.6%	−0.1%
Program 55	9	677	PME–Air Vehicle	Dollars 74.8	74.8	0.0%	1.6%	1.6%	0.0%
Program 56	5	590	PME–Air Vehicle	Dollars 6.6	6.6	0.5%	0.2%	0.2%	6.3%
Program 57	5	579	PME–Air Vehicle	Dollars 22.8	22.8	−0.1%	0.8%	0.8%	0.0%

TABLE 6B.2

Error Comparison Using Unit Theory

Program	Number of Lots	Number of Units	Component Estimated	Traditional Units RMSE	Boone RMSE	RMSE Percentage Difference	Traditional MAPE	Boone MAPE	MAPE Percentage Difference
Program 1	7	503	Airframe	Hours 4.6	3.5	23.4%	7.1%	5.0%	28.7%
Program 1	6	483	PME–Air Vehicle	Hours 5.4	1.5	72.5%	11.3%	2.9%	74.0%
Program 1	7	503	PME–Air Vehicle	Dollars 2,260.6	517.0	77.1%	12.9%	3.2%	75.2%
Program 1	7	503	Airframe	Dollars 2,383.2	857.9	64.0%	14.6%	4.9%	66.4%
Program 2	5	638	PME–Air Vehicle	Dollars 315.4	195.3	38.1%	5.8%	4.3%	26.3%
Program 3	5	500	PME–Air Vehicle	Dollars 2,984.5	1,120.2	62.5%	49.4%	17.6%	64.4%
Program 4	7	357	Airframe	Dollars 2,662.2	2,664.3	−0.1%	13.1%	13.2%	−0.1%
Program 4	9	424	PME–Air Vehicle	Dollars 9,323.3	4,999.8	46.4%	37.9%	14.1%	62.8%
Program 5	19	205	Airframe	Dollars 2,446.1	2,445.8	0.0%	12.6%	12.6%	−0.3%
Program 5	19	205	PME–Air Vehicle	Dollars 3,228.6	3,228.9	0.0%	12.4%	12.4%	0.0%
Program 6	7	459	Electronic Warfare (1)	Dollars 20.9	20.9	0.0%	30.8%	30.8%	0.0%
Program 6	7	459	PME–Air Vehicle	Dollars 1,439.9	738.1	48.7%	11.3%	5.9%	47.2%
Program 7	5	321	PME–Air Vehicle	Dollars 37.9	33.3	12.2%	3.8%	3.8%	1.1%
Program 8	6	98	Electronic Warfare (3)	Dollars 5.2	4.9	6.1%	4.8%	4.8%	1.4%
Program 8	6	98	Electronic Warfare (2)	Dollars 84.2	70.3	16.5%	11.1%	10.6%	4.7%
Program 8	6	98	PME–Air Vehicle	Dollars 375.2	339.5	9.5%	4.2%	3.7%	13.4%
Program 8	6	98	Electronic Warfare (1)	Dollars 27.5	18.7	31.9%	10.2%	5.9%	42.5%
Program 9	7	110	Electronic Warfare (5)	Dollars 102.9	99.2	3.5%	9.7%	10.4%	−6.6%
Program 9	7	110	Electronic Warfare (3)	Dollars 6.4	6.4	0.0%	4.7%	4.7%	0.0%
Program 9	7	110	Electronic Warfare (4)	Dollars 653.6	653.6	0.0%	6.2%	6.2%	0.0%
Program 9	7	110	Electronic Warfare (2)	Dollars 709.4	709.4	0.0%	6.1%	6.1%	0.0%

(Continued)

TABLE 6B.2 (Continued)
Error Comparison Using Unit Theory

Program	Number of Lots	Number of Units	Component Estimated	Traditional Units RMSE	Boone RMSE	RMSE Percentage Difference	Traditional MAPE	Boone MAPE	MAPE Percentage Difference
Program 9	7	110	PME–Air Vehicle	Dollars 668.5	668.5	0.0%	5.1%	5.1%	0.0%
Program 9	7	110	Electronic Warfare (1)	Dollars 31.6	29.1	8.0%	8.7%	8.0%	8.3%
Program 10	9	1,586	PME–Air Vehicle	Dollars 115.5	115.6	–0.2%	12.5%	12.5%	–0.2%
Program 10	10	1,796	PME–Air Vehicle	Hours 150.8	150.9	0.0%	12.5%	12.5%	–0.1%
Program 11	8	3,529	PME–Air Vehicle	Hours 0.9	0.7	21.2%	27.5%	44.9%	–63.4%
Program 11	8	3,529	PME–Air Vehicle	Dollars 97.1	97.5	–0.4%	10.1%	10.4%	–2.1%
Program 12	16	7,891	PME–Air Vehicle	Hours 520.1	525.6	–1.1%	86.2%	86.2%	0.0%
Program 12	21	1,0035	PME–Air Vehicle	Dollars 243.8	239.2	1.9%	30.1%	28.8%	4.2%
Program 13	6	3,385	EO	Dollars 12.1	9.4	22.5%	10.7%	9.6%	10.0%
Program 13	10	3,803	PME–Air Vehicle	Dollars 33.6	24.8	26.1%	10.3%	7.5%	27.1%
Program 13	10	3,803	PME–Air Vehicle	Hours 130.1	100.5	22.7%	21.5%	17.1%	20.7%
Program 14	6	180	PME–Air Vehicle	Dollars 2,249.4	1,008.9	55.2%	6.4%	2.3%	64.2%
Program 15	10	20	PME–Air Vehicle	Hours 3,430.3	3,430.4	0.0%	41.5%	41.5%	0.0%
Program 15	10	20	PME–Air Vehicle	Dollars 3,013.9	3,013.9	0.0%	17.4%	17.4%	0.0%
Program 15	7	11	Mission Computer (1)	Dollars 213.9	213.9	0.0%	11.6%	11.5%	0.6%
Program 16	5	100	Airframe	Dollars 10,807.3	7,455.4	31.0%	7.0%	4.1%	41.8%
Program 16	5	100	PME–Air Vehicle	Dollars 137,225.9	81,884.9	40.3%	51.7%	26.9%	48.0%
Program 17	5	275	PME–Air Vehicle	Dollars 8,837.5	1,396.3	84.2%	17.6%	3.3%	81.6%
Program 18	12	83	PME–Air Vehicle	Hours 266,012.8	266,015.3	0.00%	39.3%	39.3%	0.0%
Program 18	11	83	Airframe	Dollars 89,956.0	89,961.1	0.00%	39.1%	39.1%	0.0%
Program 18	10	68	Mission Computer (1)	Dollars 4,143.0	4,143.2	0.00%	68.2%	68.2%	0.0%
Program 18	11	83	PME–Air Vehicle	Dollars 82,138.6	82,143.3	0.00%	23.2%	23.2%	0.0%
Program 19	5	45	Airframe	Dollars 501.2	501.2	0.00%	53.9%	53.9%	0.0%

(Continued)

TABLE 6B.2 (Continued)
Error Comparison Using Unit Theory

Program	Number of Lots	Number of Units	Component Estimated	Traditional Units RMSE	Boone RMSE	RMSE Percentage Difference	Traditional MAPE	Boone MAPE	MAPE Percentage Difference
Program 19	5	45	PME–Air Vehicle	Dollars 649.0	649.0	0.0%	17.6%	17.6%	0.0%
Program 19	5	45	Mission Computer (1)	Dollars 61.7	59.7	3.20%	9.8%	9.7%	1.2%
Program 20	9	76	PME–Air Vehicle	Dollars 1,108.7	522.5	52.90%	7.2%	3.6%	49.9%
Program 21	5	50	PME–Air Vehicle	Dollars 24,625.3	6,362.0	74.20%	7.4%	2.3%	69.5%
Program 22	9	31	PME–Air Vehicle	Dollars 16,636.3	16,636.4	0.00%	6.6%	6.6%	0.0%
Program 23	5	14	PME–Air Vehicle	Dollars 14,475.8	14,476.0	0.00%	8.7%	8.7%	0.0%
Program 24	6	98	PME–Air Vehicle	Dollars 2,259.9	2,260.1	0.00%	3.3%	3.3%	0.0%
Program 25	7	59	Electronic Warfare (5)	Dollars 2,808.4	2,805.2	0.10%	14.8%	15.4%	-4.0%
Program 25	11	84	PME–Air Vehicle	Dollars 5,083.2	4,228.8	16.80%	8.7%	9.2%	-5.2%
Program 25	11	84	Electronic Warfare (2)	Dollars 248.9	248.6	0.10%	13.9%	14.3%	-2.9%
Program 25	7	59	Electronic Warfare (1)	Dollars 1,259.1	653.3	48.10%	16.1%	7.1%	55.6%
Program 26	7	344	Airframe	Dollars 11,474.7	8,294.90	27.70%	22.7%	21.5%	5.3%
Program 26	7	344	Avionics	Dollars 2,218.8	2,102.80	5.20%	29.5%	26.9%	8.8%
Program 26	7	344	PME–Air Vehicle	Dollars 12,898.4	8,742.10	32.20%	20.7%	16.9%	18.4%
Program 27	14	453	PME–Air Vehicle	Hours 54,142.9	53,766.40	0.70%	59.9%	63.1%	-5.4%
Program 27	14	453	Airframe	Hours 70,415.0	69,426.80	1.40%	58.8%	59.1%	-0.5%
Program 28	8	538	PME–Air Vehicle	Hours 3,828.8	3,829.80	0.00%	9.8%	9.9%	0.0%
Program 28	8	538	Airframe	Hours 3,865.3	3,866.20	0.00%	7.6%	7.6%	0.0%
Program 29	8	529	Hydraulic	Dollars 156.9	156.4	0.30%	22.3%	22.9%	-2.8%
Program 29	12	477	Airframe	Dollars 6,490.2	5,974.20	7.90%	14.2%	14.4%	-1.8%
Program 29	12	477	Wing	Dollars 712.3	712.7	-0.1%	27.8%	27.8%	-0.1%
Program 29	11	433	Electronic Warfare (1)	Dollars 57.5	57.5	0.00%	13.5%	13.5%	-0.1%

(Continued)

TABLE 6B.2 (Continued)

Error Comparison Using Unit Theory

Program	Number of Lots	Number of Units	Component Estimated	Traditional Units RMSE	Boone RMSE	RMSE Percentage Difference	Traditional MAPE	Boone MAPE	MAPE Percentage Difference
Program 29	8	309	Electrical	Dollars 230.6	230.7	–0.1%	8.2%	8.2%	0.0%
Program 29	12	1,045	Body	Dollars 1,922.2	1,826.70	5.00%	26.0%	25.9%	0.7%
Program 29	5	177	Empennage	Dollars 32.3	22	31.80%	6.1%	4.6%	24.5%
Program 29	12	477	PME–Air Vehicle	Dollars 8,218.5	5,525.30	32.80%	15.0%	10.2%	32.0%
Program 29	8	309	Alighting Gear	Dollars 205.7	42.2	79.50%	11.6%	2.0%	83.1%
Program 30	5	469	PME–Air Vehicle	Dollars 1,283.8	891.8	30.50%	13.5%	8.3%	38.3%
Program 31	10	59	PME–Air Vehicle	Dollars 11,978.9	11,979.30	0.00%	8.6%	8.6%	0.0%
Program 32	9	348	PME–Air Vehicle	Dollars 430.6	430.8	0.00%	3.5%	3.5%	–0.1%
Program 33	5	109	PME–Air Vehicle	Hours 993.9	994	0.00%	9.5%	9.5%	0.0%
Program 33	5	109	PME–Air Vehicle	Dollars 6,824.7	6,824.80	0.00%	28.2%	28.2%	0.0%
Program 34	18	631	PME–Air Vehicle	Dollars 6,926.7	2,799.90	59.60%	17.0%	6.6%	61.0%
Program 35	6	425	PME–Air Vehicle	Dollars 1,135.8	1,137.50	–0.2%	3.5%	3.5%	–0.2%
Program 35	7	522	PME–Air Vehicle	Hours 4,615.3	4,458.50	3.40%	6.3%	6.1%	3.1%
Program 35	7	522	Airframe	Hours 6,757.0	6,280.70	7.00%	5.7%	5.4%	4.8%
Program 36	9	358	PME–Air Vehicle	Hours 5,118.7	5,120.10	0.00%	6.8%	6.8%	0.0%
Program 36	9	358	Airframe	Hours 12,155.2	11,257.1	7.40%	15.50%	14.30%	7.6%
Program 37	5	204	PME–Air Vehicle	Dollars 1,468.7	921	37.30%	2.90%	1.90%	36.4%
Program 38	5	605	PME–Air Vehicle	Dollars 2,641.9	1,527.70	42.20%	14.90%	8.10%	46.0%
Program 39	5	870	PME–Air Vehicle	Dollars 310.9	311.5	–0.2%	2.30%	2.30%	–0.2%
Program 40	10	178	Electronic Warfare (3)	Dollars 751.2	551.9	26.50%	69.70%	74.70%	–7.1%
Program 40	10	712	Body	Dollars 617.6	577.6	6.50%	4.80%	5.10%	–7.6%
Program 40	10	178	Airframe	Dollars 4,251.9	4,226.40	0.60%	4.80%	4.90%	–1.0%

(Continued)

TABLE 6B.2 (*Continued*)
Error Comparison Using Unit Theory

Program	Number of Lots	Number of Units	Component Estimated	Traditional Units RMSE	Boone RMSE	RMSE Percentage Difference	Traditional MAPE	Boone MAPE	MAPE Percentage Difference
Program 40	10	178	Electronic Warfare (2)	Dollars 721.7	721.7	0.00%	393.40%	393.40%	0.0%
Program 40	10	178	Electronic Warfare (1)	Dollars 1,642.3	1,643.00	0.00%	20.70%	20.70%	0.0%
Program 40	10	178	PME–Air Vehicle	Hours 13,454.5	13,466.80	-0.1%	6.00%	6.00%	-0.1%
Program 40	10	178	Auxiliary Power Plant	Dollars 385.1	385.1	0.00%	24.9%	24.9%	0.0%
Program 40	10	178	PME–Air Vehicle	Dollars 12,231.7	12,236.60	0.00%	7.9%	7.9%	0.0%
Program 40	10	178	Alighting Gear	Dollars 233.6	233.6	0.00%	30.1%	30.1%	0.0%
Program 40	10	178	Wing	Dollars 607.4	607.6	0.00%	6.2%	6.2%	0.0%
Program 40	10	178	Empennage	Dollars 702.1	702.1	0.00%	17.4%	17.4%	0.0%
Program 40	10	178	Hydraulic	Dollars 72.2	70.2	2.80%	9.0%	8.8%	2.2%
Program 41	6	67	PME–Air Vehicle	Hours 12,741.5	12,743.80	0.00%	9.5%	9.5%	0.0%
Program 41	5	49	Empennage	Dollars 242.2	242.2	0.00%	5.8%	5.9%	0.0%
Program 41	6	67	PME–Air Vehicle	Dollars 16,643.9	16,645.60	0.00%	10.7%	10.7%	0.0%
Program 41	6	67	Surface Controls	Dollars 281.7	281.7	0.00%	7.7%	7.7%	0.0%
Program 41	6	67	EOTS	Dollars 442.3	442.4	0.00%	9.5%	9.5%	0.0%
Program 41	6	67	Wing	Dollars 1,927.0	1,927.30	0.00%	7.4%	7.4%	0.0%
Program 41	6	67	Electrical	Dollars 57.2	57.2	0.00%	2.1%	2.1%	0.0%
Program 41	6	67	Electronic Warfare (1)	Dollars 547.3	547.3	0.00%	8.1%	8.1%	0.0%
Program 41	6	67	Hydraulic	Dollars 281.5	274.6	2.40%	19.4%	19.0%	2.0%
Program 41	6	67	Mission Computer (1)	Dollars 1,698.1	1,542.40	9.20%	4.6%	3.7%	19.5%
Program 41	6	67	Airframe	Dollars 6,877.8	5,547.40	19.30%	8.7%	6.4%	26.8%
Program 41	6	67	Alighting Gear	Dollars 582.3	521.1	10.50%	28.3%	25.0%	11.6%
Program 41	6	67	EO/IR	Dollars 233.0	89.4	61.60%	9.3%	3.1%	66.8%

(*Continued*)

TABLE 6B.2 (Continued)
Error Comparison Using Unit Theory

Program	Number of Lots	Number of Units	Component Estimated	Traditional Units RMSE	Boone RMSE	RMSE Percentage Difference	Traditional MAPE	Boone MAPE	MAPE Percentage Difference
Program 41	6	201	Body	Dollars 3,431.8	2,343.20	31.70%	42.6%	29.9%	29.7%
Program 42	5	41	PME–Air Vehicle	Dollars 8,498.6	8,499.60	0.00%	6.2%	6.2%	0.0%
Program 42	5	41	PME–Air Vehicle	Hours 15,696.5	15,696.90	0.00%	10.7%	10.7%	0.0%
Program 42	5	50	EOTS	Dollars 593.3	593.3	0.00%	11.6%	11.6%	0.0%
Program 42	5	50	EO/IR	Dollars 578.4	578.4	0.00%	7.5%	7.5%	0.0%
Program 42	5	50	Hydraulic	Dollars 297.0	297	0.00%	15.4%	15.4%	0.0%
Program 42	5	50	Surface Controls	Dollars 424.9	424.9	0.00%	11.0%	11.0%	0.0%
Program 42	5	50	Radar	Dollars 733.8	733.8	0.00%	10.9%	10.9%	0.0%
Program 42	5	50	Airframe	Dollars 5,222.7	5,222.80	0.00%	5.9%	5.9%	0.0%
Program 42	5	50	Electronic Warfare (1)	Dollars 746.5	746.5	0.00%	10.7%	10.7%	0.0%
Program 42	5	50	Wing	Dollars 3,726.6	3,726.70	0.00%	16.5%	16.5%	0.0%
Program 42	5	50	Alighting Gear	Dollars 78.6	77.4	1.50%	3.6%	3.5%	2.3%
Program 42	5	150	Body	Dollars 1,588.5	892.1	43.80%	12.6%	8.7%	30.8%
Program 43	13	1,285	PME–Air Vehicle	Dollars 88.1	88.8	−0.8%	1.9%	1.9%	−1.0%
Program 44	6	432	PME–Air Vehicle	Dollars 1,621.0	1,623.30	−0.1%	10.0%	10.0%	−0.2%
Program 45	9	63	PME–Air Vehicle	Dollars 2,152.3	1,557.10	27.70%	9.5%	6.4%	33.2%
Program 46	6	44	PME–Air Vehicle	Hours 7,736.9	7,255.30	6.20%	17.6%	16.7%	4.8%
Program 46	10	113	PME–Air Vehicle	Dollars 797.9	627	21.40%	3.8%	2.9%	22.7%
Program 47	19	1,023	PME–Air Vehicle	Dollars 115.2	115.2	0.00%	4.3%	4.2%	0.2%
Program 48	5	1,725	PME–Air Vehicle	Dollars 59.8	3.1	94.90%	6.8%	0.3%	95.4%
Program 49	10	16	Data Link (1)	Dollars 470.3	470.3	0.00%	20.4%	20.4%	0.0%
Program 49	10	16	PME–Air Vehicle	Dollars 41,008.9	41,009.2	0.00%	14.1%	14.1%	0.0%

(Continued)

TABLE 6B.2 (Continued)
Error Comparison Using Unit Theory

Program	Number of Lots	Number of Units	Component Estimated	Traditional Units RMSE	Boone RMSE	RMSE Percentage Difference	Traditional MAPE	Boone MAPE	MAPE Percentage Difference
Program 50	7	577	PME–Air Vehicle	Dollars 1,674.7	1,224.7	26.90%	5.5%	4.6%	15.7%
Program 51	12	244	PME–Air Vehicle	Hours 625.6	612.8	2.00%	191.4%	191.8%	−0.2%
Program 52	11	899	Electronic Warfare (1)	Dollars 90.1	90.2	−0.1%	29.2%	29.3%	−0.1%
Program 52	11	203	PME–Air Vehicle	Dollars 2,995.1	2,992.0	0.10%	24.9%	23.6%	5.2%
Program 53	13	251	PME–Air Vehicle	Hours 4,585.2	4,585.2	0.00%	6.7%	6.7%	0.0%
Program 53	11	203	PME–Air Vehicle	Dollars 2,459.9	2,460.0	0.00%	9.6%	9.6%	0.0%
Program 54	11	184	PME–Air Vehicle	Hours 7,010.4	7,010.7	0.00%	18.0%	18.0%	0.0%
Program 54	9	134	PME–Air Vehicle	Dollars 1,907.3	970.0	49.10%	11.8%	6.5%	44.9%
Program 55	5	136	PME–Air Vehicle	Dollars 321.6	277.7	13.70%	5.5%	4.7%	14.8%
Program 56	9	155	PME–Air Vehicle	Dollars 1,356.5	1,356.6	0.00%	3.9%	3.9%	0.0%
Program 57	6	68	EO/IR	Dollars 326.0	326.0	0.00%	1,261.8%	1,261.8%	0.0%
Program 57	6	68	PME–Air Vehicle	Dollars 8,574.7	8,470.9	1.20%	4.3%	4.3%	−0.5%
Program 57	6	68	Electronic Warfare (1)	Dollars 998.8	998.9	0.00%	58.9%	58.9%	0.0%
Program 57	6	68	Electronic Warfare (2)	Dollars 750.2	750.2	0.00%	31.3%	31.3%	0.0%
Program 57	6	68	Data Link (1)	Dollars 94.8	94.8	0.00%	7.2%	7.2%	0.0%
Program 57	6	68	Electronic Warfare (4)	Dollars 1,156.3	1,156.3	0.00%	12.2%	12.2%	0.0%
Program 57	6	68	Mission Computer (1)	Dollars 1,030.6	1,030.6	0.00%	13.0%	13.0%	0.0%
Program 57	6	68	PME–Air Vehicle	Hours 6,435.9	6,435.0	0.00%	12.3%	12.3%	0.3%
Program 57	6	68	Airframe	Dollars 1,443.2	1,285.1	11.00%	6.7%	5.4%	18.5%
Program 57	6	68	Electronic Warfare (3)	Dollars 53.4	21.8	59.10%	7.2%	3.0%	58.5%
Program 58	9	36	PME–Air Vehicle	Hours 60,347.2	60,347.3	0.00%	78.2%	78.2%	0.0%
Program 58	9	36	Data Link (1)	Dollars 227.8	227.8	0.00%	29.3%	29.3%	0.0%
Program 58	9	36	PME–Air Vehicle	Dollars 4,570.2	4,570.2	0.00%	10.9%	10.9%	0.0%

(Continued)

TABLE 6B.2 (*Continued*)
Error Comparison Using Unit Theory

Program	Number of Lots	Number of Units	Component Estimated	Traditional Units RMSE	Boone RMSE	RMSE Percentage Difference	Traditional MAPE	Boone MAPE	MAPE Percentage Difference
Program 58	5	18	EO/IR	Dollars 3,488.4	3,469.8	0.50%	28.8%	28.7%	0.3%
Program 59	5	179	PME–Air Vehicle	Dollars 4,583.3	1,334.5	70.90%	8.1%	2.8%	65.4%
Program 60	6	180	PME–Air Vehicle	Dollars 1,010.5	333.9	67.00%	12.4%	4.6%	63.1%
Program 61	5	488	PME–Air Vehicle	Dollars 502.3	486.5	3.10%	9.2%	7.7%	16.3%
Program 62	6	78	PME–Air Vehicle	Hours 6,027.1	5,952.3	1.20%	33.8%	34.3%	-1.6%
Program 62	6	97	Airframe	Hours 2,648.5	2,649.0	0.00%	20.5%	20.5%	0.0%
Program 62	9	110	PME–Air Vehicle	Dollars 13,027.5	13,028.9	0.00%	24.0%	24.0%	0.0%
Program 63	6	663	PME–Air Vehicle	Dollars 23.2	21.1	9.20%	2.9%	2.6%	11.6%
Program 64	5	380	PME–Air Vehicle	Dollars 1,520.9	1,521.2	0.00%	57.4%	57.4%	0.0%
Program 65	6	749	PME–Air Vehicle	Dollars 116.6	115.9	0.60%	1.7%	1.8%	-5.1%
Program 66	8	194	PME–Air Vehicle	Dollars 128.3	119.3	7.00%	2.6%	2.4%	8.6%
Program 67	9	677	PME–Air Vehicle	Dollars 273.5	273.5	0.00%	5.1%	5.1%	0.0%
Program 68	5	590	PME–Air Vehicle	Dollars 87.1	87.2	0.00%	2.8%	2.8%	0.0%
Program 69	5	579	PME–Air Vehicle	Dollars 305.7	305.8	0.00%	9.5%	9.5%	0.0%

REFERENCES

1. United States Government Accountability Office; Oakley, S.S. Weapon Systems Annual Assessment: Knowledge Gaps Pose Risks to Sustaining Recent Positive Trends: Report to Congressional Committees; United States Government Accountability Office: Washington, DC, 2018.
2. Wright, T.P. Factors affecting the cost of airplanes. *J. Aeronaut. Sci.* **1936**, *3*, 122–128.
3. Asher, H. Cost-Quantity Relationships in the Airframe Industry. Ph.D. Thesis, The Ohio State University, Columbus, OH, 1956.
4. Boone, E.R.; Elshaw, J.J.; Koschnick, C.M.; Ritschel, J.D.; Badiru, A.B. A Learning curve model accounting for the flattening effect in production cycles. *Def. Acquis. Res. J.* **2021**. in-print.
5. Mislick, G.K.; Nussbaum, D.A. *Cost Estimation: Methods and Tools*; John Wiley & Sons: Hoboken, NJ, 2015.
6. Argote, L.; Beckman, S.L.; Epple, D. The persistence and transfer of learning in industrial settings. *Manag. Sci.* **1990**, *36*, 140–154.
7. Argote, L. Group and organizational learning curves: Individual, system and environmental components. *Br. J. Soc. Psychol.* **1993**, *32*, 31–51.
8. Jaber, M.Y. Learning and forgetting models and their applications. *Handb. Ind. Syst. Eng.* **2006**, *30*, 30–127.
9. Glock, C.H.; Grosse, E.H.; Jaber, M.Y.; Smunt, T.L. Applications of learning curves in production and operations management: A systematic literature review. *Comput. Ind. Eng.* **2019**, *131*, 422–441.
10. Jaber, M.Y.; Bonney, M. Production breaks and the learning curve: The forgetting phenomenon. *Appl. Math. Model.* **1996**, *2*, 162–169.
11. Nembhard, D.A.; Uzumeri, M.V. Experiential learning and forgetting for manual and cognitive tasks. *Int. J. Ind. Ergon.* **2000**, *25*, 315–326.
12. Sikström, S.; Jaber, M.Y. The Depletion–power–integration–latency (DPIL) model of spaced and massed repetition. *Comput. Ind. Eng.* **2012**, *63*, 323–337.
13. Anzanello, M.J.; Fogliatto, F.S. Learning curve models and applications: Literature review and research directions. *Int. J. Ind. Ergon.* **2011**, *41*, 573–583.
14. Crossman, E.R. A Theory of the acquisition of speed-skill*. *Ergonomics* **1959**, *2*, 153–166.
15. Moore, J.R.; Elshaw, J.J.; Badiru, A.B.; Ritschel, J.D. *Acquisition Challenge: The Importance of Incompressibility in Comparing Learning Curve Models*; US Air Force Cost Analysis Agency: Arlington, TX, 2015.
16. Honious, C.; Johnson, B.; Elshaw, J.; Badiru, A. *The Impact of Learning Curve Model Selection and Criteria for Cost Estimation Accuracy in the DoD*; Air Force Institute of Technology: Wright Patterson AFB, OH, 2016.
17. Baloff, N. Startups in Machine-intensive production systems. *J. Ind. Eng.* **1966**, *17*, 25.
18. Baloff, N. Startup management. *IEEE Trans. Eng. Manag.* **1970**, *4*, 132–141.
19. Corlett, E.N.; Morecombe, V.J. Straightening out learning curves. *Pers. Manag.* **1970**, *2*, 14–19.
20. Yelle, L.E. The learning curve: Historical review and comprehensive survey. *Decis. Sci.* **1979**, *10*, 302–328.
21. Hirschmann, W.B. Profit from the learning-curve. *Harv. Bus. Rev.* **1964**, *42*, 125–139.
22. Li, G.; Rajagopalan, S. A Learning curve model with knowledge depreciation. *Eur. J. Oper. Res.* **1998**, *105*, 143–154.
23. Chalmers, G.; DeCarteret, N. *Relationship for Determining the Optimum Expansibility of the Elements of a Peacetime Aircraft Procurement Program*; Stanford Research Institute: Menlo Park, CA, 1949.

24. De Jong, J.R. The effects of increasing skill on cycle time and its consequences for time standards. *Ergonomics* **1957**, *1*, 51–60.
25. Knecht, G.R. Costing, technological growth and generalized learning curves. *J. Oper. Res. Soc.* **1974**, *25*, 487–491.
26. Badiru, A.B. Computational survey of univariate and multivariate learning curve models. *IEEE Trans. Eng. Manag.* **1992**, *39*, 176–188.
27. Office of the Secretary of Defense. 1921-1 Data Item Description. Available online: https://cade.osd.mil/ content/cade/files/csdr/dids/archive/1921-1.DI-FNCL-81566B.pdf (accessed on 29 September 2020).
28. Hu, S.-P.; Smith, A. Accuracy matters: Selecting a lot-based cost improvement curve. *J. Cost Anal. Parametr.* **2013**, *6*, 23–42.
29. Cohen, J. Quantitative methods in psychology. *Nature* **1938**, *141*, 613.
30. Badiru, A.B. Half-life learning curves in the defense acquisition life cycle. *Def. Acquis. Res. J.* **2012**, *19*, 283–308.

7 A Learning Curve Model Accounting for the Flattening Effect in Production Cycles*

Evan R. Boone
US Air Force Materiel Command

John J. Elshaw, Clay M. Koschnick,
Jonathan D. Ritschel, and Adedeji B. Badiru
US Air Force Institute of Technology

CONTENTS

Many manufacturing firms today operate in a fiscally constrained and financially conscious environment. Managers throughout these organizations are expected to maximize the utility from every dollar as budgets and profit margins continue to shrink. Increased financial scrutiny adds greater emphasis on the accuracy of program and project management cost estimates to ensure acquisition programs are sufficiently funded. Cost estimating models and tools used by organizations must be evaluated for their relevance and accuracy to ensure reliable cost estimates.

* Reprinted with permission: Elshaw, J. J., Ritschel, J. D., & Badiru, A. B. (2021). A Learning Curve Model Accounting for the Flattening Effect in Production Cycles. *Defense AR Journal*, 28(1), 72–97.

DOI: 10.1201/9781003220978-9

Many of the cost estimating procedures for learning curves were developed in the 1930s (Wright, 1936) and are still in use today as a primary method to model learning. As automation and robotics increasingly replace human touch- labor in the manufacturing process, the current 80-year-old learning curve model may no longer provide the most accurate approach for estimates. New learning curve methods that incorporate automated production and other factors that lead to reduced learning should be examined as an alternative for cost estimators in the acquisition process.

> Increased financial scrutiny adds greater emphasis on the accuracy of program and project management cost estimates to ensure acquisition programs are sufficiently funded.

Since Wright's (1936) original learning curve model was developed, researchers have found other functions to model learning within the manufacturing process (Carr, 1946; Chalmers & DeCarteret, 1949; Crawford, 1944; DeJong, 1957; Towill, 1990; Towill & Cherrington, 1994). The purpose of this research is to address a gap in the literature that fails to account for the nonconstant rate of learning, which results in a flattening effect at the end of production cycles. We will investigate the learning curve estimating methodology, develop learning curve theory, and pursue the development of a new estimation model that examines learning at a nonconstant rate.

This research identifies and models modifications to a learning curve model such that the estimated learning rate is modeled as a decreasing learning rate function over time, as opposed to a constant learning rate that is currently in use. Wright's (1936) learning curve model in use today mathematically states that for every doubling of units there will be a constant gain in efficiency. For example, if a manufacturer observes a 10% reduction in labor hours in the time to produce unit 10 from the time to produce unit 5, then it should expect to see the same 10% reduction in labor hours in the time to produce unit 20 from the time to produce unit 10. We propose that more accurate cost estimates would result if a more flexible exponent were taken into consideration in developing the learning curve model. The proposed general modification would take the form:

$$\text{Cost}(x) = Ax^{f(x)} \tag{7.1}$$

where:
    $\text{Cost}(x)$ = cumulative average cost per unit
    A = theoretical cost to produce the first unit
    x = cumulative number of units produced
    $f(x)$ = learning curve effect as a function of units produced

The exponent function in Eq. 7.1 will be explored in this article. Figure 7.1 demonstrates the phenomena this research will examine. The different shades of lines in the figure represent different aspects of the learning curves. Readers should refer to the original journal publication to see the color differentiation in the learning curve lines.

To address this research gap, our study aims to model a function that has the added precision of diminishing learning effects over time by introducing a learning

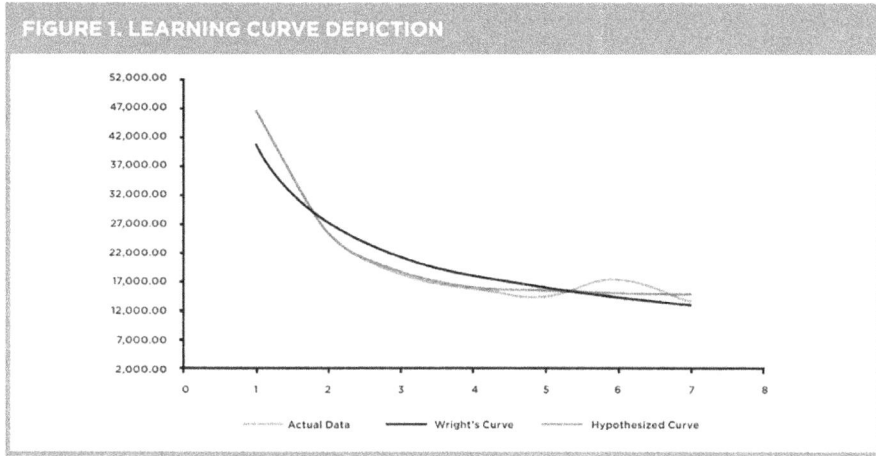

FIGURE 1. LEARNING CURVE DEPICTION

FIGURE 7.1   Learning curve depiction.

curve decay factor that more closely models actual production cycle learning. We will accomplish this by developing a new learning curve model that minimizes the amount of error compared to current estimation models. Learning curves, specifically when estimating the expected cost per unit of complex manufactured items such as aircraft, are frequently modeled with a mathematical power function. The intent of these models is to capture the expected reduction in costs over time due to learning effects, particularly in areas with a high percentage of human touch labor. Typically, as production increases, manufacturers identify labor efficiencies and improve the process. If labor efficiencies are identified, it translates to unit cost savings over time. The general form of the learning curve model frequently used today is based on Wright's theory and is shown in Eq. 7.2. Note that the structure of exponent b ensures that as the number of units produced doubles, the cost will decrease by a given percentage defined as the learning curve slope (LCS). For example, when LCS is 0.8, then the cost per unit will decrease by 80% between units 2 and 4.

$$\text{Cost}(x) = Ax^b \qquad (7.2)$$

where
   $\text{Cost}(x)$ = cumulative average cost per unit
   A = theoretical cost to produce the first unit
   X = cumulative number of units produced
   $b = \dfrac{\ln \text{Learning Curve Slope prior to additional work}}{\ln 2}$

The cost of a particular production unit is modeled as a power function that decreases at a constant exponential rate. The problem is that the rate of decrease is not likely to be constant over time. We propose that the majority of cost improvements are to be found early in the production process, and fewer revelations are made later as the manufacturer becomes more familiar with the process. As time progresses, the production process should normalize to a steady state and additional cost reductions prove less likely.

> Our study aims to model a function that has the added precision of diminishing learning effects over time by introducing a learning curve decay factor that more closely models actual production cycle learning.

For relatively short production runs, the basic form of the learning curve may be sufficient because the hypothesized efficiencies will not have time to materialize. However, when estimating production runs over longer periods of time, the basic learning curve could underestimate the unit costs of those furthermost in the future. The underestimation occurs because the model assumes a constant learning rate, while actual learning would diminish, causing the actuals to be higher than the estimate. Current models may underestimate a significant amount when dealing with high unit cost items such as those in major acquisition programs; a small error in an estimate can be large in terms of dollars. Through the use of curve fitting techniques, a comparison can be made to determine which model best predicts learning within a particular production process. The remainder of this article is organized as follows. A literature review of the most common learning curve processes is presented in the next section, followed by methodology and model formulation. We then provide an in-depth analysis of the learning curve models, followed by future research directions, conclusions, and limitations of this research.

7.1 LITERATURE REVIEW

Learning curve research dates back to 1936 when Theodore Paul Wright published the original learning curve equation that predicted the production effects of learning. Wright recognized the mathematical relationship that exists between the time it takes for a worker to complete a single task and the number of times the worker had previously performed that task (Wright, 1936). The mathematical relationship developed from this hypothesis is that as workers complete the same process, they get better at it. Specifically, Wright realized that the rate at which they get better at that task is constant.

The relationship between these two variables is as follows: as the number of units produced doubles, the worker will do it faster at a constant rate. He proposed that this relationship takes the form of:

$$F = N^x$$

Or

$$x = \frac{\log F}{\log N};$$

where F = a factor of cost variation proportional to the quantity N. The reciprocal of F then represents a direct percent variation of cost vs. quantity (Wright, 1936). The relationship between these variables can be modified to predict the expected cost of

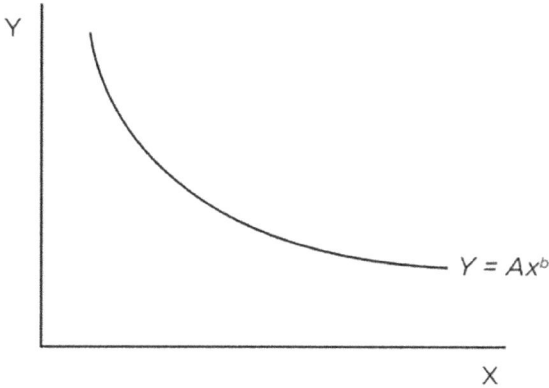

FIGURE 7.2 Wright's learning curve model (Martin, n.d.).

a given unit number in production by multiplying the factor of cost variation by the theoretical cost of the first unit produced—this relationship was stated in Eq. 7.2 and is shown in Figure 7.2. It is a log-linear relationship through an algebraic manipulation. The logarithmic form of this equation (taking the natural log of both sides of the equation) allows practitioners to run linear regression analysis on the data to find what slope best fits the data using a straight line (Martin, n.d.).

The goal of using learning curves is to increase the accuracy of cost estimates. Having accurate cost estimates allows an organization to efficiently budget while providing as much operational capability as possible because it can allocate resources to higher priorities. While the use of learning curves focuses on creating accurate cost estimates, learning curves often use the number of labor hours it takes to perform a task. When Wright originated the theory, he proposed the output in terms of time to produce, not production cost. However, many organizations perform learning curve analysis on both production cost and time to produce, depending on the data available. Nevertheless, labor-hour cost is relevant because it is based on factors such as labor rates and other associated values. The use of labor hours in learning curve development allows a common comparison over time without the effects of inflation convoluting the results. However, the same goal can be achieved by using inflation-adjusted cost values.

Wright's model has been compared to some of the more contemporary models that have surfaced in recent years since the original learning curve theory was established (Moore et al., 2015). Moore compared the Stanford-B, Dejong, and the S-Curve models to Wright's model to see if any of these functions could provide a more accurate estimate of the learning phenomenon. Both the Dejong and the S-curve models use an incompressibility factor in the calculation. Incompressibility is a factor used to account for the percentage of automation in the production process. Values of the incompressibility factor can range from zero to one where zero is all touch labor and one is complete automation. Moore found that when using an incompressibility factor between 0 and 0.1, the Dejong and S-Curve models were more accurate (Moore et al., 2015). In other words, when a production process had very little automation and high amounts of touch labor, the newer learning curve models tended to be more accurate. For all other values of incompressibility, Wright's model was more accurate.

More recently, Johnson (2016) proposed that a flattening effect is evident at the end of the production process where learning does not continue to occur at a constant rate near the end of a production cycle. Using the same models as Moore, Johnson explored the difference in accuracy between Wright's model and contemporary models early in the production process versus later in the production process. He had similar findings to Moore in that Wright's model was most accurate except in cases where the incompressibility factors were extremely low. When the incompressibility factor is low, more touch labor is involved in the process allowing for the possibility of additional learning to occur. He also found that Wright's learning curve was more accurate early in the production process whereas the Dejong and S-Curve models were more accurate later in the production process (Johnson, 2016). Another key concept in learning curve estimation and modeling is the idea of a forgetting curve (Honious et al., 2016). A forgetting curve explains how configuration changes in the production process can cause a break in learning, which leads to a loss of efficiency that had previously been gained. When a configuration change occurs, the production process changes. Changes may include factors such as using different materials, different tooling, adding steps to a process, or might even be attributed to workforce turnover. The new process affects how workers complete their tasks and cause previously learned efficiencies to be lost. If manufacturers fail to take these breaks into account, they may underestimate the total effort needed to produce a product. Honious et al. (2016) found that configuration changes significantly changed the learning curve, and that the new learning curve slope was steeper than the previous steady slope prior to a configuration change. The distinction between pre- and post-configuration change is important to ensure both types of effects are taken into account.

> When a production process had very little automation and high amounts of touch labor, the newer learning curve models tended to be more accurate. For all other values of incompressibility, Wright's model was more accurate.

The International Cost Estimating and Analysis Association (ICEAA) published learning curve training material in 2013. While presenting the basics of learning curve theory, it also presented some rules of thumb for learning. The first rule is that learning curves are steepest when the production process is touch-labor intensive. Conversely, learning curves are the flattest when the production process is highly automated (ICEAA, 2013). Another key piece of information is that adding new work to the process can affect the overall cost. ICEAA states that this essentially adds a new curve for the added work, which increases the original curve by the amount of the new curve (ICEAA, 2013). The equation is as follows:

$$\text{Cost}(x) = A_1 x^{b_1} + A_2 (x - L)^{b_2} \tag{7.3}$$

where:

Cost(x) = cumulative average cost per unit

A_1 = theoretical cost to produce the first unit prior to addition of new work

x = cumulative number of units produced

L = last unit produced before addition of new work

A_2 = theoretical cost to produce the first unit after addition of new work

$$b_1 = \frac{\ln \text{Learning Curve Slope prior to additional work}}{\ln 2}$$

$$b_2 = \frac{\ln \text{Learning Curve Slope prior to additional work}}{\ln 2}$$

(typically same as b_1)

Equation 7.3 is important to consider when generating an estimate after a major configuration change or engineering change proposal (ECP). For example, while producing the eighth unit of an aircraft, the customer realizes they need to drastically change the radar on the aircraft. Learning has already taken place on the first eight aircraft; the new radar has not yet been installed, and therefore no learning has taken place. To accurately take the new learning into account, the radar would be treated as a second part to the equation, ensuring we account for the learning on the eight aircraft while also accounting for no learning on the new radar.

Finally, Anderlohr (1969) and Mislick and Nussbaum (2015) wrote about production breaks and the effects they have on a learning curve. These production breaks can cause a direct loss of learning, which can fully or partially reset the learning curve. For example, a 50% loss of learning would result in a loss of half of the cost reduction that has occurred (ICEAA, 2013). This information is important when analyzing past data to ensure that breaks in production are accounted for.

Thus far, we have laid out the fundamental building blocks for learning curve theory and how they might apply in a production environment. Wright's learning curve formula established the method by which many organizations account for learning during the procurement process. Following Wright's findings, other methods have emerged that account for breaks in production, natural loss of learning over time, incompressibility factors, and half-life analysis (Benkard, 2000). This article adds to the discussion by examining the flattening effect and how various models predict learning at different points in the production process.

When examining learning curve theory and the effects learning has on production, it is critical to understand the production process being estimated. Since Wright established learning curve theory in 1936, factory automation and technology have grown tremendously and continue to grow. Contemporary learning curve methods try to account for this automation. To get the best understanding, we will examine the aircraft industry, specifically how it behaves in relation to the rest of the manufacturing industry.

The aircraft industry has relatively low automation (Kronemer & Henneberer, 1993), especially compared to other industries. Kronemer and Henneberer (1993) state that the aircraft industry is a fairly labor-intensive process with relatively little reliance on automated production techniques, despite it being a high-tech industry. Specifically, they list three main reasons why manufacturing aircraft is so labor-intensive. First, aircraft manufacturers usually build multiple models of the same aircraft, typically for the commercial sector alone. These different aircraft models mean different tooling and configurations are needed to meet the demand of the customer. Second, aircraft manufacturers deal with a very low unit volume when compared to other industries in manufacturing. The final reason

> Following Wright's findings, other methods have emerged that account for breaks in production, natural loss of learning over time, incompressibility factors, and half-life analysis (Benkard, 2000).

for low levels of automation is the fact that aircraft are highly complex and have very tight tolerances. To attain these specifications, manufacturers must continue to use highly skilled touch laborers or spend extremely large amounts of money on machinery to replace them (Henneberer & Kronemer, 1993). For these reasons, we should typically see or use low incompressibility factors in the learning curve models when estimating within the aircraft industry.

Although the aircraft industry remains largely unaffected by the shift to machine production from human touch labor, many industries are seeing a rise in the percentage of manufacturing processes that are automated. In a *Wall Street Journal* article posted in 2012, the author showed how companies have been increasing the amount of money spent on machines and software while spending less on manpower. They proposed that part of the reason behind this shift was a temporary tax break "that allowed companies in 2011 to write off 100% of investments in the first year" (Aeppel, 2012). Tax breaks combined with extremely low-interest rates provided industry with incentive to invest in future production. Investment in production technology increases the incompressibility factor that should be used when estimating the effects of learning. In a separate article for the *Wall Street Journal*, Kathleen Madigan also pointed out the increase in spending on capital investments in relation to labor. She stated that "businesses had increased their real spending on equipment and software by a strong 26%, while they have added almost nothing to their payrolls" (Madigan, 2011).

7.2   METHODOLOGY

7.2.1   MODEL FORMULATION

Before we can begin the process of developing a new learning curve equation, we need to examine the characteristics of the curve we expected to best fit the data. Our hypothesis is that a learning curve whose slope decreases over time would fit the data better than Wright's curve. To adjust the rate at which the curve flattens, the *b* value from Wright's learning curve, or the exponent in the power function, needs to be adjusted. Specifically, to make the curve move in a flatter direction, the exponent in the power curve must decrease as the number of units produced increases. Initially we modified Wright's existing formula by dividing the exponent by the unit number as shown in Equation 4.

$$\text{Cost}(x) = Ax^{b/x} \tag{7.4}$$

where:
   Cost(x) = cumulative average cost per unit
   A = theoretical cost of the first unit

x = cumulative number of units produced
b = Wright's learning curve constant as described in Eq. 7.2

Using Wright's learning curve, b is a negative constant that has a larger magnitude for larger amounts of learning (i.e., as LCS decreases, b becomes more negative). Therefore, in Eq. 7.4, when b is divided by x, the amount of learning is reduced. In fact, the flattening effect is fairly drastic. For example, when applying Eq. 7.4, a standard 80% Wright's learning curve exhibits 90% learning by the second unit and flattens to 97% by the fourth unit. To implement a learning curve that has the flexibility to not flatten as quickly, we instead divide b by $1+x/c$ where c is a positive constant (see Eq. 5). The term $1+x/c$ is always greater than 1 and is increasing as x increases; therefore, a flattening effect always occurs (i.e., learning decreases as the number of units produced increases). The choice of the constant c is critical in determining how quickly the learning decreases. For example, when $c=4$, a standard 80% Wright's learning curve exhibits 86% learning by the second unit and approximately 89% learning by the fourth unit. For the same standard 80% curve when $c=40$, the learning decreases to 80.9% by the second unit and to 81.6% by the fourth unit. The new equation (which we also refer to as Boone's learning curve hereafter) took the form:

$$\text{Cost}(x) = Ax^{b/(1+x/c)} \tag{7.5}$$

where:
Cost(x) = cumulative average cost per unit
A = theoretical cost of the first unit
x = cumulative number of units produced
b = Wright's learning curve constant as described in Eq. 7.2
c = decay value (positive constant)

The function that modifies the traditional learning curve exponent in Eq. 7.5—i.e., $1+x/c$ – has a key characteristic—ensures that the rate of learning associated with traditional learning curve theory decreases as each additional unit is produced. Specifically, $1+x/c$ is always greater than 1 since x/c is always positive. Note that c is an estimated parameter and x increases as more units are produced, so the term x/c is decreasing. When c is large, Boone's learning curve would effectively behave like Wright's learning curve. For example, if the fitted value of c is 5,000, then $1+x/c$ equals 1.0002 after the first unit has been produced and 1.004 after the 20th unit has been produced. This equates to a decrease in the learning rate of the traditional theory (i.e., b) of less than 0.07%. More formally, as c goes to infinity, $b/(1+x/c)$ goes to b.

Note that the previous discussion assumed that b was the same value for both Wright's and Boone's learning curve to help demonstrate the flattening effect. In practice, nothing precludes each of the learning curves from having different b values. For instance, if we desire a learning curve that possesses more learning early in production and less learning later in production (compared to Wright's curve), then the b parameters could be different—this was shown in Figure 7.1. In this case,

Boone's curve would have a b value less than Wright's curve (i.e., a more negative value representing more learning). Then the flattening effect of dividing by $1+x/c$ as production increases would eventually result in a curve with less learning than Wright's curve. For example, consider an 80% Wright's learning curve and a Boone's learning curve that initially has 70% learning and a decay value of 8; by the eighth production unit, Boone's curve would be at 82% learning.

7.2.2   POPULATION AND SAMPLE

To test the new learning curve in Eq. 7.5, we looked at quantitative data from several DoD airframes to gain a comprehensive understanding of how learning affects the cost of lot production. The costs used in this analysis are the direct lot costs and exclude costs for items such as Research, Development, Test, & Evaluation (RDT&E), support items, and spares. These data specifically include Prime Mission Equipment (PME) only as these costs are directly related to labor and can be influenced directly through learning. To ensure we are comparing properly across time, we used inflation and rate-adjusted PME cost data for each production lot of the selected aircraft systems. The PME cost data were adjusted using escalation rates for materials using Office of the Secretary of Defense (OSD) rate tables, when applicable. We used data from fighter, bomber, and cargo aircraft, as well as missiles and munitions. This diverse dataset allowed comparison among multiple systems in different production environments.

7.2.3   DATA COLLECTION

Data used were gathered from the Cost Assessment Data Enterprise (CADE). CADE is a resource available to DoD cost analysts that stores historical data on weapon systems. Some of the older data also came from a DoD research library in the form of cost summary reports. The data used can be broken out by Work Breakdown Structure (WBS) or Contract Line Item Number (CLIN). For this research, the PME cost data were broken out by WBS element, then rolled up into top line, finished product elements and used for the regression analysis. In total, 46 weapon system platforms were analyzed (see Table 7.1).

7.2.4   ANALYSIS

Regression analysis was used to test which learning curve model was most accurate in estimating the data. The goal is to minimize the sum of squared errors (SSE) in the regression to examine how well a model estimates a given set of data. The SSE is calculated by taking the vertical distance between the actual data point (in this case lot midpoint PME cost) and the prediction line (or estimate) (Mislick & Nussbaum, 2015). This error term is then squared and the sum of these squared error terms is the value for comparing which model is a more accurate predictor. However, since an extra parameter is available in fitting the regression for the new model, it should be able to maintain or decrease the SSE in most cases. As previously mentioned, as the decay parameter in Eq. 7.5 approaches infinity, Boone's learning curve approaches

TABLE 7.1
Results

Program	Wright's SSE	Wright's MAPE (%)	Boone SSE	Boone MAPE (%)	SSE Difference (%)	MAPE Difference (%)
Platform A	2.78E+08	5.3	2.17E+08	4.8	−22	−10
Platform B	4.88E+08	5.4	4.90E+08	5.6	0	5
Platform C	1.58E+07	10.8	4.51E+05	2.1	−97	−80
Platform D	6.56E+10	22.1	6.02E+10	24.5	−8	11
Platform E	1.14E+09	6.2	1.10E+09	5.6	−4	−9
Platform F	1.94E+06	4.6	1.95E+06	4.6	0	1
Platform G	7.14E+08	13.6	6.28E+08	12.9	−12	−5
Platform H	5.49E+06	4.6	5.00E+06	4.0	−9	−13
Platform I	1.30E+09	18.6	1.21E+09	23.8	−7	28
Platform J	7.90E+06	3.9	6.12E+06	3.6	−23	−8
Platform K	2.18E+07	6.0	7.48E+06	3.2	−66	−47
Platform L	1.06E+08	9.6	1.05E+08	9.7	−1	0
Platform M	1.49E+07	10.7	1.48E+07	13.4	0	26
Platform N	9.92E+08	16.3	7.67E+07	10.0	−92	−39
Platform O	1.81E+08	13.0	1.78E+08	14.0	−1	7
Platform P	1.71E+07	6.3	7.96E+06	4.7	−53	−26
Platform Q	8.00E+06	10.1	4.11E+06	7.6	−49	−25
Platform R	1.48E+09	18.8	1.31E+09	18.3	−12	−2
Platform S	5.00E+07	6.2	4.89E+07	6.1	−2	−2
Platform T	4.01E+07	11.1	5.45E+06	6.5	−86	−41
Platform U	1.19E+06	8.8	1.34E+06	7.8	13	−11
Platform V	1.60E+09	10.6	1.74E+02	0.0	−100	−100
Platform W	1.39E+09	6.4	1.38E+09	6.4	−1	0
Platform X	7.61E+08	18.1	3.18E-01	0.0	−100	−100
Platform Y	6.81E+05	3.3	1.10E+06	4.1	4.1	26
Platform Z	2.12E+06	7.5	1.57E+06	6.8	6.8	−9
Platform AA	2.66E+07	5.0	2.73E+07	5.5	5.5	10
Platform AB	1.48E+09	18.8	1.31E+09	18.3	18.3	−2
Platform AC	3.81E+07	5.9	2.45E+07	4.5	4.5	−24
Platform AD	3.03E+11	21.9	1.34E+11	16.7	16.7	−24
Platform AE	1.04E+09	10.0	1.03E+09	10.3	10.3	3
Platform AF	9.01E+05	5.1	6.94E+05	4.0	4.0	−23
Platform AG	8.20E+06	5.9	1.77E+06	3.7	3.7	−37
Platform AH	6.40E+06	10.8	6.11E+06	9.8	9.8	−9
Platform AI	1.47E+07	8.2	5.22E+06	5.4	5.4	−35
Platform AJ	4.95E+07	10.0	4.98E+07	10.7	10.7	6
Platform AK	5.99E+07	19.8	5.69E+07	20.4	20.4	3
Platform AL	1.50E+10	12.9	1.43E+10	14.8	14.8	15
Platform AM	1.29E+07	5.5	1.28E+07	5.4	5.4	−3
Platform AN	4.99E+06	3.7	3.02E+06	3.4	3.4	−9

(Continued)

TABLE 7.1 (Continued)
Results

Program	Wright's SSE	Wright's MAPE (%)	Boone SSE	Boone MAPE (%)	SSE Difference (%)	MAPE Difference (%)
Platform AO	9.63E+07	21.9	9.45E+07	21.5	21.5	−2
Platform AP	1.18E+06	3.1	1.22E+06	3.4	3.4	7
Platform AQ	2.77E+03	3.4	1.19E-05	0.0	0.0	−100
Platform AR	1.84E+06	17.3	1.82E+06	18.0	18.0	4
Platform AS	3.27E+06	1.3	1.09E+00	0.0	0.0	−100
Platform AT	1.98E+03	2.8	1.19E+03	1.7	1.7	−40

Note: The actual names of each system and contractor have been removed and replaced with a designator of Platform A…Platform AT.

Wright's learning curve formula. With this in mind, we also examined the Mean Absolute Percentage Error (MAPE). MAPE takes the same error term as the SSE calculation but then divides it by the actual value; then the mean of the absolute value of these modified error terms is calculated. By examining the error in terms of a percentage, comparisons between different types and sizes of systems are more robust. If Boone's curve reduces both SSE and MAPE when compared to the SSE and MAPE of Wright's curve, it would indicate the new model may be better suited for modeling learning and the associated costs.

As stated previously, Wright's learning curve is suitable for a log-log model. A log-log model is used when a logarithmic transformation of both sides of an equation results in a model that is linear in the parameters. As Wright proposed, this linear transformation occurs because learning happens at a constant rate throughout the production cycle. If learning happens at a nonconstant rate (as in Boone's learning curve), then the curve in log-log space would no longer be linear. This constraint means typical linear regression methods would not be suitable for estimating Boone's learning curve; therefore, we had to use nonlinear methods to fit these curves.

Specifically, we used the Generalized Reduced Gradient (GRG) nonlinear solver package in Excel to minimize the SSE by fitting the A, b, and c parameters from Eq. 5. To use this solver, bounds for the three parameters had to be established. These are values that are easy to obtain for any dataset, as they are provided by Microsoft Excel when fitting a power function or by using the "linest()" function in Excel. We used this as a starting point because Wright's curve is currently used throughout the DoD. For the A variable, the lower bound was one-half of Wright's A and the upper bound was 2 times Wright's A. These values were used to give the solver model a wide enough range to avoid limiting the value but small enough to ease the search for the optimal values. Neither of these limits was found to be binding. For the exponent parameter b, we chose values between 3 and −3 times Wright's exponent value. In theory, the value of the exponent should never go above 0 due to positive learning leading to a decrease in cost, but in practice, some datasets go up over time and we wanted to be able to account for those scenarios, if necessary. Again, these values between 3 and −3 times Wright's exponent value were never found to be binding limits for the model. Finally, for the decay parameter c, fitted values were bounded between 0 and 5,000; the 5,000 upper bound was found to be a binding constraint in the solver on several occasions. In practice, analysts could bound the value as high as possible to reduce error, but in the case of this research, we used 5,000 as no significant change was evidenced from 5,000 to infinity—relaxing this bound would have only further reduced the SSE for Boone's learning curve.

7.2.5 STATISTICAL SIGNIFICANCE TESTING

Once the SSE and MAPE values were calculated for each learning curve equation, we tested for significance to determine whether the difference between the error values for the two equations was statistically different.

Specifically, we conducted t-tests on the differences in error terms between Wright's and Boone's learning curve equations. This t-test was conducted for both SSE and MAPE values separately. A nonsignificant t-test indicates no statistically significant difference between the two learning curves.

7.3 ANALYSIS AND RESULTS

The Table shows the SSE and MAPE values for both Wright's and Boone's learning curve for each system in the dataset. The last two columns are the percentage difference in SSE and MAPE between the two learning curve methods. This percentage was calculated by taking the difference of Boone's error term minus Wright's error term divided by Wright's error term. Negative values represent programs where Boone's learning curve had less error than Wright's learning curve, and positive values represent programs where Wright's curve had less error than Boone's curve.

Based on this analysis, we observed that Boone's learning curve reduced the SSE in approximately 84% of programs and reduced MAPE in 67% of programs. The mean reduction of SSE and MAPE was 27% and 17%, respectively. As previously mentioned, these values were based on using both learning curve equations to minimize the SSE for each system in the dataset. This is standard practice in the DoD as prescribed by the U.S. Government Accountability Office (GAO, 2009) *Cost Estimating and Assessment Guide* when predicting the cost of subsequent units or subsequent lots.

> The incompressibility factor represents the amount of automation in the production process, which limits how much learning can occur (Badiru et al., 2013).

We conducted additional tests to determine if a statistical difference existed between the means of both curve estimation techniques. On average, programs estimated using Boone's learning curve had a lower error rate ($M = 4.73$, SD $= 2.15$) than those estimated using Wright's learning curve ($M = 8.64$, SD $= 4.55$). Additionally, the difference between these two error rates expressed as a percentage and compared to a hypothesized value of 0 (no difference) was significant, $t(46) = -4.87$, $p < .0001$, and represented an effect of $d = 1.10$. We then applied the same test to the difference in the MAPE values from Boone's learning curve and Wright's learning curve.

On average, programs estimated using Boone's learning curve had a lower MAPE value ($M = .08$, $SD = .07$) than those estimated using Wright's MAPE value ($M = .10$, $SD = .06$). The difference between these two estimates has a mean of $-.17$, which translates to Boone's curve reducing MAPE by 17% more on average. Additionally, the difference between these two error rates expressed as a percentage and compared to a hypothesized value of 0 (no difference) was significant, $t(46) = -3.48$, $p < .0005$, and represented an effect of $d = .22$. The results indicate that in both SSE and MAPE, Boone's learning curve reduced the error, and that each of those values was statistically significant when using an alpha value of 0.05.

7.4   DISCUSSION

As stated previously, an average of a 27% reduction in the SSE resulted from among the 46 programs analyzed. These results were statistically significant. Also, a 17% reduction in the MAPE resulted from among the programs analyzed, which was also found to be statistically significant. Based on these results, we can conclude that Boone's learning curve equation was able to reduce the overall error in cost estimates for our sample. This information is critical to allow the DoD to calculate more accurate cost estimates and better allocate its resources. These conclusions help answer our three guiding research questions. Specifically, we were looking for the point where Wright's model became less accurate than other models. We found that adding a decay factor caused the learning curve to flatten out over time, which resulted in less error than Wright's model. Additionally, we found that Boone's learning curve was more accurate throughout the entire production process, not just during the tail end when production was winding down. Boone's learning curve was steeper during the early stages of production when it's hypothesized that the most learning occurs. Toward the end of the

> Future research should identify decay values for different types of weapon systems –similar to the way learning curve rates are established for different categories of programs.

production process, Boone's curve flattens out more than Wright's curve, supporting our contention that learning toward the end of the production cycle yields diminishing returns. While Wright's curve assumes constant learning throughout the entire process, Boone's curve treats learning in a nonlinear fashion that slows down over time. By reducing the error in the estimates and properly allocating resources, the DoD could potentially minimize risk for all parties involved. The benefit of Boone's learning curve is accuracy in the estimation process. If labor estimates aren't accurate in the production process, risks escalate, such as schedule slip, cost overruns, and increased costs for all involved. Accuracy in the cost estimate should be the goal of both the contractor and government, thereby facilitating the acquisition process with better data.

7.4.1   LIMITATIONS

One limitation of this study is that all 46 of the weapon systems analyzed were U.S. Air Force systems. While the list included many platforms spanning decades, we hesitate to draw conclusions outside of the U.S. Air Force without further research and analysis. That said, we see no reason our model wouldn't apply equally well in any aircraft production environment, both within and outside the DoD. Another limitation in this research is the use of PME cost as opposed to labor hours. Labor-hour data are not readily available across many platforms, which led to the use of PME cost. Contractor data provided to the government normally come in the form of lots, which is the lowest level tracked by cost estimators. To compare learning curves across multiple platforms, the same level of analysis is required to ensure a fair comparison. Future research should attempt to examine data at the individual level of analysis between systems and exclude those where only a lot of data are available. Because there are inherently less lots than units, this may affect how the equation behaves when applied at the unit level. For this research, we used the lot midpoint formula/method (Mislick & Nussbaum, 2015), but further research should be conducted to evaluate the performance of Boone's learning curve with unitary data. Finally, we only performed a comparison to Wright's learning curve since that is a primary method of estimation in the DoD. A comparison with other learning curve models may yield different results, although previous research found those curves were not statistically better than Wright's.

7.4.2   RECOMMENDATIONS FOR FUTURE RESEARCH

Data outside of the U.S. Air Force should be examined to test whether this equation applies broadly to programs, and not just to Air Force programs. Also, conducting the analysis with unitary data could confirm that this works for predicting subse-quent units as well as subsequent lots, while reducing error over Wright's method. We also made an attempt to select weapon systems that had minimal automation in the production process. However, DeJong's Learning Formula is another deri-vation from Wright's original in which an incompressibility factor is introduced. The incompressibility factor represents the amount of automation in the production process, which limits how much learning can occur (Badiru et al. 2013). Other models such as the S-Curve model (Carr, 1946) and a more recent version (Towill, 1990; Towill & Cherrington, 1994) also account for some form of incompressibility. Additional research could also include modifications to Boone's formula to try and further reduce the error types listed in this research. Furthermore, fitting Boone's curve in this analysis was based on past data whereas cost estimates are used to project future costs. Therefore, future research should identify decay values for dif-ferent types of weapon systems—similar to the way learning curve rates are estab-lished for different categories of programs. Finally, further research could examine whether the incorporation of multiple learning curve equations at different points in the production process would be beneficial in reducing additional error in the estimates.

We developed a new learning curve equation utilizing the concept of learning decay. This equation was tested against Wright's learning equation to see which equa-tion provided the least amount of error when looking at both the SSE and MAPE. We found that Boone's learning curve reduced error in both cases and that this reduction in error was statistically significant. Follow-on research in this field could lead to further discoveries and allow for broader use of this equation in the cost community.

REFERENCES

Aeppel, T. (2012, January 17). Man vs. machine: Behind the jobless recovery. *Wall Street Journal.* http://www.wsj.com/articles/SB10001424052970204468004577 164710231081398.

Anderlohr, G. (1969). What production breaks cost. *Industrial Engineering, 1*(9), 34–36.

Badiru, A., Elshaw, J., & Mack, E. (2013). Half-life learning curve computations for airframe life-cycle costing of composite manufacturing. *Journal of Aviation and Aerospace Perspective, 3*(2), 6–37.

Benkard, C. L. (2000). Learning and forgetting: The dynamics of aircraft production. *American Economic Review, 90*(4), 1034–1054. https://www.aeaweb.org/articles?id=10.1257/aer.90.4.1034.

Carr, G. W. (1946). Peacetime cost estimating requires new learning curves. *Aviation, 45*(4), 220–228.

Chalmers, G., & DeCarteret, N. (1949). *Relationship for Determining the Optimum Expansibility of the Elements of a Peacetime Aircraft Procurement Program.* Wright-Patterson AFB, OH: USAF Air Materiel Command.

Crawford, J. R. (1944). *Estimating, Budgeting and Scheduling.* Burbank, CA: Lockheed Aircraft Co.

De Jong, J. R. (1957). The effects of increasing skill on cycle time and its consequences for time standards. *Ergonomics, 1*(1), 51–60. https://doi. org/10.1080/00140135708964571.

Honious, C., Johnson, B., Elshaw, J., & Badiru, A. (2016, May 4–5). The impact of learning curve XE model selection and criteria for cost estimation accuracy in the DoD. In Maj Gen C. Blake, USAF (Chair), *Proceedings of the 13th Annual Acquisition Research Symposium*, Monterey, CA. https://apps.dtic.mil/dtic/tr/fulltext/u2/1016823.pdf.

International Cost Estimating and Analysis Association (2013). *Cost Estimating Body of Knowledge* [PowerPoint Slides]. ICEAA. https://www.iceaaonline.com/cebok/.

Johnson, B. J. (2016). *A Comparative Study of Learning Curve Models and Factors in Defense Cost Estimating Based on Program Integration, Assembly, and Checkout* [Master's Thesis]. Air Force Institute of Technology. https://apps.dtic.mil/dtic/tr/ fulltext/u2/1056447.pdf.

Kronemer, A, & Henneberger, J. E. (1993). Productivity in aircraft manufacturing. *Monthly Labor Review, 16*(6), 24–33. http://www.bls.gov/mfp/mprkh93.pdf.

Madigan, K. (2011, September 28). It's man vs. machine and man is losing. *Wall Street Journal.* https://blogs.wsj.com/economics/2011/09/28/its-man-vs-machine-andman-is-losing/.

Martin, J. R. (n.d). *What is a learning curve?* Management and Accounting Web. http://maaw. info/LearningCurveSummary.htm.

Mislick, G. K., & Nussbaum, D. A. (2015). *Cost Estimation: Methods and Tools.* Wiley. https://www.academia.edu/24430098/Cost_Estimation_Methods_and_Tools_by_Gregory_K_Mislick_and_Daniel_A_Nussbaum_1st_Edition.

Moore, J. R., Elshaw, J. Badiru, A. B., & Ritschel, J. D. (2015, October). Acquisition challenge: The importance of incompressibility in comparing learning curve models. *Defense Acquisition Research Journal, 22*(4), 416–449. https://www.dau. edu/library/arj/ARJ/ARJ75/ARJ75-ONLINE-FULL.pdf.

Towill, D. R. (1990). Forecasting learning curves. *International Journal of Forecasting, 6*(1), 25–38. https://www.sciencedirect.com/science/article/abs/ pii/016920709090095S.

Towill, D. R., & Cherrington, J. E. (1994). Learning curve models for predicting the performance of AMT. *The International Journal of Advanced Manufacturing Technology, 9*, 195–203. https://doi.org/10.1007/BF01754598.

U.S. Government Accountability Office. (2009). *GAO cost estimating and assessment guide (GAO-09-3SP).* Government Printing Office. https://www.gao.gov/new.items/d093sp.pdf.

Wright, T. P. (1936). Factors affecting the cost of airplanes. *Journal of the Aeronautical Sciences, 3*(4), 122–128. https://arc.aiaa.org/doi/10.2514/8.155.

Section 3

Space Technologies

8 Shifting Satellite Control Paradigms
Operational Cybersecurity in the Age of Megaconstellations*

Carl Poole
US Space Force

Robert Bettinger and Mark G. Reith
US Air Force

CONTENTS

DISCLAIMER

The following articles are written by Department personnel or drawn from Department information available to non-affiliated researchers, both operating under the principles of academic freedom. Article selection is based on upcoming Secretary or Chiefs calendar events, topical or enduring service issues, or relevant geopolitical commentary and analysis. The entries reflect ongoing discussions on ideas and their implications, rather than news events. The articles are for informational purposes only and should not be interpreted as endorsements by the Department of the Air Force or services.

Synopses and full text for Weekend Reads articles follow.

* Reprinted with permission: Poole, C., Bettinger, R., & Reith, M. (2021). Shifting Satellite Control Paradigms: Operational Cybersecurity in the Age of Megaconstellations. *Air & Space Power Journal*, *35*(3), 46–56.

DOI: 10.1201/9781003220978-11

SYNOPSES

This article explains how the use of automated satellite control systems in a space-mission environment historically dominated by human-in-the-loop operations will change the cybersecurity measures that ensure space system safety and security. The author argues that both the defense and commercial space sectors must adapt to the rapidly changing digital landscape of future space operations to promote cybersecurity in an increasingly competitive, contested, and congested space domain.

Space and cybersecurity professionals will need increased interactive cooperation and mission understanding to address new potential cybersecurity issues presented by emerging commercial space applications and automation.

FULL TEXT

The introduction of automated satellite control systems into a space-mission environment historically dominated by human-in-the-loop operations will require a more focused understanding of cybersecurity measures to ensure space system safety and security. On the ground-segment side of satellite control, the debut of privately owned communication antennas for rent and a move to cloud-based operations or mission centers will bring new requirements for cyber protection for both Department of Defense (DOD) and commercial satellite operations alike. It is no longer a matter of whether automation will be introduced to satellite operations, but how quickly satellite operators can adapt to the onset of control automation and promote cybersecurity in an increasingly competitive, contested, and congested space domain.

8.1   INTRODUCTION

Control automation has spread from industrial manufacturing and self-driving cars to home and household appliances. Control automation has also moved into the realm of satellite-control operations, with the focus of satellite-control automation being driven on two fronts. First, the ability to incorporate cost-effective, highly capable equipment in the satellite design allows for an increase in onboard control processing. Second, the proliferation of space operations in various orbital regimes—this article will focus on low-Earth orbit (LEO)—is pushing complex tasks, such as satellite-link scheduling and conjunction-avoidance maneuvers, beyond the control of human operators.

An additional operational distinction is made between satellite automation—the self-contained system process of conducting repetitive tasks—and satellite autonomy, which gives the satellite the ability to implement changes with limited to no human-in-the-loop actions.[1] This distinction will add a level of complexity to the cybersecurity of satellite control. Placing tasks previously controlled by humans under the control of a computer-executed algorithm may be the only viable way to manage the development of future megaconstellations and enable effective space-traffic management.[2] But the prospect of improved space-traffic safety and collision avoidance via control automation raises several concerns.

While increasing the levels at which LEO constellations can interact and cooperate, the needed hardware infrastructure and data-exchange alterations that will allow for such interoperability will introduce new entry points that, in turn, will likely increase cybersecurity risks. The introduction of software-defined equipment, cloud-based mission-control centers, and Ground Stations as a Service (GSaaS) are prime examples. Space and cybersecurity professionals will need increased interactive cooperation and mission understanding to address new potential cybersecurity issues presented by emerging commercial space applications and automation.

8.2 CURRENT SATELLITE CONTROL OPERATIONS

The control architecture for satellites has remained nearly constant since the beginning of the Space Age in the mid-twentieth century. Starting with the launch of the first artificial satellites, each on-orbit system has mostly featured a unique design, function, and mode of operation. This uniqueness has led to self-contained and independent operating procedures controlled by the satellite owner. In the typical satellite-control structure, a satellite downlinks information such as payload data and spacecraft state-of-health information when it is within view of a ground-based receiver. From the receiver, the information is processed and passed to the satellite operations center (SOC), which reviews it for faults and assesses the need for required operating adjustments and/or new system instructions.

In the case of orbital maneuvers to correct for position or to change location (such as slewing, station keeping, or collision avoidance with another object), one member of the operations team scripts the commands for the prescribed maneuver. Several operation team members then review the script before passing it to the human-in-the-loop satellite operator for processing. During the next scheduled uplink opportunity with the satellite, the commands are sent from the SOC to the transceiver and then to the satellite for processing and command execution. This type of hands-on approach developed due to constraints in the onboard systems, specifically, limited computing power and proprietary operating structures.

The emphasis on human control ostensibly meant reduced concerns for cybersecurity and an increased sense of command situational awareness due to the human use of protected ground communications systems and owner-controlled data links. Despite its benefits, this process can be very time-consuming, and task scheduling becomes increasingly complex with the addition of new satellites to the satellite-control architecture. Consequently, this human-in-the-loop satellite-control architecture will be unable, without a substantial increase in infrastructure, manning, and funding, to effectively manage the size of megaconstellations in the near future.

8.3 ANATOMY OF MEGACONSTELLATIONS

The development of constellations consisting of thousands of individual satellites controlled by one operator is no longer a wistful dream of science fiction or avant-garde technologists. With the introduction of LEO constellations such as "Starlink" or "OneWeb," the concept of megaconstellations is becoming a reality, precipitating the rise of megaconstellations as a potential means to provide regional and global telecommunications services.[3]

In Asia, China Telecom reportedly plans to create a 10,000-satellite megaconstellation called "China StarNet" in the next 5–10 years.[4] In late 2020, the European Union revealed plans to initiate a program to develop a telecommunications megaconstellation to establish "European digital sovereignty."[5] The proliferation of LEO with tens of thousands of satellites will require increasing levels of automation to handle intraconstellation operations and to enable future constellation growth and system safety in a given orbital altitude regime.

The creation of megaconstellations is the result of two factors. First, the shift in the commercial space industry to create standardized, rapidly produced, and high-volume space-capable vehicles has caused both the size and cost of individual satellites to decrease drastically.[6] The ability to buy commercial-off-the-shelf components instead of making proprietary hardware lowers the cost of research and development, thus accelerating system production.

The second factor is a function of satellite size. As the satellite form factor decreases, more satellites can fit inside the payload fairing of a single launch vehicle, which, in turn, drives down the cost per satellite to reach orbit. Overall, the costs of satellite design, production, and space launch are decreasing, thus allowing for the nearly exponential proliferation of near-Earth orbital regimes. Consequently, the increase in satellites will lead to an escalation of costs associated with operations if the current satellite control paradigm does not evolve to meet the challenges of proliferated orbits.

The evolution of satellite control from human-in-the-loop commands to automation will require the megaconstellation, in concert with the ground communications networks, to deconflict satellite pass times over receiver antennas at specified ground stations.[7] By definition, a "pass time" is the time each satellite needs to downlink, or transmit, data to the ground antenna, as well as to uplink, or receive, commands from the ground station. Depending on the mission and amount of information transmitted, timing is critical.

In addition, the orbital altitude of a given satellite determines the access durations to each ground antenna: the lower the satellite altitude, the faster the satellite passes over a given point on the ground. This planning will be increasingly important as the communication bandwidths become more crowded due to more satellites flying within the ground receiver's view.

Since the early twenty-first century, an increase in CPU power has enabled the addition of programmable capabilities to onboard satellite subsystems.[8] A growing number of satellites are now being equipped with onboard systems that resemble a standard personal computer.[9] This design architecture, in turn, increases reliability. A satellite's onboard system can now identify and correct for faults and adapt to changing parameters much faster than a human-in-the-loop system.[10] A human-in-the-loop system is comparatively slower due to data transmission and analysis delays and the need for an extra layer of review to verify the correctness and validity of planned operations before command uplink.

One of the most common satellite-control tasks is that of station keeping or maintaining a satellite's predetermined, mission-centric orbital attitude and position.

For megaconstellations, an attitude determination and control system may control all station-keeping operations. Due to an increase in ground station demand resulting from a vastly greater number of contacts, each satellite will have to determine the correct orbital attitude and position deviations autonomously to ensure continued constellation stability and mission functionality and to reduce the likelihood of satellite collisions.[11]

Shifting such attitude and orbit maintenance tasks away from the ground segment, however, will require the introduction of a robust fault- and error-alert architecture to identify and notify the human satellite operators of any anomalous events. Ultimately, raising more housekeeping commands into the purview of control automation will shift the satellite maintenance workload from continuous hands-on, day-to-day human operations to an oncall, human-response control structure. Greater automation will also remove the likelihood of an incomplete command sent by human operators or the need to check for unsafe commands before data uplink.[12]

8.4   SATELLITE-CONTROL EVOLUTION

While automation will play a large role in handling satellite functions, the main changes for cybersecurity will come from the evolutionary shifts made in the ground-control segments and associated security implementation requirements. In the 2020 Space Capstone Publication *Spacepower: Doctrine for Space Forces*, the foundation for cybersecurity is defined in the cyber operations spacepower discipline as the "knowledge to defend the global networks upon which military space power is vitally dependent," the "ability to employ cybersecurity and cyber defense of critical space networks and systems," and the "skill to employ future offensive capabilities."[13]

The future of security implementation is already being felt on the manufacturing side for DOD contracts. The recently introduced Cybersecurity Maturity Model Certification (CMMC) program pushes the level of responsibility for cybersecurity down, starting with the industries providing the components and systems, then to the Department of Defense by requiring it to use the published National Institute of Standards and Technology rating system.[14] The CMMC is also rooted in the Federal Acquisition Regulation, Federal Information Processing Standards, and general industry collaboration.[15]

The CMMC does have several caveats such as not requiring compliance for commercial-off-the-shelf systems.[16] This shift will ensure the hardware and software introduced for future satellite-control needs will be primed for cyberdefense. Another aspect that will play a role in the coming changes focuses on the protection of potential dual-use technologies. "Entrepreneurs with innovative and potentially dual-use technologies must improve the protection of their intellectual property from unintended foreign assimilation, including protecting their networks from cyber exfiltration attempts, and avoiding exit strategies that transfer intellectual property to foreign control hostile to U.S. interests."[17]

Some of these dual-use technologies can come in the form of software-defined components that will allow for greater flexibilities in upgrading the on-orbit and ground-control segments, especially in the area of communication systems.[18] Though software-defined systems will add increased flexibility and allow for faster fixes if damaged (for example, there is no need to replace expensive parts if the component can be simply reprogrammed), it will also introduce a new level of security require-ments and response capabilities due to the inherent vulnerabilities in all software control systems.

Unlike traditional cybersecurity training provided to most Airmen, Guardians may require enhanced cyber skills to manage risk in the space environment. The Space Force chief technology and innovation officer describes USSF as a digital ser-vice; accordingly, Guardians will likely need to understand how digital engineering intersects with cybersecurity in order to model complex systems and cyber threats.[19] Guardians will need to be able to conceptualize how existing hardware and evolving software components interact as well as how they may be exploited by threat actors.

Furthermore, they will likely benefit from development, security, and operations training that will help them craft new software components that not only meet mis-sion needs but are continuously hardened in response to evolving threats. Advanced digital twin modeling—a one-for-one virtual model tested in an operationally accu-rate simulated environment—may provide a feedback loop to inform operators of how well these new software components perform across a risk spectrum.

Another area of evolving satellite control relates to the use of flexible ground-control systems, more specifically, the ground antennas used to transmit commands and receive data. Commercial entities such as Microsoft are introducing GSaaS to increase capabilities and offset costs associated with satellite command and control.[20] These systems will need to be diverse in operational software and equipment to cover the wide range of satellite technologies currently used. Alternatively, future satellite designs that intend to use this emerging method of ground control can establish a form of technological standardization. In either case, commercializing the ground segment will help handle the increased volume and bolster networked capabilities.

Despite these benefits, however, current satellite programs base network security on the legacy assumption that ground stations and the associated ground network are program- or owner-controlled, system-specific, and isolated from other net-works. A new control structure is only half of the required change—the other half involves changing how and where some of the satellite-control operations and tasks are conducted.

This second change is coming in the form of cloud-based SOCs. As with the software-defined component and commercialized ground stations, cloud-based con-trol will provide a more robust and flexible answer for growing constellations without the need to build costly new mission-specific "brick and mortar" operations cen-ters.[21] This area already has several working examples, such as the "Major Tom" sys-tem—produced by the commercial firm Kubos—that is implemented by the Planet company for use in its Dove constellation consisting of approximately 250 small satellites.[22]

Cloud-based systems will have the added benefit of being accessible from any "secure" networked computer. In concert with the aforementioned commercial ground stations, cloud-based systems could enable megaconstellation control from any location on the globe featuring a proper access point.

In the emerging satellite-control dynamic, an example of potential operations starts with a customer satellite sending spacecraft state-of-health or other data to a configured service receiver. The receiver then uploads the data to a cloud-based

SOC that is accessible by satellite operators from any networked computer system. Even with this control flexibility, the use of increasingly networked systems owned by third parties, rather than the satellite or constellation operator, can introduce new entry points and areas for cyber vulnerability.[23] To ensure the cyber protection of all US and Allied space-based assets, satellite programs and control architectures directly in touch with these evolving systems will need to change just as drastically as the systems themselves.[24]

8.5 SATELLITE SURVIVABILITY CONSIDERATIONS

The goal of any satellite system is to maintain mission functionality for the planned mission lifetime; this requires satellite survivability. Satellite survivability is a function of three time-separated phases: susceptibility, vulnerability, and recoverability. Survivability is promoted if a system's susceptibility and vulnerability to natural and/or manmade threats are minimized while the prospect of recoverability is maximized. From a manmade-threat perspective, susceptibility analysis focuses on the threat system and its ability to successfully detect, be employed, intercept, and finally function as intended vis-à-vis the target satellite system.

Similarly, a satellite's vulnerability relates to its ability to survive the threat's intended weapon effects. Finally, recoverability is the ability of a satellite (and the satellite operators), following damage from a threat system, to take emergency action to prevent the loss of the satellite and/or to regain a level of satellite mission capability.[25] These components of survivability can be extrapolated to megaconstellations as a system-of-systems due to their interconnected internal communications and mission architecture.

The Venn diagram (Figure 8.1) depicts survivability considerations for megaconstellations, outlining the aspects of susceptibility, vulnerability, and recoverability. Overall, the high number of satellites comprising megaconstellations and the use of emerging autonomy and network technologies represent both primary strengths and weaknesses for megaconstellations. While the risk of satellite collision and debris impact constitute a constant environmental risk to operations, megaconstellations are increasingly susceptible to cybersecurity threats due to the use of commercial GSaaS and cloud-based satellite operations.

Cybersecurity threats are varied based on the source of origin and damage mechanism. Satellite operators must maintain a proper understanding of the cyber-threat landscape and the digital and networked functionality of the megaconstellation in order to secure continued mission effectiveness and survivability.

8.6 THE NETWORKED OPERATIONS CENTER

With anticipated shifts in both methods and infrastructure for space-control operations, there should be an equal shift in the cadre structure and training for satellite-operation teams in the Department of Defense and commercial sectors. On the satellite-operations floor, operators often reach out to fellow team members when anomalous situations arise with satellite systems. But this consultation only works well if the members on both ends of the conversation talk the same language.

Commercial Ground Stations-as-Service
Cloud-Based Operations
Networked Satellite Control Architecture

Survivability

Susceptibility

Recoverability   Vulnerability

Autonomous Satellite Operations   Satellite Collision Risks
Rapid Constellation Replenishment   Man-made Debris Hypervelocity Impacts
Capability for Degraded Performance   Networked Satellite Control Architecture

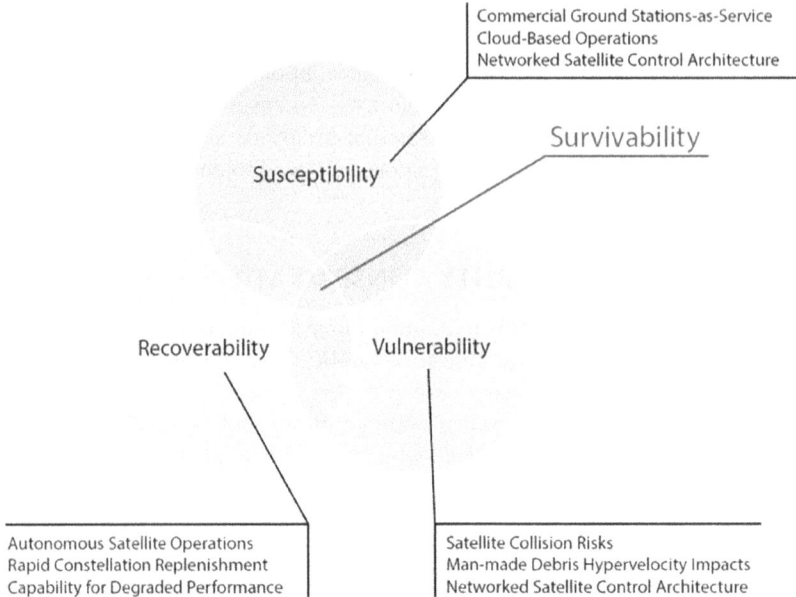

FIGURE 8.1   Megaconstellation survivability.

As the transition to increasingly networked centers interfacing with highly auto-mated systems progresses, space operations and cybersecurity professionals should learn and understand more of the other members' skill sets and technical terminol-ogy. Ideally, the formal training for satellite-operations team members will evolve to include a space- and cyber-centric curriculum. This training could be in the form of introductory classes in cyberdefense for space professionals, and satellite mission design and communications for cyber professionals.

The USSF is in a crucial position to make this happen starting at the ground level. As mentioned in the Space Capstone Publication, increased education will add to the understanding of the "network dimension."[26] Optimally, this education would result in embedding cyberoperations members at key SOCs, in addition to having increased cybersecurity and monitoring training at all levels of satellite operations. This approach will facilitate a highly digitally capable satellite-operations cadre.[27]

Building a cyber-minded and space-proficient space-control foundation will ensure space and cyberspace professionals will have the tools needed to tackle any future growth in satellite capabilities and space mission execution. It will also empower members with the abilities and confidence to react rapidly and even pre-emptively to future threats.

8.7   CONCLUSION

Satellite systems and control architectures are in a rapid state of change. Satellite automation could significantly alter the current hands-on satellite-operations mis-sion to one of key-event monitoring, with a consolidated human-in-the-loop team present to react to and resolve issues that cannot be directly handled by the satellite itself or by the megaconstellation. Additionally, the introduction of a more capable and increasingly flexible mission-operations system, one using emerging technologies such as cloud-based networks and services like privately owned and networked ground stations, will make it possible for true 24/7 global access to and control of satellite systems.

To ensure the continued safety and security of on-orbit satellite systems, both the defense and commercial space sectors must adapt to the rapidly changing digital landscape of future space operations. The introduction of the CMMC has already demonstrated such an adaptation, along with the alignment of emergent USSF doctrine and strategy with cyber-mindedness. The final step will be to shape the future of the USSF and USAF space and cyberspace cadre to be better prepared as a digital force synergistically working to remain at the forefront of protection in the increasingly competitive, contested, and congested domain of space.

As LEO becomes more congested and the mission sets for megaconstellations expand beyond telecommunications, the operating altitudes for megaconstellations will also expand. As a result, the space and cyberspace cadre – Airmen and Guardians alike – must be poised to handle considerations of autonomy and cybersecurity in LEO, geosynchronous Earth orbit, and beyond into the cislunar realm.

NOTES

1. J. B. Hartley and P. M. Hughes, "Automation of Satellite Operations: Experiences and Future Directions at NASA GSFC," in *Space Mission Operations and Ground Data Systems - SpaceOps '96, Proceedings of the Fourth International Symposium held 16–20 September 1996 in Munich, Germany*, ed. T. D. Guyenne (Paris: European Space Agency, 1996): 1262–69.
2. Steven J. Butow et al., *State of the Space Industrial Base 2020: A Time for Action to Sustain US Economic & Military Leadership in Space*, (Washington, DC: USSF, Air Force Research Laboratory, and Defense Innovation Unit, July 2020), http://aerospace.csis.org/.
3. Jonathan C. McDowell, "The Low Earth Orbit Satellite Population and Impacts of the SpaceX Starlink Constellation," *Astrophysical Journal Letters* 892, no. 2 (2020): 1–18, https:// iopscience.iop.org/.
4. Dan Swinhoe, "China's Moves into Mega Satellite Constellations Could Add to the Space Debris Problem," Data Center Dynamics, April 20, 2021, https://www.datacenterdynamics.com/.
5. Jonathan O'Callaghan, "Europe Wants to Build Its Own Satellite Mega Constellation to Rival SpaceX's Starlink," *Forbes*, December 23, 2020, https://www.forbes.com/.
6. Mohamed Khalil Ben-Larbi et al., "Towards the Automated Operations of Large Distributed Satellite Systems, Part I: Review and Paradigm Shifts," *Advances in Space Research* 67, no. 1 (June 1, 2021), https://www.sciencedirect.com/; and Ben-Larbi et al., "Towards the Automated Operations of Large Distributed Satellite Systems, Part II: Classification and Tools," *Advances in Space Research* 67, no. 1 (June 1, 2021), https://www.sciencedirect.com/.
7. Ben-Larbi et al., "Paradigm Shifts"; Ben-Larbi et al., "Classification and Tools"; Michael J. Bentley, Alan C. Lin, and Douglas D. Hodson, "Overcoming Challenges to Air Force Satellite Ground Control Automation," in *Proceedings of the IEEE Multi-Disciplinary Conference on Cognitive Methods in Situational Awareness and Decision*

Support (CogSIMA) (Curran Associates, June 2017), https://ieeexplore.ieee.org/; and Jun Tominaga, José Demísio Simões da Silva, and Mauricio Goncalves Vieira Ferreira, "A Proposal for Implementing Automation in Satellite Control Planning" (paper, SpaceOps 2008 Conference, Heidelberg, Germany, May 1216, 2008), https:// arc.aiaa. org/.

8. Misa Iovanov et al., "Automation of Daily Tasks Necessary for the Management of a Large Satellite Constellation" (paper, American Institute of Aeronautics and Astronautics (AIAA) Space 2003 Conference & Exposition, Long Beach, CA, September 23–25, 2003), https://arc.aiaa.org/.

9. Ben-Larbi et al., "Paradigm Shifts"; and Ben-Larbi et al., "Classification and Tools."

10. Gilles Kbidy, "Flying Large Constellations Using Automation and Big Data" (paper, SpaceOps 2016 Conference, Daejeon, South Korea, May 13, 2016), https://arc.aiaa.org/.

11. Jérôme Thomassin, Maxime Ecochard, and Guillaume Azema, "Predictive Autonomous Orbit Control Method for Low Earth Orbit Satellites" (paper, International Symposium on Space Flight Dynamics, Matsuyama, Japan, June 6–9, 2017), https://issfd.org/; and Byoung-Sun Lee, Yoola Hwang, and Hae-Yeon Kim, "Automation of the Flight Dynamics Operations for Low Earth Orbit Satellite Mission Control," in *Proceedings of the 2008 International Conference on Control, Automation, and Systems* (Curran Associates, April 2009), https://ieeexplore.ieee.org/.

12. Ben-Larbi et al., "Paradigm Shifts"; and Ben-Larbi et al., "Classification and Tools."

13. John W. Raymond, *Spacepower: Doctrine for Space Forces*, Space Capstone Publication (Washington, DC: USSF, June 2020), 52, https://www.spaceforce.mil/.

14. Barry Rosenberg, "'Start of a New Day': DoD's New Cybersecurity Regs Take Effect Today,'" Breaking Defense, December 1, 2020, https://breakingdefense.com/.

15. "Understanding the CMMC Fundamentals," Cybersecurity Maturity Model Certification Center of Excellence, November 26, 2020, https://cmmc-coe.org/.

16. Rosenberg, "New Day."

17. Butow et al., "Space Industrial Base."

18. M. Manulis et al., "Cyber Security in New Space," *International Journal of Information Security* 20 (2020): 287–311, https://link.springer.com/.

19. Space Force Chief Technology and Innovation Office, *U.S. Space Force Vision for a Digital Service*, (Washington, DC: USSF, May 2021), https://media.defense.gov/.

20. Theresa Hitchens, "Microsoft Boosts Space Services, Partnerships," Breaking Defense, October 20, 2020, https://breakingdefense.com/.

21. Ben-Larbi et al., "Paradigm Shifts"; and Ben-Larbi et al., "Classification and Tools."

22. Kubos, "Major Tom"; and Ben-Larbi et al., "Paradigm Shifts."

23. J. D. Scanlan et al., "New Internet Satellite Constellations to Increase Cyber Risk in Ill-Prepared Industries," (paper, 70th International Astronautical Congress, Washington, DC, October 21–25, 2019).

24. Department of Defense (DOD), *Defense Space Strategy Summary* (Washington, DC: DOD, June 2020), https://media.defense.gov/.

25. Andrew J. Lingenfelter, Joshuah A. Hess, and Robert A. Bettinger, "From Sanctuary to Warfighting Domain: A Space System Survivability Framework," *Aircraft Survivability*, Summer 2021, 7–16.

26. Raymond, *Spacepower, 7.*

27. Charles Pope, "Driven by 'a Tectonic Shift in Warfare' Raymond Describes Space Force's Achievements and Future," SpaceForce News, September 15, 2020, https://www.spaceforce.mil/.

9 Using a CubeSat Reference Architecture for Accelerated Model Development and Analysis[*]

Major Sean Kelly, David Jacques,
Brad Ayres, Richard Cobb, and Thomas Ford
US Air Force Institute of Technology

CONTENTS

9.1 INTRODUCTION

The CubeSat class of nanosatellites has lowered the barrier of entry to space and has rapidly gained popularity over recent years. The lower development cost, small form factor, and use of commercial off-the-shelf (COTS) components (Karvinen et al., 2015) make the CubeSat form factor an ideal platform for university teams, where

[*] Reprinted with permission: Kelly, M. S., Jacques, D., Ayres, B., Cobb, R., & Ford, T. Using a CubeSat Reference Architecture for Accelerated Model Development and Analysis. *Journal of Small Satellites* *10(3)*, 1097–1108.

budget and development time are extremely limited. Many academic institutions have embraced this field for research and have developed their own space programs (Pradhan and Cho, 2020). To successfully design a CubeSat system in a rapid cycle conducive to academic timelines, a reference architecture geared toward university CubeSat development would be helpful. A reference architecture would further speed up the development process by providing a template, capturing previous work and lessons learned from subject matter experts, and providing the framework to focus on the design rather than the intricacies of modeling software. A reference architecture can also add functionality that student teams could use and improve over time, such as pre-built analysis functions and a library of components to choose from. This chapter discusses the need for a CubeSat reference architecture and explores features of one developed at the Air Force Institute of Technology (AFIT) for their space program.

9.2 MODEL-BASED SYSTEMS ENGINEERING

The International Council on Systems Engineering (INCOSE) defines Systems Engineering as *An interdisciplinary approach and means to enable the realization of successful systems* (Walden et al., 2015). The system, comprised of a collection of hardware, software, people, facilities, and procedures, begins as a theoretical concept in the eyes of users or stakeholders, and from that idea, needs are defined, a system is developed and used operationally, and finally retired or disposed of (Buede and Miller, 2016). Systems Engineering is all about addressing this complete life cycle, and there are many strategies or techniques to accomplish this. The Department of Defense and NASA have traditionally used a linear, document-based approach, but they are currently transitioning to a Model-Based Systems Engineering (MBSE) approach.

Documents are the primary artifacts available to stakeholders (Delligatti, 2014) in the traditional approach, including requirement and traceability matrices, interface documents, concept of operation documents, and other unique documents in a wide variety of formats. As systems become more complex, the traditional document-based approach becomes challenging to maintain. Each document is manually generated, so file management and version control are problematic. It is difficult to know for sure if a file is current or if it has been subsequently updated but located on some other file system or storage drive. Furthermore, any changes in one document, drawing, etc., must be made in any other document that contains items affected by the change, or risk multiple versions of the same document being presented. This system is prone to errors, inconsistencies, and difficulties in maintaining an accurate representation of the entire system. MBSE provides a solution to these increasingly relevant problems. In MBSE, a system model represents the system and any information traditionally needed for documents can be found within this model. The model becomes the source of truth instead of the documentation. When in doubt, the model always has the most current information, making it easier to stay consistent. If the modeler updates a component or interface in one area, it will be updated throughout the system as appropriate. Acquisition program reviews may still require paper documents, but the necessary information for those can still be found within the system model.

MBSE requires a modeling language, a modeling method, and a modeling tool (Delligatti, 2014). This CubeSat reference architecture was developed using the Systems Modeling Language and NoMagic's Cameo Systems Modeler (CSM) tool.

SysML is a standard modeling language that added systems engineering functionality to the Unified Modeling Language (UML) which has been used extensively in Software Engineering for decades (Delligatti, 2014). SysML provides a language, or the definitions and notations for nine different diagram types to describe a complex system, many of which will be used in this reference architecture. Model elements are expressed graphically through those diagrams, with SysML defining what those model elements are and how they are expressed. For example, a Block Definition Diagram (bdd) expresses system structure, and an Activity Diagram can show specific system behaviors. Within blocks, further detail can be expressed on an Internal Block Diagram (ibd). The modeling tool implements the SysML language, which allows for custom extensions if needed.

The modeling method is the specific methodology used to ensure important design tasks have been accomplished and provides the general guidance, processes, or steps for the system design. This chapter focuses on the Object-Oriented Systems Engineering Method (OOSEM), but there are other popular methods, such as the Weilkiens System Modeling (SYSMOD) method (Weilkiens, 2016) and the IBM Telelogic Harmony-SE method (Hoffman, 2020).

OOSEM uses SysML in a top-down, model-based approach that leverages object-oriented concepts with traditional systems engineering methods to architect more flexible and extensible systems that can evolve with technology and changing requirements (Estefan, 2008). OOSEM was developed in part by Lockheed Martin Corporation as a method to capture and analyze requirements of complex systems, integrate with object-oriented software and hardware, and support system-level reuse and design evolution (Walden et al., 2015).

OOSEM includes the following steps in an iterative fashion (Object Management Group, 2011), all of which are incorporated into the reference architecture.

1. *Analyze Stakeholder Needs*: Capture the "as-is" system and mission enterprise and identify gaps or issues. The "as-is" depiction helps develop the "to-be" system, and the gaps or issues can help drive mission requirements for the new system. OOSEM frequently uses measures of effectiveness for the primary mission objectives identified in this step.

2. *Define System Requirements*: Once the "as-is" system is defined and produces Mission Requirements, the system is modeled as a "black box" in a Mission Enterprise model. For example, instead of going deep into subsystem-level detail on a CubeSat, the entire CubeSat will be a "black box" that interacts with ground stations, other satellites, and the environment. This "black box" model allows for system-level activity diagrams and uses cases to show how the "to-be" system will support the mission enterprise. This step helps derive system-level functional, performance, and interface requirements.

3. *Define Logical Architecture*: A "logical" architecture is created that captures key functions in logical blocks, allowing for specific components to be chosen later in place of the logical depiction.

4. *Synthesize Candidate Allocated Architectures*: From the logical architecture, create potential physical instantiations using value properties and selected components. Each component at this stage is then traced to system requirements in table or matrix form.
5. *Optimize and Evaluate Alternatives*: Trade studies or other analysis is conducted at this step among the candidate architectures. Parametric diagrams within the model or integrating other tools can simulate system performance with the chosen components so alternative solutions can be compared.
6. *Validate and Verify System*: Once a candidate architecture has been chosen from the alternatives, the system needs to be validated and verified to ensure the requirements are being met and that stakeholder needs are satisfied. This step uses inspection, demonstration, analysis, and test activities to validate and verify the system.

Finally, the modeling tool is how the language, developer, and method work together. The modeling tool is a critical piece of software that maintains an underlying model of the system that can be used to display many different viewpoints or diagrams, depending on what is needed. The system model in a modeling tool is comprised of model elements and relationships between those elements, and from those, diagrams can be generated and displayed. When the source element or relationship is modified or deleted, that change gets carried out throughout the entire model, in all diagrams those elements or relationships appeared. The authors of this chapter used the CSM tool from No Magic Inc., but other tools are available on the market to accomplish the same goals with different user interfaces and feature sets.

9.3   REFERENCE ARCHITECTURES

Complex systems require a well-thought-out architecture early on in the design process. The Department of Defense recognized this issue for their complex systems, so they published the Department of Defense Architecture Framework (DoDAF) to establish "Enterprise-level Architectures" and "Solution Architectures" throughout the department. DoDAF defined an architecture as a "fundamental organization of a system embodied in its components, their relationships to each other and to its environment, and the principles governing its design and evolution over time (OASD/NII, 2010a)." This framework works well for major Defense Acquisition Programs, but small University teams that turn over every academic cycle are not able to take full advantage of this framework. Reference architectures can help alleviate that problem by consolidating subject matter expertise and previous relevant architectures into digestible models that system designers can benefit from when creating a solution architecture (Cloutier et al., 2010). The DoD saw the benefits of reference architectures and put out a reference architecture description in 2010, describing them as *an authoritative source of information about a specific subject area that guides and constrains the instantiations of multiple architectures and solutions* (OASD/NII, 2010b). A reference architecture should be an "elaboration of company (enterprise) or consortium mission, vision, and strategy, facilitating a shared understanding about the current architecture and the vision on the future direction" (Cloutier et al., 2010).

Furthermore, it should be continuously developed and improved over time as more teams use the architecture.

Finally, reference architectures should have at least the following elements (Cloutier et al., 2010):

1. *Strategic Purpose*: Goals, objectives, and a specific purpose or problem to be addressed
2. *Principles*: High-level foundational statements of rules, culture, and values that drive technical positions and patterns
3. *Technical Positions*: Technical guidance and standards that must be followed by solution architectures (maybe data vocabulary/ data model)
4. *Patterns (Templates)*: Generalized representations (e.g., Viewpoints, Views, Diagrams, Products, Artifacts) showing relationships between elements specified in the Technical Position
5. *Vocabulary*: acronyms, terms, definitions

Reference architectures are being used in many industries, and at least one has been developed for CubeSats already. Kaslow, Ayres, et al., as part of the INCOSE Space Systems Working Group (SSWG), drafted a CubeSat Reference Model (CRM) to help promote and institutionalize the practice of MBSE for CubeSat development. Their CRM provides a reusable logical architecture for a generic CubeSat and provides a model to create a physical architecture (Kaslow et al., 2017). The SSWG's CRM did not meet some specific needs for students though. For example, the CRM is not designed to generate traditional documents for system-level reviews. There is no easy way to generate a Concept of Operations document or Operational Requirements Document, for example, and that is a desire for a CubeSat reference architecture. Second, the CRM does not appear to have a component library or a generic, intuitive system that can be easily adapted by students new to MBSE. Finally, the CRM does not appear to have sufficiently detailed value properties for the system to be useful for detailed mission analysis using MATLAB and STK. Students in university courses must design down to a greater level of detail with many value properties for each subsystem in order to perform the required analysis and calculations. The CRM was quite useful though in examining what subject matter experts deem important for a CubeSat model and for their various subsystem internal block diagrams.

In summary, reference architectures can help systems engineers by providing a template, developed from years of experience, to aid in the systems engineering process. From the literature, it is clear that a reference architecture would be particularly useful for teams designing a CubeSat in a university setting, and this chapter will address that need.

9.4 RAPID DESIGN ENVIRONMENT

Between the compressed schedule, the distraction of other courses and projects, and the lack of modeling experience for most students, designing a satellite in a short timeframe is a challenge. Using AFIT as an example, the space vehicle design sequence lasts just nine months. Students start with a Mission Capabilities Document

(MCD), outlining the stakeholders' required capabilities and design constraints, and from there, they're expected to derive mission, system, and subsystem-level requirements, design the physical architecture, simulate that design, and ultimately test physical hardware to verify the requirements. Throughout the process, they build a system model from scratch and use the model to create traditional stakeholder documents, such as a Concept of Operations, Space Vehicle Requirements Document, etc. They must also demonstrate traceability throughout the model, from that original MCD through the tiers of requirements and to the physical components themselves.

Clearly, developing brand new components within this short time period is not feasible, so COTS components or components developed by the University are used to accomplish their objectives. A component library within the reference architecture would aid this process so teams can reuse or improve previous model elements if applicable. Students could copy and paste existing component blocks and simulate their system using those blocks to quickly assess mission feasibility with the chosen parts. Additionally, the primary mission stakeholders usually prefer traditional documents instead of a complex system model, so the model should aid in that process. In the past, students would copy and paste diagram images or transcribe requirement text into other tools, but this reference architecture will automate this process to rapidly generate deliverable documentation while avoiding the version control issues discussed previously.

In the university setting, the students generally come from a wide range of experience levels, with some having industry experience and others who have zero experience with satellites or with MBSE. Furthermore, students generally need to collaborate remotely due to their schedule demands. To address this, a cloud-based collaborative environment would be useful, and including examples and guidance will aid the less experienced team members. Template tables and diagrams should be provided so students can focus more on the design choices instead of the details of model organization, structure, stereotypes, etc. In the end, a reference architecture should support rapidly designing, simulating, prototyping, and testing a system by members of all experience levels.

9.4.1 Developing a CubeSat Reference Architecture

AFIT currently provides a template model to get students started with Stakeholder Analysis and developing a Concept of Operations, but support stops when students move onto future courses that build upon that foundation. Teams quickly diverge from using the model after the focus shifted from MBSE to design presentations and reports, so the need for a reference architecture to assist students became apparent.

The primary goal of this reference architecture is to encourage the use of MBSE throughout the entire design sequence, all the way through the testing of hardware while incorporating faculty input to meet the needs of three courses. In this course series, students use the textbook "Space Mission Engineering: The New SMAD" (Wertz, 2011), which was used as the primary source for equations, subsystem details, and mission activity descriptions.

The reference architecture opens with an overview diagram to show the organization of the model as shown in Figure 9.1. It shows the top-level package structure and

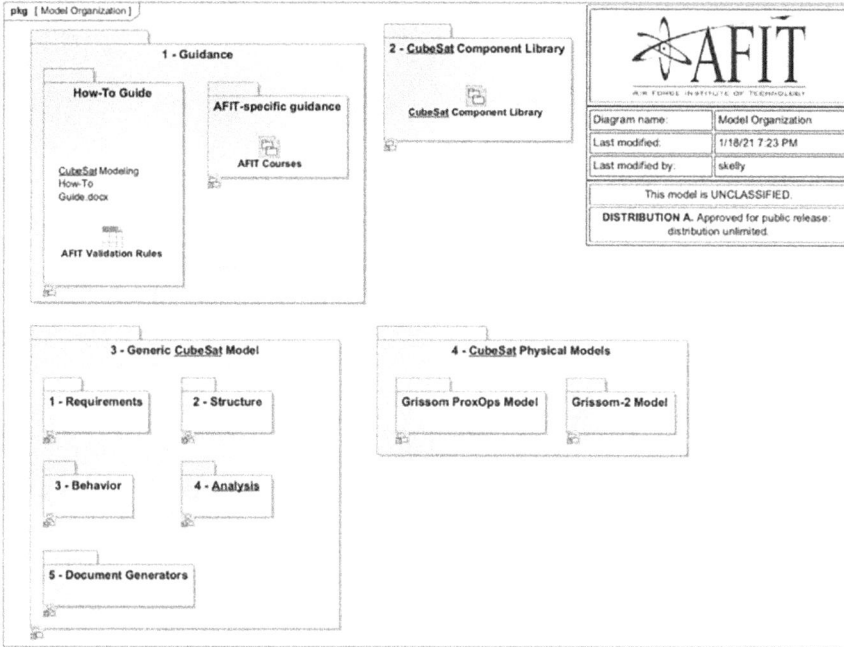

FIGURE 9.1 Top-level model organization.

includes hyperlinks to additional organizational pages for each model section. This allows for intuitive navigation, instead of always digging into the directory structure (called the "containment tree" in CSM) to search for sections. The first package contains guidance for students, with a how-to guide and example diagrams for those that are new to MBSE.

The Component Library is a new feature inspired by an AFIT-developed Small Unmanned Aircraft System (SUAS) reference architecture (Jacques and Cox, 2019). This feature is still a work in progress, but the goal is to have a library of components for each subsystem that can be reused in new models for rapid prototyping. For example, if an engineer wants to quickly test how different antenna options affect the radio frequency link analysis, they can use the available antenna options in the component library, each having value properties that affect the calculations in the Analysis section of the model. Each subsystem has starter components contained within the respective packages, and the intent is for this library to be updated as new CubeSat designs are created using the reference architecture.

The third package (Generic CubeSat Model) is the core of the reference architecture. The "Generic CubeSat Model" is the template that teams will start from. It has a pre-built, generic CubeSat model with diagrams, tables, and matrices provided with template data that is meant to be replaced by the design teams. It also contains the Document Generator tools that will be discussed later. Each package within the "Generic CubeSat Model" is hyperlinked to an informative diagram linking to all of the included tools and instructions for how to navigate them. An example diagram is shown in Figure 9.2, with links to all relevant requirement-related diagrams to fill out.

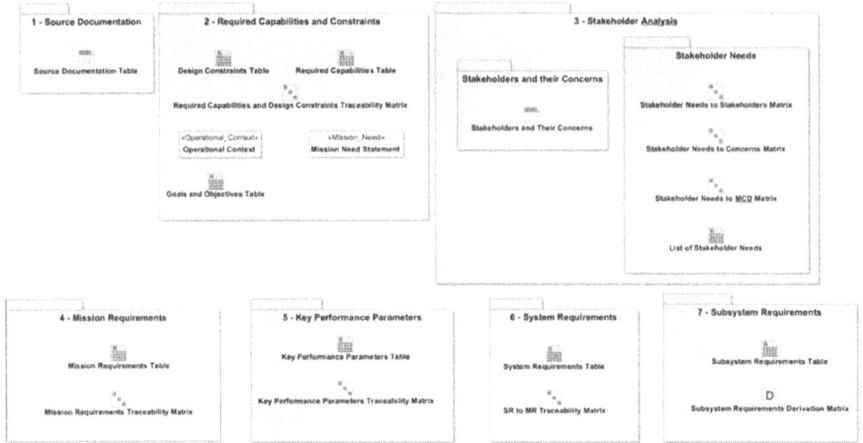

FIGURE 9.2 Requirements organization.

The CubeSat Physical Models package contains the various physical instantiations of the reference architecture. This could contain past projects to reference if needed, but for the purposes of the reference architecture development, it was used as the testbed to validate the model. This is also where teams will place their starting template to build from, keeping the generic CubeSat model for future use.

9.5 USE OF CUBESAT REFERENCE ARCHITECTURE FOR REQUIREMENTS VERIFICATION AND VALIDATION

One of the key functions of this CubeSat reference architecture is the verification and validation section. Figure 9.3 shows the analysis portion of the reference architecture, which helps guide teams through the requirement verification and validation processes. Because the CubeSat component blocks include predefined value properties, parametric diagrams could be created to perform a variety of calculations based on these values. An example of the included parametric analysis is associated with the thermal properties of the system, shown in Figure 9.4. This thermal analysis parametric diagram includes a constraint block with MATLAB code. This MATLAB code uses value properties from the "Thermal Subsystem" block (radiating area, emissivity, absorptivity, specific heat capacity, etc.), some parameters from the "Orbit" block (altitude, period, etc.), and any other value properties needed to perform the calculations. The MATLAB script then displays graphs that change automatically if the user swaps out a new thermal subsystem block, changes the altitude, or otherwise modifies the value properties. The constraint blocks can integrate engineering analysis into SysML modeling, and several analysis patterns were included in the reference architecture to assist future teams in performing rapid analysis while keeping all work inside the system model.

The objective for the Analysis section of the reference architecture is to keep as much analysis contained within the model as possible, using the actual value properties to perform the calculations. Instead of moving values to other tools, the analysis

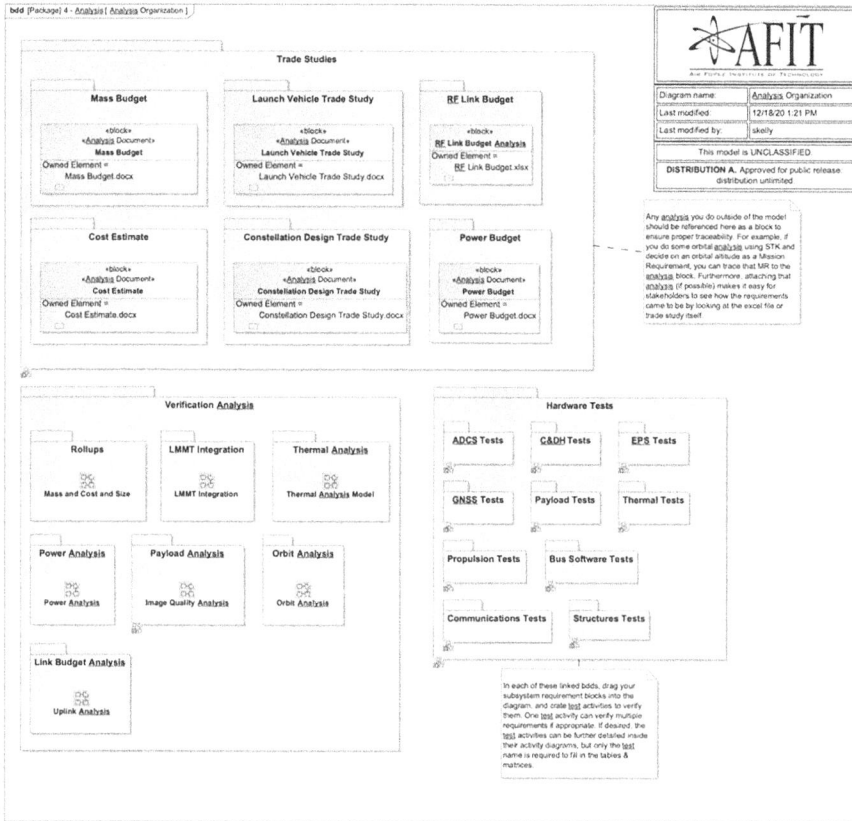

FIGURE 9.3 Analysis organization.

calculations are kept within the model. Additional functionality can be added as well, depending on the requirements. For example, Figure 9.5 shows how a requirement for a Near Infrared (NIR) Ground Sample Distance (GSD) of less than 4 m could be tested using the same methodology. In this parametric diagram, the NIR GSD is calculated based on the imager's value properties and the CubeSat's altitude. This result is compared to the requirement and will automatically flag the result as green or red, depending on if it meets the requirement's constraint block or not, as shown in Figure 9.6. In this example, an engineer could tweak the design variables and instantly see how the GSD is affected, which would be very useful in the early stages of design.

In addition to the parametric diagrams, teams will need to perform hardware tests in the lab, and the reference architecture accounts for that as well. There is a package for defining hardware tests for each subsystem with features to help keep everything organized. Figure 9.7 shows the testing diagram for the Electrical Power Subsystem (EPS). The user can access the applicable requirements in the linked subsystem requirement table, create test activities to verify requirements, and include descriptions of each test in the included test description table. The linked subsystem

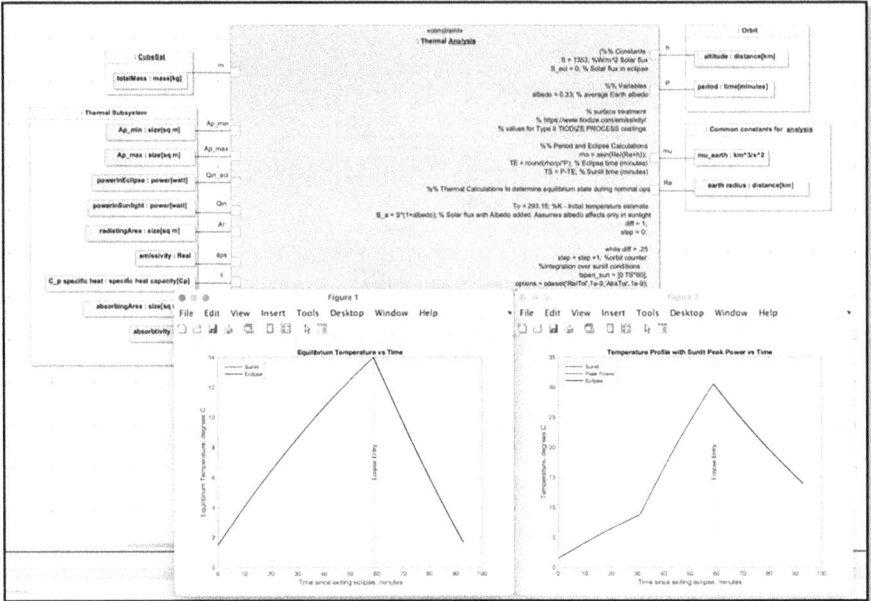

FIGURE 9.4   Thermal analysis.

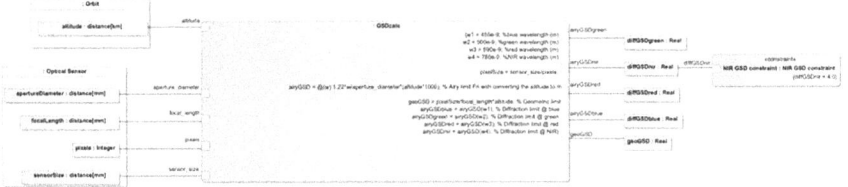

FIGURE 9.5   Image quality parametric diagram.

Name	Value
Image Quality Analysis	Image Quality Analysis@62a1c657
diffGSDblue : Real	1.8503
diffGSDgreen : Real	2.0333
diffGSDnir : Real	3.1720
diffGSDred : Real	2.3993
geoGSD : Real	0.0018
: Optical Sensor	Optical Sensor@18638d7d
: Orbit	Orbit@15816cce
: GSDcalc {w1 = 455e-9; %blue wavelength (m)w2...	GSDcalc@19a1763b
NIR GSD constraint : NIR GSD constraint {diffGSDnir...	NIR GSD constraint@58591fd7
diffGSDnir	3.1720

FIGURE 9.6   Image quality results.

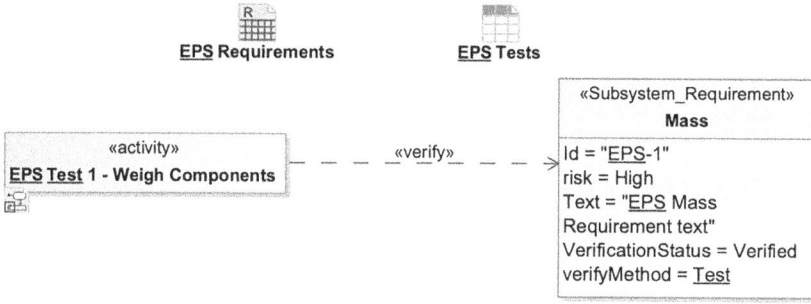

FIGURE 9.7 EPS tests.

requirement tables are automatically generated and also include color coding to high-light testing status. As tests are completed, the user can choose a verification status from a dropdown menu (such as Requirement Verified, Test Not Completed, Testing in Progress, etc.) to keep track of the test campaign in one standardized location.

9.6 USE OF CUBESAT REFERENCE ARCHITECTURE FOR GENERATING TRADITIONAL DOCUMENTATION

This reference architecture includes a polished document generator to create traditional documents for stakeholders. CSM can use Apache's Velocity Template Language (VTL) to export model elements into external tools such as Microsoft Word and Microsoft PowerPoint, so by including Microsoft Word file templates with tailored VTL code, documents can be automatically generated as needed as the model is updated. Even narrative text, such as introduction paragraphs or other lead-in text, can be included within the model as notes. The reference architecture includes document generators for a Stakeholder Analysis Report (SAR), Concept of Operations (CONOPS), Mission Requirements Document (MRD), Operational Requirements Document (ORD), Space Vehicle Requirements Document (SRD), MCD, and a document to help start Test Plans. The templates were based on AFIT course requirements, but they may be tailored if different sections are required or if the order of sections needs to be modified. Additionally, the reference architecture includes a template for a master document, which exports all model elements that the team generated in an intuitive and visual format. This is useful as it contains all code that a user may wish to use if they want to create a new document template that was not provided. The generic model document can be used as a template to build custom documents. The generic template is commented on so users can know how it works and what to copy for a new document, as this template language is not well explained in the software user's manual.

The goal of these document generators is to encourage the use of the model throughout the design and build process. Historically, teams would copy and paste model elements into reports and transcribe requirement text into Microsoft Word or PowerPoint tables for reviews or presentations. This led to version control issues, such as requirement text being updated in the PowerPoint table but not in the underlying model. By keeping everything entirely within the model, these document generators take most of the manual work out of the process. Teams can focus on ensuring the model is accurate instead of needing to cross-check every diagram each time a change is made.

9.7   MISSION MODELING ON THE AFIT CUBESAT BUS

AFIT's CubeSat reference architecture is not only designed for classroom projects. AFIT has their own space program, and this reference architecture is intended to assist with its mission modeling. AFIT uses a 6U-sized bus called the "Grissom Bus" for its upcoming missions, and this reference architecture has that bus included in the component library along with other payloads and subsystem components that AFIT has in stock. As future Grissom-based CubeSats are designed at AFIT, this reference architecture can be used to rapidly prototype design configurations using the available components and the built-in analysis tools. Two upcoming missions, Grissom-1 and Grissom-P, have already been modeled using this reference architecture as test cases. Their requirements and structure were created, and stakeholder documentation was generated to ensure the process and tools worked for other models. Additional research is being done using this baseline reference architecture to easily simulate multiple AFIT payloads using the Grissom bus, and it all relies on the built-in structure and value properties associated with components. The goal is to be able to quickly test multiple design configurations using the same bus. For example, two payloads on the same bus may have conflicting pointing requirements or a combined power draw that exceeds the power budget requirement, and the simulations should highlight these conflicts. As missions, such as the Grissom-based missions, are fully modeled in this architecture, the analysis tools can be used to model various mission phases or activities. For example, the relevant orbital parameters can be used to automatically create an STK scenario, displaying a visual simulation of a ground station contact using parametric diagrams in the reference architecture.

Grissom technicians at AFIT were consulted during the development of this reference architecture to determine what value properties should be included, and those mission modeling tools are actively being developed using this reference architecture as a baseline.

9.8   FUTURE WORK AND CONCLUSION

To date, the CubeSat reference architecture effort has been to establish a working template to test with the next cohort of Space Systems Engineering students at AFIT. As the course sequence curriculum changes and as more people look at and use this reference architecture, improvements can and should be made for the benefit of future teams. The reference architecture as-is has been tested by prior students, but every team has different styles and preferences, and the reference architecture can reflect those differences.

The Component Library will be expanded as the reference architecture is used by design teams. After several iterations, there will be several different types of payloads to choose from, multiple propulsion systems, chassis sizes, etc. The Component

Library also contains other non-physical blocks for reuse as well, such as constraint blocks for analysis, object flows, value types, and custom stereotypes.

The Analysis section includes several working examples that work with the generic component blocks, but as future teams add working MATLAB code or STK configurations to their analysis, those can be saved in the Component Library as well. Future teams can copy any relevant constraint blocks to use in their own parametric diagrams, and over time, a wealth of working analysis can be saved and continuously improved upon.

Finally, the verification functionality built into the model primarily addresses technical performance, not programmatic requirements. Further work should be done to add functionality to validate top-level mission requirements such as schedule or regulatory requirements.

In summary, this first attempt at a CubeSat reference architecture is designed to be improved over time. Currently, it will guide teams and constrain them in a way that makes the modeling effort easier so they can focus on the design details and technical analysis. Students new to MBSE should be able to use this model, and those more familiar with the tool can add new features for future teams to take advantage of. It will also be the platform upon which future mission modeling tools are based to integrate STK or similar tools for more in-depth analysis. The goal here is to keep system data within the model, and this reference architecture will help encourage teams to do that.

ACKNOWLEDGMENTS

The author would like to thank the Center for Space Research and Assurance (CSRA) at AFIT for their guidance throughout this research.

REFERENCES

Buede, D. and Miller, W. (2016): *The Engineering Design of Systems: Models and Methods* (3rd ed.). Hoboken, NJ: John Wiley and Sons.

Cloutier, R., Muller, G., Verma, D., Nilchiani, R., Hole, E. and Bone, M. (2010): The Concept of Reference Architectures. Systems Engineering. Vol 13: 14–27. doi 10.1002/sys.20129.

Delligatti, L. (2014): *SysML Distilled: A Brief Guide to the Systems Modeling Language*, Boston, MA: Addison-Wesley Professional.

Estefan, J. (2008): Survey of Model-Based Systems Engineering (MBSE) Methodologies (rev. B). Seattle, WA, USA: International Council on Systems Engineering (INCOSE). INCOSE-TD-2007-003-02. Available at http://www.omgsysml.org/MBSE_Methodology_Survey_RevB.pdf (accessed Nov 5, 2020).

Hoffmann, H.P. (2020): Systems Engineering Best Practices with the Rational Solution for Systems and Software Engineering. Deskbook release 3.1.2 ed. IBM.

Jacques, D. and Cox, A. (2019): The Use of MBSE and a Reference Architecture in a Rapid Prototyping Environment, Tech. rep., Air Force Institute of Technology.

Karvinen, M., Tikka, T., and Praks, J. (2015): Using Hobby Prototyping Boards and Commercial-Off-The-Shelf (COTS) Components for Developing Low-Cost, Fast-Delivery Satellite Subsystems. *Journal of Small Satellites*, vol. 4(1), pp. 301–314.

Kaslow, D., Ayres, B., Cahill, P. et al. (2017): Developing a CubeSat Model-Based Systems Engineering (MBSE) Reference Model - Interim Status 3, Tech. rep., IEEE.

Office of the Assistant Secretary of Defense, Networks and Information Integration (2010a): The DoDAF Architecture Framework Version 2.02. Available at https://dodcio.defense. gov/library/dod-architecture-framework/ (accessed Nov 5, 2020).

Office of the Assistant Secretary of Defense, Networks and Information Integration (2010b): Reference Architecture Description. Available at dodcio.defense.gov/Portals/0/ Documents/Ref_Archi_Description_Final_v1_18Jun10.pdf (accessed Nov 5, 2020).

Object Management Group (2011): INCOSE Object-Oriented Systems Engineering Method (OOSEM). Available at https://www.omgwiki.org/MBSE/doku.php?id=mbse:incoseoosem (accessed Oct 15, 2020).

Pradhan, K. and Cho, M. (2020): Shortening of Delivery Time for University-Class Lean Satellites. *Journal of Small Satellites*. Vol. 9(1), pp. 881–896.

Walden, D., Roedler, G., Forsberg, K. et al. (2015): *Systems Engineering Handbook: A Guide for System Life Cycle Processes and Activities* (4th ed.). San Diego, CA: INCOSE.

Weilkiens, T. (2016): *SYSMOD - The Systems Modeling Toolbox* (2nd ed.). Fredesdorf, Germany: MBSE4U Booklet.

Wertz, J. R., Everett, D. F., and Puschell, J. J. (2011): *Space Mission Engineering: The New SMAD* (1st ed.). Torrance, CA: Microcosm Press.

10 Cislunar Debris Propagation Following a Catastrophic Spacecraft Mishap*

Nathan R. Boone and Robert A. Bettinger
US Air Force Institute of Technology

CONTENTS

This research involves theoretical analysis of the short- and long-term motion of space debris in cislunar trajectories following a spacecraft catastrophic mishap. Specifically, the research formulates a debris propagation model using four-body dynamics and determines debris trajectories following breakup events for a variety of different initial orbital positions. A spacecraft survivability model is then used to quantify the risks from the debris to other cislunar spacecraft. Three cislunar debris case studies are

---

* Reprinted with permission: Boone, N., & Bettinger, R. A. (2021). Cislunar Debris Propagation Following a Catastrophic Spacecraft Mishap. In *AIAA Scitech 2021 Forum* (p. 0102).

DOI: 10.1201/9781003220978-13

examined, which include catastrophic spacecraft mishaps during an Apollo-like Earth-Moon transfer, during a transfer between L_1 and Earth along the L_1 manifold, and at the stable Lagrange points L_4 and L_5. Risks to current operational spacecraft near Earth were found to be greatest for the Apollo-like transfer case study, and slight risks to spacecraft in cislunar orbits were found for the other case studies. Overall, research into cislunar debris propagation will enhance operational planning outside the traditional near-Earth paradigm of spacecraft mission operations and increase understanding of the debris-related consequences of mishaps within this orbital regime.

10.1   NOMENCLATURE

A_1 = Distance from the Sun to the Sun-Earth-Moon barycenter
A_2 = Distance from the Sun-Earth-Moon barycenter to the Earth-Moon barycenter
A_P = Surface area of spacecraft facing debris fragment cloud
A_S = Surface area of debris fragment cloud facing a spacecraft
C = Jacobi's Constant
E = Expected number of debris hits on a spacecraft
M = Number of debris fragments
m_S = Mass of the Sun
$P_{K|H}$ = Probability of spacecraft kill with a debris particle hit
P_{HZ} = Probability of spacecraft hazard
t = Time in terms of the period of the Earth-Moon system about its barycenter
v = Speed
μ = Mass of the Moon in terms of the mass of the Earth-Moon system, equal to 0.01215
ρ = Debris cloud density
ρ_E = Distance from simulated particle to the Earth
ρ_M = Distance from simulated particle to the Moon
ρ_S = Distance from simulated particle to the Sun
ω_1 = Angular velocity of the Earth-Moon barycenter about the Sun-Earth-Moon barycenter
ω_2 or ω = Angular velocity of the Earth-Moon system about the Earth-Moon barycenter

10.2   INTRODUCTION

Traditionally, space operations for most emerging or heritage spacefaring nations were notionally limited to near-Earth space, with mission altitudes extending to geosynchronous or highly elliptical orbits. During the 2010s, however, space operations began moving beyond this approximate altitude limit to encompass the space between the Earth and the Moon. This space, known as cislunar space,[†] offers to serve as a new "high ground" for space operations, thus allowing a positional and logistic advantage over other space-based assets. In terms of these cislunar space

---

[†] The term "cislunar" refers to the spatial volume between the Earth and Moon. By comparison, the term "translunar" refers to specific trajectories and/or missions between the Earth and Moon. For the present research, "cislunar" will be used exclusively in conjunction with all analysis showcased herein.

operations, there are reinvigorated U.S. initiatives to return to the Moon, planned commercial space projects, and cislunar injection trajectories for geosynchronous-orbiting satellites. Additionally, an increase in international projects has pushed space domain awareness and space operations considerations beyond near-Earth space. Recent activity includes Israel's attempted lunar surface mission (2019) [1] and China's Chang'e-4 far side Moon mission and the accompanying Queqiao relay satellite orbiting the Earth-Moon L_2 Lagrange point. China eventually intends to industrialize the Moon to make economic use of its resources, potentially creating an enormous advantage to Chinese operations in the cislunar region [2]. International competition in cislunar space is likely to increase the number of spacecraft operating in this region.

As more spacecraft operate in cislunar space, a catastrophic spacecraft mishap in this region becomes more likely, with each mishap generating a significant number of debris fragments. Many studies have investigated the threat of debris events in lower Earth orbits. Concerns over debris near Earth have grown with recent anti-satellite (ASAT) tests performed by China in 2007 [3] and by India in 2019 [4], which generated debris clouds that continue to threaten other spacecraft. Several satellites have also suffered failures that led to breakup events, including battery explosions on board the U.S. Air Force's Defense Meteorological Satellite Program (DMSP) satellite and the NOAA 16 weather satellite in 2015 [5]. The 2009 collision of Cosmos 2,251 and Iridium 33 also created two large fragment clouds [5]. These fragmentation events could lead to a cascading series of collisions, eventually leaving certain Earth orbits so full of debris that they are unusable, a phenomenon known as "Kessler Syndrome" [6].

While the threat of debris in orbits near Earth has been heavily studied, few prior studies have investigated the effects of a similar debris event in cislunar space. The unique dynamics of this environment make it difficult to predict the motion of the debris over time. Namazyfard [7] discussed the effects of lunisolar perturbations on the Eccentric Geophysical Observatory (EGO) satellite, which was launched in 1964 and expected to return to Earth in 16 years but remains in space over 50 years later. The unexpected longevity of this satellite demonstrates the difficulties in properly modeling the perturbations in cislunar space, and the potentiality for objects to remain there indefinitely under certain conditions.

One spacecraft mishap has already occurred in cislunar space, namely the explosion of the oxygen tank onboard the Apollo 13 spacecraft during its transfer to the Moon. This explosion generated so many fragments that the swarm of debris surrounding the spacecraft made navigation by the stars impossible, forcing mission control to instead rely on navigation using the Sun [8]. The debris field could also be imaged from the ground, and images following the explosion show a debris cloud extending about 60 km from the spacecraft [9]. Following the explosion, the debris likely moved back toward Earth, possibly intersecting lower Earth orbits at very high relative velocities. A similar debris event today could potentially threaten spacecraft in the crowded Low Earth Orbit (LEO) and Geostationary (GEO) environments if the debris intersects those orbits. Therefore, the first cislunar debris case study presented in this research involves the study of a catastrophic spacecraft mishap during a transfer to the Moon much like the Apollo 13 scenario.

The second case study presented in this research involves a catastrophic space-craft mishap during a transfer to the Moon using a low-energy trajectory. Trajectories that utilize three-body dynamics are an ongoing topic of research and can enable transfers between Earth orbits and the Moon or the Lagrange points for less fuel than conventional direct transfers. These types of transfers are discussed in detail by Parker and Anderson [10]. The Gravitational Recovery and Interior Laboratory (GRAIL) mission, launched in 2011, was the first spacecraft to utilize a low-energy transfer to reach the Moon [10]. These types of trajectories are expected to become more common due to their fuel savings as lunar exploration increases, increasing the need to understand the motion of debris following a catastrophic spacecraft mis-hap that occurs while a spacecraft is utilizing these techniques. The stable nature of these trajectories, and their great distance from Earth that negates the effect of atmospheric drag, could lead to long-term debris threats in cislunar space.

Finally, some types of cislunar debris events could lead to a hazardous accumula-tion of debris in certain regions. Like observed natural phenomena, debris particles produced by a catastrophic spacecraft mishap may begin to "pool" in certain orbits in cislunar space, eventually creating significant debris hazards in those orbits. Several papers have investigated the existence of "Kordylewski Clouds," which are accumu-lations of cosmic dust that have been observed near the stable Earth-Moon Lagrange points L_4 and L_5. The dynamics of these clouds may suggest debris risks to space-craft operating near these points. Slíz-Balogh, Barta, and Horváth noted in their study of the Kordylewski clouds that rock-sized particles should be able to circulate in the vicinity of the L_5 Lagrange point for a long time [11]. If a significant debris event occurs near the L_4 or L_5 points, then it could impact the reliability of future spacecraft that operate at those points. Therefore, the third case study presented in this research identifies the risks to spacecraft at L_4 and L_5 following a catastrophic spacecraft mishap at each Lagrange point.

In terms of structure, the first section of the paper reviews existing literature related to cislunar debris. Next, the methodology for the simulations is presented, including the trajectory generation model, the models for the spacecraft trajectories prior to the catastrophic mishaps, the model for the velocity and mass distributions that result from the spacecraft catastrophic mishap, and the spacecraft survivability model used to quantify risks from the debris to other spacecraft. Finally, the results of the simulations for each of the three case studies are provided.

10.3 BACKGROUND

10.3.1 Natural Debris in Cislunar Space

The ambient meteoroid[‡] environment in cislunar space has been a subject of study since the 1960s. These studies can provide a baseline for the natural particle flux in cislunar space. Burbank, Cour-Palais, and McAllum [12] developed expressions for the flux of meteoroid particles by mass in cislunar space in a 1965 NASA report.

[‡] Meteoroids are rocky or metallic objects in space that range in size from small, dust-sized grains to rock-sized objects less than about 1 m in diameter.

This chapter noted that the size of meteoroid particles is always greater than 0.3 μm for metallic particles because smaller particles are swept out of the solar system by solar radiation pressure. Davidson and Sandorff [13] studied the sources of meteoroids, the structural damage meteoroid impacts could cause to spacecraft and spacecraft shielding methods in a 1963 NASA report. Singer [14] analytically predicted a dust shell around the Earth with a concentration of dust a few times higher than the concentration in interplanetary space in a 1961 paper. Hyde and Alexander [15] studied ejecta from the lunar surface and found that some intersected the Earth's magnetic field and obtained geocentric orbits. More recently, Altobelli, Grün, and Landgraf [16] studied data collected from the Helios spacecraft to analyze the density, composition, and interaction with solar radiation pressure of interstellar dust particles near Earth. In this chapter, the ratio of solar radiation pressure to the gravitational force, β, was calculated as a function of particle mass. This ratio is extremely small for particles heavier than about 10^{-10} kg, indicating that gravity has a much more significant influence on the motion of larger particles.

10.3.2 COLLINEAR LAGRANGE POINT DEBRIS

Some prior literature exists on catastrophic spacecraft mishaps at collinear Lagrange points. Landgraf and Jehn [17] studied the effects of a spacecraft explosion in the Sun-Earth L_2 point. The explosion creates a cloud of debris in the stable manifold, and about 56% of the fragments in the simulation moved toward Earth, with almost all approaching within the Moon's orbit. About 7% of the fragments came within geosynchronous altitude and about 2% reached the LEO environment. Bandyopadhyay, Sharma, and Tewari [18] conducted a similar study on a fragmentation event at the Earth-Moon L_1 point, and found that about 1.6% of the fragments came within geosynchronous altitude. The authors emphasized the need for a more detailed study of fragmentation events at the collinear Earth-Moon Lagrange points.

10.3.3 KORDYLEWSKI CLOUDS

Of the five stationary Lagrange points where the gravitational forces balance in the three-body problem, two, L_4 and L_5, are stable. Objects orbiting the L_4 or L_5 points, located 60° ahead and behind the central body, respectively, are in stable equilibrium and will tend to return to those points if perturbed. Any object orbiting at one of the collinear Lagrange points L_1, L_2, or L_3 is in unstable equilibrium and will tend to fall away from these points if perturbed. Therefore, natural satellites at the L_4 and L_5 points are common. For example, over 7,000 asteroids have been discovered at the Sun-Jupiter L_4 and L_5 points [18].

Polish astronomer Kazimierz Kordylewski first began searching for large objects near the Earth-Moon L_4 and L_5 points with a telescope in 1951. After the initial search was unsuccessful, Kordylewski instead began looking for a cloud of small dust particles too small to be seen individually, but collectively visible with the naked eye on dark and clear nights. He first observed large patches (about four times the size of the Moon) near the L_5 point in 1956, then succeeded in photographing the dust patches in 1961. There have since been numerous attempts, both successful and unsuccessful, to observe these clouds, now known as the "Kordylewski clouds." The Japanese Hiten space probe intended to confirm the existence of the Kordylewski clouds when it passed through the L_4 and L_5 points in 1993, but no obvious increase in dust concentration was detected [18].

Several papers have sought to explain the dynamics of the hypothesized Kordylewski clouds and demonstrate that they can exist. Pohle used the CR3BP to analyze the motion of the clouds in 1962 [20,21], concluding that the dynamics allowed cloud shapes like those observed by Kordylewski. Salnikova and Stepanov in Ref. [18,22–24] have extensively studied the Kordylewski clouds using the planar BCR4BP and a model for the force exerted by solar radiation pressure, showing that dust is stable at the L_4 and L_5 points even with gravitational perturbations from the Sun. The optimal times for maximum visibility of the clouds as seen from Earth are discussed in [22], the motion of charged dust particles in the clouds is studied in [18], a probabilistic model for the distribution of dust is developed in [23], and the possibility of multiple dust clouds is examined in [24].

In 2018, a detailed study by Slíz-Balogh, Barta, and Horváth [11,25] was reported to have confirmed the existence of the Kordylewski clouds. In the first part of the study [11], a three-dimensional restricted four-body problem was used to simulate the motion of particles in the vicinity of the L_5 Lagrange point. The positions and velocities of the Sun, Earth, and Moon were taken from the JPL HORIZONS database. To determine if interplanetary dust would become trapped near the L_5 point, 1,860,000 particles moving slowly relative to L_5 were simulated using four-body dynamics in the vicinity of L_5. Solar radiation pressure was also considered; however, because solar radiation ejects particles with sizes between 0.1 and 0.5 μm from the L_5 point while having a negligible effect on larger particles, it was not needed in the simulation. The resulting summed distribution of particle positions matched well with optical analysis conducted using ground-based imaging polarimetry in the second part of the study [25].

The existence of the Kordylewski clouds could suggest hazards to spacecraft operating near the L_4 and L_5 points, particularly if rock-sized particles can circulate in the L_4 and L_5 for long periods as suggested in Ref. [11]. This issue could be exacerbated if artificial debris begins to collect in these points. Methods very similar to those used to study the Kordylewski clouds are used in the present research to determine if large fragments of artificial debris from a catastrophic spacecraft mishap in cislunar space could begin to collect near the stable Lagrange points.

10.4 METHODOLOGY

10.4.1 THE CIRCULAR RESTRICTED THREE-BODY PROBLEM

The Circular Restricted Three-Body Problem (CR3BP) may be used to model the motion of a spacecraft of negligible mass moving in the gravitational influence of the Earth and Moon, assuming that the Earth and Moon move in coplanar circular orbits about their barycenter with a constant angular velocity. The equations of motion are given in the coordinates of the Earth-Moon barycentric rotating reference frame shown in Figure 10.1.

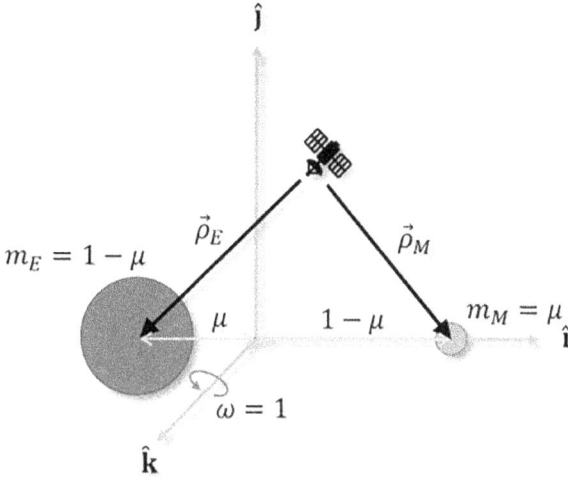

FIGURE 10.1 Earth-moon-spacecraft CR3BP in a Barycentric rotating reference frame.

The non-dimensionalized second-order equations of motion in the rotating reference frame are:

$$\ddot{x} = x + 2\dot{y} - \frac{(1-m)(x+m)}{\rho_E^3} - \frac{(x-1+m)}{\rho_M^3} \tag{10.1}$$

$$\ddot{y} = y - 2\dot{x} - \frac{(1-m)y}{\rho_E^3} - \frac{my}{\rho_M^3} \tag{10.2}$$

$$\ddot{z} = -\frac{(1-m)z}{\rho_E^3} - \frac{mz}{\rho_M^3} \tag{10.3}$$

where

$$\rho_E = \sqrt{(x+m)^2 + y^2 + z^2} \tag{10.4}$$

$$\rho_M = \sqrt{(x-1+m)^2 + y^2 + z^2} \tag{10.5}$$

The value of the parameter μ is 0.01215. These equations of motion are in terms of the characteristic quantities given in Table 10.1. The Distance Unit (DU) is equal to the distance between the Earth and the Moon, and the Time Unit (TU) is the time unit that makes the time for the system to complete one orbit about its barycenter 2π.

Equilibrium points in the CR3BP are determined by setting the velocities and accelerations in the equations of motion equal to zero. Solving for positions with

TABLE 10.1

Characteristic Quantities for the Earth-Moon System

Characteristic Length (l^*)	Characteristic Mass (m^*)	Characteristic Time (t^*)
384,400 km	6.0458×10^{24} kg	4.3425 days

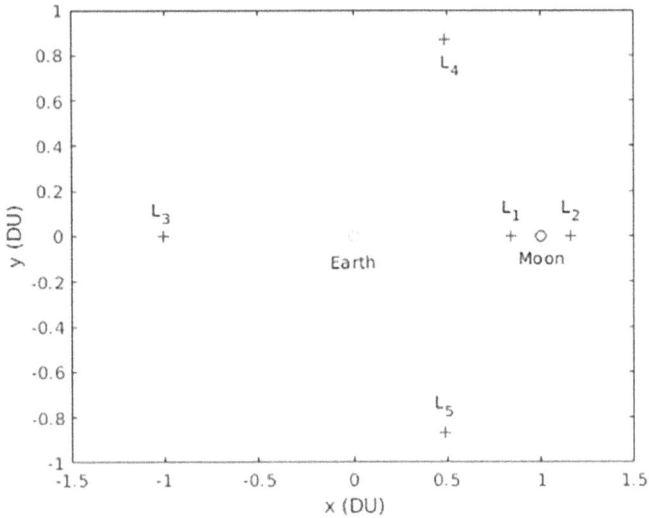

FIGURE 10.2   Equilibrium points in the Earth-Moon CR3BP.

TABLE 10.2

Coordinates of Lagrange Points in the Earth-Moon System

Lagrange Point	x Position (DU)	y Position (DU)
L_1	0.837	0
L_2	1.156	0
L_3	−1.005	0
L_4	0.488	0.866
L_5	0.488	−0.866

these conditions gives the five Lagrange points shown in Figure 10.2. The two-dimensional coordinates for these points are given in Table 10.2.

In the rotating frame, a constant of the motion known as Jacobi's Constant exists. This constant is expressed as:

$$C = 2U - n^2 \tag{10.6}$$

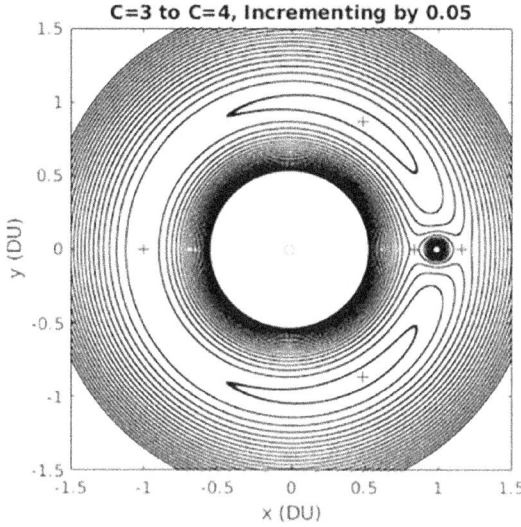

FIGURE 10.3 ZVCs for varying Jacobi Constants.

where v is the magnitude of the total spacecraft velocity, and U is given by:

$$U = \frac{1-m}{\rho_E} + \frac{m}{\rho_M} + \frac{1}{2}(x^2 + y^2) \tag{10.7}$$

The Jacobi Constant defines regions of cislunar space that are accessible to a particle with a given energy level. Positions associated with zero velocity for a particular value of Jacobi's Constant may be plotted to generate energy contour curves known as Zero Velocity Curves (ZVCs) as shown in Figure 10.3.

10.4.2 THE BI-CIRCULAR RESTRICTED FOUR-BODY PROBLEM

Although the CR3BP provides significant insight into the dynamics of the Earth-Moon system, a model that incorporates the gravitational influence of the Sun is preferable for long-term trajectory modeling. The Bi-circular Restricted Four-Body Problem (BCR4BP) is an extension of the CR3BP that incorporates solar gravity. In the formulation used for this research, the Earth and Moon revolve in circular orbits about their barycenter, which revolves in a circular orbit about the Sun-Earth-Moon barycenter. All orbits are assumed to lie in the same plane. A diagram of this problem is shown in Figure 10.4. The equations of motion are again derived in the Earth-Moon barycentric rotating reference frame shown in Figure 10.1. These equations form the main trajectory model for the present research.

The second-order non-dimensional equations of motion in the Earth-Moon rotating frame are:

$$\ddot{x} = x + 2\dot{y} + A_2\omega_1^2 \cos(t(\omega_1 - 1)) - \frac{m_s(x - x_S)}{\rho_S^3} - \frac{(1-m)(x+m)}{\rho_E^3} - \frac{m(x-1+m)}{\rho_M^3} \tag{10.8}$$

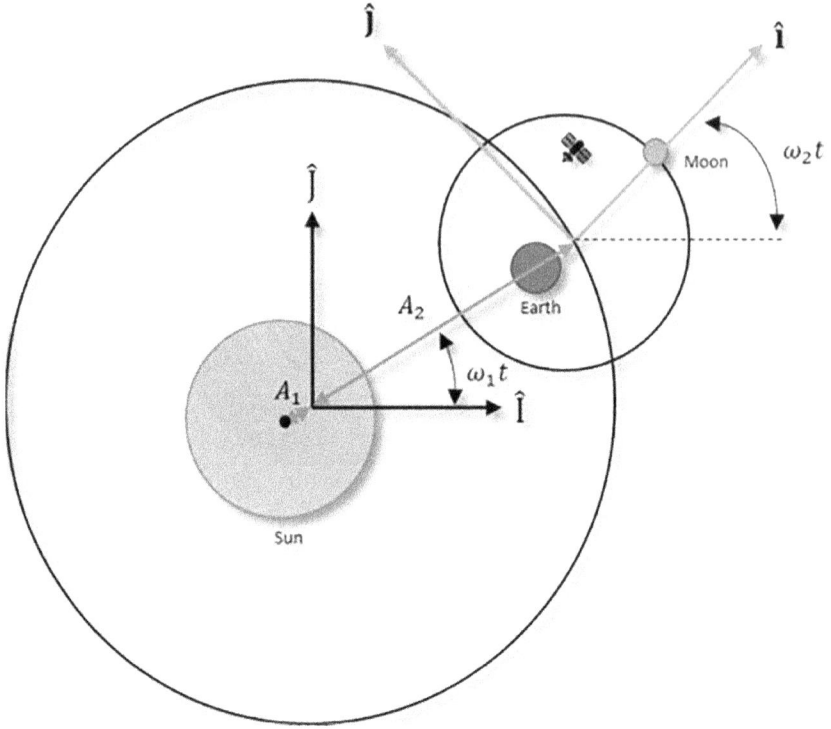

FIGURE 10.4   Coordinate frames in the BCR4BP.

$$\ddot{y} = y + 2\dot{x} + A_2\omega_1^2 \sin\left(t(\omega_1 - 1)\right) - \frac{m_s(y - y_S)}{\rho_S^3} - \frac{(1-m)y}{\rho_E^3} - \frac{my}{\rho_M^3} \qquad (10.9)$$

$$\ddot{z} = -\frac{m_s z}{\rho_E^3} - \frac{(1-m)z}{\rho_M^3} - \frac{mz}{\rho_M^3} \qquad (10.10)$$

Numerically integrating these equations will yield the trajectory. All values are again in terms of the characteristic quantities in Table 10.1. The angular velocity of the system about the Sun-Earth-Moon barycenter, ω_1, is given by:

$$\omega_1 = \sqrt{\frac{m_s + 1}{(A_1 + A_2)^3}} \qquad (10.11)$$

The coordinates of the Sun are:

$$x_S = -(A_1 + A_2)\cos\left(t(\omega_1 - 1)\right) \qquad (10.12)$$

$$y_S = -(A_1 + A_2)\sin\left(t(\omega_1 - 1)\right) \qquad (10.13)$$

The distance to the Sun is given by:

$$\gamma_S = \sqrt{(x - x_S)^2 (y - y_S)^2 + z^2} \qquad (10.14)$$

The distances to the Earth and Moon were given in Eqs. (10.4) and (10.5) respectively.

10.4.3 SPACECRAFT PRE-MISHAP TRAJECTORIES

The BCR4BP trajectory model was used to generate the trajectories of spacecraft prior to the catastrophic mishaps. The spacecraft that suffer catastrophic mishaps include a spacecraft that transfers from the Earth to the Moon using a conventional direct transfer, much like the Apollo missions; a spacecraft that transfers from the Earth to the L_1 Lagrange point using the L_1 stable manifold; and spacecraft fixed at L_4 and L_5.

The Apollo spacecraft utilized a three-day transfer to reach the Moon. This transfer is slightly less efficient than a standard Hohmann transfer, but shortens the mission by about 1.5 days, requiring less consumables for the humans on board. The Apollo missions began with a parking orbit in LEO, followed by a Trans-Lunar Injection (TLI) burn that placed the spacecraft on a free-return trajectory. The free-return trajectory enables a return to Earth in the event of a problem. As the Moon approached, a series of small mid-course correction burns were conducted to lower the spacecraft's perilune. Finally, a Lunar Orbit Insertion (LOI) burn was conducted at perilune to bring the spacecraft to a parking orbit around the Moon. Following the completion of the lunar landing mission, a Trans-Earth Injection (TEI) burn was conducted to bring the spacecraft back to Earth, where it splashed down in the Pacific Ocean [26]. The approximation of the Apollo trajectory used for the simulations is shown in the rotating Earth-Moon reference frame in Figure 10.5. The blue circles mark the locations of the simulated catastrophic mishap for each run of the simulation and the red dots mark the locations of burns. Note that only one mid-course correction burn was modeled for simplicity, and the time spent in lunar orbit was not modeled.

The next spacecraft trajectory modeled for this research was a theoretical future spacecraft that transfers from the Earth to L_1 along the stable manifold of the Earth-Moon CR3BP. The manifolds of the Earth-Moon system can enable transfers between Earth orbits and Lagrange point or lunar orbits or less fuel than conventional methods and are an ongoing topic of research. Manifolds have been used before for missions to the Sun-Earth Lagrange points, and the L_1 manifold of the Sun-Earth system is especially useful because intersects the Earth. Unlike the Sun-Earth L_1 manifold, the Earth-Moon L_1 manifold does not intersect the Earth, requiring the use of a trajectory segment that connects LEOs to the manifold [10].

The stable and unstable manifolds that extend from L_1 toward Earth are shown in Figure 10.6 along with the ZVCs associated with the Jacobi Constant of the L_1 point. In this figure, the arrows extending from L_1 represent the stable/unstable eigenvectors of the CR3BP plant matrix linearized about the L_1 point. The manifolds are then determined by propagating the trajectory of a particle displaced slightly along the stable/unstable eigenvectors. The unstable manifold extending toward Earth is found by propagating forwards in time, while the stable manifold that brings particles toward the L_1 point is found by propagating backward in time. The arrows on

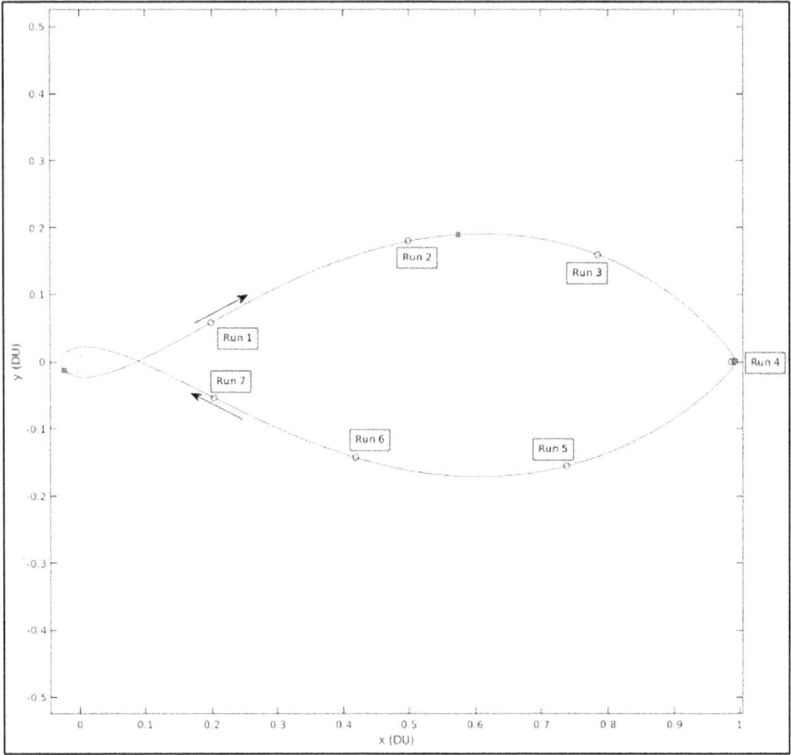

FIGURE 10.5   Apollo trajectory model in the rotating Earth-Moon reference frame.

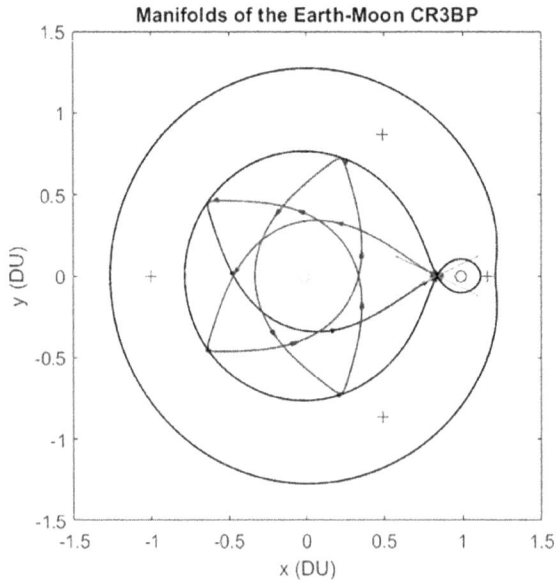

FIGURE 10.6   Manifolds of the Earth-Moon L_1 point.

the manifold lines indicate the direction of travel in forward time. The stable and unstable manifolds eventually connect.

To simulate the trajectory of a spacecraft that transfers between the Earth and the Moon, the direct transfer method to L_1 described by Parker and Anderson [10] was used as a guide. This trajectory involves the use of a "bridge" segment that connects a parking orbit around the Earth to the stable manifold. Two burns are required; the first to exit the parking orbit, and the second to enter the stable manifold. For the present research, only the manifold segment was simulated to study the effects of the manifold on the debris distribution. The trajectory of the spacecraft that suffers the catastrophic mishap begins with the spacecraft on the stable manifold at the stable manifold's lowest y-position and ends after the spacecraft has passed L_1 and returned the vicinity of Earth as shown in Figure 10.7. The circles mark the locations of the simulated catastrophic mishap for each simulation run. Note that a small maneuver at or before reaching L_1 would enable a spacecraft that follows this trajectory to continue through the L_1 point toward the Moon.

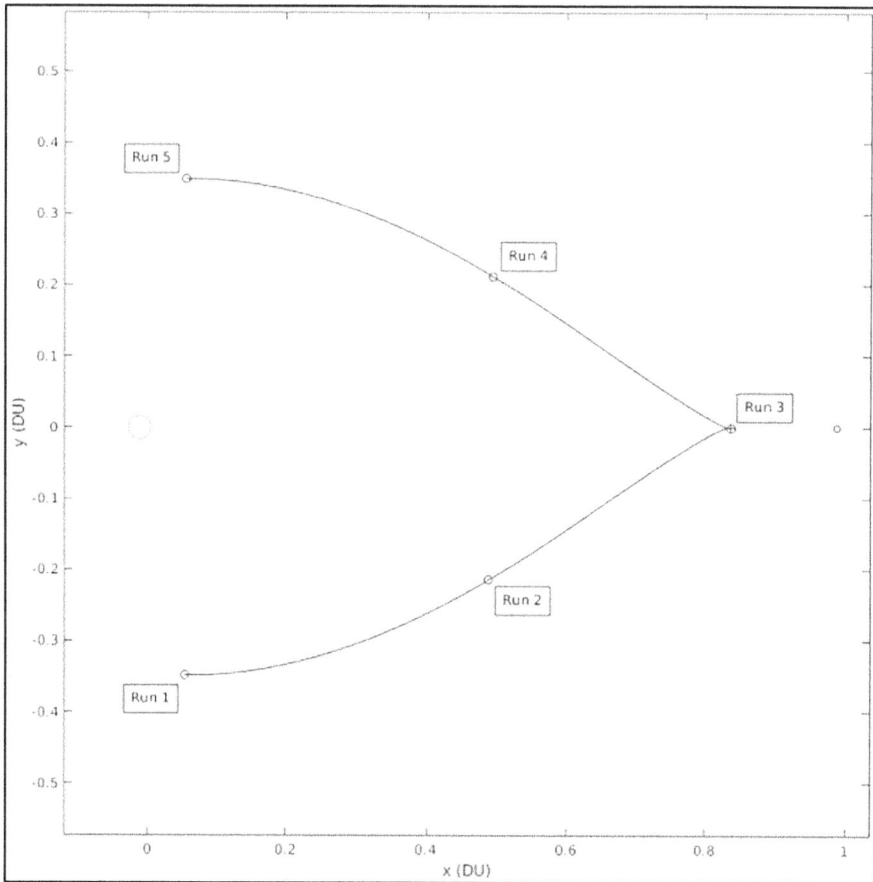

FIGURE 10.7 Trajectory for spacecraft transferring along the L_1 manifolds in the rotating reference frame.

For the final spacecraft catastrophic mishap case study, mishaps at L_4 and L_5, the spacecraft were assumed to be fixed at the L_4 or L_5 points prior to the explosions. Since the spacecraft remain fixed, the BCR4BP trajectory model was not used to generate the trajectory of the spacecraft. One run of the simulation was conducted for a spacecraft mishap at L_4 and one run was conducted for a spacecraft mishap at L_5.

10.4.4 CATASTROPHIC MISHAP MODEL

The NOAA 16 satellite battery explosion was used as the model for the spacecraft catastrophic mishap [5,27]. The two parameters of interest in the simulation are the mass distribution of particles released in the explosion and the change in velocity (ΔV) given to each particle by the explosion. These parameters are determined according to the statistical models discussed in this section.

The particle masses were determined in the simulation by fitting a probability distribution to the observed mass distribution in the NOAA 16 explosion, then selecting random numbers from this distribution to assign each particle mass. Each tracked debris particle contains a measure of the size of the particle, the Radar Cross Section (RCS). The approximate masses of the particles were calculated based on the following formula from [28] for the mass of a debris particle given the RCS:

$$M = 62 \times 10^3 (RCS)^{1.13} \tag{10.15}$$

The masses of 135 tracked debris particles were calculated and plotted in a histogram, and then a lognormal distribution was fit to the data. The parameters of the resulting lognormal distribution are $\mu = -1.7286$ and $\sigma = 1.4511$. The mass histogram, with the lognormal probability distribution overlaid, is shown in Figure 10.8.

In the simulation, particle masses were selected randomly from the lognormal distribution until the combined mass of the particles matched the mass of the original NOAA 16 satellite, 1,457 kg. Therefore, the total number of particles simulated is random and changes with each simulation.

After determining the masses of each debris particle, the ΔV given to each particle was calculated by determining the amount of kinetic energy released in the NOAA 16 explosion. The Two-Line Element sets (TLEs) for 135 debris particles tracked within a month of the explosion were used to propagate the particles back to the explosion time, approximately 07:20 UTC on 25 November 2015 [27]. The velocity vectors in the Earth Centered Inertial (ECI) reference frame were calculated at the time of the explosion and then subtracted from the velocity vector of the NOAA 16 satellite to determine the relative velocities of the debris particles. This relative velocity was then used to calculate the relative kinetic energies of all particles. The average kinetic energy per particle was 6,678 J. For the simulation, all particles are assumed to be given the same amount of kinetic energy by the explosion. Therefore, given a particle's mass, the scalar change in velocity of that particle is given by:

$$v = \sqrt{\frac{2}{m} KE} = \sqrt{\frac{2}{m} (6678 \, J)} \tag{10.16}$$

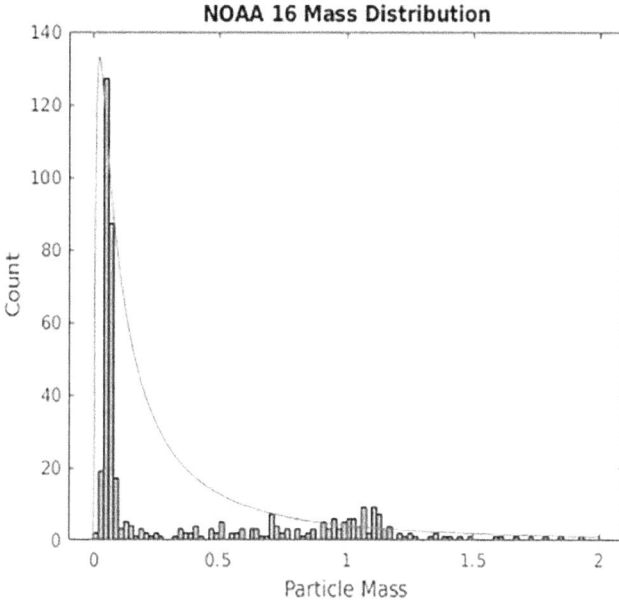

FIGURE 10.8 Particle mass histogram with fitted probability distribution.

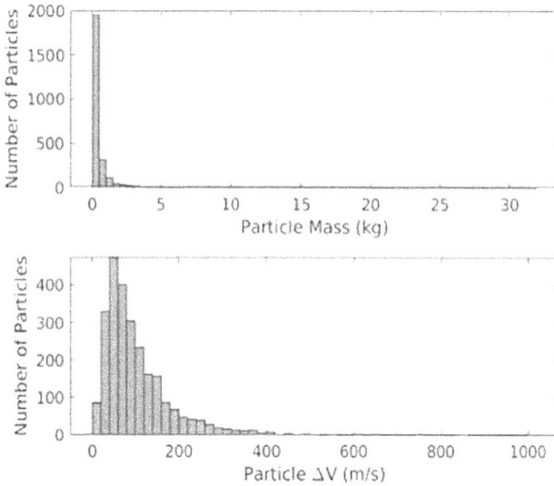

FIGURE 10.9 Example histogram of simulated particle masses and changes in velocity.

Example histograms of particle mass and ΔV determined using this method for one run of the simulation are shown in Figure 10.9. Note that each run of the simulation will involve slightly different particle masses and velocities due to the random selection of particle masses. The simulated mass and velocity distributions match well with the data from the NOAA 16 satellite. The average particle ΔV for the simulation and the actual NOAA 16 explosion are both approximately 100 m/s.

Explosion Velocity Distribution

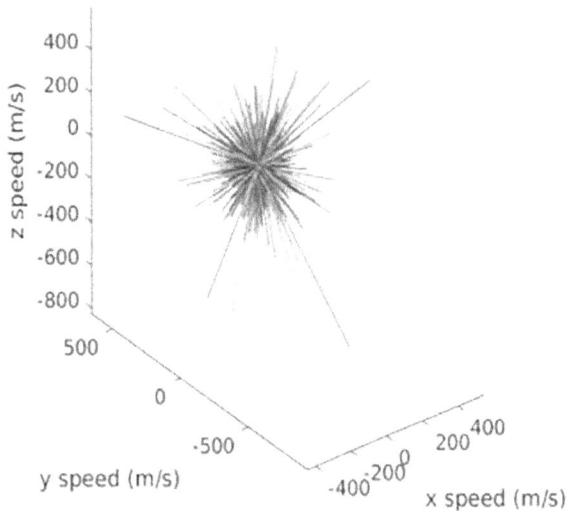

FIGURE 10.10   Example simulated explosion velocity vectors.

A direction for the velocity vector was assigned by picking a random direction on a sphere to represent an omnidirectional explosion. The velocity vectors obtained using this method from one run of the simulation are shown in Figure 10.10. Again, note that each simulation run will result in a slightly different explosion velocity distribution due to the random selection of particle masses and the random assignment of velocity directions.

10.4.5   SPACECRAFT SURVIVABILITY

A method for determining the survivability of an aircraft moving in a cloud of particles is the Poisson approach developed by Ball [29]. A similar method was applied to spacecraft for the present research. In the Poisson approach, the number of hits to the spacecraft is a random variable, with an expected number of hits E. If the cloud of debris fragments is treated as a spray of M penetrators over a volume V_S, then the penetrator spray density ρ is:

$$\gamma = M/V_S \tag{10.17}$$

where V_S is the volume of a spherical "danger zone" that surrounds the spacecraft, within which the particle density is calculated. For the present research, this danger zone is assumed to have a radius of 10,000 km. Once ρ has been calculated through debris propagation modeling, the expected number of hits on a spacecraft's "hazard zone," with hazard zone volume V_{HZ}, is given by:

$$E = \rho V_{HZ} \tag{10.18}$$

Like the spacecraft fragmentation study by Bettinger and Hess [30], V_{HZ} defines a sphere around the spacecraft such that any particle that enters this volume is considered to have hit the spacecraft. For the present research, the hazard zone is a sphere of radius of 500 m. This volume is like an error ellipsoid around the spacecraft and avoids numerical precision issues related to determining the exact location of the spacecraft and debris particles at cislunar scales in the computer simulation.

With a probability of kill with a hit of $P_{K|H}$, or the probability a particle will cause critical damage to the spacecraft if it strikes it, the instantaneous probability of spacecraft hazard is given by:

$$P_{HZ} = 1 - e^{-E P_{K|H}} \tag{10.19}$$

The probability of spacecraft hazard P_{HZ} represents the probability that a particle that would cause critical damage to the spacecraft will enter the 500 m hazard zone sphere. In a real-world application, this may represent the probability that a spacecraft will be significantly threatened by debris and should perform an avoidance maneuver. For each simulation, the value of P_{HZ} is tracked over time. The total probability of hazard during a time interval t_0 to t_f is the area under the P_{HZ} curve with respect to time:

$$\bigcup_{t_0}^{t_f} P_{HZ}(t)\,dt \tag{10.20}$$

As shown in Eq. (10.19), the probability of hazard P_{HZ} depends on the chosen probability of kill with a hit $P_{K|H}$. Once the particle masses have been determined through the spacecraft catastrophic mishap model, the $P_{K|H}$ can be determined based on the severity of damage impacts from particles of those masses are likely to cause to a spacecraft. Due to the wide range of particle masses in this research, a variable $P_{K|H}$ that depends on particle mass was used. A logistic curve was used to model the dependence of the $P_{K|H}$ on the mass of particles. Logistic curves are often used to measure vulnerabilities in other applications, including the likelihood of bridge collapse in earthquakes of varying magnitudes [31,32]. Studies of the damage to spacecraft caused by particle impacts of various sizes, such as that by Elvidge [33] can aid in determining the parameters of the logistic curve for the present research. The assumed model for the logistic curve for $P_{K|H}$ is shown in Figure 10.11. The equation for this logistic curve is:

$$P_{K|H} = A + \frac{K - A}{(C + Qe^{-B*m})^{1/v}} \tag{10.21}$$

where m is particle mass, and values chosen for the parameters A, B, C, K, Q, and v are given in Table 10.3.

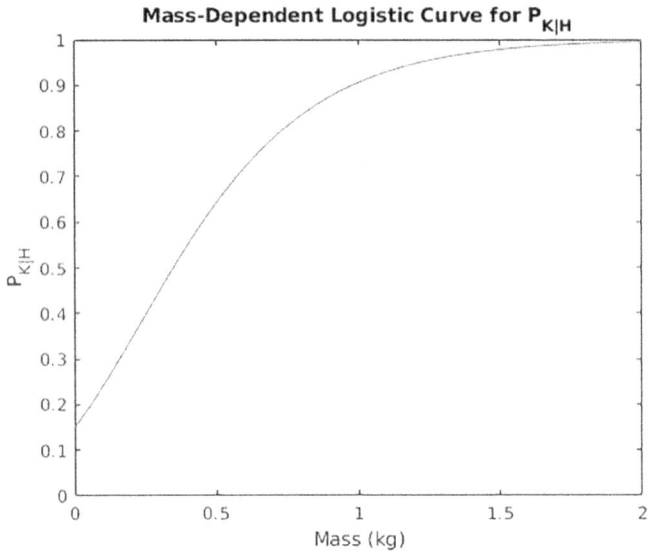

FIGURE 10.11   Logistic curve model for probability of kill with a hit.

TABLE 10.3
Logistic Curve Parameters

Parameter	Value
A	0
B	3
C	1
K	1
Q	0.1
v	0.05

10.5   ANALYSIS AND RESULTS

The results of the cislunar debris case studies are presented in the following sections. Each case study involves a spacecraft on a different cislunar trajectory before it suffers a catastrophic mishap. The models of these cislunar trajectories were discussed in Section IV Part B. For each simulation, $t=0$ refers to when the Sun, Earth, and Moon are aligned (i.e., $\omega_1 t = 0$ in Figure 10.4).

10.5.1   CASE STUDY 1: APOLLO-LIKE TRANSFER

For each run location along the Apollo-like transfer, particles were simulated from the time of the catastrophic spacecraft mishap to $t=50$ days. There was no notional spacecraft for this case, and the survivability model was not applied. Instead, the

TABLE 10.4
Number of Particles with Perigee within GEO Altitude, Apollo Transfer Simulations

Run	Mishap Time (Days)	Number of Particles with Final Perigee within GEO Altitude
1	0.25	655
2	1.00	657
3	2.00	566
4	2.97	657
5	4.00	630
6	5.00	364
7	5.49	156

Run Locations

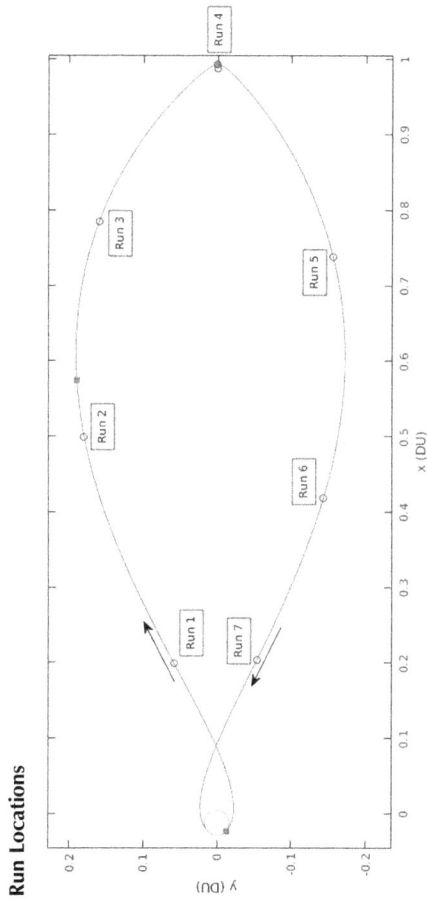

TABLE 10.5

Status of Particles at Simulation End, Apollo Transfer Simulations

Run Number	Escaped Earth-Moon System (%)	Impacted Moon (%)	Impacted Earth (%)	End in Lunar Sphere of Influence (%)	End in Earth Sphere of Influence (%)
1	9.62	5.27	25.89	0	59.21
2	8.22	6.72	23.21	0	61.85
3	6.14	16.43	15.46	0	61.97
4	5.36	0.66	41.43	0	52.26
5	1.68	0	73.49	0	24.80
6	2.67	0	82.98	0	14.25
7	5.94	0	87.73	0	6.25

risks to spacecraft near Earth are evaluated based on the number of particles that have a perigee that is within GEO altitude at the end of the simulation. The number of particles that have a perigee within GEO altitude for each run of the simulation is shown in Table 10.4. The locations of all particles in the Earth-Moon system at the end of the simulation are shown in Table 10.5. Note that Run 3 is most like the Apollo 13 mishap, since the Apollo 13 incident took place about two days into the mission, following the maneuver to exit the free-return trajectory [8].

Snapshots of the simulation during selected runs are shown in Figure 10.12. In these snapshots, the green circle marks the location of the Earth, the black circle marks the location of the Moon, the red crosses mark the locations of the Lagrange points, the yellow line indicates the current direction to the Sun, and the black dots represent debris particles.

In Run 1 (shown in Figure 10.12a), Run 2, and Run 3 (shown in Figure 10.12b), most of the debris particles are thrown out of the Earth-Moon plane after passing the Moon and begin orbiting at an angle of approximately 90° relative to the Earth-Moon line. In addition, a cardioid-like shape forms at apogee as particles on the outer edges of the particle cloud orbit are farther from the Earth. This shape is especially apparent in Run 3 in Figure 10.12b and may be due to the effects of the Moon on the particle cloud as the particle cloud passes the Moon. Because the orbits for Runs 1 and 2 prior to the explosion started in a LEO parking orbit, many particles return to within GEO when they reach their perigee following the explosion. These particles would likely continue to intersect lower Earth orbits at each perigee passage for a very long time following the explosion due to the high apogee height, which would make it difficult for atmospheric drag to decay the orbits of the particles enough for the particles to re-enter the Earth's atmosphere. This could cause long-term debris threats to Earth-orbiting spacecraft.

In Run 3, shown in Figure 10.12b, the catastrophic mishap occurs at a similar location along the spacecraft's trajectory as the Apollo 13 disaster. Run 3 is also the first run that takes place following the course-correction maneuver that lowers the spacecraft's perilune. As expected, this causes many more particles to impact the Moon. Overall, the results are very similar to Runs 1 and 2 due to the small ΔV in the course-correction maneuver. The particles enter an orbit that is at an angle of

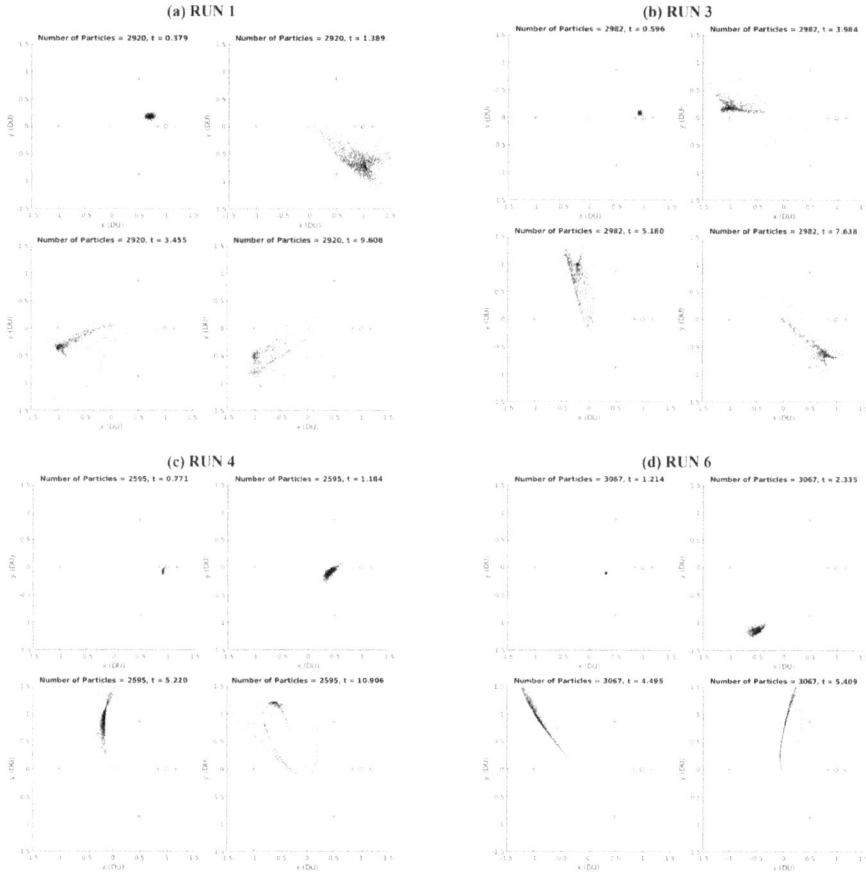

FIGURE 10.12 Snapshots of selected runs, Apollo transfer simulations: (a) Run 1; (b) Run 3; (c) Run 4; (d) Run 6.

approximately 90° from the Earth-Moon line and many have a perigee within GEO altitude. This may provide an indication of the motion of the debris from the Apollo 13 incident in the 50 days following the explosion and suggest risks to spacecraft near Earth if a similar event occurred today.

The spacecraft mishap in Run 4, shown in Figure 10.12c, occurs at perilune, resulting in different behavior from the previous three runs. After passing the Earth for the first time, the particles enter a highly elliptical orbit with an apogee well beyond the Moon. Unlike the previous runs, the motion is mostly in the Earth-Moon plane. However, after passing the Moon again, many of the particles enter an orbit at an angle of about 90° to the Earth-Moon line, similar to the previous runs, while other particles continue in an orbit has an apogee well beyond the Moon. This behavior is shown in the bottom right corner of Figure 10.12c. Particles in each of the two main particle cloud orbits still pass close to the Earth, potentially creating long-term threats to Earth-orbiting spacecraft.

Significantly more particles impact the Earth in Runs 5–7 because the spacecraft is on a trajectory that would bring it into the Earth's atmosphere prior to the explosion. The debris particles that avoid the Earth's atmosphere then travel well beyond the Moon's distance before returning to a perigee close to Earth. Particles in these runs would intersect the LEO environment at each perigee passage. However, the threat would diminish within about 50 days as particles return to Earth. Progressively fewer particles survive each perigee passage as more and more particles enter the Earth's atmosphere. The thinning of the particle cloud as the particles enter the atmosphere is shown for Run 6 in the bottom right corner of Figure 10.12d. Note that atmospheric drag was not modeled in this simulation, and particles are terminated when they pass within 80 km of the Earth.

10.5.2 CASE STUDY 2: MANIFOLD TRANSFER

The simulations for a spacecraft transferring along the L_1 manifold apply the spacecraft survivability model to analyze the threats to a notional spacecraft operating at the L_1 point. The simulations prior to the L_1 point start at negative time because the trajectory that brings a spacecraft toward L_1 was obtained by propagating the trajectory backward in time. The simulation begins at the time of the catastrophic spacecraft mishap and continues to $t = 50$ days. Five runs were conducted for catastrophic mishaps at different locations along the L_1 manifold. The survivability results for each run are shown in Table 10.6. The final locations of all particles at the end of the simulation are given in Table 10.7. No particles end within GEO altitude, and the primary threats for a catastrophic mishap in the manifolds are to a spacecraft at L_1 and perhaps to future spacecraft that utilize the L_1 manifold. However, as the very low probability of hazards indicates, the threat to any one spacecraft is very low due to the vast distances in cislunar space.

Snapshots of selected simulation runs are shown in Figure 10.13. In all runs, the debris particle cloud stretches to fill the L_1 manifolds that were shown in Figure 10.6. For runs that occur prior to the L_1 point, particles with large enough energy levels can cross the L_1 point and continue toward the Moon, while the remaining particles travel along the unstable manifold back toward Earth. The particles that continue toward the Moon could suggest possible hazards to lunar orbiting spacecraft since many of the particles can remain near the Moon for at least 50 days. The particles that continue back toward Earth could also create some risk because they can remain in the manifold long enough to eventually return to the L_1 point. This occurs for particles in Runs 4 and 5, which started after L_1, as the non-zero P_{HZ} in Table 10.6 indicates. The particles that return to L_1 do so at much greater velocities, likely due to perturbations that have increased the particle velocity relative to L_1. Overall, a mishap at any location along the L_1 manifold could create small but measurable threats to a spacecraft operating at L_1, a spacecraft traveling through the L_1 manifold, or a spacecraft operating in a distant lunar orbit.

10.5.3 CASE STUDY 3: STABLE LAGRANGE POINTS

A catastrophic mishap at L_4 or L_5 is of particular interest due to the stability of these points in the CR3BP, which could lead to an accumulation of debris if these points

TABLE 10.6

Survivability Results for a Spacecraft at L_1, Manifold Transfer Simulations

Run Number	Mishap Time (Days)	Closest Approach to Notional Spacecraft at L_1 (km)	Mean Relative Speed (m/s)	Total P_{HZ} (%)
1	-13.04	186	140	2.676e-9
2	-11.43	233	118	3.872e-9
3	0	52	75	1.806e-8
4	11.94	736	376	4.668e-12
5	13.59	832	480	4.731e-12

Run Locations

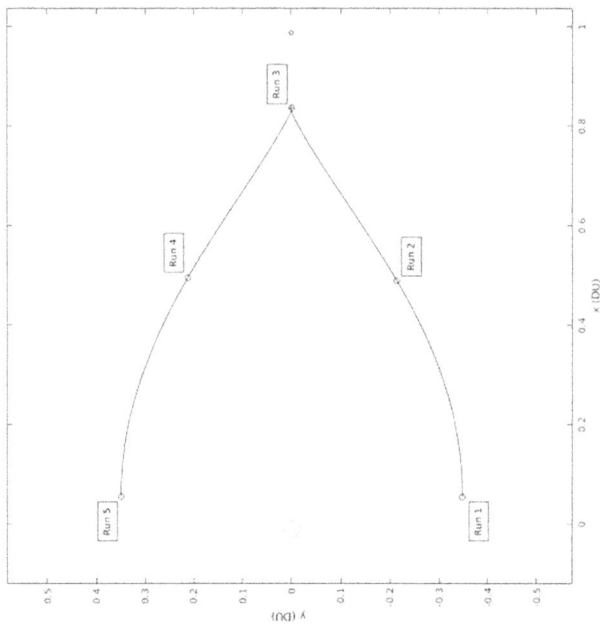

TABLE 10.7

Status of Particles at Simulation End, Manifold Transfer Simulations

Run Number	Escaped Earth-Moon System (%)	Impacted Moon (%)	Impacted Earth (%)	End in Lunar Sphere of Influence (%)	End in Earth Sphere of Influence (%)
1	16.94	3.77	0	2.13	77.16
2	20.76	4.33	0	2.73	72.18
3	1.99	5.19	0	20.73	72.08
4	0.80	0.30	0	0.27	98.63
5	1.21	1.37	0	0	97.42

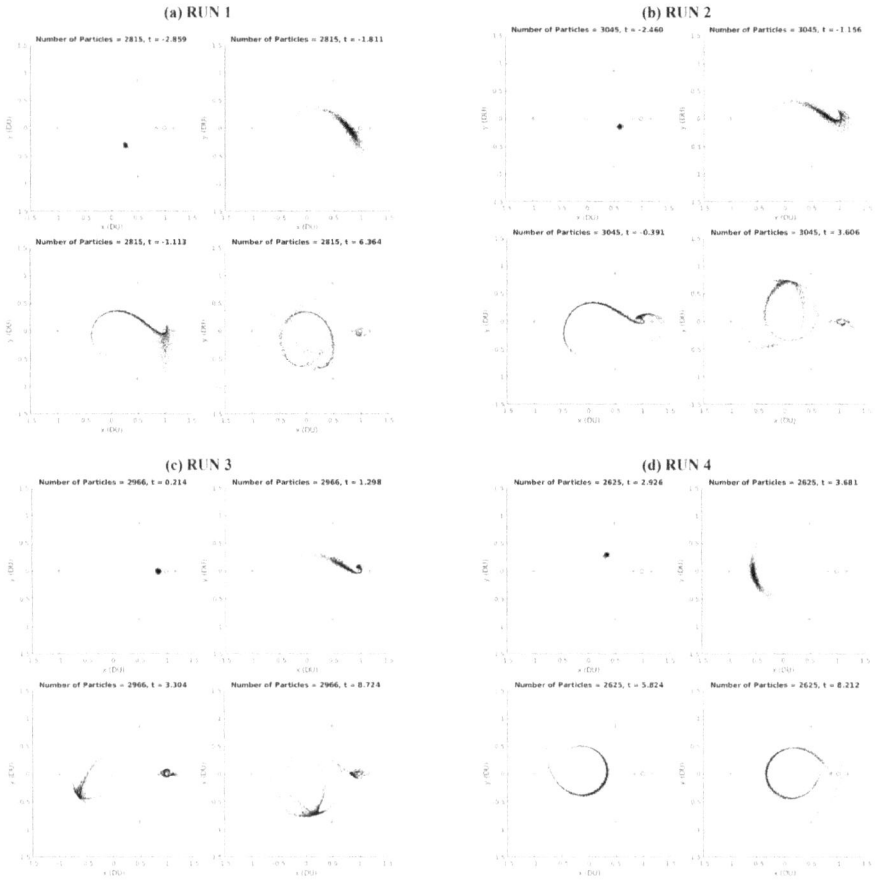

FIGURE 10.13 Snapshots of selected runs, manifold transfer simulations: (a) Run 1; (b) Run 2; (c) Run 3; (d) Run 4.

TABLE 10.8

Survivability Results for Spacecraft at L_4/L_5, Stable Lagrange Point Simulations

Run	Mishap Time (Days)	Closest Approach to Notional Spacecraft (km)	Mean Relative Speed (m/s)	Total P_{HZ} (%)
L_4	0	270	38	6.377e-8
L_5	0	462	40	4.314e-8

TABLE 10.9

Status of Particles at Simulation End, Stable Lagrange Point Simulations

Run	Escaped Earth-Moon System (%)	Impacted Moon (%)	Impacted Earth (%)	End in Lunar Sphere of Influence (%)	End Within 200,000 km of L_4/L_5 (%)	End Elsewhere in Earth Sphere of Influence (%)
L_4	27.03	5.90	0	0.50	33.80	32.76
L_5	34.44	4.09	0	0.17	29.81	31.49

are still stable in the presence of the Sun's gravity. The simulation for a catastrophic spacecraft mishap at L_4 applies the survivability model to a different notional spacecraft also operating at L_4, while the simulation for a catastrophic spacecraft mishap at L_5 applies the survivability model to a notional spacecraft operating at L_5. Both simulations start with the catastrophic mishap at $t = 0$ and continue for 365 days. The survivability results for the runs at L_4 and L_5 are shown in Table 10.8. The total P_{HZ} is higher for these cases than any of the previously discussed L_1 manifold cases due to the stability of L_4 and L_5. As shown in Table 10.9, about a third of the particles remain within a 200,000-radius of the Lagrange point after a year for the runs at L_4 and L_5. This creates a continuous hazard to the notional spacecraft, although the risk remains low due to the vast distances in cislunar space.

Figure 10.14 shows snapshots of the L_4 and L_5 simulation runs, and the debris particles move similarly for each case. The particles initially expand rapidly due to the explosion, then many begin to fall back toward their respective Lagrange point. The particles that do not escape are initially confined to an area roughly equivalent to the ZVCs associated with the Jacobi Constants of the L_4 and L_5 points. These ZVCs were shown in Figure 10.3, and look similar to the particle distributions shown in the upper right corners of Figure 10.14a and b. As solar perturbations act on the particles, the particles begin to expand and contract about L_4 and L_5 on a regular cycle, like what was noted in the study of the Kordylewski clouds by Slíz-Balogh, Barta, and Horváth [11]. The particles spin about the Lagrange point in the rotating frame with pinwheel-like arms that periodically collapse and expand. The velocities of the trapped particles that pass within the 10,000 km danger zone of the notional spacecraft relative to the spacecraft also vary according to a regular cycle between about 20 and 70 m/s. This cyclical pattern matches the cycle in the masses of particles within

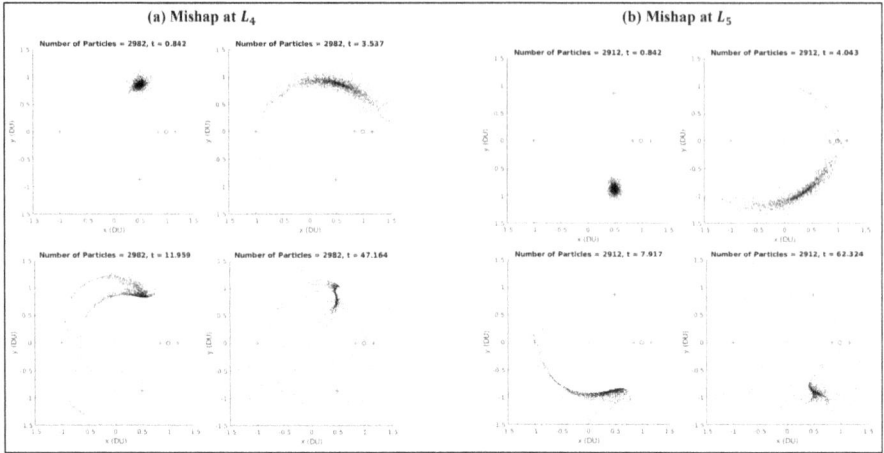

FIGURE 10.14 Snapshots of selected runs, stable Lagrange point simulations: (a) Mishap at L_4 (b) Mishap at L_5.

the danger zone, with lighter particles, which are given more ΔV from the explosion, tending to enter the danger zone with higher speeds relative to the notional spacecraft.

Over time, particles are ejected from L_4 and L_5 due to the destabilizing effect of solar gravity, but nearly 1,000 particles remain trapped at the end of the year-long simulation in each case. In a real-world mission, the significant number of particles that remain near these points may increase the need for spacecraft operators to monitor the area around the notional spacecraft for debris. Difficulties tracking debris at such great distances from Earth may make it more likely that the spacecraft will have to maneuver to avoid debris, despite the low debris density relative to orbital environments closer to Earth.

10.6   CONCLUSION

Three cislunar debris case studies have been presented in this research. These case studies analyzed the debris-related consequences of spacecraft mishaps during an Apollo-like Earth-Moon transfer, during a transfer to or from L_1 using the L manifolds, and at the stable Earth-Moon Lagrange points L_4 or L_5. Each scenario threatens unique regions of cislunar space; however, the overall risk to spacecraft is very low in each case. See Salnikova and Stepanov (2019) [19].

A catastrophic spacecraft mishap during the Apollo-like transfer resulted in the greatest risk to spacecraft near Earth, with hundreds of debris particles coming within geosynchronous altitude at high relative velocities during each perigee passage. The particles would likely remain in space for a very long time (much longer than the 50-day simulation) due to their very high apogees that would decay slowly due to atmospheric drag. Therefore, this type of debris event may be the most concerning of the case studies due to the potential for long-term risks to crowded orbits near Earth.

Catastrophic spacecraft mishaps in the L_1 manifold would pose little threat to spacecraft closer to Earth, but they could pose some risk to future spacecraft that utilize the unique dynamics of cislunar spacecraft for mission operations. The L_1 manifold can enable transfers to the lunar region for less fuel than a conventional Hohmann transfer, and the L_1 point could be an appealing location for future spacecraft due to its high vantage point over the Earth and lower orbital altitudes. Debris would circulate through these regions indefinitely if a mishap occurred in the L_1 manifold. In addition, some debris could threaten spacecraft in lunar orbits. However, the risk to any particular spacecraft would be extremely low due to the vast distances in cislunar space.

Finally, catastrophic spacecraft mishaps at L_4 and L_5 would pose risks to other spacecraft operating at those points. L_4 and L_5 could be an appealing location for future spacecraft due to their stability. This case study showed that much of the debris from a catastrophic spacecraft mishap at either point would be also stable at those points for at least a year following the explosion. This results in a much higher probability of hazard to the notional spacecraft than any of the L_1 simulation runs. Debris circulating these points would also likely be difficult to track from Earth due to the great distance, making it difficult to maneuver a spacecraft to avoid debris. However, like the other case studies, the risk is still low, and debris would only become problematic if these points become more crowded in the future or if repeated mishaps occur.

Future studies of cislunar debris could analyze other types of case studies, such as catastrophic mishaps in lunar orbits. In addition, while the $P_{K|H}$ model in the present research incorporated only the mass of the particle in determining its potential to destroy a spacecraft, future studies could incorporate both the mass and velocity of the particle in computing this value. Both the L_1 manifold and the stable Lagrange point case studies resulted in widely varying relative velocities, so changing the risk to a spacecraft depending on the velocities of debris particles would result in a more realistic simulation.

REFERENCES

1. Maria Temming, "Israel's First Moon Mission Lost Moments before Landing," *ScienceNews*, April 2019, https://www.sciencenews.org/article/israel-moon-mission -spacecraft-crash.
2. David, Leonard. "US Military Eyes Strategic Value of Earth-Moon Space," *Space.com*, 29 August 2019. URL https://www.space.com/us-military-strategic-value-earth-moon-space.html.
3. Kelso, T. "Analysis of the 2007 Chinese ASAT Test and the Impact of its Debris on the Space Environment." Paper presented at the *8th Advanced Maui Optical and Space Surveillance Technologies Conference*, Maui, Hawaii, 12–15 September 2007.
4. Henry, C. "India ASAT debris spotted above 2,000 kilometers, will remain a year or more in orbit," *SpaceNews*, 9 April 2019. URL https://spacenews.com/india-asat-debris-spotted-above-2200-kilometers-will-last-a-year-or-more/.
5. Braun, V., S. Lemmens, B. Reihs, H. Krag, and A. Horstmann. "Analysis of Breakup Events." Paper presented at the *7th European Conference on Space Debris*, Darmstadt, Germany, 18–21 April 2017.

6. Kessler, Donald J. and Burton G. Cour-Palais. "Collision Frequency of Artificial Satellites: The Creation of a Debris Belt," *Journal of Geophysical Research* 83, no. A6: 2637–2646 (1978).

7. Namazyfard, Hossein. "Computational Exploration of the Cislunar Region and Implications for Debris Mitigation." MS Thesis: Department of Aerospace Engineering, The University of Arizona, Tucson, Arizona, April 2019.

8. Dunbar, Brian, "Apollo 13," *NASA.gov*, July 2009, https ://www.nasa.gov/mission_pages/apollo/missions/apollo13.html

9. Keel, William C., "Telescopic Tracking of the Apollo Lunar Missions," Bill Keel's Space Bits, *Department of Physics and Astronomy at the University of Alabama*, https://pages.astronomy.ua.edu/keel/space/apollo.html

10. Parker, Jeffrey S. and Rodney L. Anderson. *Low-Energy Lunar Trajectory Design.* Hoboken, NJ: Wiley, 2014.

11. Slíz-Balogh, Judit, András Barta, and Gábor Horváth, "Celestial Mechanics and Polarization Optics of the Kordylewski Dust Cloud in the Earth-Moon Lagrange Point L5, Part I: 3D Celestial Mechanical Modelling of Dust Cloud Formation," *Monthly Notices of the Royal Astronomical Society* 480: 5550–5559 (2018).

12. Burbank, Paige B., Burton G. Cour-Palais, and William E. McAllum. "A Meteoroid Environment for Near-Earth, Cislunar, and Near-Lunar Operations," *NASA TN D-2747.* Houston, Texas: National Aeronautics and Space Administration, 1965.

13. Davidson, John R. and Paul E. Sandorff. "Environmental Problems of Space Flight Structures: II. Meteoroid Hazard," *NASA TN D-1493.* Langley Station, Hampton, Virginia: NASA Langley Research Center, 1963.

14. Singer, S.F. "Interplanetary Dust Near the Earth," *Nature* 192, no. 4800: 321–323 (1961).

15. Hyde, T.W. and W.M. Alexander. "Transport Dynamics Calculated Under the Full Mie Scattering Theory for Micron and Submicron Lunar Ejecta in Selenocentric, Cislunar, Geocentric Space." Paper presented at the *19th Lunar and Planetary Science Conference*, Houston, Texas, 14–18 March 1988.

16. Altobelli, N., E. Grün, and M. Landgraf. "A New Look into the Helio Dust Experiment Data: Presence of Interstellar Dust Inside the Earth's Orbit," *Astronomy and Astrophysics* 448: 243–252 (2006).

17. Landgraf, M. and R. Jehn. "Space Debris Hazards from Explosions in the Collinear Sun-Earth Lagrange Points." Paper presented at the *Third European Conference on Space Debris*, Darmstadt, Germany, 19–21 March 2001.

18. Bandyopadhyay, Priyankar, Ram Krishan Sharma, Ashish Tewari. "Space Debris Hazards from Fragmentations in Collinear Earth-Moon Points." Paper presented at the *5th European Conference on Space Debris*, Darmstadt, Germany, 30 March to 2 April 2009.

19. Salnikova, T. and S. Stepanov. "Existence of the Elusive Kordylewski Cosmic Dust Clouds," *Acta Astronautica* 163, Part A: 138–141 (2019).

20. Pohle, Frederick V. "The Least Density of a Spherical Swarm of Particles, with an Application to Astronomical Observations of K. Kordylewski," *MRC Technical Summary Report #351.* Madison, WI: Mathematics Research Center, United States Army, The University of Wisconsin, 1962.

21. Pohle, Frederick V. "A Dynamical Model for Kordylewski Cloud Satellites," *AIAA Journal* 2, no. 10: 1818–1820 (1964).

22. Salnikova, T.V., S. Ya. Stepanov. "Mathematical Model of Formation of Kordylewski Cosmic Dust Clouds," *Doklady Physics* 60, no. 7: 323–326 (2015).

23. Salnikova, T.V., S. Ya. Stepanov, and A. I. Shuvalova. "Probabilistic Model of Kordylewski Clouds," *Doklady Physics* 61, no. 5: 243–246 (2016).

24. Salnikova, Tatiana and Sergey Stepanov. "On the Kordylewski Cosmic Dust Clouds." Paper presented at the *2016 International Conference "Stability and Oscillations of Nonlinear Control Systems" (Pyatnitskiy's Conference)*, Moscow, Russia, 1–3 June 2016.

25. Slíz-Balogh, Judit, András Barta, and Gábor Horváth, "Celestial Mechanics and Polarization Optics of the Kordylewski Dust Cloud in the Earth-Moon Lagrange Point L5, Part II: Imaging Polarimetric Observation: New Evidence for the Existence of the Kordylewski Dust Cloud," *Monthly Notices of the Royal Astronomical Society* 482: 762–770 (2019).
26. Riché, L. J., G. M. Colton, and T. A. Guillory. *Apollo 11 Flight Plan.* Houston, Texas: National Aeronautics and Space Administration, Manned Spaceflight Center, 1969.
27. "NOAA 16 Satellite Breakup Leaves Dozens of Debris in Orbit," *SpaceFlight 101*, 5 December 2019. URL https://spaceflight101.com/noaa-16-satellite-breakup-leaves-dozens-of-debris-in-orbit/
28. Dickey, Michael R. and Robert D. Culp. "Determining Characteristic Mass for Low-Earth-Orbiting Debris Objects," *Journal of Spacecraft and Rockets* 26, no. 6: 460–464 (2012).
29. Ball, Robert E. *The Fundamentals of Aircraft Combat Survivability Analysis and Design.* AIAA (American Institute of Aeronautics & Astronautics), 2003.
30. Bettinger, Robert A. and Joshuah A. Hess. "Fractionated Spacecraft Survivability following a Catastrophic Explosion." Paper presented at the *AIAA SciTech 2020 Forum*, Orlando, FL, 6–10 January 2020.
31. Lee, Seongkwan Mark, Tschangho John Kim, and Seung Lim Kang. "Development of Fragility Curves for Bridges in Korea," *KSCE Journal of Civil Engineering* 11, No. 3: 165–174 (2007).
32. Bangerjee, S. and M. Shinozuka. "Nonlinear Static Procedure for Seismic Vulnerability Assessment of Bridges," *ComputerAided Civil and Infrastructure Engineering* 22, No. 4: 293–305 (2007).
33. Elvidge, Michael E. "A Stochastic Method for Calculating Spacecraft Vulnerability to Low Earth Orbit Debris." MS Thesis: Department of Mechanical and Aerospace Engineering, Ottawa-Carleton Institute for Mechanical and Aerospace Engineering, Ottawa, Canada, August 1994.

11 Elliptical Orbit Proximity Operations Differential Games*

Eric R. Prince, Joshuah A. Hess, and Richard G. Cobb
US Air Force Institute of Technology

Ryan W. Carr
US Air Force Research Laboratory

CONTENTS

Differential games are formulated and solved for an inspector satellite operating nearby a resident space object (RSO) in an elliptical orbit. For each differential game, the goal of the inspector satellite is to minimize the time to achieve an inspection goal, whereas the goal of the RSO is to delay that condition as long as possible. Thus, it is assumed that the RSO can maneuver, and that both the inspector satellite and

* Reprinted with permission: Prince, E. R., Hess, J. A., Cobb, R. G., & Carr, R. W. (2019). Elliptical orbit proximity operations differential games. Journal of Guidance, Control, and Dynamics, 42(7), 1458–1472.

DOI: 10.1201/9781003220978-14

the RSO use a constant, steerable thruster (e.g., an electric engine) during each game. The following games are formulated and solved via an indirect heuristic method, where each game corresponds to an inspection goal of the inspector satellite: (i) intercept; (ii) rendezvous; (iii) obtain sun vector; (iv) match energy; (v) obtain sun vector and match energy; and (vi) match energy and remain in close proximity during the ensuing orbit. The open-loop strategies obtained for these games represent worst-case scenarios, and they may inform requirements for future satellites as well as be actual open-loop strategies for a given scenario.

11.1   INTRODUCTION

For an inspector satellite conducting proximity operations about an uncooperative RSO, it may be desirable to achieve an inspection goal in an optimal manner in order to characterize the RSO in an effective way. Typically, the RSO of interest is passive and remains in its given orbit. However, there may be times when the RSO maneuvers away from its reference orbit, possibly due to a faulty thruster or due to mission operations. In the worst-case scenario, the RSO may maneuver in a way that makes achieving the inspection goal the most difficult. If it is assumed that the RSO indeed maneuvers optimally to evade or prolong an inspection goal of the inspector satellite while the inspector satellite maneuvers optimally to achieve its inspection goal, then the scenario may be treated as a differential game.

There are various techniques that have been used to solve related differential games. Horie and Conway [1] solved a fighter pursuit–evasion game via a method they called the semidirect collocation with nonlinear programming (SDCNLP). With this method, the optimal control for one player is found numerically where a genetic algorithm (GA) is used to produce an initial guess [2] for the nonlinear programming problem solver, and the optimal control for the other is found based on the analytic necessary conditions of the two-sided optimization problem. Pontani and Conway [3] used the SDCNLP method and developed an algorithm for the numerical solution of a three-dimensional (3-D) orbital pursuit–evasion game in which the objective of the pursuer is to minimize the time to intercept while the evader's objective is to maximize it. Shen et al. [4] and Blasch et al. [5] also studied the pursuit–evasion orbital game in which the pursuer minimizes the time to interception while the evader attempts to maximize it for collision avoidance. In their formulation, the pursuer rotated its orbit to the same plane as that of the evader, and then the game was solved via a MATLAB-based optimization environment called TOMLAB.

Differential game solution techniques applied to relative satellite motion problems include the work of Sun et al. [6], in which they solved a pursuit–evasion game of two spacecraft in low Earth orbit using the Hill–Clohessy–Wiltshire (HCW) equations [7] via two methods they introduced: the semidirect control parameterization (SDCP) method and a hybrid method that combine the SDCP method with a multiple shooting method. A constant mass was assumed for the games, and the function to be minimized (or maximized) was the separation distance between the two spacecraft at a given terminal time instead of the final time with intercept terminal constraints. Selvakumar and Bakolas [8] solved the pursuit–evasion game using the HCW equations, in which the pursuer satellite aims to minimize time to capture while the evader attempts to prolong it. Both spacecraft had control constraints, and the problem was solved by transforming the free final time problem into a fixed final time problem with a terminal cost by first transforming the relative motion equations into Levi-Civita coordinates [9], in which the equations of motion become decoupled harmonic oscillators, and then employing the Gutman et al. state transformation [10]. A semi-analytic solution was presented for the time of capture, and a closed-form expression was developed for the optimal control inputs. However, the problem was limited to the orbital plane, and the mass was assumed constant.

Stupik et al. [11] and Stupik [12] solved a differential game using the HCW equations in which the objective of the pursuer was to minimize the time to capture and the objective of the evader was to maximize it. The control variables for the players were the three-dimensional directions of each player's constant thrust, and mass loss was accounted for. The solution was obtained by applying the analytic necessary conditions for a differential game solution and solving the resulting system with particle swarm optimization (PSO) where only 3 of the initial 12 costates were needed as PSO variables. This technique is basically what Pontani and Conway called the indirect heuristic method (IHM) [13] but applied to a pursuit–evasion game instead of to a one-sided optimization problem.

The research herein intends to use the IHM and the approach of Stupik et al. [11] to solve various types of differential games for an inspector satellite operating nearby an RSO in an elliptical orbit. Specifically, this research will address multiple types of games in addition to the intercept game, where each game corresponds to an inspection goal of the inspector satellite. Instead of using the HCW equations of motion, the Tschauner–Hempel (TH) equations of motion [14] will be used to allow the RSO of interest to be in an elliptical orbit. Like Stupik et al. [11], both players will use a constant, steerable thrust, which may be especially applicable for electric engines in such a game. It is desired to use the IHM due to the relative simplicity of the method if it can be applied. Thus, the IHM will be used to formulate and solve multiple types of games in which the objective of the pursuer is to minimize the time to achieve an inspection goal and the objective of the evader is to maximize it. The inspection goals considered in this paper are the following: intercept, rendezvous, obtain sun vector, match energy, obtain sun vector and match energy, and match energy and remain in close proximity. The resulting solutions will represent open-loop strategies generated from known initial conditions.

This chapter is organized as follows. First, the equations of motion and control definition are presented. Then, the necessary conditions for a zero-sum pursuit–evasion game are put forth, and the general game formulations including how the IHM is used to formulate and solve the games is explained. Next, the six games considered in this chapter are developed, in which each game corresponds to a specific inspection goal for the inspector satellite. The techniques used to solve the resulting boundary value problem for each game, again following the IHM, are then detailed. Finally, simulation results are presented, as well as some validation efforts to add confidence that the differential game solutions have indeed been found.

11.2 EQUATIONS OF MOTION AND CONTROL DEFINITION

The equations describing the motion of an inspector satellite (or deputy) relative to an RSO (or chief) have been derived under a variety of assumptions. Typically, the equations are expressed in a noninertial, rotating, coordinate frame fixed to the RSO with the RSO at its center, in which the $\hat{x}$ axis points in the direction from the center of the Earth to the RSO, the $\hat{z}$ axis points in the same direction as the specific angular momentum vector, and the $\hat{y}$ axis completes the right-handed coordinate system. This frame is commonly called the local-vertical/local-horizontal frame. Using this coordinate system, the equations of relative motion for two-body, or Keplerian, motion can be derived under the additional assumption that the distance between the deputy and the chief is much smaller than the distance between the center of the Earth and the chief [15]:

$$\ddot{x} - 2\dot{f}\dot{y} - \ddot{f}y - \dot{f}^2x - 2\mu x\left(\frac{\dot{f}}{h}\right)^{(3/2)} = a_x \tag{11.1}$$

$$\ddot{y} + 2\dot{f}\dot{x} + \ddot{f}x - \dot{f}^2y + \mu y\left(\frac{\dot{f}}{h}\right)^{(3/2)} = a_y \tag{11.2}$$

$$\ddot{z} + \mu z\left(\frac{\dot{f}}{h}\right)^{(3/2)} = a_z \tag{11.3}$$

where f is the true anomaly of the RSO; $\mu = 398;600.5\,\mathrm{km^3/s^2}$ is the Earth's gravitational parameter; h is the constant specific angular momentum of the RSO; and a_x, a_y, and a_z are the acceleration terms resulting from any non-Keplerian forces acting on the inspector satellite. If the independent variable is changed from time to the true anomaly f of the RSO and the variables are transformed according to

$$\begin{bmatrix} \tilde{x} \\ \tilde{y} \\ \tilde{z} \end{bmatrix} = (1 + e\cos f)\begin{bmatrix} x \\ y \\ z \end{bmatrix} \tag{11.4}$$

where e is the eccentricity of the RSO's orbit, then a simplified set of equations can be derived, known as the Tschauner–Hempel equations of motion [14]:

$$\tilde{x}'' = \frac{3}{k}\tilde{x} + 2\tilde{y}' \tag{11.5}$$

$$\tilde{y}'' = -2\tilde{x}' \tag{11.6}$$

$$\tilde{z}'' = -\tilde{z}' \tag{11.7}$$

where a_x, a_y, and a_z have been set equal to zero, $\dfrac{d()}{df} = ()'$ and $k = 1 + e\cos f$.

Equations (11.5–11.7) form a linear, nonautonomous set of differential equations since k is a function of the true anomaly of the RSO, and they can be written as:

$$\tilde{X}' = A(f)\tilde{X} \tag{11.8}$$

where $\tilde{X} = \left[\tilde{x}, \tilde{y}, \tilde{z}, \tilde{x}', \tilde{y}', \tilde{z}'\right]$. Yamanaka and Ankersen (YA) [16] found a relatively simple and singularity-free solution to this set of differential equations, where the solution can be described by

$$\tilde{X}(f) = \Theta(f; f_0)\tilde{X}(f_0) \tag{11.9}$$

and the state transition matrix (STM) is [16]

$$
\Theta(f, f_0) =
\begin{bmatrix}
s & 2 - 3esI & 0 & c & 0 & 0 \\
c\left(1 + \dfrac{1}{k}\right) & -3k^2I & 0 & -s\left(1 + \dfrac{1}{k}\right) & 1 & 0 \\
0 & 0 & \cos f & 0 & 0 & \sin f \\
s' & -3e\left(s'I + \dfrac{s}{k^2}\right) & 0 & c' & 0 & 0 \\
-2s & -3(1 - 2esI) & 0 & e - 2c & 0 & 0 \\
0 & 0 & -\sin f & 0 & 0 & \cos f
\end{bmatrix}
$$

$$
\times \dfrac{1}{\eta^2}
\begin{bmatrix}
-3s\dfrac{k + e^2}{k^2} & 0 & 0 & c - 2e & -s\dfrac{k + 1}{k} & 0 \\
3k - \eta^2 & 0 & 0 & es & k^2 & 0 \\
0 & 0 & \eta^2 \cos f & 0 & 0 & -\eta^2 \sin f \\
-3\left(e + \dfrac{c}{k}\right) & 0 & 0 & -s & -\left(c\dfrac{k + 1}{k} + e\right) & 0 \\
-3es\dfrac{k + 1}{k^2} & \eta^2 & 0 & -2 + ec & -es\dfrac{k + 1}{k} & 0 \\
0 & 0 & \eta^2 \sin f & 0 & 0 & \eta^2 \cos f
\end{bmatrix}_{f = f_0}
$$

$$\tag{11.10}$$

where, for clarity, it is emphasized that the first matrix in Eq. (11.10) is evaluated at f and the second matrix is evaluated at f_0. The terms inside Eq. (11.10) are defined as follows:

$$k = 1 + e \cos f \tag{11.11}$$

$$s = k \sin f \tag{11.12}$$

$$c = k \cos f \tag{11.13}$$

$$I = \frac{h}{p^2} (t - t_0) \tag{11.14}$$

$$\eta = \sqrt{1 - e^2} \tag{11.15}$$

where p is the semilatus rectum of the RSO,

$$p = a\eta^2 \tag{11.16}$$

and a is the semimajor axis of the RSO.

Several of the game conditions or inspection goals addressed in this chapter require the pursuer to match the energy of the evader. This is called the energy-matching condition and has been shown for the TH equations of motion to be [17]

$$(2 + 3e \cos f_f + e^2)\tilde{x}_f + e \sin f_f(1 + e \cos f_f)\tilde{x}_f' + (1 + e \cos f_f)^2 \tilde{y}_f' = 0 \tag{11.17}$$

where the subscript "f" denotes the final value of a given variable after a given maneuver. If these conditions hold at the end of a maneuver, and thus at the beginning of the ensuing natural motion, then the motion of the inspector satellite will be bounded with respect to the RSO. Given this condition, Sengupta and Vadali [17] parameterized the solution to the TH equations of motion in order to more clearly describe the geometry of the periodic relative motion:

$$\tilde{x} = \frac{\tilde{n}_1}{p} \sin(f + \alpha_0)(1 + e \cos f) \tag{11.18}$$

$$\tilde{y} = \frac{\tilde{n}_1}{p} \cos(f + \alpha_0)(2 + e \cos f) + \frac{\varrho_2}{p} \tag{11.19}$$

$$\tilde{z} = \frac{\tilde{n}_3}{p} \sin(f + \beta_0) \tag{11.20}$$

where the relative orbit parameters $\tilde{n}_1, \ldots, _3, \alpha_0$, and β_0 are as follows:

$$\tilde{n}_1 = \frac{a}{\eta}(\eta^2 \delta e^2 + e^2 \delta M_0^2)^{(1/2)} \tag{11.21}$$

$$\tilde{n}_2 = p\left(\delta\omega_\alpha + \delta\Omega\cos i + \frac{1}{\eta^3}\delta M_0\right) \tag{11.22}$$

$$\tilde{n}_3 = p(\delta i^2 + \delta\Omega^2 \sin^2 i)^{(1/2)} \tag{11.23}$$

$$\alpha_0 = \tan^{-1}\left(-\frac{\eta}{e}\frac{\delta e}{\delta M_0}\right) \tag{11.24}$$

$$\beta_0 = \tan^{-1}\left(-\frac{\delta\Omega\sin i}{\delta i}\right) + \omega_\alpha \tag{11.25}$$

and where the differential orbital element vector $\delta_\alpha = [\delta a, \delta e, \delta i, \delta\Omega, \delta\omega_\alpha, \delta M_0]$ contains the orbital element differences (to the first order) with respect to the reference orbit of the semimajor axis, eccentricity, inclination (i), right ascension of the ascending (Ω), argument of perigee (ω_α), and initial mean anomaly (M_0). Relating these differential orbital elements back to the TH states, Sengupta and Vadali showed the following [17]:

$$\delta a = \frac{2a}{\eta^2}c_3 \tag{11.26}$$

$$\delta M_0 = \frac{\eta^3}{e}c_2 \tag{11.27}$$

$$\delta e = -\eta^2 c_1 \tag{11.28}$$

$$\delta i = \sin(\omega_\alpha + f_f)\tilde{z}_f + \cos(\omega_\alpha + f_f)\tilde{z}'_f \tag{11.29}$$

$$\delta\Omega = \frac{-\left[\cos(\omega_\alpha + f_f)\tilde{z}_f - \sin(\omega_\alpha + f_f)\tilde{z}'_f\right]}{\sin i} \tag{11.30}$$

$$\delta\omega_\alpha = c_4 - \frac{\delta M_0}{n^3} - \delta\Omega\cos i \tag{11.31}$$

where

$$c_1 = -\frac{3}{\eta^2}(e + \cos f_f)\tilde{x}_f - \frac{1}{\eta^2}\sin f_f(1 + e\cos f_f)\tilde{x}'_f - \frac{1}{\eta^2}(2\cos f_f + e + e\cos^2 f_f)\tilde{y}'_f \tag{11.32}$$

$$c_2 = -\frac{3}{\eta^2}\frac{\sin f_f(1 + e\cos f_f + e^2)}{1 + e\cos f_f}\tilde{x}_f \tag{11.33}$$

$$+ \frac{1}{\eta^2}(\cos f_f - 2e + e\cos^2 f_f)\tilde{x}'_f - \frac{1}{\eta^2}\sin f_f(2 + e\cos f_f)\tilde{y}'_f$$

$$c_3 = (2 + 3e\cos f_f + e^2)\tilde{x}_f + e\sin f_f(1 + e\cos f_f)\tilde{x}'_f + (1 + e\cos f_f)^2\tilde{y}'_f \tag{11.34}$$

$$c_4 = -\frac{1}{\eta^2}(2+e\cos f_f)\left[\frac{3e\sin f_f}{1+e\cos f_f}\tilde{x}_f + (1-e\cos f_f)\tilde{x}_f' + e\sin f_f\tilde{y}_f'\right] + \tilde{y}_f \qquad (11.35)$$

and where c_3 has already been set equal to zero in Eq. (11.17) by enforcing the energy-matching condition. The size of the unforced relative motion can thus be prescribed by choosing ϱ_1 and ϱ_3, the bias can be set with ϱ_2, and the phase angles can be set with α_0 and β_0.

The equations of motion used by each player are the TH equations of motion [Eqs. (11.5–11.7)], but now with acceleration terms added due to a constant, steerable thrust, with mass loss accounted for. The acceleration terms are thus composed of a time-varying acceleration magnitude at, which increases with mass loss:

$$a(t) = \frac{a_0}{1 - t(a_0 / c)} \qquad (11.36)$$

where a_0 is the initial acceleration value due to a constant thrust and the initial mass of the spacecraft, and c is the effective exhaust velocity. This magnitude increases with time, and its direction is defined by the in-plane and out-of-plane thrust angles, α and ϕ, as shown in Figure 11.1. These two angles are the control variables for each player and are bounded as follows:

$$a \in -\pi, \pi \qquad (11.37)$$

$$\phi \in \left[-\frac{\pi}{2}, \frac{\pi}{2}\right] \qquad (11.38)$$

Thus, the acceleration terms can be written as

$$a_x = a(f)\sin\alpha(f)\cos\phi(f) \qquad (11.39)$$

$$a_y = a(f)\cos\alpha(f)\cos\phi(f) \qquad (11.40)$$

$$a_z = a(f)\sin\phi(f) \qquad (11.41)$$

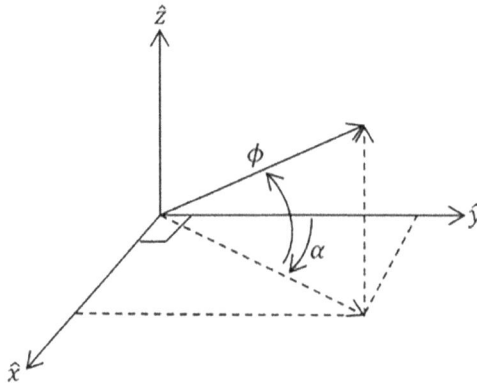

FIGURE 11.1   Thrust direction.

where the acceleration magnitude and the angles are now functions of the true anomaly f of the reference orbit because that is the independent variable when using the TH equations of motion. When these acceleration terms are retained in the development of the TH equations of motion, they become multiplied by a scalar coefficient, as in Ref. [18]:

$$\beta(f) = \frac{p^3 a(f)}{\mu k(f)^3} \tag{11.42}$$

where $a(f)$ has been included in the scalar coefficient. With the defined acceleration terms and the coefficient $\beta(f)$, the TH equations of motion become

$$\tilde{x}'' = 2\tilde{z}' + \beta(f)\sin\alpha\cos\phi \tag{11.43}$$

$$\tilde{y}'' = -\tilde{y} + \beta(f)\cos\alpha\cos\phi \tag{11.44}$$

$$\tilde{z}'' = \frac{3\tilde{z}}{k(f)} - 2\tilde{x}' + \beta(f)\sin\phi \tag{11.45}$$

where it is understood that the control variables α and ϕ, as well as the TH states $\tilde{X} = [\tilde{x}, \tilde{y}, \tilde{z}, \tilde{x}', \tilde{y}', \tilde{z}']^T$, are also functions of the true anomaly. These equations can be written in state space form as follows:

$$\tilde{X}' = A(f)\tilde{X} + B(f)U(\alpha,\phi) \tag{11.46}$$

where

$$B(f) = \begin{bmatrix} 0 & 0 & 0 \\ 0 & 0 & 0 \\ 0 & 0 & 0 \\ \beta(f) & 0 & 0 \\ 0 & \beta(f) & 0 \\ 0 & 0 & \beta(f) \end{bmatrix} \tag{11.47}$$

and

$$U(\alpha,\phi) = \begin{bmatrix} \sin\alpha\cos\phi & \cos\alpha\cos\phi & \sin\phi \end{bmatrix}^T \tag{11.48}$$

Both the pursuer and the evader use Eq. (11.46) as their equations of motion, where the dynamics of the pursuer will be represented with a subscript "p" and the evader with a subscript "e":

$$\tilde{X}'_p = A(f)\tilde{X}_p + B(f)U_p \tag{11.49}$$

$$\tilde{X}'_e = A(f)\tilde{X}_e + B(f)U_e \tag{11.50}$$

11.3   NECESSARY CONDITIONS FOR A DIFFERENTIAL GAME SOLUTION

Differential games were formally introduced by Isaacs [19] and treat the two-sided optimization of two uncooperative or competing players. This chapter addresses two-player zero-sum differential games, in which the strategies obtained are open-loop strategies based on the knowledge of each player's initial conditions. Following Bryson and Ho [20], the games considered herein can be described by starting with the dynamic system of the two players:

$$\dot{X} = f(X,u,v,t), \qquad X(t_0) = X_0 \tag{11.51}$$

where X is the vector containing the states of both players, u is the control of the minimizing player (the pursuer), v is the control of the maximizing player (the evader), and any terminal constraints can be described by

$$\psi(X(t_f),t_f) = 0 \tag{11.52}$$

The general performance index for the zero-sum game may be written in the Bolza form as

$$J = \phi(X(t_f),t_f) + \int_{t_0}^{t_f} L(X,u,v,t)\,dt \tag{11.53}$$

where the problem is to find u^* and v^* such that

$$J(u^*,v) \leq J(u^*,v^*) \leq J(u,v^*) \tag{11.54}$$

where the pursuer wishes to minimize and the evader wishes to maximize the same cost functional. Still following Ref. [20], it is desired to find the game-theoretic saddle-point solution by applying the first-order necessary conditions. The necessary conditions can be described by first forming the Hamiltonian:

$$\mathcal{H} = \lambda^T f + L \tag{11.55}$$

where λ is the vector of costates, f is the vector containing the system dynamics, and L is the Lagrangian. The function Φ is also formulated:

$$\Phi = \phi + v^T \psi \tag{11.56}$$

where ϕ is the function of terminal values in the cost function, and v is the vector of Lagrange multipliers associated with the terminal constraints. Given H and Φ, the necessary conditions for stationarity are

$$\dot{\lambda} = -\left[\frac{\partial \mathcal{H}}{\partial X}\right]^T \tag{11.57}$$

$$\lambda_f = -\left[\frac{\partial \Phi}{\partial X_f}\right]^T \qquad (11.58)$$

with

$$\frac{\partial \mathcal{H}}{\partial u} = 0, \qquad \frac{\partial \mathcal{H}}{\partial v} = 0 \qquad (11.59)$$

or, if the control is bounded,

$$\mathcal{H}^* = \max_{v} \min_{u} \mathcal{H} \qquad (11.60)$$

The transversality condition may also apply:

$$\mathcal{H}_f + \frac{\partial \Phi}{\partial t_f} = 0 \qquad (11.61)$$

and resulting equations typically form a two-point boundary value problem.

11.4 DIFFERENTIAL GAME FORMULATIONS USING INDIRECT HEURISTIC METHOD

The boundary value problems resulting from applying the necessary conditions may be solved with the IHM if certain properties apply. Pontani and Conway solved several one-sided optimization problems with the IHM [13,21,22], as well as the intercept pursuit–evasion game with Stupik et al. [11]. The IHM transforms the optimal control or differential game problem into a parameter optimization problem in which the initial (or final) costates can be sought for in an arbitrary range in order to determine the optimal control. This can only be done, however, if the resulting costate equations [Eq. (11.57)] and costate boundary conditions [Eq. (11.58)] are homogeneous with respect to the costates. Also, the optimal control equations resulting from applying Eq. (11.59) or Eq. (11.60) must only depend on the relative magnitude of the costates. If this is the case, then the transversality condition [Eq. (11.61)] is ignorable and the optimal control may be found via the IHM.

Given the equations of motion and control for both players and the necessary conditions for a solution, the differential games can now be formulated. Six different inspection goals or games are examined in this paper, where, in all games, the pursuer has a higher initial acceleration value than the evader (i.e., $a_{0_p} > a_{0_e}$) and both players have the same effective exhaust velocity. This ensures that the pursuer should achieve its inspection goal, if given enough time.

For all games, the cost function is

$$J = f_f \qquad (11.62)$$

which is equivalent to the final time of the game. This means that the pursuer wishes to minimize the time to achieve a specific inspection goal, whereas the evader wishes to prolong that inspection goal as long as possible. This is a zero-sum game because the sum of the two players' costs is zero.

Given the dynamic system of the two players [Eqs. (11.49) and (11.50)], their initial conditions, the terminal constraints for each game,

$$\psi(\tilde{X}_p(f_f), \tilde{X}_e(f_f), f_f) = 0 \tag{11.63}$$

and the performance criterion [Eq. (11.62)], the goal is to find u^* and v^* such that

$$J(u^*, v) \le J(u^*, v^*) \le J(u, v^*) \tag{11.64}$$

where

$$u = \begin{bmatrix} \alpha_p & \phi_p \end{bmatrix}^T \tag{11.65}$$

$$v = \begin{bmatrix} \alpha_e & \phi_e \end{bmatrix}^T \tag{11.66}$$

Following the IHM, it is desired to find the game-theoretic saddle point by first applying the first-order necessary conditions. The Hamiltonian for all games considered is

$$\mathcal{H} = \lambda_p^T \tilde{X}_p' + \lambda_e^T \tilde{X}_e' = \lambda_p^T \left[A(f)\tilde{X}_p + B(f)U_p \right] + \lambda_e^T \left[A(f)\tilde{X}_e + B(f)U_e \right] \tag{11.67}$$

where λ_p and λ_e are the vectors of costates for each player:

$$\lambda_p = [\lambda_1, \lambda_2, \lambda_3, \lambda_4, \lambda_5, \lambda_6]^T \tag{11.68}$$

$$\lambda_e = [\lambda_7, \lambda_8, \lambda_9, \lambda_{10}, \lambda_{11}, \lambda_{12}]^T \tag{11.69}$$

The costate equations for all games considered are

$$\lambda_p' = -A(f)^T \lambda_p \tag{11.70}$$

$$\lambda_e' = -A(f)^T \lambda_e \tag{11.71}$$

where, like Stupik et al. [11], it is desirable to find the STM for the costate equations. Because the YA STM [16] [Θ in Eq. (11.10)] already exists for the system

$$\tilde{X}' = A(f)\tilde{X} \tag{11.72}$$

then it can be used to find the STM for the costate equations when $A(f)$ is $-A(f)^T$. The STM for the costate equations, Ξ, can be found symbolically by using MATLAB's Symbolic Toolbox via

$$\Xi(f, f_0) = (\Theta(f; f_0)^{-1})^T \tag{11.73}$$

The optimal control for each player must satisfy

$$[u^{*^T} \quad v^{*^T}]^T = \arg \max_v \min_u \mathcal{H} \tag{11.74}$$

since the controls are bounded. It can be seen that the Hamiltonian is separable, thus

$$u^* = [\alpha_p^* \quad \phi_p^*]^T = \arg \min_{(\alpha_p, \phi_p)} (\lambda_4 \sin\alpha_p \cos\phi_p + \lambda_5 \cos\alpha_p \cos\phi_p + \lambda_6 \sin\phi_p) \tag{11.75}$$

This can be written as

$$u^* = [\alpha_p^* \quad \phi_p^*]^T = \arg \min_{(\alpha_p, \phi_p)} (\cos\phi_p(\lambda_4 \sin\alpha_p + \lambda_5 \cos\alpha_p) + \lambda_6 \sin\phi_p) \tag{11.76}$$

which is in a form where Isaac's lemma on circular vectograms [19] may be applied to obtain the following:

$$\sin\alpha_p^* = \frac{-\lambda_4}{\sqrt{\lambda_4^2 + \lambda_5^2}} \tag{11.77}$$

$$\cos\alpha_p^* = \frac{-\lambda_5}{\sqrt{\lambda_4^2 + \lambda_5^2}} \tag{11.78}$$

$$\sin\phi_p^* = \frac{-\lambda_6}{\sqrt{\lambda_4^2 + \lambda_5^2 + \lambda_6^2}} \tag{11.79}$$

Similarly, the optimal control equations for the evader can be obtained from

$$v^* = [\alpha_e^* \quad \phi_e^*] = \arg \max_{(\alpha_e, \phi_e)} (\cos\phi_e(\lambda_{10} \sin\alpha_e + \lambda_{11} \cos\alpha_e) + \lambda_{12} \sin\phi_e) \tag{11.80}$$

and thus

$$\sin\alpha_e^* = \frac{\lambda_{10}}{\sqrt{\lambda_{10}^2 + \lambda_{11}^2}} \tag{11.81}$$

$$\cos\alpha_e^* = \frac{\lambda_{11}}{\sqrt{\lambda_{10}^2 + \lambda_{11}^2}} \tag{11.82}$$

$$\sin\phi_e^* = \frac{\lambda_{12}}{\sqrt{\lambda_{10}^2 + \lambda_{11}^2 + \lambda_{12}^2}} \tag{11.83}$$

These optimal control equations are the same for all games examined in this paper. The costate boundary conditions are different for each game considered. For each game, the function Φ is generated:

$$\Phi = f_f + v^T \psi \tag{11.84}$$

where ψ is different for each game or inspection goal. The next section describes the specific terminal constraints and costate boundary conditions for each game. To use the IHM, as discussed, the costate boundary conditions must be homogeneous with respect to the costates in conjunction with the homogeneity of the costate equations. The developed costate equations [Eqs. (11.70) and (11.71)] are homogeneous, and thus it is desirable that the costate boundary conditions developed in the next section are homogeneous as well. If these conditions apply, then the transversality condition ($\mathcal{H}_f + 1 = 0$ for all games considered) is ignorable, and if an initial value of λ for the pursuer and evader can be found such that $\lambda_0 = k_\lambda \lambda_0^*(k_\lambda > 0)$, then the same proportionality holds between λ and the optimal λ (λ^*) at any time. And because the optimal control equations [Eqs. (11.77–11.79) and (11.81–11.83)] depend only on the relative magnitude of the costates and not the optimal costates themselves, then by satisfying all of the necessary conditions except for the transversality condition, the optimal control law can be found with a scaled version of the optimal costates.

In addition to the homogeneity of the boundary conditions, it is desirable that the terminal constraints (developed in the next section) are linear with respect to the states because this simplifies the resulting boundary value problem and makes the problem more tractable to solve.

For most games examined, as will be seen, the following property holds, as in the work of Stupik et al. [11]:

$$\lambda_{pf} + \lambda_{ef} = 0 \qquad (11.85)$$

Using the STM for the costates, it can be seen that

$$\lambda_{pf} + \lambda_{ef} = 0 = \Xi\lambda_{p0} + \Xi\lambda_{e0} = \Xi(\lambda_{p0} + \lambda_{e0}) \qquad (11.86)$$

which, assuming Ξ is invertible, means that

$$\lambda_{p0} = -\lambda_{e0} \qquad (11.87)$$

and thus, at any true anomaly,

$$\lambda_p = -\lambda_e \qquad (11.88)$$

meaning that the pursuer and evader costates are always equal and opposite. Plugging this relationship into the optimal control equations, it can be seen that, as shown in the work of Stupik et al. [11],

$$\alpha_P^* = \alpha_e^* \qquad (11.89)$$

$$\phi_P^* = \phi_e^* \qquad (11.90)$$

11.5 GAME CONDITIONS

For each game, a different set of terminal constraints is considered. Therefore, the following subsections present the terminal constraints for each game and the corresponding costate boundary conditions. All of the corresponding costate boundary conditions are homogeneous and allow the use of the IHM. This means that all $m \, \nu(\nu_1 - \nu_m)$, where m is the number of terminal constraints, can be eliminated from the set of 12 costate boundary condition equations [e.g., from Eq. (11.92)] and the 12 equations can be reduced to a set of $12 - m$ equations, which can be written in matrix form as $A\lambda = 0_{[12-m \times 1]}$, where $\lambda = [\lambda_1, \lambda_2, \lambda_3, \lambda_4, \lambda_5, \lambda_6, \lambda_7, \lambda_8, \lambda_9, \lambda_{10}, \lambda_{11}, \lambda_{12}]^T$. In other words, there is no effect on the $12 - m$ equations if λ is scaled by any constant.

11.5.1 INTERCEPT

In the intercept game, the goal of the pursuer is to match the final position of the evader in minimum time, whereas the goal of the evader is to delay that condition as long as possible. Thus, the vector of terminal constraints is simply

$$\psi = \begin{bmatrix} \tilde{x}_p - \tilde{x}_e \\ \tilde{y}_p - \tilde{y}_e \\ \tilde{z}_p - \tilde{z}_e \end{bmatrix} \tag{11.91}$$

and the costate boundary conditions become

$$
\begin{aligned}
&\lambda_{1_f} = \nu_1 \quad &\lambda_{7_f} = -\nu_1 \\
&\lambda_{2_f} = \nu_2 \quad &\lambda_{8_f} = -\nu_2 \\
&\lambda_{3_f} = \nu_3 \quad &\lambda_{9_f} = -\nu_3 \\
&\lambda_{4_f} = 0 \quad &\lambda_{10_f} = 0 \\
&\lambda_{5_f} = 0 \quad &\lambda_{11_f} = 0 \\
&\lambda_{6_f} = 0 \quad &\lambda_{12_f} = 0
\end{aligned}
\tag{11.92}
$$

If given $\lambda_{1_f} - \lambda_{3_f}$, all final costates can be determined.

11.5.2 RENDEZVOUS

In the rendezvous game, the goal of the pursuer is to match the final position and velocity of the evader in minimum time, whereas the goal of the evader is to delay that condition as long as possible. Thus, the vector of terminal constraints is simply

$$\psi = \begin{bmatrix} \tilde{x}_p - \tilde{x}_e \\ \tilde{y}_p - \tilde{y}_e \\ \tilde{z}_p - \tilde{z}_e \\ \tilde{x}'_p - \tilde{x}'_e \\ \tilde{y}'_p - \tilde{y}'_e \\ \tilde{z}'_p - \tilde{z}'_e \end{bmatrix} \qquad (11.93)$$

and the costate boundary conditions become

$$\begin{aligned} \lambda_{1_f} &= v_1 & \lambda_{7_f} &= -v_1 \\ \lambda_{2_f} &= v_2 & \lambda_{8_f} &= -v_2 \\ \lambda_{3_f} &= v_3 & \lambda_{9_f} &= -v_3 \\ \lambda_{4_f} &= v_4 & \lambda_{10_f} &= -v_4 \\ \lambda_{5_f} &= v_5 & \lambda_{11_f} &= -v_5 \\ \lambda_{6_f} &= v_6 & \lambda_{12_f} &= -v_6 \end{aligned} \qquad (11.94)$$

If given $\lambda_{1_f} - \lambda_{6_f}$, all final costates can be determined.

11.5.3 Obtain Sun Vector

In this game, the goal of the pursuer is to align its position with the vector from the sun to the evader at a point between the sun and the evader such that the evader is lit with respect to the pursuer, in minimum time. The goal of the evader is to delay that condition as long as possible. Figure 11.2 shows the general terminal conditions for this game, in which the goal of the pursuer is to end on the vector from the sun to

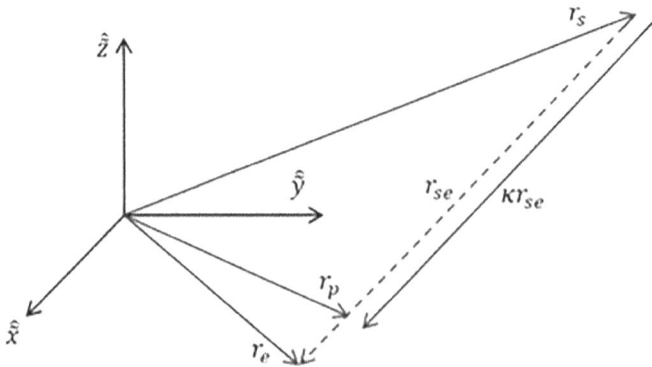

FIGURE 11.2   General terminal conditions for obtaining the sun vector.

the evader r_{se} at some distance from the evader, thus at κr_{se}, where $\kappa \in 0;1$ and should be very close to one. This formulation is linear with respect to the states. However, the final range cannot be exactly prescribed, and some errors must be allowed from the prescribed final range. For example, if a final range of d km is desired between the pursuer and evader, then κ may be estimated as

$$\kappa = \frac{-d\|r_s(f_0)\|_2 + \|r_s(f_0)\|_2^2}{\|r_s(f_0)\|_2^2} \tag{11.95}$$

where r_s is the vector to the sun in the relative frame. This should produce a final range relatively close to d as long as the motion of the pursuer and evader remains close to the evader's original orbit (the reference orbit) and the game does not last for too long with respect to the period of the reference orbit. Given an estimate for κ to achieve d, then the terminal constraint may be expressed as

$$r_p = r_s + \kappa(r_{se}) \tag{11.96}$$

where r_p is the relative position of the pursuer, r_p is the relative position of the evader, and $r_{se} = r_e - r_s$. Thus, in standard form, the vector of terminal constraints is

$$\psi = \begin{bmatrix} \tilde{x}_p - \tilde{x}_s - \kappa(\tilde{x}_e - \tilde{x}_s) \\ \tilde{y}_p - \tilde{y}_s - \kappa(\tilde{y}_e - \tilde{y}_s) \\ \tilde{z}_p - \tilde{z}_s - \kappa(\tilde{z}_e - \tilde{z}_s) \end{bmatrix} \tag{11.97}$$

where r_s does not depend on the pursuer and evader states and can be calculated for any given final true anomaly via Vallado's sun algorithm [23]. This formulation of the terminal constraints produces the following costate boundary conditions:

$$
\begin{aligned}
&\lambda_{1_f} = v_1 \quad \lambda_{7_f} = -\kappa v_1 \\
&\lambda_{2_f} = v_2 \quad \lambda_{8_f} = -\kappa v_2 \\
&\lambda_{3_f} = v_3 \quad \lambda_{9_f} = -\kappa v_3 \\
&\lambda_{4_f} = 0 \quad \lambda_{10_f} = 0 \\
&\lambda_{5_f} = 0 \quad \lambda_{11_f} = 0 \\
&\lambda_{6_f} = 0 \quad \lambda_{12_f} = 0
\end{aligned} \tag{11.98}
$$

which are not a function of the players' states or the position of the sun, thus simplifying the resulting boundary value problem.

If given $\lambda_{1_f} - \lambda_{3_f}$, all final costates can be determined.

If a more accurate or exact d is desired, κ can be changed in an iterative fashion in an outer loop if the resulting final range is not close enough to the value desired. The game can be solved multiple times until some tolerance is reached and the resulting κ value produces the desired terminal range between the pursuer and evader.

11.5.4 MATCH ENERGY

The match energy game is where the goal of the pursuer is to match the energy of the evader in minimum time such that, if each player stopped thrusting at the end of the game, then the relative motion between the two players would remain bounded after one period. The goal of the evader is to delay that condition as long as possible. To ensure the energy of the two players becomes matched, the change in the semimajor axes of both players should be the same. Thus, using Eq. (26) for each player, the following should apply:

$$\delta a_p = \frac{2a}{\eta^2}c_{3_p} = \delta a_e = \frac{2a}{\eta^2}c_{3_e} \tag{11.99}$$

where a and η are those of the reference orbit, and the coefficients c_{3_p} and c_{3_e} from Eq. (11.34) are referenced to the reference orbit as well but are functions of the relative states of the pursuer and evader, respectively. Expressing the terms multiplied by the pursuer or evader states inside Eqs. (11.32–11.35) as b_{ij}, where $i = 1, \ldots, 4$ for each c_i and where $j = 1, \ldots, 3$ for the three terms multiplied by the relative states inside each c_i, the vector of terminal constraint can be written as

$$\psi = [b_{31}(\tilde{x}_p - \tilde{x}_e) + b_{32}(\tilde{x}'_p - \tilde{x}'_e) + b_{33}(\tilde{y}'_p - \tilde{y}'_e)] \tag{11.100}$$

where each b_{ij} is a function of the final true anomaly. This formulation of the terminal constraints produces the following costate boundary conditions:

$$
\begin{aligned}
\lambda_{1_f} &= b_{31}v_1 & \lambda_{7_f} &= -b_{31}v_1 \\
\lambda_{2_f} &= 0 & \lambda_{8_f} &= 0 \\
\lambda_{3_f} &= 0 & \lambda_{9_f} &= 0 \\
\lambda_{4_f} &= b_{32}v_1 & \lambda_{10_f} &= -b_{32}v_1 \\
\lambda_{5_f} &= b_{33}v_1 & \lambda_{11_f} &= -b_{33}v_1 \\
\lambda_{6_f} &= 0 & \lambda_{12_f} &= 0
\end{aligned}
\tag{11.101}
$$

If given λ_{1_f} and the final true anomaly, all final costates can be determined.

11.5.5 OBTAIN SUN VECTOR AND MATCH ENERGY

In this game, the goal of the pursuer is a combination of two previous conditions: to align itself with the sun vector at an approximate range while also matching the energy of the evader in minimum time, whereas the goal of the evader is to delay that combination of conditions as long as possible. The vector of terminal constraints is thus

$$\psi = \begin{bmatrix} \tilde{x}_p - \tilde{x}_s - \kappa(\tilde{x}_e - \tilde{x}_s) \\ \tilde{y}_p - \tilde{y}_s - \kappa(\tilde{y}_e - \tilde{y}_s) \\ \tilde{z}_p - \tilde{z}_s - \kappa(\tilde{z}_e - \tilde{z}_s) \\ b_{31}(\tilde{x}_p - \tilde{x}_e) + b_{32}(x'_p - \tilde{x}'_e) + b_{33}(\tilde{y}'_p - \tilde{y}'_e) \end{bmatrix} \tag{11.102}$$

$$\lambda_{1_f} = v_1 + b_{31}v_4 \qquad \lambda_{7_f} = -\kappa v_1 - b_{31}v_4$$

$$\lambda_{2_f} = v_2 \qquad \lambda_{8_f} = -\kappa v_2$$

$$\lambda_{3_f} = v_3 \qquad \lambda_{9_f} = -\kappa v_3$$

$$\lambda_{4_f} = b_{32}v_4 \qquad \lambda_{10_f} = -b_{32}v_4 \qquad (11.103)$$

$$\lambda_{5_f} = b_{33}v_1 \qquad \lambda_{11_f} = -b_{33}v_4$$

$$\lambda_{6_f} = 0 \qquad \lambda_{12_f} = 0$$

If given $\lambda_{1_f} - \lambda_{4_f}$ and the final true anomaly, all final costates can be determined.

11.5.6 MATCH ENERGY AND REMAIN IN CLOSE PROXIMITY

In this last game considered, the goal of the pursuer is to both match the energy of the evader and also to remain within a prescribed range of the evader during the ensuing period and to accomplish these goals in minimum time. The goal of the evader is to delay that combination of conditions as long as possible. To remain close during the ensuing period, Sengupta and Vadali's [17] relative orbit parameters for periodic, elliptical relative motion [ϱ_1, ϱ_2, and ϱ_3 in Eqs. (11.18–11.20)] can be constrained to place the pursuer into the same orbit as the evader, at a prescribed distance away, and with a prescribed amplitude for the desired relative crosstrack motion.

To do this, the same energy-matching constraint must be enforced as before, namely, Eq. (11.100). Next, ϱ_1 must be set equal to zero, where this is one of the two relative motion size parameters for the motion of the pursuer relative to the evader, and not relative to the reference orbit. Thus, examining Eq. (11.21), δe and δM_0 of the pursuer relative to the evader must both be equal to zero. Therefore, δe_p and δe_e (the change in eccentricity of each player with respect to the reference orbit or the original orbit of the evader) must be equal. Therefore, the second constraint is

$$\delta e_p - \delta e_e = -\eta^2(c_{1_p} - c_{1_e}) = 0 \qquad (11.104)$$

Likewise, δM_{0_p} and δM_{0_e} must be equal, and the third constraint is

$$\delta M_{0_p} - \delta M_{0_e} = \frac{\eta^2}{e}(c_{2_p} - c_{2_e}) = 0 \qquad (11.105)$$

These two constraints ensure that $\varrho_1 = 0$ for the motion of the pursuer with respect to the evader. Next, ϱ_2 / p_e (where p_e is the semilatus rectum of the evader's final orbit since this is for the motion of the pursuer relative to the evader) must be set equal to the desired intrack placement of the pursuer relative to the evader, or d_y. Therefore, the following condition must apply:

$$\varrho_2 = p_e d_y \qquad (11.106)$$

It can be seen by examining Eq. (11.22) that, because δM_0 of the pursuer relative to the evader is already equal to zero, then a way to enforce Eq. (11.106) is to set $\delta\Omega$ of the pursuer relative to the evader equal to zero, which means the following must apply:

$$\varrho_2 = p_e(\delta\omega_{\alpha_p} - \delta\omega_{\alpha_e}) \tag{11.107}$$

Therefore, setting this equation equal to Eq. (11.106),

$$\delta\omega_{\alpha_p} - \delta\omega_{\alpha_e} = d_y \tag{11.108}$$

and thus the fourth constraint, using Eq. (11.31), is

$$\delta\omega_{\alpha_p} - \delta\omega_{\alpha_e} = \left(c_{4_p} - \frac{\delta M_{0_p}}{\eta^3} - \delta\Omega_p \cos i\right)$$
$$- \left(c_{4_e} - \frac{\delta M_{0_e}}{\eta^3} - \delta\Omega_e \cos i\right) = c_{4_p} - c_{4_e} = d_y \tag{11.109}$$

As mentioned, $\delta\Omega$ of the pursuer relative to the evader must also be set equal to zero, and thus the fifth constraint is

$$\delta\Omega_p - \delta\Omega_e = -(\cos(\omega_\alpha + f)\tilde{z}_p - \sin(\omega_\alpha + f)\tilde{z}'_p)$$
$$- \frac{-(\cos(\omega_\alpha + f)\tilde{z}_e - \sin(\omega_\alpha + f)\tilde{z}'_e)}{\sin i} = 0 \tag{11.110}$$

or

$$\cos(\omega_\alpha + f)(\tilde{z}_e - \tilde{z}_p) + \sin(\omega_\alpha + f)(\tilde{z}'_p - \tilde{z}'_e) = 0 \tag{11.111}$$

And, finally, the sixth and final constraint prescribes the amplitude of the ensuing crosstrack motion of the pursuer relative to the evader. To do this, ϱ_3 / p_e must be set equal to the desired amplitude, or d_z.
Therefore,

$$\varrho_3 = p_e d_z \tag{11.112}$$

and by examining Eq. (11.23) and realizing that $\delta\Omega$ of the pursuer relative to the evader is already set equal to zero, then the following condition must apply:

$$\varrho_3 = p_e(\delta i_p - \delta i_e) \tag{11.113}$$

Thus, setting the two previous equations equal to one another,

$$\delta i_p - \delta i_e = d_z \tag{11.114}$$

and the sixth and final constraint can be written as

$$\delta i_p - \delta i_e = (\sin(\omega_\alpha + f)\tilde{z}_p + \cos(\omega_\alpha + f)\tilde{z}'_p)$$
$$- (\sin(\omega_\alpha + f)\tilde{z}_e + \cos(\omega_\alpha + f)\tilde{z}'_e) = d_z \tag{11.115}$$

or

$$\sin(\omega_\alpha + f)(\tilde{z}_p - \tilde{z}_e) + \cos(\omega_\alpha + f)(\tilde{z}'_p - \tilde{z}'_e) - d_z = 0 \tag{11.116}$$

The vector of terminal equality constraints can thus be written as

$$\psi = \begin{bmatrix} b_{31}(\tilde{x}_p - \tilde{x}_e) + b_{32}(x'_p - \tilde{x}'_e) + b_{33}(\tilde{y}_p - \tilde{y}'_e) \\ b_{11}(\tilde{x}_p - \tilde{x}_e) + b_{12}(x'_p - \tilde{x}'_e) + b_{13}(\tilde{y}_p - \tilde{y}'_e) \\ b_{21}(\tilde{x}_p - \tilde{x}_e) + b_{22}(x'_p - \tilde{x}'_e) + b_{33}(\tilde{y}_p - \tilde{y}'_e) \\ b_{41}(\tilde{x}_p - \tilde{x}_e) + b_{42}(x'_p - \tilde{x}'_e) + b_{43}(\tilde{y}_p - \tilde{y}'_e) + \tilde{y}_p - \tilde{y}_e - d_y \\ \cos(\omega_\alpha + f)(\tilde{z}_e - \tilde{z}_p) + \sin(\omega_\alpha + f)(\tilde{z}'_p - \tilde{z}'_e) \\ \sin(\omega_\alpha + f)(\tilde{z}_p - \tilde{z}_e) + \cos(\omega_\alpha + f)(\tilde{z}'_p - \tilde{z}'_e) - d_z \end{bmatrix} \tag{11.117}$$

These constraints ensure that the pursuer matches the orbit of the evader at a prescribed intrack distance away d_y and with a relative crosstrack amplitude of d_z. They have been developed to be linear with respect to the states such that the resulting boundary value problem is more tractable to solve. The corresponding costate boundary conditions are

$$\lambda_{1_f} = b_{31}v_1 + b_{11}v_2 + b_{21}v_3 + b_{41}v_4 \qquad \lambda_{7_f} = -\lambda_{1_f}$$

$$\lambda_{2_f} = v_4 \qquad \lambda_{8_f} = -\lambda_{2_f}$$

$$\lambda_{3_f} = -\cos(\omega_\alpha + f)v_5 + \sin(\omega_\alpha + f)v_6 \qquad \lambda_{9_f} = -\lambda_{3_f}$$

$$\lambda_{4_f} = b_{32}v_1 + b_{12}v_2 + b_{22}v_3 + b_{42}v_4 \qquad \lambda_{10_f} = -\lambda_{4_f} \tag{11.118}$$

$$\lambda_{5_f} = b_{33}v_1 + b_{13}v_2 + b_{23}v_3 + b_{43}v_4 \qquad \lambda_{11_f} = -\lambda_{5_f}$$

$$\lambda_{6_f} = \sin(\omega_\alpha + f)v_5 + \cos(\omega_\alpha + f)v_6 \qquad \lambda_{12_f} = -\lambda_{6_f}$$

If given $\lambda_{1_f} - \lambda_{6_f}$ and the final true anomaly, all final costates can be determined.

11.6 BOUNDARY VALUE PROBLEM FORMULATIONS VIA INDIRECT HEURISTIC METHOD

Following the IHM, each boundary value problem resulting from each game condition is solved by formulating a parameter optimization problem. Each optimization problem is then solved with a metaheuristic optimization method, where in this study either MATLAB's PSO or GA is used. The purpose of the optimizer is solely to satisfy the terminal constraints contained in ψ for each game. For all games, the minimum number of final costates required to determine all final costates (which is equal to the number of terminal constraints m) becomes the optimization variables. Stupik et al. [11] used initial costates; however, the final costates are used here to make the problem simpler to formulate and solve. Because the properties required to apply the IHM are satisfied, only the relative magnitude of the costates is of importance to find the optimal control; thus, following the IHM, the m final costates for each problem can be bounded as follows:

$$\lambda_{i_f} \in -1; 1 \qquad i = 1; \ldots; m \qquad (11.119)$$

The final time (which can be converted to the final true anomaly) is also included as an optimization variable in each problem and is bounded:

$$t_f \in \left[0; t_{f_{max}}\right] \qquad (11.120)$$

where $t_{f_{max}}$ should be large enough to ensure a solution can be found. The optimization variables for each problem must then be found to satisfy the terminal constraints. This is accomplished as follows:

1. Given the value of each of the minimum number of final costates required for the current problem and the value of the final true anomaly, determine the rest of the final costates via the costate boundary condition equations for the current problem.
2. Propagate the costates backward analytically from the given final time (final true anomaly) using the costate STM Ξ in Eq. (11.73) to find the initial costates.
3. Propagate the states forward numerically (for example, with MATLAB's *ode45*) using Eqs. (11.49) and (11.50) where, at each time step, the control is calculated via Eqs. (11.77–79) [and Eqs. (11.1–11.83) if necessary], which are functions of the costates and can be propagated forward analytically from the initial costates, again with Ξ.
4. Once the states have been propagated to the given final time (or final true anomaly), evaluate the terminal constraints contained in ψ for the current problem.
5. Iterate (using the metaheuristic optimization algorithm), repeating steps 1 through 4 until the prescribed tolerance is met.
6. Extract the differential game solution.

For each problem, the static optimization problem is solved with either MATLAB's PSO or MATLAB's GA, depending on the amount and type of constraints. If the PSO is used, the constraints are satisfied by summing the square of each constraint in the cost function:

$$J_{PSO} = \sum_{i}^{m} \psi_i^2 \qquad (11.121)$$

If the set of terminal constraints for a particular problem is a function of the positions ($\tilde{x}$, $\tilde{y}$, or $\tilde{z}$) and not the velocities ($\tilde{x}'$, $\tilde{y}'$, or $\tilde{z}'$) or, likewise, a function of the velocities and not the positions, then the PSO performs well in driving the constraint violations to zero. On the other hand, if the set of terminal constraints for a particular problem is a function of both positions and velocities, then the GA performs fairly well to satisfy all constraints. If the GA is used, then the cost function is set equal to zero:

$$J_{GA} = 0 \tag{122}$$

and the equality constraints are satisfied by using the nondefault nonlinear constraint algorithm called the penalty algorithm. Table 11.1 shows the following for each game: the number of equality constraints, m, contained in ψ, the optimization parameters, and the solver to be used.

11.7 SIMULATION RESULTS

The solutions presented in this section use the simulation parameters shown in Table 11.2. The same initial conditions from Stupik et al. [11] are used in order to compare the intercept case when the eccentricity is set equal to zero. Note that $a_{0_p} > a_{0_e}$ to ensure the games can be completed.

TABLE 11.1
Differential Game Problem Formulations

Inspection Goal	Size of Ψ, m	Optimization Parameters	Solver
Intercept	3	$\chi = [\lambda_{1f}, \lambda_{2f}, \lambda_{3f}, t_f]$	PSO
Rendezvous	6	$\chi = [\lambda_{1f}, \lambda_{2f}, \lambda_{3f}, \lambda_{4f}, \lambda_{5f}, \lambda_{6f}, t_f]$	GA
Obtain sun vector	3	$\chi = [\lambda_{1f}, \lambda_{2f}, \lambda_{3f}, t_f]$	PSO
Match energy	1	$\chi = [\lambda_{1f}, t_f]$	PSO
Obtain sun vector and match energy	4	$\chi = [\lambda_{1f}, \lambda_{2f}, \lambda_{3f}, \lambda_{4f}, t_f]$	GA
Match energy and remain in close proximity	6	$\chi = [\lambda_{1f}, \lambda_{2f}, \lambda_{3f}, \lambda_{4f}, \lambda_{5f}, \lambda_{6f}, t_f]$	GA

TABLE 11.2
Simulation Parameters

Time and Sun	Reference Orbit	Pursuer Properties	Evader Properties
$Date_0$ = 17 March 2018	a_c = 42, 164.137 km	r_{po} = $[-38.93, -100,0]^T$ km	r_e = $[0, 0, 0]^T$ km
$Time_0$ = 13:00.00	e = 0.2	vpo = $[0, 0, 0]^T$ m/s	Ve = $[0,0,0]^T$ m/s
K = 0.9999999328126	f_o = 0°	a_0 = 0.0686 m/s^2	$a0_e$ = 0.0343 m/s^2
—	i = 5°	c_p = 3 km/s	Ce = 3 km/s
—	ω_{oe} = 10°	—	—
—	Ω = 45°	—	—

It was desired to compare the intercept solution with $e = 0$ to the first solution obtained by Stupik et al. [11] to see if they were approximately the same and to verify that the developed method reduced to the HCW solution when the eccentricity was set equal to zero. Indeed, the same solution was obtained, with a final time of $t_f = 41.32$ min and the same final position. This solution took 38.5 s to compute. All solutions obtained in this section used MATLAB's parallel processing with four "workers" (MATLAB computational engines) and were obtained on a laptop with eight 2.4 GHz processors and 16 GB of RAM. The value of the cost function, or the sum of the constraint violations squared, was 1.15×10^{-8}. Because the solutions match those in Ref. [11], they are not shown here.

With the zero-eccentricity intercept solution matching the work of Stupik et al. [11], the intercept game was run with $e = 0.2$, which the rest of the games in this section use to showcase the ability to solve games that take place with respect to an elliptical reference orbit. The resulting costates and control are shown in Figure 11.3, where the evader costates are equal and opposite the pursuer costates, and the control for both players is the same. Note that the last control values are a result from the velocity costates being zero, and they can be ignored. The states and trajectory are shown in Figure 11.4. This solution took 36.4 s to compute, with a final time of $t_f = 41.53$ min, and a PSO constraint violation of 3.79×10^{-8}.

The rendezvous game costates and control are shown in Figure 11.5, where the evader costates are equal and opposite the pursuer costates, and the control for both players is the same. The states and trajectory are shown in Figure 11.6. The states take an interesting shape, in which the optimal states for the evader prolong the rendezvous condition as long as possible. This solution took 8.64 min to compute because the GA was used instead of the PSO. The final time was $t_f = 57.61$ min, and the maximum GA constraint violation was 5.12×10^{-10}.

The obtain sun vector game control and trajectory are shown in Figure 11.7. The controls for both players are approximately the same, but not exactly, because $\lambda_{p_f} + \lambda_{e_f} \neq 0$. For this game, the desired range at the end of the game was set to $d = 10$ km. Using Eq. (11.95) for κ to obtain d, the resulting range is 10.0006 km. The trajectory shows how, at the end of the game, the pursuer is approximately 10 km away from the evader, at a point along the sun vector with respect to the evader, such that the

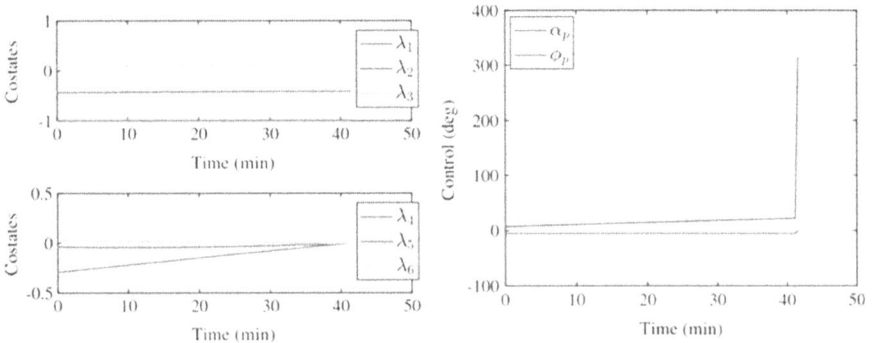

FIGURE 11.3 Intercept game: costates and control.

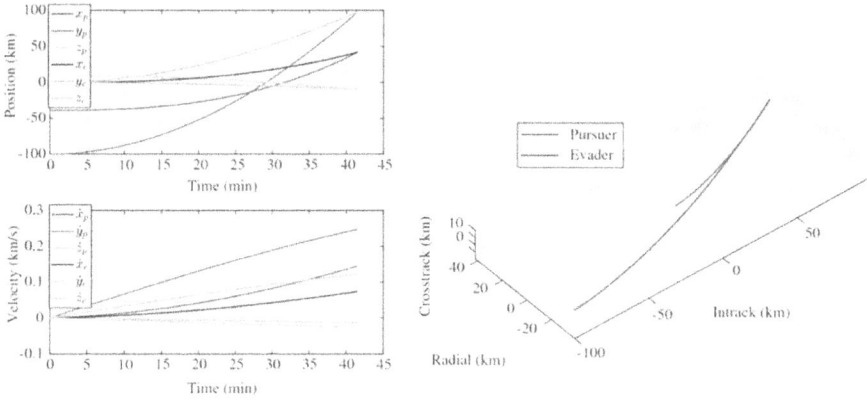

FIGURE 11.4 Intercept game: states and trajectory.

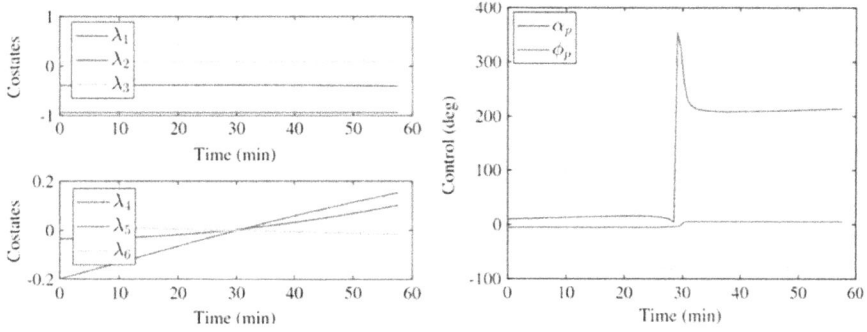

FIGURE 11.5 Rendezvous game: costates and control.

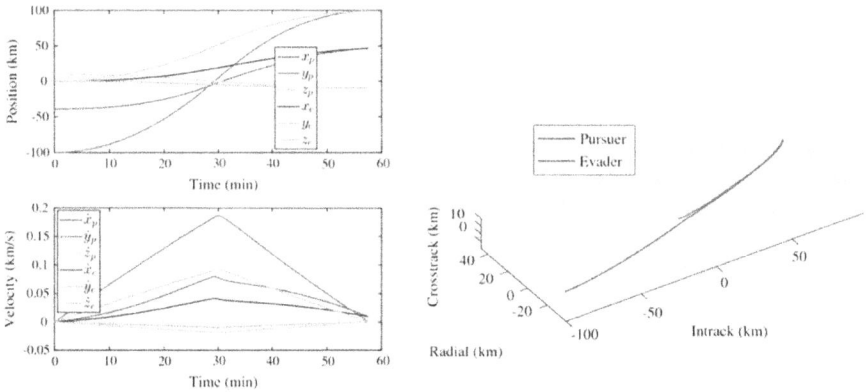

FIGURE 11.6 Rendezvous game: states and trajectory.

(a)

(b)

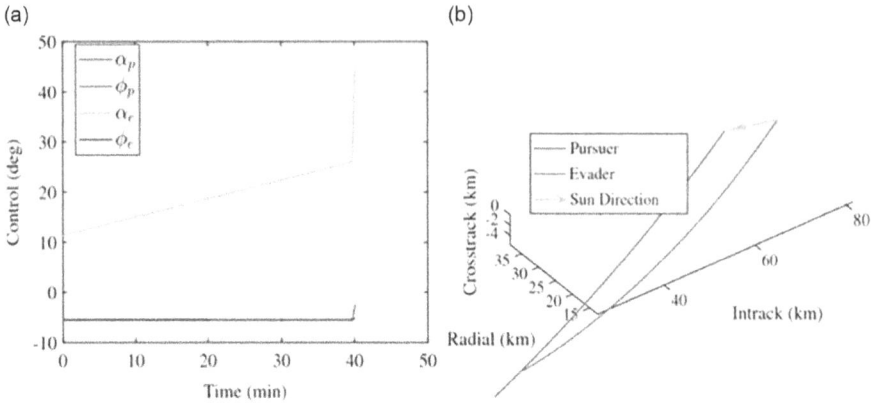

FIGURE 11.7 Obtain sun vector game: control and trajectory (a) Pursuer and evader control (b) 3-D trajectory.

evader is illuminated with respect to the pursuer. This solution took 51.3 s to compute, with a final time $t_f = 40.08$ min, and a PSO constraint violation of 6.97×10^{-7}.

The match energy game ends relatively quickly, with a final time of $t_f = 3.86$ min. This is quite possible because no other constraint is being enforced and the pursuer simply needs to match the semimajor axis of the evader. The resulting motion after the game concludes is shown in Figure 11.8, in which Figure 11.8a shows the relative positions of both players with respect to the original reference orbit, and Figure 11.8b shows the position of the pursuer with respect to the evader over the course of a couple of orbits. Thus, as can be seen, the two players travel far from the original reference orbit. Also, they do not necessarily remain close to each other, but the motion of the pursuer is bounded with respect to the evader because the energy has been matched. This solution took just 13.2 s to compute, with a PSO constraint violation of 1.00×10^{-10}.

The obtain sun vector and match energy game results are shown in Figure 11.9, in which Figure 11.9a shows that the desired lighting is obtained at the end of the game, and Figure 11.9b shows that the energy has been matched. With the estimate for κ, the

(a)

(b)

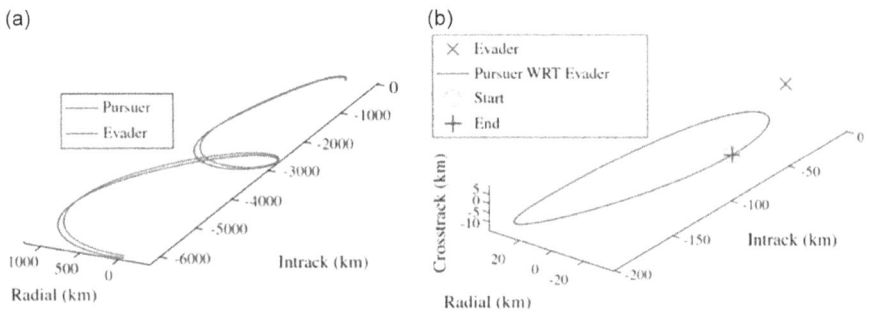

FIGURE 11.8 Match energy game: trajectories after game concludes (WRT denotes with respect to) (a) 3-D trajectory (b) Pursuer relative to evader.

(a)

(b)

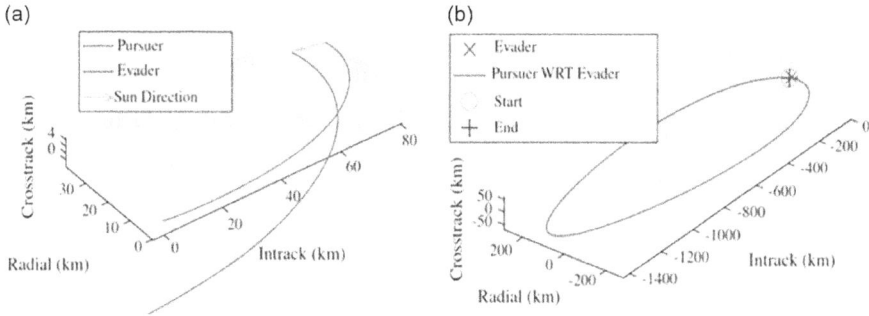

FIGURE 11.9 Obtain sun vector and match energy game: 3-D trajectories (a) 3-D trajectory during game (b) Pursuer relative to evader after game.

final range of 10.0008 km is very close to the desired range (10 km). It is interesting to note, however, that although the energy has been matched, the pursuer does not remain close to the evader throughout the ensuing motion and gets even farther away from the evader than in the previous game. This solution took 2.73 min to compute, with a final time of $t_f = 51.80$ min and a maximum GA constraint violation of $1.58 - 10^{-10}$.

For the match energy and remain in close proximity game, the desired intrack distance of the pursuer from the evader for the motion after the game concludes is $d_y = 10$ km, and the desired crosstrack amplitude for the motion after the game ends is $d_z = 10$ km. It is assumed that both players cease to maneuver at the conclusion of the game, and thus the energy-matched state is naturally maintained during the ensuing motion. The results can be seen in Figure 11.10, in which Figure 11.10a shows the trajectory during the game, and Figure 11.10b shows that the desired motion after the game concludes has been achieved. Due to the elliptic nature of the orbit, the intrack separation distance oscillates about d_y throughout the orbit. This solution

(a)

(b)

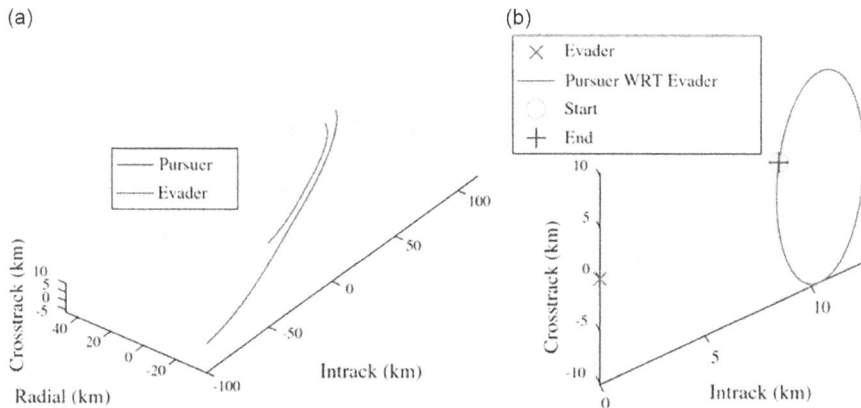

FIGURE 11.10 Match energy and remain close game: 3-D trajectories (a) 3-D trajectory during game (b) Pursuer relative to evader after game.

TABLE 11.3
Simulation Results

Game Condition	t_f, min	Solver	CPU Time	Constraint Violation
Intercept (with $e = 0$)	41.32	PSO	38.5s	1.15×10^{-8}
Intercept	41.53	PSO	36.4s	3.79×10^{-8}
Rendezvous	57.61	GS	8.64 min	5.12×10^{-10}
Obtain sun vector	40.08	PSO	51.3s	6.97×10^{-7}
Match energy	3.86	PSO	13.2s	1.00×10^{-10}
Obtain sun vector and match energy	51.80	GA	2.73 min	1.58×10^{-10}
Match energy and remain in close proximity	59.56	GA	13.10 min	1.26×10^{-10}

took 13.10 min to compute, with a final time of $t_f = 59.56$ min and a maximum GA constraint violation of 1.26×10^{-10} (see Table 11.3).

11.8 SOLUTION VALIDATION

This section is presented to add confidence that the solutions obtained in the previous section are in fact the differential game solutions. If a differential game solution has been found, that is, if u^* and v^* have been found such that

$$J(u^*, v) \leq J(u^*, v^*) \leq J(u, v^*) \tag{11.123}$$

then if any suboptimal v is used while u^* is used, the pursuer should be able to further minimize the performance index. Likewise, if any suboptimal u is used while v^* is used, then the evader should be able to further maximize the performance index. Thus, this section uses that logic to formulate and solve one-sided optimal control problems, again using the IHM, where the control of one of the players is determinGistic and known beforehand, to add confidence that the differential game solutions have indeed been found. Both the intercept game and rendezvous game solutions are analyzed, where these two have been chosen to be analyzed since one used the PSO to solve the resulting boundary value problem and the other used the GA.

To analyze both the intercept game and rendezvous game solutions, multiple one-sided problems are formulated and solved. The first one-sided optimal control problem uses v^* for the control of the evader, and a problem is formulated to see if the pursuer can intercept (or rendezvous with) the evader any sooner than the final time obtained from the differential game solution. If the solution obtained in the previous section is indeed the differential game solution, then the final time from this one-sided optimal control problem should match. To do this for the intercept case, approximating polynomials are fit to $\tilde{x}_e^*$, $\tilde{y}_e^*$, and $\tilde{z}_e^*$ as functions of the true anomaly. This was done to avoid numerically propagating the deterministic position of the evader, and instead use the polynomial functions in the terminal constraint vector. For the rendezvous case, the approximating polynomials were not accurate enough, thus the evader state is numerically propagated based on the optimal evader cosGtates found from the differential game. The objective for this first one-sided optimal control problem is to minimize the time to intercept (or rendezvous with) the deterministic position (or state) of the evader. Thus, the cost function is

$$J = f_f \tag{11.124}$$

subject to

$$\psi = \begin{bmatrix} \tilde{x}_{p_f} - \tilde{x}_e^*(f_f) \\ \tilde{y}_{p_f} - \tilde{y}_e^*(f_f) \\ \tilde{z}_{p_f} - \tilde{z}_e^*(f_f) \end{bmatrix} \tag{11.125}$$

for the intercept case and

$$\psi = \begin{bmatrix} \tilde{x}_{p_f} - \tilde{x}_e^*(f_f) \\ \tilde{y}_{p_f} - \tilde{y}_e^*(f_f) \\ \tilde{z}_{p_f} - \tilde{z}_e^*(f_f) \\ \tilde{x}'_{p_f} - \tilde{x}_e'^*(f_f) \\ \tilde{y}'_{p_f} - \tilde{y}_e'^*(f_f) \\ \tilde{z}'_{p_f} - \tilde{z}_e'^*(f_f) \end{bmatrix} \tag{11.126}$$

for the rendezvous case, where it is emphasized that the terminal states of the evader are solely functions of the final true anomaly. These boundary conditions still allow for the use of the IHM for these one-sided optimal control problems because, after applying the necessary conditions, the costate boundary conditions are the same as they were for the pursuer in the differential game for both the intercept and rendezvous cases. Applying the necessary conditions also produces the same control, costate, and state equations that the pursuer had in the differential games.

For the rest of the one-sided optimal control problems, the evader uses a control other than v^*. Thus, the pursuer, in a one-sided optimal control problem, should be able to intercept the evader sooner. If there is a v that outperforms v^* [i.e., makes the pursuer take longer to intercept (or rendezvous with) the evader as compared to the differential game solution], then the differential game solution was not found. For these one-sided optimal control problems in which a v other than v^* is used for the evader, the first strategy for v is to point the thruster in the direction of r_{p_0} such that the evader accelerates continually away from r_{p_0}. For the rest of the suboptimal v strategies, the thruster is pointed in a constant, random direction during each of the problems. These formulations still allow the use of the IHM, and the are developed to show that the pursuer can intercept the evader equal to or sooner than the differential game final time for any of the suboptimal strategies tested. For these one-sided optimal control problems, the evader states are numerically propagated from the initial conditions with the constant control laws. The cost function is the same as before, and the terminal constraints are the same, except that the evader states at a given final true anomaly are not the optimal ones from the differential game solution. Again, for all these one-sided optimal control problems, the IHM can be used, and the same equations from applying the necessary equations are obtained as before.

The intercept game analysis results, for which MATLAB's PSO is used to solve the resulting boundary value problems for the one-sided optimal control problems, can be seen in Figure 11.11. Figure 11.11a shows the PSO exit flags, the constraint violations, and the CPU times for all 50 one-sided problems tested. As can be seen, constraints were satisfied for every problem, and most problems were solved in less than 1 min. Figure 11.11b shows the final time for each problem. The first final time obtained matches the differential game final time, as it should, because v^* was used for the evader's control. The rest of the final times obtained are less than the differential game final time, as they should be, because a v other than v^* is used and the pursuer is able to further minimize the performance index.

The rendezvous game analysis results, in which MATLAB's GA is used to solve the resulting boundary value problems for the one-sided optimal control problems, can be seen in Figure 11.12. Figure 11.12a shows the GA exit flags, the constraint violations, and the CPU times for all 50 one-sided problems. As can be seen, constraints were satisfied for every problem, and most problems were solved in about 6 min. Figure 11.12b shows the final time for each problem. The first final time obtained matches the differential game final time, as it should. The rest of the final times obtained are less than the differential game final time, also as they should be.

11.9   CONCLUSIONS

The IHM is a desirable method to solve the differential games and one-sided problems herein due to (i) no need for an initial guess for the costates; (ii) the ability to search for the final (or initial) costates in an arbitrary range; (iii) the relatively fast computation times; (iv) the global nature of the method; and (v) the fact that an entire

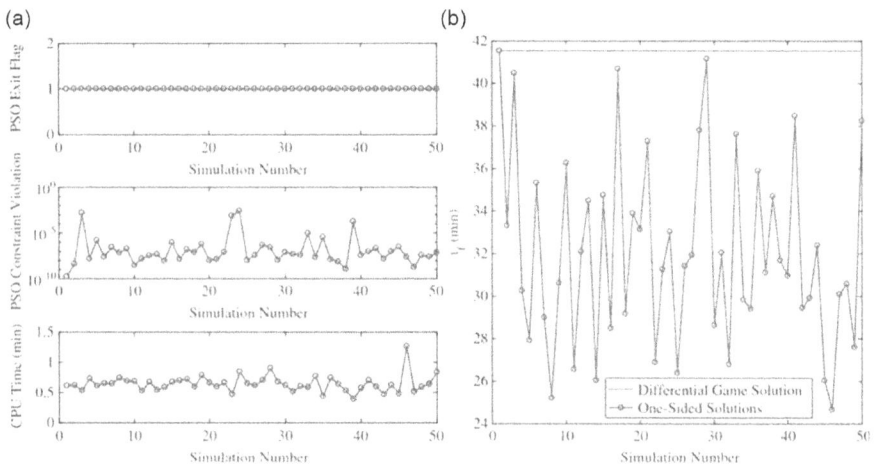

FIGURE 11.11 Intercept game: solution validation (a) Exit flag, violation, and CPU time (b) t_f for each one-sided problem.

(a)   (b)

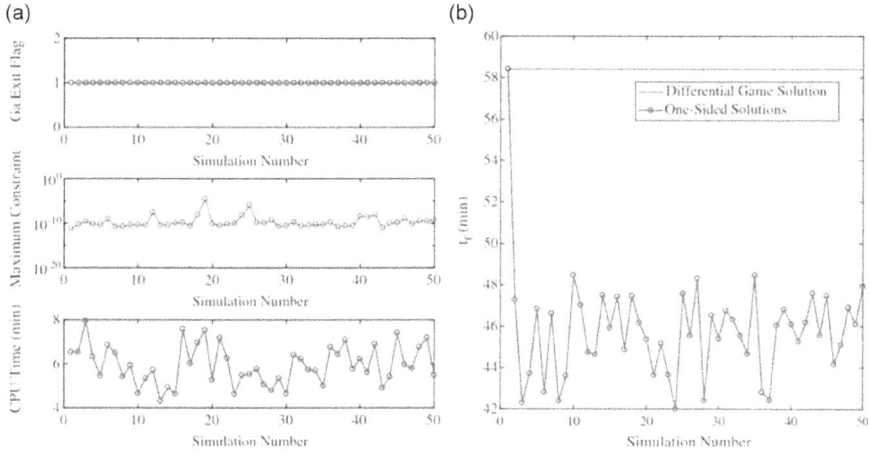

FIGURE 11.12 Rendezvous game: solution validation (a) Exit flag, violation, and CPU time (b) t_f for each one-sided problem.

dynamic, differential game can be reduced to a static optimization problem composed of very few optimization variables. Regarding point (iii), the fast computation times are partly due to the analytic propagation of the costates. Thus, it is desirable that the costates can be propagated analytically. If not, the computation times would be slower, and the costates may become unstable from the numerical propagation.

Pontani and Conway's IHM [13] and the framework of Stupik et al. [11] for solving a differential game can be successfully applied to multiple types of zero-sum pursuit–evasion games using the Tschauner– Hempel equations of motion and the Yamanaka–Ankersen STM. Thus, the developed method can be used for games with respect to a noncircular player or reference orbit, and the costates can still be propagated analytically. Pursuit–evasion games can be formulated and solved for not only the intercept case but also for the following games: (i) rendezvous, (ii) obtain sun vector, (iii) match energy, (iv) obtain sun vector and match energy, and (v) match energy and remain in close proximity. The terminal constraints for these games can be formulated in such a way that the IHM can still be used, and such that the problem remains tractable. Also, the differential game solutions can be verified to a certain degree by solving multiple one-sided optimal control problems and comparing the solutions to the differential game solution.

The solutions to these games provide the open-loop strategies for the pursuer and evader and may help to answer several questions. First, the solutions may provide the worst-case time (and corresponding fuel) for the pursuer to accomplish its inspection goal since it is based on the evader playing optimally. If the evader behaves any differently, then the pursuer may ideally be able to accomplish its inspection goal sooner, and thus with less fuel. Second, the game solutions may provide the actual strategy for the evader to implement in such a scenario, regardless of the strategy of the pursuer. If the evader uses the open-loop strategy generated from these solutions, then it may have the best chance at evading the pursuer, forcing the pursuer to use the most time (and corresponding fuel) to chase the evader. Implementing such a strategy may force the pursuer to discontinue its pursuit, if the pursuer realizes the time and fuel that would be required to continue pursuing the optimally evading evader. Third, these game solutions may aid in the design process. Based on the expected capabilities of the other player, the propulsion properties of the player of interest may be designed to be capable of performing at a certain level in such a scenario.

REFERENCES

1. Horie, K., and Conway, B. A., "Optimal Fighter Pursuit-Evasion Maneuvers Found via Two-Sided Optimization," *Journal of Guidance, Control, and Dynamics*, Vol. 29, No. 1, 2006, pp. 105–112. doi:10.2514/1.3960.
2. Horie, K., and Conway, B. A., "Genetic Algorithm Preprocessing for Numerical Solution of Differential Games Problems," *Journal of Guidance, Control, and Dynamics*, Vol. 27, No. 6, 2004, pp. 1075–1078. doi:10.2514/1.3361.
3. Pontani, M., and Conway, B. A., "Numerical Solution of the ThreeDimensional Orbital Pursuit-Evasion Game," *Journal of Guidance, Control, and Dynamics*, Vol. 32, No. 2, 2009, pp. 474–487. doi:10.2514/1.37962.
4. Shen, D., Pham, K., Blasch, E., Chen, H., and Chen, G., "PursuitEvasion Orbital Game for Satellite Interception and Collision Avoidance," *Proceedings of SPIE: Sensors and Systems for Space Applications IV*, Vol. 8044, May 2011, Paper 80440B. doi:10.1117/12.882903.
5. Blasch, E. P., Pham, K., and Shen, D., "Orbital Satellite Pursuit-Evasion Game-Theoretical Control," *2012 11th International Conference on Information Science, Signal Processing and their Applications (ISSPA), Montreal*, 2012, pp. 1007–1012. doi:10.1109/ISSPA.2012.6310436.
6. Sun, S., Zhang, Q., Loxton, R., and Li, B., "Numerical Solution of a Pursuit-Evasion Differential Game Involving Two Spacecraft in Low Earth Orbit," *Journal of Industrial and Management Optimization*, Vol. 11, No. 4, 2015, pp. 1127–1147. doi:10.3934/jimo.
7. Clohessy, W. H., and Wiltshire, R. S., "Terminal Guidance System for Satellite Rendezvous," *Journal of the Aerospace Sciences*, Vol. 27, No. 9, 1960, pp. 653–658. doi:10.2514/8.8704.
8. Selvakumar, J., and Bakolas, E., "A Pursuit-Evasion Game in the Orbital Plane," AAS/AIAA Spaceflight Mechanics Meeting, AAS Paper 17–474, San Antonio, TX, Feb. 2017.
9. Hernandez, S., and Akella, M. R., "Lyapunov-Based Guidance for Orbit Transfers and Rendezvous in Levi-Civita Coordinates," *Journal of Guidance, Control, and Dynamics*, Vol. 37, No. 4, 2014, pp. 1170–1181. doi:10.2514/1.62305.
10. Gutman, S., Esh, M., and Gefen, M., "Simple Linear Pursuit-Evasion Games," *Computers and Mathematics with Applications*, Vol. 13, Nos. 1–3, 1987, pp. 83–95. doi:10.1016/0898-1221(87)90095-2.
11. Stupik, J., Pontani, M., and Conway, B., "Optimal Pursuit/Evasion Spacecraft Trajectories in the Hill Reference Frame," AIAA/AAS Astrodynamics Specialist Conference, AIAA Paper 2012–4882, 2012. doi:10.2514/6.2012-4882.
12. Stupik, J. M., "Optimal Pursuit/Evasion Spacecraft Trajectories in the Hill Reference Frame," M.S. Thesis, Univ. of Illinois, Urbana– Champaign, Champaign, IL, 2013.
13. Pontani, M., and Conway, B., "Minimum-Fuel Finite-Thrust Relative Orbit Maneuvers via Indirect Heuristic Method," *Journal of Guidance, Control, and Dynamics*, Vol. 38, No. 5, 2015, pp. 913–924. doi:10.2514/1.G000157.
14. Tschauner, J., and Hempel, P., "OptimaleBeschleunigeungsprogramme fur das Rendezvous-Manover," *Astronautica Acta*, Vol. 10, 1964, pp. 296–307.
15. Alfriend, K. T., Vadali, S. R., Gurfil, P., How, J. P., and Breger, L. S., Spacecraft Formation Flying, 1st ed., Elsevier, New York, 2010. doi:10.1016/C2009-0-17485-8.

16. Yamanaka, K., and Ankersen, F., "New State Transition Matrix for Relative Motion on an Arbitrary Elliptical Orbit," *Journal of Guidance, Control, and Dynamics*, Vol. 25, No. 1, 2002, pp. 60–66. doi:10.2514/2.4875.

17. Sengupta, P., and Vadali, S. R., "Relative Motion and the Geometry of Formations in Keplerian Elliptic Orbits," *Journal of Guidance, Control, and Dynamics*, Vol. 30, No. 4, 2007, pp. 953–964. doi:10.2514/1.25941.

18. Li, J., and Xi, X.-n., "Fuel-Optimal Low-Thrust Reconfiguration of Formation-Flying Satellites via Homotopic Approach," *Journal of Guidance, Control, and Dynamics*, Vol. 35, No. 6, 2012, pp. 1709–1717. doi:10.2514/1.57354.

19. Isaacs, R., *Differential Games*, Wiley, New York, 1965.

20. Bryson, A. E., and Ho, Y.-C., Applied Optimal Control: Optimization, Estimation, and Control, Hemisphere, Washington, DC, 1975.

21. Pontani, M., and Conway, B.A., "ParticleSwarmOptimizationApplied to Space Trajectories," *Journal of Guidance, Control, and Dynamics*, Vol. 33, No. 5, 2010, pp. 1429–1441. doi:10.2514/1.48475.

22. Pontani, M., and Conway, B. A., "Optimal Finite-Thrust Rendezvous Trajectories Found via Particle Swarm Algorithm," *Journal of Spacecraft and Rockets*, Vol. 50, No. 6, 2013, pp. 1222–1234. doi:10.2514/1.A32402.

23. Vallado, D. A., *Fundamentals of Astrodynamics and Applications*, 3rd ed., Microcosm Press, Hawthorne, CA, 2007.

Section 4

Human Factors

12 Systems Modeling Language Extension to Support Modeling of Human-Agent Teams[*]

Michael E. Miller, John M. McGuirl,
Michael F. Schneider, and Thomas C. Ford
US Air Force Institute of Technology

CONTENTS

12.1 INTRODUCTION

Artificial intelligence has once again gained attention due to significant advances in deep learning technology,[1] adding credence to Marvin Minsky's vision of artificial intelligence as a superior replacement to human cognition.[2] While the use of artificial intelligence to replace human intelligence has been demonstrated to be effective in well-defined problem spaces, such as strategy games, the utility of the resulting systems often degrades over time in complex environments as they lack resiliency. In contrast, systems which incorporate the adaptive capacity of humans are better able to respond to changes in the environment and other external pressures.[3–6] Therefore, the vision of human-machine or human-agent teaming has been proposed which extends Licklider's vision of cooperative interaction between humans and artificial cognitive systems, resulting in human-computer symbiosis.[7]

---

[*] Reprinted with permission: Miller, M. E., McGuirl, J. M., Schneider, M. F., & Ford, T. C. (2020). Systems modeling language extension to support modeling of human-agent teams. *Systems Engineering, 23*(5), 519–533.

DOI: 10.1201/9781003220978-16

Unfortunately, current system modeling techniques do not easily support the design and specification of systems that embody cooperative interaction between humans and systems which include artificial agents. Most models constructed using standardized modeling languages, including the Unified Modeling Language (UML) or Systems Modeling Language (SysML),[8,9] depict the user as an actor external to the system. Although these modeling languages can be used to represent the user interface and human interaction within the system,[10,11] these languages were not designed to permit the analysis and representation of human-agent teams. More specifically, SysML was designed to capture specific system implementations with prescribed sequences of actions, rather than the variable, goal-directed behavior of interdependent agents which adapt their behavior in response to environmental or other unpredictable external influences.

Structured analytic techniques have also been applied to analyze and describe human processes within the human factors literature, beginning with the Gilbreths' Therblings and continuing through modern work analysis.[12] Some of these methods provide mechanisms that permit the description of multiple potential sequences of actions to achieve some higher-level goal. For example, some forms of the Goals, Operators, Methods, and Selection rules (GOMS) models, originally developed by Card and colleagues,[13] acknowledge that humans may employ one of the multiple methods (ie, procedures or activities) to accomplish goals. This approach of modeling various sequences of actions to achieve the desired goal is important as an operator's goals are unlikely to change with the introduction of technology. However, automation often changes how a goal is achieved, that is, the sequences of actions necessary to achieve the goal, and who or what accomplishes the actions to achieve the goals.[14,15] Therefore, by modeling and decomposing goals, we can model the effect of multiple system implementations to achieve a stable set of human goals. Thus, centering a system analysis on the goals permit the analysis of one or more system implementations that can be employed to achieve the desired goals.

The study of how teams of humans and artificial cognitive agents can be designed to adapt to external influences has also gained significant attention within the robotics and human factors literature.[16,17] For example, a recent approach explores the interdependence between human and artificial cognitive agents, attempting to design the behavior of the artificial cognitive agents to be observable, predictable, and direct-able by human team members to permit the system to exploit various activities which achieve the same goal.[18,19] These three attributes are known to facilitate teamwork for highly collaborative interaction within human-human teams and are believed to be important attributes for the design of effective human-agent teams.[20]

Furthermore, within the Human Factors and Human-Computer Interface design literature, languages have been developed for describing human interaction with software interface elements and more recently with software agents.[21–25] This work has been expanded to include tools for simulating the behavior of human and agents to facilitate task allocation,[26,27] as well as, performing formal analysis on human-computer interaction to attempt to define potential safety hazards.[28–30] While these approaches permit the modeling of human-agent interaction, they are often concerned with the interface between an individual or small team of humans and one or a small team of artificial agents.

Similar influences also exist within the agent-based software and multiagent systems literature.[31–34] Although these languages focus on the design and interaction of software agents and consider humans as entities external to the system, the terminology and the concepts applied within this literature mimic the terminology that is often applied to describe the structure of human-human teams within the psychology, organizational behavior, and human factors literature. Although not explicitly stated, it would appear that some of the constructs, for example, the definition of goals, roles, and protocols within these languages, are intended to provide flexibility to the procedures an artificial agent employs to achieve a goal in response to environmental variability.

Our interest in the current research is to explore an approach to model, both structurally and dynamically, the design of teams of humans and artificial cognitive agents as they interact within a complex system to uncover requirements driven by intrateam interdependencies. We are particularly interested in models in which humans and artificial cognitive agents have common, shared goals within an interaction model, as discussed in the goal models provided by Sterling and Taveter.[33] Furthermore, the modeling approach should be able to exploit the interdependence of team members to provide the flexibility necessary to permit the teams to achieve these goals under various, unpredictable external influences. Finally, we seek to define a modeling language that can be employed to augment, rather than replace existing modeling approaches. Therefore, we seek a modeling language extension that describes the high-level goals, structure, and behavior within the human-agent team and that integrates with the use of the existing modeling approaches, for example, UML and SysML, to describe the design of systems in which these human-agent teams function; in other words, a UML or SysML profile.

12.2 HUMAN-AGENT TEAMING MODELING LANGUAGE PROFILE OVERVIEW

We first provide the basic assumptions and definitions to provide a summary of the structure of the Human-Agent Teaming Modeling Language (HAT-ML) which is a profile of SysML, similar to existing profiles, such as SoaML,[35] IFML,[36] and MARTE.[37] We then propose a human-agent teaming metamodel to the language, which we view as an extension to existing system models, that is useful in designing and specifying human-agent teams. Finally, a modeling method, referred to as Human-Agent Teaming Design Method (HAT-DM), is discussed which is useful in the application of the proposed modeling language extension.

It is worth noting that HAT-ML and HAT-DM have been developed primarily as an instructional tool within a course in human-agent team design offered at the US Air Force Institute of Technology. Variations of this modeling language and process have been employed in nearly a dozen offerings of this course to perform the conceptual design of these systems. It has been shown that the modeling language and method lead the students to consider many factors during the conceptual design of human-agent teams that they would not consider without such a language and method. While the language is likely incomplete and not fully specified, our experience in the classroom indicates that it is useful during conceptual designs of systems that include teams of humans and artificial agents.

12.2.1 ASSUMPTIONS AND DEFINITIONS

As stated earlier, our interest is to explore an approach to model the design of teams of humans and artificial agents as they apply to automation. "Automation" is defined as the process of substituting an activity originally performed by a human with an activity performed by a man-made artifact or system.[38] This definition is important as the goals defined within this approach must be goals that are relevant to the humans within the human-agent team for the approach to be meaningful. Automation differs from "autonomy," which is the capability and authority to be self-directing with respect to some set of activities. As a result, automation and the capability to perform autonomously are characteristics of system design while autonomy is an attribute of agents within a particular application context.

Agents are entities that continuously monitor their environment and perform activities consistent with pursuing one or more goals within that environment. More specifically, in the current research, an "agent" is defined as a persistent entity that is capable of: (i) perceiving the environment to obtain state information, (ii) applying this information to engage in reasoning relative to a set of goals, and (iii) applying this reasoning to drive activities and actions it performs in the environment based upon granted authority. It is important that humans fulfill the requirements of an agent according to our definition. This definition is consistent with the typical definition of an artificial agent, which is an entity capable of perceiving its environment, processing the perceived environment to determine a desired action, and performing the desired action in response to the environment.[39] Although not strictly required, artificial agents will typically take action to achieve a goal in response to perceiving multiple attributes of the environment. In the current context, artificial agents may be embodied and encased within a single physical entity, as a robot, or disembodied and distributed, where the sensors or actuators are located remotely from the reasoning engine. These distributed components can logically perform as an agent. An example of a logical agent might be a tsunami warning system with distributed sensors to continually gather information such as geologic activity and changes in water level, reason about this information with regard to the goal of detecting dangerous tsunamis, and automatically issuing a warning. For clarity, we will only refer to agents as either human or artificial when it is important to distinguish human from artificial cognitive agents. The generalization of humans and artificial agents as agents is not a new concept and has been included in previous modeling approaches.[33,40] It is noted, however, that by this definition of agent, the term human-agent team is redundant. However, this term is applied in this chapter as it is common within the literature.

In HAT-ML, a "machine" is a man-made apparatus used to apply energy to matter to change its properties within an environment. Conceptually, this differs from man-made apparatus for generating, communicating, or manipulating symbols that enable sensing, commanding actuation, and reasoning, which are within the purview of the agents within the system. However, components that generate, communicate, or manipulate symbols and are not logically part of an artificial agent within the system, will be considered part of the machine.

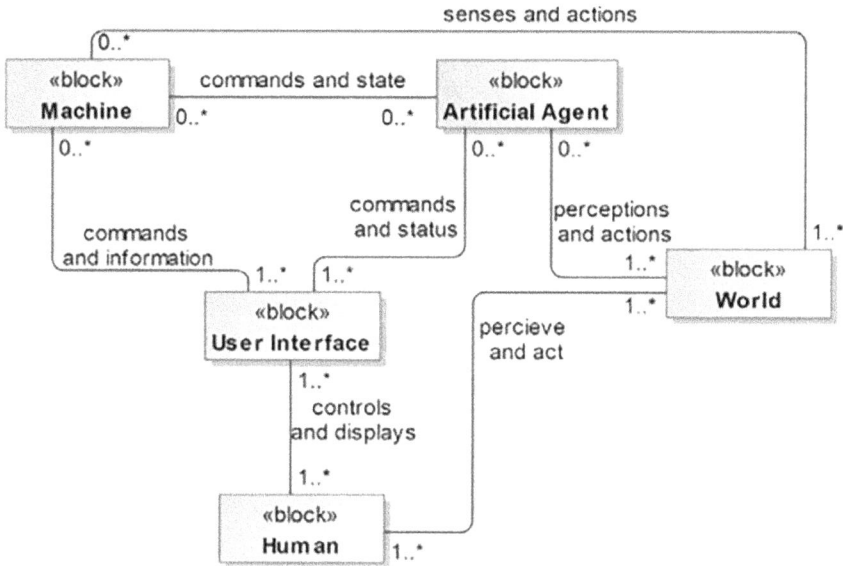

FIGURE 12.1 Block definition diagram showing the interactions of the humans and agents with a machine which is situated in the world.

Figure 12.1 provides a block diagram of the assumed logical system architecture. As shown, one or more human agents interact with one or more machines and one or more artificial agents through one or more user interfaces. Importantly, artificial agents can share a user interface with a machine or can have dedicated interface elements which permit them to interact directly with one or more humans. As a result, an individual human might interact with one or more machines, one or more artificial agents and one or more other humans through common or individual human interfaces. However, it is important that the artificial agents and the machine are viewed as separate components as it may be important for the human to direct either the machine or the agent, which might then direct the machine. For example, an automobile driver using adaptive cruise control may wish to reduce the following distance in a single instance by pressing the gas pedal to increase the speed of his or her car relative to the car he or she is following. In this example, the human is interacting with a user interface element to convey commands directly to the machine through the user interface. Alternatively, the driver may direct the adaptable cruise control agent to reduce the following distance for all cars within a busy metropolitan area through a different interface control. In this example, the human issues commands to an artificial agent through the user interface and the artificial agent commands the behavior of the machine.

Figure 12.1 also depicts the fact that the human can perceive and act directly on the world, as can artificial agents. However, it is also possible that the machine can sense and take actions in the world that are commanded by a human or an artificial agent. Additionally, a human or an artificial agent can obtain state information about the world through the machine.

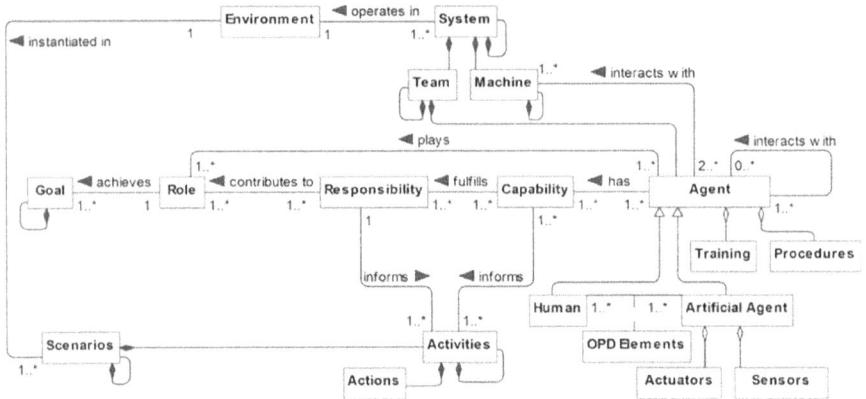

FIGURE 12.2 Block definition diagram depicting the proposed human-artificial agent teaming metamodel. Note that composition (filled diamonds), aggregation (open diamonds), and generalization (open arrows) connector types are adopted from SysML and therefore these connector types are not labeled.

12.2.2 Proposed Human-Agent Teaming Metamodel

To explain the proposed modeling approach, it is useful to begin by describing the metamodel which is graphically represented in Figure 12.2. Block definition diagram depicting the proposed human-artificial agent teaming metamodel. Note that composition (filled diamonds), aggregation (open diamonds), and generalization (open arrows) connector **types are adopted from SysML** and therefore these connector types are not labeled. The terms applied in the metamodel are applied in the Systems Engineering, Agent-Based Software Engineering, Human Factors Engineering, and Cognitive Systems Engineering literature. Unfortunately, these connector types are not labeled terms are not applied consistently across or sometimes within each of these disciplines. To define the model, this metamodel includes the object or element types, connector types, and multiplicity relationships. To more completely describe the language each of the elements is defined in this section and a notation example is provided in the following section. These components are common in the description of existing visual modeling languages.[41]

Beginning at the top of Figure 12.2, we begin with a "System" which can be defined as "an integrated set of elements, subsystems, or assemblies that accomplish a defined objective".[42] In the current chapter, we refer to the "Objective" as the top-level goal. This system is deployed in an "Environment." Consistent with the definition of a system, this system is comprised of other systems, a "Machine," such as an aircraft, and a "Team" which controls the system. A "Team" refers to a group of two or more agents who cooperate in a structured way, adjusting their behaviors dynamically to achieve one or more common goals.[43] As shown, the team is potentially comprised of other teams and multiple "Agents," as defined earlier. These agents can include both "Humans" and "Artificial Agents" functioning to interact with the machine. To analyze systems without automation, models can be constructed which include a human and any human teammates as human agents. As automation is designed into the system, artificial agents should be added to represent the complete human-agent team as elements of the "System."

Within this metamodel, all agents require "Procedures" (ie, an established or official process for completing an activity or action) and "Training" (ie, the action of teaching an agent to perform an activity or action) to enable "Capabilities" (ie, the power or ability to do some activity or action). As the model is specified, procedural or training requirements can be captured for the artificial agents to inform the design of these agents as required. As indicated earlier, each of the artificial agents, at least logically, includes one or more "Sensors" for sensing the environment of the system under control. These artificial agents additionally include at least one "Actuator" which gives the agent the capability to perform actions that control the machine to influence the environment.

Within this model, agents play the "Roles" necessary within the human-agent team. A role is defined as a set of "Responsibilities" necessary to fulfill a "Goal" or a set of goals. In human teams, roles provide a mechanism to compartmentalize the work in such a way that it reduces human training and often the communication between team members. These roles are designed to partition the high-level activities necessary to fulfill the goals within the goal hierarchy for the human-agent team.[44,45] Responsibilities are functional descriptions of the necessary actions, and the authority to act, required to contribute to a role. They are fulfilled through applying agent capabilities, with appropriately designed authority, wherein the capabilities require knowledge, skills, and abilities which are unique from the knowledge skills and abilities needed to perform other roles.

In this context "goals" are desirable states that must be achieved by the human-agent team.[46] Importantly, goals describe what needs to be achieved, but do not attempt to describe how these goals are achieved. As a result, by defining and deriving the system function from the goals, various system implementations can be explored which support successful goal completion.[14] In the goal model, high-level goals are completely decomposed into lower-level goals to provide the goal hierarchy.

One or more roles play a part in completing each of the goals. At least one agent must play each role, however, multiple agents can play a part in each role. It is also possible for an agent to participate in multiple roles. As such, roles provide a method to map agents and their capabilities to responsibilities which are associated with goals through these roles, permitting a more abstract mapping between agents and goals. Within most of the agent-based software literature, only artificial agents can be directly mapped to goals, as such the goals are not shared between the human and artificial agents.[31,46] However, by defining agents such that they are a generalization of humans and artificial agents, the roles can be shared between these two classes of agents, as discussed by Sterling and Tavateer.[33] The shared goals then provide a starting point from which to design the interaction between humans and automated agents, permitting the designer to consider the personnel selection criteria; procedures and training of the human; the required capabilities of the artificial agents; and requirements for the user interface between the humans, artificial agents, and any machine that they collaborate to control. Note that this metamodel interface between agents, human and artificial, is partially specified through the Observability, Predictability, and Direct-ability (OPD) elements within

Figure 12.2. OPD elements of the interface between the agents within the system permit their behaviors to be observed, future behaviors to be predicted, and new behaviors directed when necessary. These three behaviors are commonly cited as being necessary to facilitate teamwork.[47]

As the model shows, each role will require one or more responsibilities to accomplish a goal. Furthermore, one or more capabilities of one or more agents are required to fulfill a responsibility, where each capability is allocated to one or more individual agents. This is important as the capabilities can be employed in a flexible manner by the team of agents to fulfill the requirements within a role to accomplish the goals. Importantly agents possess the capabilities and enable the team of agents to successfully complete the goals. Therefore, the model permits the capabilities possessed or designed into the agents to be matched to the required responsibilities through design. It is also important to recognize that the capabilities of a human agent are heavily influenced by the user interface. Thus, if one is to apply a human capability to fulfill a required capability, it will be necessary to add functionality to the user interface to support this capability.

While it is important to understand the capabilities of each of the agents, it is also important to understand that each agent will also have certain "Constraints" which limit the conditions under which these capabilities can be applied. These constraints should be captured in the model as they are understood to facilitate later design. In this context, constraints are defined as rules which limit the actions of the agents. These constraints may exist at the component, system, or protocol level. Alternately, they may exist due to interactions among these levels.[48] As these constraints and the fact that an agent can fulfill a responsibility under at least some circumstances exist, it must be recognized that the agent may not be capable of fulfilling the responsibility under all circumstances. Therefore, the exercise of mapping capabilities to responsibilities should include all possible methods of fulfilling a responsibility. Having multiple agents with capabilities to fulfill a responsibility, permits backup behaviors to be designed into the team to improve the resilience of the system to a broader range of circumstances.

The capabilities permit the human-agent team to enable the activities and actions necessary to achieve a goal in a given scenario context.

It should be noted that we have adopted the word "action" to represent a discrete step taken by an agent to execute a higher-level action or activity to fulfill a goal and an "activity" is a type assigned to an action. The terms activity and action are consistent with SysML. Therefore, an action is a discrete step of type "activity," the time-ordered sequence of which can be shown in an activity diagram. These time-ordered sequences of "actions" can be allocated to an entity and required for the completion of a goal.[8] We purposefully avoid the use of the term task as this term is, at times, defined either with actions performed by a human or the outcomes that people are trying to achieve.[49] This later definition is more consistent with goals within our language. Of course, as these actions are conducted, they influence the machine, which influences the environment. Although not indicated directly in Figure 12.2, changes in the environment and machine affect the information sensed and perceived by the agents. This influence is necessary to recognize as it completes the feedback loop which is required to assess progress toward goal attainment.

To facilitate team communication, the system should be designed to facilitate at least the transition and action processes discussed by Marks and colleagues.[50] These processes include mission formulation and planning, goal specification and strategy formulation as transition processes and monitoring, backup and coordination activities which are conducted during the mission. Many of these are addressed through the use of an analysis approach similar to that proposed by Johnson and colleagues.[18] It is additionally hoped that this approach will aid conflict management within the interpersonnel (interagent in the language of the current chapter) processes discussed by Marks and colleagues. However, this approach does not necessarily address motivation or affect management, which are elements of backup and coordination activities within Marks and colleague's taxonomy.

While we have discussed a process referred to as "Interdependence Analysis[18]", we have not yet indicated how this process is applied to inform the model. In practice, the construction of the model begins with understanding the goals of the system to view the system using a top-down approach. Definition of the agents can also be performed using a bottom-up approach. Within our model, "scenarios" for applying the system within a specified environment are created. These "scenarios" represent envisioned applications of the systems. These scenarios are analyzed to determine a time sequential group of activities and actions which can be undertaken by the system to achieve the goals within each scenario. We then apply existing agents to the resulting actions to document how each agent contributes to the cognitive activity necessary to fulfill the activities and actions within each scenario. These scenarios can be selected to represent a broad range of operating conditions, including scenarios where the design is likely to excel, as well as scenarios where the design is likely to be challenged. Design at this level includes evaluating each agent's ability to not only perform the necessary activities but to support other agents through cooperative activities. As such the design can be evaluated and refined using an extension of Interdependence Analysis,[16,18,51,52] insuring that the behaviors of the agents are designed to be observable, predictable, and direct-able as necessary to facilitate teamwork.

12.2.3 HAT-DM OVERVIEW

The HAT-DM for applying the HAT-ML is depicted in the activity diagram shown in Figure 12.3. This figure illustrates a three-step process. The initial step is system characterization which ensures a comprehensive examination of Human-Agent Teams by integrating a top-down goal role-responsibility decomposition analysis and a bottom-up agent and capability synthesis within the initial characterization. These results of these two techniques are then structurally mapped to one another in the second step, responsibilities-to-capabilities, to define all potential means by which a team could achieve any given goal. Finally, the agents and their capabilities are examined situationally, via an extension to Interdependence Analysis, to validate the structural completeness and explore the team dynamics. As the process is completed, it provides insight into requirements for the system, the human interface, artificial agent definitions, as well as human training and procedures.

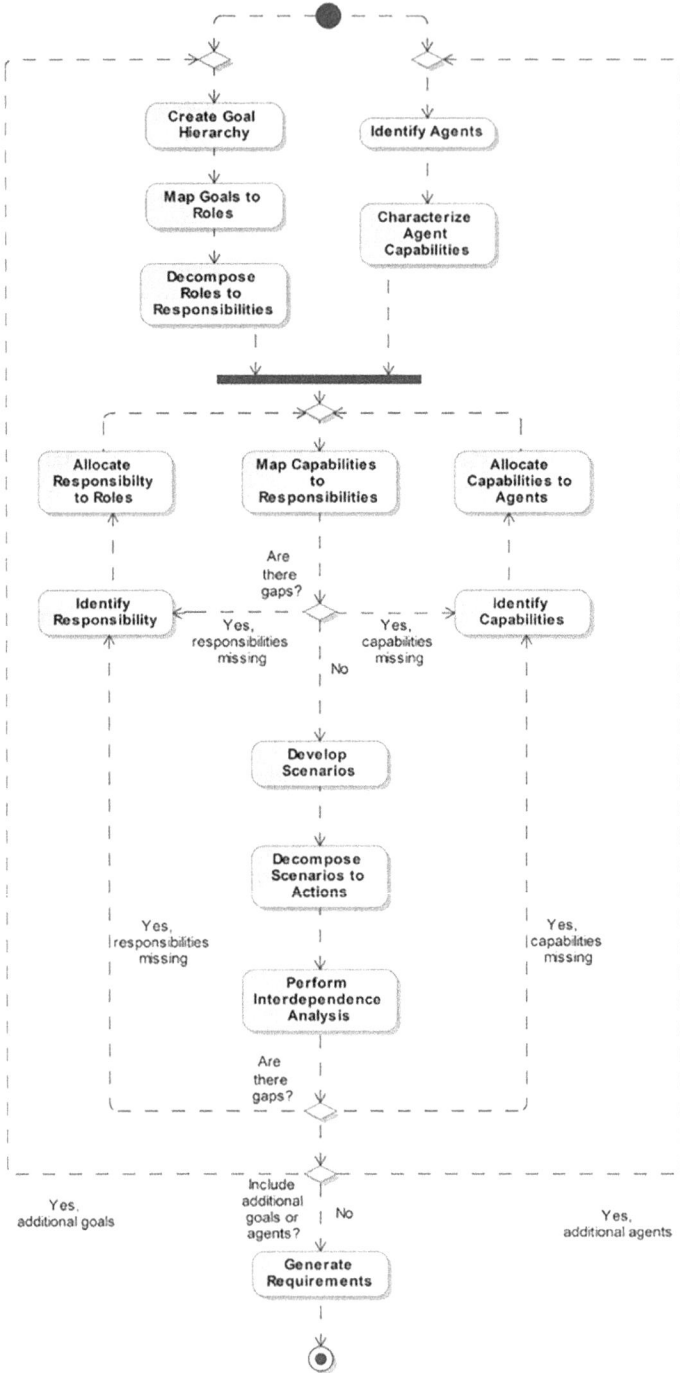

FIGURE 12.3   Activity diagram depicting the Design of Agent teams through Analysis of Goals method for employing the proposed metamodel.

The human-agent team modeling process begins with analyzing the work as it is performed to create a goal hierarchy the team must be able to achieve. In this process, one or more goals are mapped to each role, where roles are defined such that a cohesive set of knowledge or skills are required within each role. Each of the roles is then decomposed into a list of responsibilities that must be undertaken to fulfill the goals associated with each role, to include team coordination which is necessary when multiple roles contribute toward a goal. The synthesis process begins by defining an initial set of agents, including both human and artificial. The agents identified either currently accomplish the work or are proposed to be included in the new design. The specific, known capabilities of these agents are characterized by their means to sense the world, reason about it, and take action. This initial understanding of the work and the system provides a basis for refinement of the design.

A list of the initial capabilities, already allocated to agents, are then mapped to the responsibilities they fulfill during the structural analysis. Although there is no mandatory mapping between roles through responsibilities and capabilities to agents, it is generally desired that a minimum number of agents are necessary to complete each role. This ensures that each agent has a unique set of knowledge and skills, reducing the training requirements for any single agent. By tracing the structure of the responsibilities and capabilities, we can identify any gaps, missing capabilities, or responsibilities that were not previously identified. These missing elements must then be allocated to roles or agents appropriately to ensure that the team can accomplish the goals. However, as we seek to understand interdependence relationships between humans and artificial agents, many of these capabilities may be held by at least one human and one artificial agent. It is also useful to understand that when analyzing an "as-is" system without automation, that only human agents may initially exist in the system and the analysis required to identify capabilities and to determine which agents have specific capabilities may suggest artificial agents which can be used to support the humans within the system. A complete mapping of at least one capability to each responsibility models all the permutations and combinations of agent effort which could accomplish the goals. While only a subset of these potential team configurations may be considered desirable, understanding them all may help identify undesirable anomalous conditions which can be precluded in design. Implicit in each step of this process is capturing requirements for agent actuators, sensors, procedures, and training which could be used to populate a supporting SysML model.

Once a human-agent team design is structurally sound, that is, agents have all the capabilities to fulfill the responsibilities, the situational analysis can now be performed. "Sunny day" and "rainy day" scenarios are developed to support this analysis. The "Sunny day" scenarios represent mission segments which we expect to commonly occur while the "Rainy day" scenarios represent mission segments in which resources or environmental factors prevent one or more agents to perform as expected. For example, we might reasonably assume that fog or freezing rain may reduce the visibility of information collected by a sensor which is utilized by an artificial agent, degrading its performance. Analysis of such a "Rainy day" scenario may illuminate new interaction methods which are required to permit the team of agents to perform successfully within the imagined conditions. Our proposed modeling method is flexible and does not prescribe a specific scenario development methodology, although many sources can be consulted for help in developing scenarios.[53–55]

Scenarios need not be exhaustive, however, several requirements pertaining to scenario development must be met. First, they should be selected to represent best or nominal case situations, extremely challenging yet achievable situations, and situations of specific interest or criticality. Second, it must be possible to decompose these scenarios to actions which we can reasonably allocate in whole to capabilities of individual agents. This decomposition might include multiple action layers within a hierarchy. These actions are used in place of the leaf tasks of the hierarchical task analysis which is typically applied in Interdependence Analysis, to perform an extended version of this technique. Generally, this process comprises examining various allocations of the actions to the agents, eventually selecting an agent who is primarily responsible. It is important to note that criteria for allocation have been discussed extensively in the human factors and system design literature[56–61] with existing literature providing a number of useful heuristics in this area. However, the focus of the current approach is not to allocate tasks to individual agents, but to design cooperative behaviors where select agents can support each other during task performance.[62,63] Therefore, additional supporting agents are also selected whenever possible to aid the primary agent. As these backup behaviors and agents are developed, interface requirements supporting mutual OPD among the agents are recorded in support of interaction requirements. As before, this analysis will likely lead to the discovery of additional responsibilities and capabilities which were not included in the original analysis and the previous portions of the model, particularly the agent definitions and responsibilities, will require revision. When the process is completed, requirements for artificial agents, their interaction with human agents and requirements for human training and procedures can be developed based upon the model of the human-agent team interaction. Finally, scenarios must be complete enough to ensure that the set of capabilities required to meet the goals can be performed in the context of the scenario.

It is worth noting that it is not absolutely necessary to perform this analysis for a complete system. Instead it may be performed for subsets of human goals within a system or for the inclusion of partial sets of artificial agents. Once performed for a subset of the system, the process can be expanded to include additional system elements. This permits the application of this modeling process through an Agile approach as resources, including subject matter experts from which goals can be elicited, are available to support the analysis.

12.3   A SIMPLIFIED UNMANNED AERIAL VEHICLE EXAMPLE

To illustrate this model and method, we provide a simplified example of a human-agent team managing a machine within a particular domain. Specifically, we begin by describing a goal for an Unmanned Aerial Vehicle (UAV) team, restricting our unit of analysis to the activities performed by the two individuals most closely aligned with controlling a midaltitude UAV. Furthermore, in this example, we restrict our attention to the goal of "locate target," as depicted in Figure 12.4. The reader should note that this example is not intended to be complete or illustrate "the" solution to the proposed problem. Instead, the example is simply intended to illustrate some of the

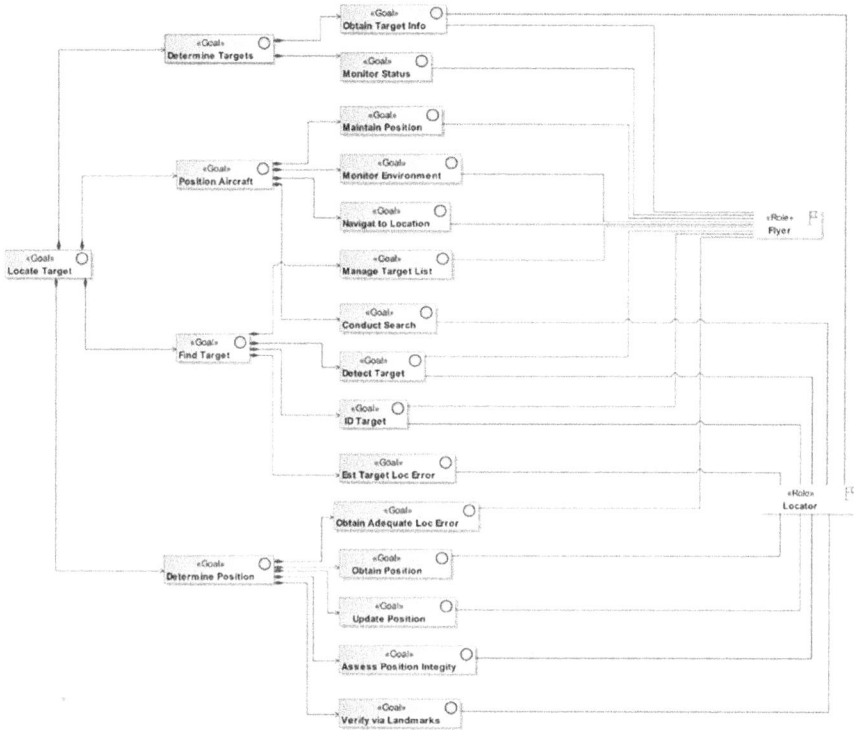

FIGURE 12.4 Hierarchical goal diagram, depicting the goals of the human-agent team and roles used to subdivide the responsibilities necessary to support these goals, providing a decomposition of the operational activities to be completed by the human-agent team.

important artifacts which are created through the application of the proposed Agent-Oriented SysML and the Design of Agent Teams through Analysis of Goals method.

As shown, the goal "locate target" can be decomposed into lower-level goals, which can be further decomposed. Each of the lower-level goals is then associated with a role. In this example, two roles are shown, including the "Flyer" role, which is primarily conducted by the UAV Pilot and the "Locator" role, which is primarily conducted by the Sensor Operator. Note that as shown, many of the goals are performed by an agent acting in a single role. However, some goals, such as detecting and identifying target are completed by more than one role, with each role having specific training, skills, and capabilities which contribute to the reliable completion of this goal. It is also useful to note that in this specific example, the two roles each correspond to a specific agent acting within the "as is" system. However, this is not necessary and it is common that a single agent will perform multiple roles within a system. A goal hierarchy of this type can be derived using a traditional goal-directed task analysis, which has been shown to be useful in describing the goals of large teams and to be useful when incorporating automation into real-world systems.[64,65]

Once the goals have been decomposed to a level that permits one to assign roles and determine specific responsibilities that must be completed to complete each goal,

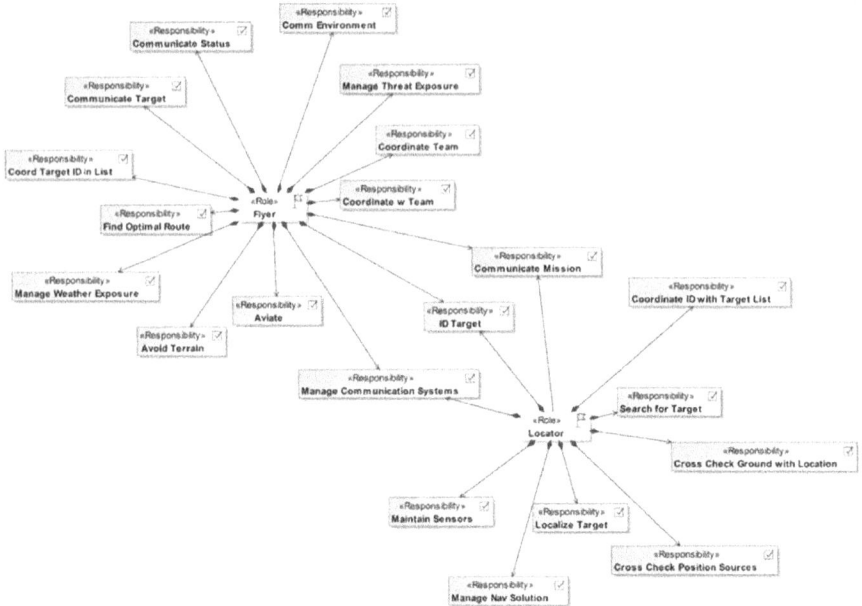

FIGURE 12.5   Role diagram, depicting the roles within the system and the responsibilities to be performed within each role to support the completion of the goals. This diagram further decomposes the operational activities to be completed by the human-agent team.

these responsibilities can be determined for each role. The results can be depicted in a Role Diagram, such as the diagram shown in Figure 12.5.

This diagram attempts to identify each of the responsibilities that each agent must fulfill to complete the goals. These responsibilities should permit the completion of each of the goals within the goal hierarchy. Therefore, there may be responsibilities such as "Search for Target," which are necessary to complete the low-level goal "Find the Target." However, there might be higher-level responsibilities such as "Coordinate Team" which is not associated with any specific goal but is required for successful completion of the highest level goal. Once again, although we expect most of these responsibilities to be associated with a single role, there will also likely be responsibilities, such as "Manage Communication Systems" which are shared among multiple roles.

Returning to Figure 12.3, once responsibilities have been determined, agents in support of the system are defined. Furthermore, the capabilities necessary to fulfill the responsibilities are defined and these capabilities are allocated to the human and artificial agents. This process may use traditional allocation processes,[56,57] identifying areas where automation could improve the performance or reliability of the human agents in the "as is" system and specifying artificial agents which will assist the human agents in the "to be" system. This can be a highly iterative process as considerations such as human workload, technology readiness, and the relative performance of the human and any artificial agent should be assessed and used to guide allocation.

Note that this specification includes not only specifying the capabilities of each agent but specifying requirements for the sensors, actuators, procedures, and knowledge to be trained for each of the agents as is necessary to enable them to fulfill the required capabilities. Furthermore, traceability can be established between these requirements and specific capabilities, which are derived from responsibilities, which are decomposed from roles, which enable the fulfillment of the goals.

Figure 12.6 illustrates a portion of an Agent Diagram. In this diagram, Agents are shown with associated capabilities, which are linked to the responsibilities. As these capabilities are defined for each of the agents it becomes possible to attempt to define the information needs of each agent. For the human operators, this step may specify information elements to be portrayed within the operator workstation. For the artificial agents, the information elements imply sensed information which must be determined and therefore the class of sensors which will be required to collect this information. In addition, it becomes possible to identify the items the human agents will need to control and the actuators which will be necessary for the artificial agents. Furthermore, we can begin to develop requirements for procedures and training for each type of agent. These can be captured in requirements which are linked to the agents and to specific required capabilities.

An initial high-level design of our human-agent team is now defined as it functions within our system to accomplish the goals we have identified. However, at this stage, we have thought about the function of the system without applying any constraints, other than perhaps constraints of the human or artificial agent capabilities. As we progress to the next step, we will need to consider constraints, such as time, environmental factors, and limits of the system under control. Furthermore, referring to Figure 12.6, we can see that multiple agents often can have overlapping capabilities, which permit them to contribute toward many of the responsibilities. At this stage, we have not provided any consideration of the interdependence between these agents as they employ the capabilities to meet the team's responsibilities and goals.

To address constraints and interactions, we begin by developing a series of "Scenarios." These scenarios ideally include both "sunny day" scenarios where the technology functions as designed with a range of typical environmental conditions. In addition, "rainy day" can include both system and automation failures, as well as more extreme environmental conditions, potentially including adversarial actions which stress the performance of the team in foreseen ways. These scenarios can be decomposed to provide a temporal list of activities and actions. The decomposition should extend until there is a single primary agent that can be identified for each action, other agents may provide support. Once defined and decomposed, these scenarios, the agents, and their capabilities can be used as inputs to an Interdependence Analysis as discussed by Johnson and colleagues as described earlier.

In this process, the scenarios and actions to be taken are arranged within a table. For each action, a primary agent is defined and the capabilities of the agent which are necessary for completion of the action are listed. An assessment is made regarding the reliability of the agent in performing the action. The designer then considers whether any agent can provide assistance to the primary agent and assigns one or more agents as the backup agent, once again capturing the necessary capabilities of this agent. If, during this process, the designer discovers that additional capabilities are necessary, the earlier list of capabilities, responsibilities, agents, and roles are updated to maintain model integrity. Finally, the designer considers whether it would

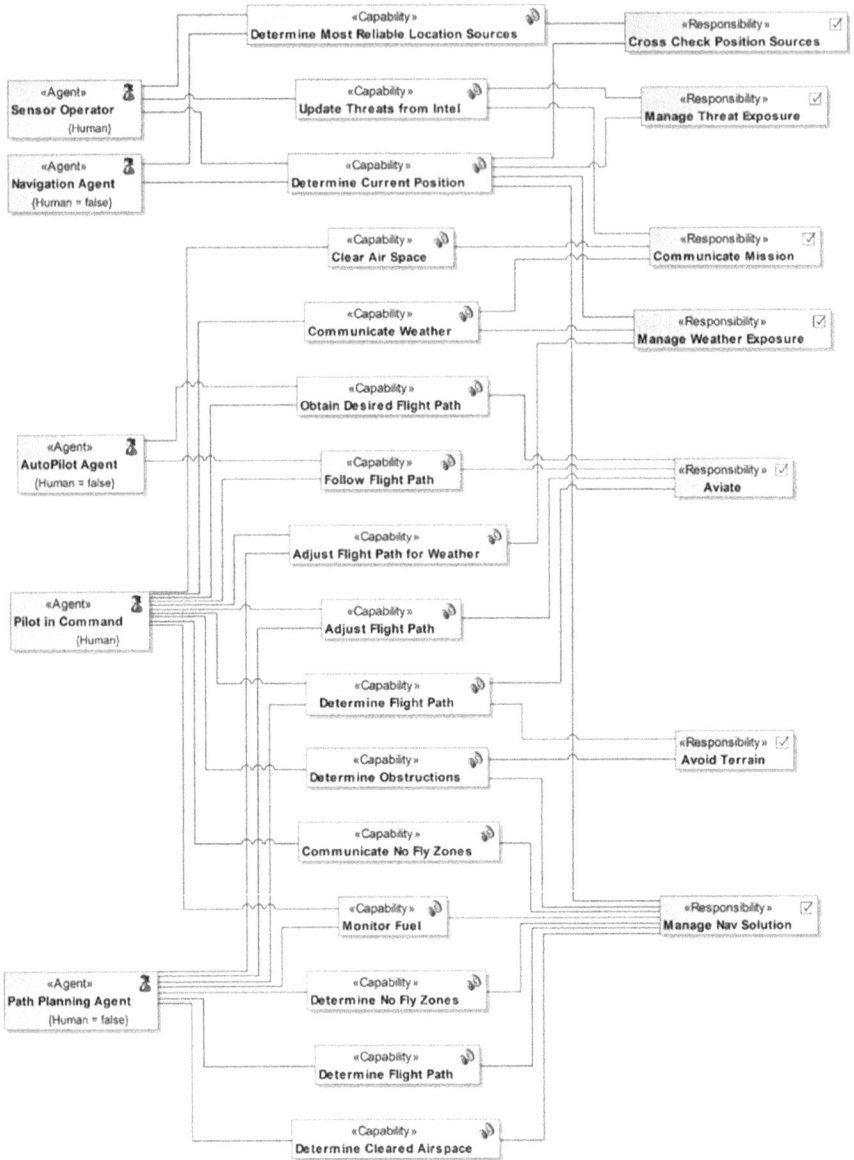

FIGURE 12.6 A depiction of an Agent-Capability diagram, illustrating the members of the human-agent team, their required capabilities and the allocation of these capabilities the members of the team where the capabilities are defined to address a portion of the responsibilities from Figure 12.5.

aid the two agents to share any information to improve the mutual OPD of the interaction between the two or more agents during the interaction.

A brief example of the output from such an analysis is shown in Table 12.1. Example interdependence analysis table for a sample scenario. As shown, the table

TABLE 12.1

Example Interdependence Analysis Table for a Sample Scenario

Scenario	Actions	Primary Agent	Primary Capabilities	Secondary Agent	Secondary Capabilities	OPD Requirements
Fly to Check Point 1	Input Data for Flight Plan	Pilot in Command	Clear Air Space Comm NoFlv Determine Obstructions	Agent	Monitor Fuel	Pilot shall be able to convey spatial and temporal locations of cleared air space, no fly zones, obstructions to the Path Planning Agent
	Create Flight Plan	Path Planning Agent	Determine FP Obtain FP	Pilot in Command	Determine FP Obtain FP Determine Risk	Path Planning Agent shall be able to show one or more potential paths and information elements pertaining to decision, Pilot must be able to direct alternate course and understand implications
	Communicate Flight Plan	Pilot in Command	Communicate Flight Path			Pilot shall be able to provide flight plan to external entities.
	Initiate Flight Plan	Pilot in Command	Determine FP			Pilot shall be able to command AutoPilot Agent to follow Flight Path
	Follow Flight Plan	AutoPilot Agent	Obtain FP Follow FP	Pilot in Command	Obtain FP Follow FP	Pilot shall be able to assume control if AutoPilot Fails while monitoring flight plan. Autopilot Agent shall convey rationale for any deviation from Flight Plan
	Gain New Intelligence	Sensor Operator	Update Threats from Intel			Sensor Operator shall be able to communicate with external entities regarding threats
	Input New Data	Sensor Operator	Update Threats from Intel	Pilot in Command		Sensor Operator shall be able to communicate threats to pilot in charge and Path Planning Agent
	Update Flight Plan	Path Planning Agent	Adjust Flight Path	Pilot in Command	Adjust Flight Path	Path Planning Agent shall notify pilot if updates in flight path are warranted
	Alter Flight Plan	Pilot in Command	Determine FP	Pilot in Command		Pilot shall be capable of understanding impact of FP change on mission.
	Follow Flight Plan	AutoPilot Agent	Obtain FP Follow FP	Pilot in Command	Obtain FP Follow FP	Pilot shall be able to command AutoPilot Agent to follow Flight Path

includes a scenario name and a list of actions necessary to complete the scenario. Primary and Secondary Agents, together with the necessary capabilities are listed. Finally, the last column provides a text-based OPD requirement as derived from this analysis.

As this analysis is completed, it is likely that capabilities will be discovered which were overlooked in our initial analysis. For example, the capabilities of "Determine Risk" and "Communicate Flight Path" did not exist within the initial model. However, the need for these capabilities becomes obvious while completing this analysis table.

It is also clear that we should be able to identify a primary agent for each action. The agent background is color coded in this example to indicate its robustness at completing the action, where this color code is an attribute of the connector between the capability and action. For example, green may mean that the agent is highly reliable in completing the action, yellow may signal the ability to perform the action with reasonable reliability, and red may signal the inability to complete the action reliably without assistance. Secondary agents are also shown for many actions, effectively allocating the action to the agent as the secondary agent. These secondary agents can be color coded to indicate whether they are able to enhance the performance of the primary agent, indicated in green, whether they can perform the action reasonably if required, indicated in yellow, or whether they have the authority and accountability to aid in the performance of the action but perhaps are not equipped, indicated in orange. Once again these color codes represent an attribute of the connector between the capability and the action. In the example, we see that the Pilot in Command is often shown in yellow as it is the designer's belief that the pilot in charge will have limited time and computing resources to complete most of these activities and therefore may not perform the action as well as desired for successful system operation. There is also one example where the Pilot in Command is shown in orange. This indication is provided because the design currently has no capability or procedure which permits the pilot to communicate with intelligence and therefore they have not been equipped to fully support this action.

The OPD requirements provide requirements for communication among the agents which allow them to observe, predict, and at times, direct the other's behavior. Again, these requirements are part of the design. One might theoretically want the agents to always be aware of each other's actions and to be able to understand and direct some changes. However, this may be an unrealistic goal due to computational limitations, among other reasons. Therefore, this list provides an acceptable set which is judged to provide adequate communication among the agents in the team by the designer and these requirements are captured to be enabled in the subsequent design.

12.4   DISCUSSION AND CONCLUSION

We have proposed and used a brief example to illustrate HAT-ML and HAT-DM for describing and capturing the structure and assumed interaction for human-agent teams. This modeling language and method have been conceived as an approach to the early stage, conceptual design of human-agent teams. It permits the derivation of requirements, structure, and high-level activity which may be used to inform system-specific system models. An initial implementation of this modeling language has been constructed within Cameo Systems Modeler, permitting the model to be integrated with and extend SysML and UML modeling artifacts, which are also supported by this modeling environment. While the language is likely incomplete and underspecified at the current time, we suggest that the application of the proposed model and method may facilitate the conceptual design of human-agent teams within systems. It is our belief that establishing traceability from software and hardware models to the model artifacts present in this language may provide greater insights to designers and developers at later stages of the design cycle.

Table 12.2 illustrates some of the perceived benefits and potential opportunities for improving the HAT-ML. Observing interactions with students, there is a natural tendency for designers to focus on the interface between a human and a system (eg, an agent) as the default unit of analysis. As this modeling language is introduced into the classroom, the students quickly begin to define integrated human teams to analyze as the unit of analysis. As such, they quickly begin to define teams of humans and multiple agents. This is especially true as the advantages of modular artificial agents are discussed within the course. The ability to capture the interaction among members of this team appears to be important in most systems of interest to the Department of Defense (DoD) and many consumer-oriented products, based on the student projects which have applied this modeling approach. Furthermore, by using a method that begins with an analysis of the common goals of this team, students define human-agent teams which generally fit within the workflow of the existing

TABLE 12.2
Observed Benefits and Known Areas for Improvement

Observed Benefits	Known Improvement Opportunities
Encourages consideration of full multiagent team	Information needs and transfer requirements are not currently modeled.
Encourages development of coordination	Does not explicitly capture decision logic.
Provides a method for capturing capabilities	Does not provide automated facilities or tools for deriving artificial agents or allocations or insuring completeness of the capability set.
Permits traceability of capabilities and requirements to agents and to the goals in current environment, forcing thought about how automation will fit in existing environment	Design begins with an analysis of the extent environment, which has the potential to limit the perceived design space.
Defines a method for establishing Observability, Predictability, and Direct-ability requirements	Currently difficult to consider or explore multiple completion paths with true collaborative behavior.
Provides a method for determining training requirements for humans and artificial, machine-learning-based agents	Application of the model and method is time-consuming and requires access to domain experts
Encourages thought on appropriate sensors and actuators	
Illustrates certain personnel and procedure requirements	

teams. Although not shown in the example, many of the students who are familiar with SysML quickly begin constructing Block Definition Diagrams which define the environment and the composition of extent system, clearly defining the system boundary before developing the goal diagram.

Creation of the goal diagram creates interaction with the user community, providing a deeper understanding of the user needs and constraints. Occasionally, individual students begin to capture not only the goals but data models which capture information needs, as well as, use cases, activity, or sequence diagrams which begin to capture some of the decision logic used by the team. While these artifacts have not been represented in the current metamodel, it is likely

that these artifacts will be useful in defining the detailed behavior of the human-agent team and can be observed in existing modeling approaches.[27,33,64,66] Therefore, future revisions of HAT-ML should consider integration with these components. It is also clear that model validation approaches could be useful, certainly to insure completeness but also to help in the detection of possible human or artificial agent-induced error and approaches such as those discussed by Bolton and Bass may be useful in this regard.[28,29]

It is also worth noting that, like many modeling approaches, value is often derived from the process of modeling. In one iteration of a class, students were asked to create an initial conceptual design before this modeling language and method were introduced. When applying this modeling language and method, one student realized that their initial concept enabled their entry-level employees to be replaced with automation and therefore their design did not permit the development of experienced employees. In multiple other examples within the same course, individuals broadened their thought to include a service provider within their team definition, realizing that much more significant changes could be made in team behavior if they could create a tighter coupling with the service provider.

Also as shown in Table 12.2, faithful application of the model and method can be time-consuming to apply and will require access to domain and technology experts. However, the model and method can assist the design team in structuring the knowledge necessary for the description of systems including human-agent teams. Application of the model in an iterative fashion permits the designer to model the system in the abstract and add fidelity to the model and the modeled elements with successive iterations. Specifically, the modeling method may be applied iteratively to subsets of system goals, as long as the goals chosen enable meaningful scenarios. The ability to share model artifacts with stakeholders, analyze portions of the model, and adjust the model within each iteration makes the approach compatible with existing Agile processes. We suggest that complex systems which require interaction with multiple individuals and multiple artificial agents to perform highly interdependent work are likely to benefit from such a structured description and analysis method. Without such a structured analysis method and language, integration of the software and the human processes and training to employ this software may be difficult to achieve.

While the Interdependence Analysis, which this modeling method extends, has been shown to provide systems with more robust human-agent teaming in design competitions, the model and method described here are yet to be fully demonstrated within a practical system development process. Therefore, the utility of the HAT-ML and the HAT-DM is yet to be demonstrated outside of the classroom and is a topic for future research. While attempting to improve the model to address the improvement opportunities discussed earlier, we believe the current language extension and method are ready for experimental application and we have begun to explore these applications. However, we believe the modeling language extension and method would benefit from broader use to determine if they are useful in creating more robust human-agent teams.

12.5 DISCLAIMER AND ACKNOWLEDGMENT

The views expressed in this article are those of the authors and do not necessarily reflect the official policy or position of the Department of the US Air Force, US Department of Defense, nor the US Government. The authors gratefully acknowledge the financial support of the Air Force Office of Scientific Research, Computational Cognition, and Machine Intelligence Program.

REFERENCES

1. Goodfellow I, Bengio Y, Courville A. *Deep Learning*. Cambridge, MA: MIT Press; 2016.
2. Minsky M. Steps toward artificial intelligence. *Proc IRE*. 1961;49(1):8–30. https://doi.org/10.1109/JRPROC.1961.287775.
3. Guilford G, GM's decline truly began with its quest to turn people into machines. https://qz.com/1510405/gms-layoffs-can-be-tracedto-its-quest-to-turn-people-into-machines/.
4. Randall KR, Resick CJ, DeChurch LA. Building team adaptive capacity: the roles of sensegiving and team composition. *J Appl Psychol*. 2011;96(3):525–540. https://doi.org/10.1037/a002 2622.
5. Smith K, Hancock PA. Situation awareness is adaptive, externally directed consciousness. *Hum Factors*. 1995;37(1):137–148. https://doi. org/10.1518/001872095779049444.
6. Woods DD, Branlat M. Basic patterns in how adaptive systems fail. In: Nemeth CP, Hollnagel E, eds. *Resilience Engineering in Practice, Volume 2: Becoming Resilient*. Burlington, VT: Ashgate Publishing; 2015: 1–21.
7. Licklider JCR. Man-computer symbiosis. *IRE Trans Hum Factors Electron*. 1960;HFE-1(1):4–11. https://doi.org/10.1109/THFE2.1960.4503259.
8. Delligatti L. *SysML Distilled: A Brief Guide to the Systems Modeling Language*. Upper Saddle River, NJ: Addison-Wesley; 2013.
9. Friedenthal S, Moore A, Steiner R. *A Practical Guide to SysML*. Amsterdam: Elsevier Science and Technology; 2014.
10. Watson M, Rusnock CF, Colombi JM, Miller ME. Human-centered design using system modeling language. *J Cogn Eng Decis Mak*. 2017;11(3):252–269. ttps://doi.org/10.1177/1555343417705255.
11. Watson M, Rusnock CF, Miller ME, Colombi JM. Informing system design using human performance modeling. *Syst Eng*. 2017;20(2):173187. https://doi.org/10.1002/sys.
12. Naikar N. Cognitive work analysis: an influential legacy extending beyond human factors and engineering. *Appl Ergon*. 2017;59:528–540. https://doi.org/10.1016/j.apergo.2016.06.001.
13. Card SK, Moran TP, Newell A. *The Psychology of Human-Computer Interaction*. Hillsdale, NJ: Lawrence Erlbaum Associates, Inc; 1983.

14. Endsley MR, Jones DG. *Designing for Situation Awareness: An Approach to User-Centered Design*. Boca Raton, FL: CRC Press, Inc; 2012.

15. Geddes ND. *Understanding human operators' intentions in complex systems* [dissertation]. Georgia Tech; 1989.

16. Johnson M, Vera AH. No AI is an island: the case for teaming intelligence. *AI Mag.* 2019;40(1):16–28.

17. Christoffersen K, Woods DD. How to make automated systems team players. In: Salas E, ed. *Advances in Human Performance and Cognitive Engineering Research.* 1st ed. Bingley, UK: Emerald Group Publishing Limited; 2002:1–12. https://doi.org/10.1016/S1479–3601(02) 02003–9.

18. Johnson M, Bradshaw JM, Feltovich PJ, Jonker CM, Van Riemsdijk MB, Sierhuis M. Coactive design: designing support for interdependence in joint activity. *J Hum Robot Interact.* 2014;3(1):43–69. https://doi.org/10.5898/JHRI.3.1.Johnson.

19. Johnson M, Bradshaw JM, Feltovich PJ. Tomorrow's human–machine design tools: from levels of automation to interdependencies. *J Cogn Eng Decis Mak.* 2018;12(1):77–82. https://doi.org/ 10.1177/1555343417736462.

20. Bradshaw JM, Hoffman RR, Woods DD, Johnson M. The seven deadly myths of "autonomous systems". *IEEE Intell Syst.* 2013;28(3):54–61. https://doi.org/10.1109/MIS.2013.70.

21. Paterno F, Mancini C, Meniconi S. ConcurTaskTrees: a diagrammatic notation for specifying task models. *Human-Computer Interact INTERACT '97*; 1997:362–369. https://doi.org/10.1007/978-0-387-351759_58.

22. Paternò F. Towards a UML for interactive systems. In: *Lect Notes Comput Sci (including Subser Lect Notes Artif Intell Lect Notes Bioinformatics).*2001;2254:7–18. https://doi.org/10.1007/3-540-45348-2_4.

23. Martinie C, Palanque P, Barboni E, Ragosta M. Task-model based assessment of automation levels: application to space ground segments. *Conference Proceedings - IEEE International Conference on Systems, Man, and Cybernetics*; 2011:3267–3273. https://doi.org/10.1109/ ICSMC.2011.6084173.

24. Martinie C, Palanque P, Ragosta M, Fahssi R. Extending procedural task models by systematic explicit integration of objects, knowledge and information. *ACM International Conference Proceeding Series;* 2013. https://doi.org/10.1145/2501907.2501954.

25. Bindewald JM, Miller ME, Peterson GL. A function-to-task process model for adaptive automation system design. *J Hum Comput Stud.* 2014;72(12):822–834. https://doi.org/10.1016/j.ijhcs.2014. 07.004.

26. Ijtsma M, Ma LM, Feigh KM, Pritchett AR. Demonstration of the "work Models that Compute" simulation framework for objective function allocation. *Proceedings of the Human Factors and Ergonomics Society*, 2018;1:321–324. https://doi.org/10.1177/1541931218621074.

27. Ma LM, Feigh KM. Jumpstarting modelling systems design a generalized xml abstraction of simulation model. *Proceedings of the Human Factors and Ergonomics Society*, 2017:718–722. https://doi.org/10.1177/ 1541931213601665.

28. Bass EJ, Bolton ML, Feigh KM, et al. Toward a multi-method approach to formalizing human-automation interaction and human-human communications. *Conference Proceedings - IEEE International Conference on Systems, Man, and Cybernetics.* 2011:1817–1824. https://doi.org/10.1109/ICSMC.2011.6083935.

29. Bolton ML, Bass EJ, Siminiceanu RI. Using formal verification to evaluate human-automation interaction: a review. *IEEE Trans Syst Man Cybern A Syst Hum.* 2013;43(3):488–503. https://doi.org/10.1109/ TSMCA.2012.2210406.

30. Bolton ML, Siminiceanu RI, Bass EJ. A systematic approach to model checking human-automation interaction using task analytic models. *IEEE Trans Syst Man Cybern A Syst Hum.* 2011;41(5):961–976. https://doi.org/10.1109/TSMCA.2011.2109709.

31. Deloach SA. The MaSE methodology. In: Bergenti F, Gleizes MP, Zambonelli F, eds. *Methodologies and Software Engineering for Agent Systems*. New York: Springer; 2004: 107–125.

32. Deloach SA, García-Ojeda JC. O-MaSE: a customisable approach to designing and building complex, adaptive multi-agent systems. *Int J Agent-Oriented Softw Eng*. 2010;4(3):244–280. http://citeseerx.ist.psu. edu/viewdoc/download?doi=10.1.1.460.728 7&rep=rep1&type=pdf.

33. Sterling LS, Taveter K. *The Art of Agent-Oriented Modeling*. Boston, MA:MIT Press; 2009.

34. Damacharla P, Javaid AY, Gallimore JJ, Devabhaktuni VK. Common metrics to benchmark Human-Machine Teams (HMT): a review. *IEEE Access*. 2018;6:38637–38655. https://doi.org/10.1109/ACCESS.2018.2853560.

35. Object Management Group. Service oriented architecture Modeling Language (SoaML) specification. *Language (Baltim)*. 2012;(March):1144.

36. Brambilla M, Fraternali P. Implementation of applications specified with IFML. *Interact Flow Model Lang*. 2015:279–334. https://doi.org/10.1016/b978-0-12-800108-0.00010–2.

37. Object Management Group. UML profile for MARTE: modeling and analysis of real-time embedded systems v.1.0. *Engineering*.2009;15(November):738. http://www.omg. org/spec/MARTE/1.1/.

38. Parsons MH. Automation and the individual: comprehensive and comparative views. *Hum Factors*. 1985;27(1):99–112. https://doi.org/10.1177/001872088502700109.

39. Weiss G. *Multiagent Systems*. Cambridge, MA: MIT Press; 2013.

40. Madni A, Madni C. Architectural framework for exploring adaptive human-machine teaming options in simulated dynamic environments. *Systems*. 2018;6(4):44. https://doi. org/10.3390/systems6040044.

41. Bork D, Karagiannis D, Pittl B. A survey of modeling language specification techniques. *Inf Syst*. 2020;87:101425. https://doi.org/10.1016/ j.is.2019.101425.

42. Walden DD, Roedler GJ, Forsbert KJ, Hamelin RD, Shortell TM. *Systems Engineering Handbook: A Guide for System Lifecycle Processes and Activities*. San Diego, CA: International Council on Systems Engineering (INCOSE); 2015.

43. Rasmussen J, Pejtersen AM, Goodstein LP. *Cognitive Systems Engineering*. New York, NY: Wiley; 1994.

44. Wright P, Dearden A, Fields B. Function allocation: a perspective from studies of work practice. *Int J Hum Comput Stud*. 2000;52:335–355. https://doi.org/10.1006/ijhc.1999.0292.

45. Dearden A, Harrison M, Wright P. Allocation of function. *Int J Hum Comput Stud*. 2000;52:289–318.

46. Garcia-Ojeda JC, DeLoach SA. The O-MaSE process: a standard view. *CEUR Workshop Proc*. 2010;627(39):55–66.

47. Woods DD, Hollnagel E. *Joint Cognitive Systems: Patterns in Cognitive Systems Engineering*. Boca Raton, FL: CRC Press; 2006.

48. Alderson DL, Doyle JC. Contrasting views of complexity and their implications for network-centric infrastructures. *IEEE Trans Syst Man Cybern A Syst Hum*. 2010;40(4):839–852. file:///2009/Unknown/ 2009ContrastingViewsofComplexityandTheirImplications ForNetworkCentricInfrastructures.pdf.

49. Crandall B, Klein GA, Hoffman RR. *Working Minds: A Practioner's Guide to Cognitive Task Analysis*. Cambridge, MA: MIT Press; 2006.

50. Marks MA, Mathieu JE, Zaccaro SJ. A temporally based framework and taxonomy of team processes. *Acad Manag Rev*. 2001;26(3):356–376.

51. Johnson M, Vignati M, Duran D. Understanding human-autonomy teaming through interdependence analysis. In: Kerstholt J, Barnhoorn J, Hueting T, Schuilenborg L, eds. *Automation as an Intelligent Teammate: Social Psychological Implications*. North Atlantic Treaty Organization Science and Technology Organization; 2019: 1–20.

52. Johnson M, Bradshaw JM, Feltovich P, Jonker C, Van Riemsdijk B, Sierhuis M. Autonomy and interdependence in human-agent-robot teams. *IEEE Intell Syst.* 2012;27(2):43–51. https://doi.org/10.1109/MIS.2012.1.

53. Chermack TJ. Studying scenario planning: theory, research suggestions, and hypotheses. *Technol Forecast Soc Chang.* 2005;72(1):59–73. https://doi.org/10.1016/j.techfore.2003.11.003.

54. Schnaars SP. How to develop and use scenarios. *Long Range Plann.* 1987;20(1):105–114. https://doi.org/10.1016/0024–6301(87) 90038–0.

55. Miller R. Futures literacy: a hybrid strategic scenario method. *Futures.* 2007;39(4):341–362. https://doi.org/10.1016/j.futures.2006.12.001.

56. Fitts P, ed. *Human Engineering for an Effective Air Navigation and Traffic Control System.* Washington, DC: National Research Council; 1951.

57. Price HE. Allocation of functions in systems. *Hum Factors.* 1985;27(1):33–45. https://doi.org/10.1177/001872088502700104.

58. Pritchett AR, Kim SY, Feigh KM. Modeling human-automation function allocation. *J Cogn Eng Decis Mak.* 2014;8(1):33–51. https://doi.org/10.1177/1555343413490944.

59. Feigh KM, Pritchett AR. Requirements for effective function allocation: a critical review. *J Cogn Eng Decis Mak.* 2014;8(1):23–32. https://doi.org/10.1177/1555343413490945.

60. Feigh KM, Dorneich MC, Hayes CC. Toward a characterization of adaptive systems: a framework for researchers and system designers. *Hum Factors.* 2012;54(6):1008–1024. https://doi.org/10.1177/ 0018720812443983.

61. de Winter JCF, Dodou D. Why the Fitts list has persisted throughout the history of function allocation. *Cogn Technol Work.* 2014;16(1):1–11. https://doi.org/10.1007/ s10111-011-0188–1.

62. Cummings MM. Man versus machine or man + machine? *IEEE Intell Syst.* 2014;29(5):62–69. https://doi.org/10.1109/MIS.2014.87.

63. Bindewald JM, Miller ME, Peterson GL. Creating effective automation to maintain explicit user engagement. *Int J Hum Comput Interact.* 2020; 36(4):341–354. https://doi.org/10.1080/10447318.2019.1642618.

64. Humphrey CM, Adams JA. Analysis of complex team-based systems: augmentations to goal-directed task analysis and cognitive work analysis. *Theor Issues Ergon Sci.* 2011;12(2):149–175. https://doi.org/10.1080/14639221003602473.

65. Kaber DB, Segall N, Green RS, Entzian K, Junginger S. Using multiple cognitive task analysis methods for supervisory control interface design in high-throughput biological screening processes. *Cogn Technol Work.* 2006;8(4):237–252. https://doi.org/10.1007/ s10111-0060029–9.

66. Feigh KM, Pritchett AR. Modeling work for cognitive work support system design in operational control centers. *J Cogn Eng Decis Mak.* 2010;4(1):1–26. https://doi.org/10.1518/155534310X495564.

13 Effects of Agent Timing on the Human-Agent Team[*]

Tyler J. Goodman, Michael E. Miller,
Christina F. Rusnock, and Jason M. Bindewald
US Air Force Institute of Technology

CONTENTS

13.1 INTRODUCTION

Although powered machines have been applied to automate difficult, energy-intensive tasks for centuries (Smeaton & De Moura, 1751), the utility of these systems has been limited by their ability to adapt to environmental changes. More recently, systems that adapt to environmental stimuli have been discussed, modeled, and demonstrated (Licklider, 1960; Rouse, 1977). However, these environmentally-adaptive systems typically allocated functions based on technical feasibility or cost-effectiveness.

---

[*] Reprinted with permission: Goodman, T. J., Miller, M. E., Rusnock, C. F., & Bindewald, J. M. (2017). Effects of agent timing on the human-agent team. *Cognitive Systems Research*, 46, 40–51.

DOI: 10.1201/9781003220978-17

This allocation was thought to be particularly desirable when the machine could perform the function more efficiently, reliably or accurately than the human operator. Thus, the human was left to perform higher-level monitoring functions, to detect and correct automation failures, and to perform functions that were difficult to automate due to their complexity (Parasuraman & Riley, 1997). Due to human vigilance loss, mental switching delays, and skill decay, this allocation of responsibility was found to be brittle as machine errors cascaded, resulting in system failure (Wiener & Curry, 1980). In response, the concept of adaptive automation was proposed. In this paradigm, the level of automation changes as a function of human, mission, environment, or system state (Hancock & Chignell, 1988). While adaptive automation changes the allocation of tasks dynamically, both traditional automation and adaptive automation specifically allocate tasks to the human or machine. Adaptive automation provides one or more state variables that permit this allocation to change dynamically. However, at any one moment in time either the human or the machine is responsible for completion of specific tasks.

Recently, the concept of human-agent teaming has been proposed, in which the human and one or more artificial autonomous agents (i) share goals (Hoc, 2001), (ii) are interdependent (Arthur et al., 2005), and (iii) dynamically allocate roles (Bruemmer, Marble, & Dudenhoeffer, 2002). In this paradigm, task allocation is not fixed. Unlike systems employing automation, autonomous systems possess the capacity and have the authority to decide which goals they are going to pursue and the tasks they are going to perform to accomplish those goals. Thus, in human-agent teams, both the agent and the human can select which actions to take to achieve a higher-level goal.

The concept of human-agent teaming is not new (Rouse, 1981; Rouse, Edwards, & Hammer, 1993; Scerbo, 1996). However, few examples of human-agent teams having all three of these attributes have been discussed in the literature. Further, the existing literature on human-agent teaming rarely explores the effect of artificial agent behavior on team performance. For example, timing of artificial agents in human-agent teams has received little attention as an important design parameter, despite timing often being considered a significant design parameter in traditional user interface design (Miller et al., 2001). In traditional human interface design, timing is typically a concern when the system exhibits unacceptable delays due to slow system response to human input, requiring design changes to reduce the perceived delay. In human-agent teams, the artificial agents do not necessarily respond to human input and can respond more rapidly to environmental or mission-based stimuli than a human teammate. This fast response permits an agent to improve system performance when system performance is limited by human response time.

Human information processing is often modeled as a four-stage process, including sensory processing, perception, decision-making, and response selection (Parasuraman, Sheridan, & Wickens, 2000). These stages are typically assumed to be performed as a serial sequence, with each of the four stages requiring a finite time period to complete. The time required to perform each stage is dependent upon the complexity of the task (Fitts & Peterson, 1964; Hyman, 1953) as well as human motivation and skill level (Dixon & Wickens, 2003). In multi-task environments, other concerns, such as task switching may introduce additional delays. An autonomous

agent may perform a series of processes analogous to the four-stage human information process. Depending upon the system design, the autonomous agent may perform
any of these four stages significantly faster or slower than a typical human operator,
with the potential to perform these processes so rapidly as to appear instantaneous to
the human operator. Therefore, a rapid response on the part of the autonomous agent
may reduce the human's opportunity to respond to environmental stimuli.

Rouse constructed a queuing model to understand the effect of the relative speed
of an autonomous agent on human involvement in a human-agent team (Rouse, 1977).
This research illustrated that the proportion of decisions made by the autonomous
agent within the team should increase as the autonomous system's speed increases
with respect to the time required for the human to make a similar decision. The
model indicated that the proportion of decisions performed by the autonomous agent
was particularly high when the event rate was low. This research implies that during
times of relatively low activity, a rapidly responding agent will perform the majority
of actions, relegating the human to the role of a supervisor. Thus, the human will be
forced to perform a vigilance task. As it is known that humans perform poorly in this
role (Warm, Parasuraman, & Matthews, 2008), it would be expected that the performance of the team may well suffer when active human tasks are converted to passive
activity due to the agent's rapid response.

The effect of varying the cycle time of adaptive automation has also been discussed in the autonomy literature (Hilburn, Molloy, Wong, & Parasuraman, 1993).
This research indicated that time triggers have limited applicability. However, in this
example, the automated system relied solely upon triggers that altered system state
between fully manual and fully automatic for predetermined lengths of time; thus,
requiring the human to perform all or none of the functions within any epoch (Feigh,
Dorneich, & Hayes, 2012; Hilburn et al., 1993). More recently, automated systems
have implemented safety features that activate when time is not available to permit
a satisfactory human response to an impending vehicle collision (Bice, Skoog, &
Howard, 1990; Rump, Steiner, & Douglas, 1996). These systems illustrate the utility
of rapid automated response when the human is unable to respond in a timely fashion.

Despite early modeling research indicating the impact of agent timing on decision-making and performance in human-agent teams, the impact of autonomous
agent timing has not been investigated through human-in-the-loop research. As
autonomous agents continue to be incorporated into dynamic and evolving environments, the effect of agent timing deserves further investigation. The moment the
agent executes an action determines its timing when responding to an environmental
or system-generated event. For both humans and agents, the timing of an action is
constrained by, but not determined by, the time required for decision-making. That
is an action cannot be undertaken before a decision is made but the action can be
deliberately delayed.

The current research sought to investigate the effect of agent delay time on human
workload, as well as, team behavior and performance within a shared environment.
To contribute toward the team objective, we recognize that the agent must consider
multiple objectives and these objectives may influence the agent's desired timing.
Specifically, the agent must complete tasks consistent with the operator's goal, while
avoiding operator overload and keeping the human engaged and aware of the task

environment to mitigate the agent's failings. Therefore, this research specifically employed an autonomous agent capable of providing an apparently instantaneous response to environmentally-generated objects within a highly dynamic environment. A purposeful delay was implemented in the agent at a stage analogous to the period between the appearance of the stimuli and the onset of a response action. This delay was then varied to vary the human's opportunity to respond to the presence of objects in the environment. The agent's and human actions were displayed in real-time within a common workspace, providing the human the opportunity to respond to the environmental stimuli and override previous responses (the autonomous agent's or their own) to the environmental stimuli.

13.2   METHOD

A combination of system modeling and human experimentation was employed, where the system models include aspects of the human, the agent, and the environment. Specifically, this research employed the method depicted in Figure 13.1. As shown, the approach includes six phases: development of a static system model, baseline human-in-the-loop (HITL) experimentation, development of a dynamic system model (e.g. simulation development), simulation execution, validation HITL experiment, and further investigation of the environment through further simulation extension. These phases were completed in sequential order with feedback from the experiments to the models. Construction of the models requires the elicitation of assumptions necessary to model human behavior. Therefore, the models provide hypotheses of human behavior and performance. Experimental validation of the models provides support for the assumptions. Thus, this modeling approach provides insight into human decision-making processes within the environment, influencing the content and research questions of the HITL experiments.

13.2.1   EXPERIMENTAL ENVIRONMENT

The experimental environment was a tablet-based, air-traffic management-style game called Space Navigator. Space Navigator provides a controlled representation of a highly dynamic, event-driven environment. In these environments, the operator has little, if any, control of the event rate. Further, there is no guarantee that the human will be capable of responding in a timely fashion to all necessary events during high event rates. The game involves space ships, representing aircraft, which must be given approach routes to planets, representing runways at an airport. As such the

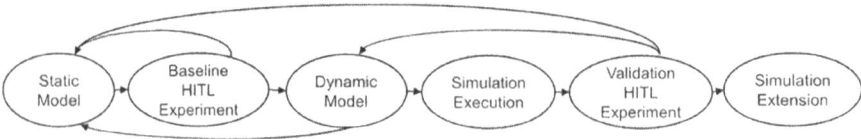

FIGURE 13.1   Methodology process, employing iterative modeling and human experimentation.

cognitive tasks performed by the participants resemble the cognitive tasks performed by an air-traffic controller performing terminal radar approach control. The ability to control the spawn rate of space ships permits task load to be manipulated similar to changes in the number of aircraft, while bonuses and no-fly zones, which appear at random locations within the environment, permit traffic complexity to be manipulated. Similar manipulations have been shown to significantly influence air-traffic controller workload within simulated terminal radar approach control (Brookings, Wilson, & Swain, 1996). While the environment does not provide a high-fidelity simulation of all aspects of these environments, most notably the consequences of decisions, it permits the control of several potentially confounding variables and logs human response. Additionally, the environment includes a single, clearly-defined, top-level goal (i.e., score the most points), as opposed to other games which provide multiple, often conflicting, goals (e.g., leveling up and score). The use of the relatively intuitive game environment simplifies participant recruitment and training.

Figure 13.2 includes a screen capture from the game and identifies several key objects within the game. Spaceships appear at set intervals from the screen edges. The player directs each spaceship to a destination planet of the same color while avoiding no-fly zones and collecting bonuses. Ships are directed by drawing a line on a touchscreen with a finger to indicate a ship trajectory. Points accumulate when a ship encounters its destination planet or a bonus is collected. Points decrement when spaceships collide, additionally resulting in the loss of each spaceship involved in the collision. Points are also lost when a spaceship traverses a no-fly zone. These no-fly zones move throughout the play area to random locations at a set time interval. An autonomous agent produces trajectories for ships that have not been assigned a trajectory. Trajectories drawn by the agent are displayed in a different color than trajectories drawn by the human. The spaceship will follow the most recent trajectory. The game ends after 5 min.

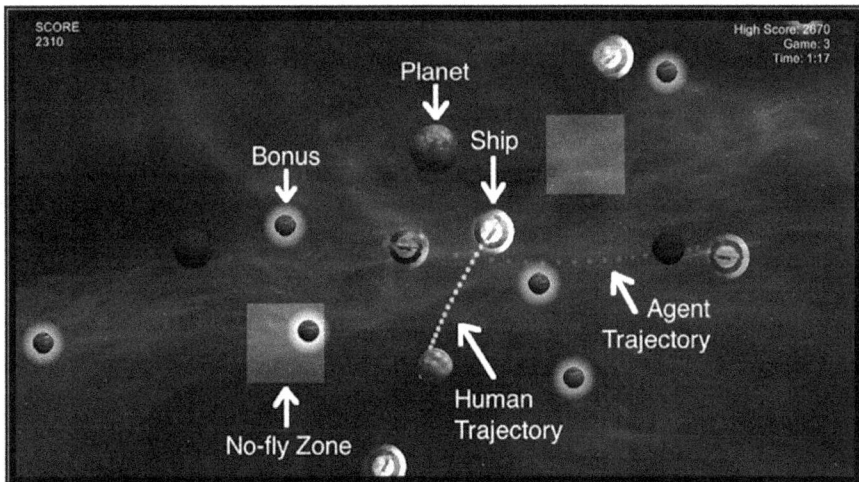

FIGURE 13.2 Screen capture from Space Navigator, highlighting spaceships, planets, trajectories, bonuses and no-fly zones.

The automated agent presented in this experiment draws straight-line trajectories from ships to their corresponding planets, ignoring the presence of other spaceships, bonuses, and no-fly zones. The trigger used for the agent considers the arrival of a new ship, human response to this event, and human inactivity. Specifically, the automated agent draws a route for a ship if the human operator does not assign the ship an initial route after a specified time. This design presumed the agent would aid the user when the user is unable to respond to the environmentally-generated event in a timely fashion. Thus, if the user was task saturated or unable to complete trajectories for other reasons, the agent would help the user by designating trajectories for the ships left unattended. In the current study, the time that a ship was displayed on screen before the agent selects and draws a trajectory (referred to as delay time) was the independent variable of interest. Therefore, its value could be varied from nearly instantaneous to a time significantly slower than the typical human response time.

13.2.2 STATIC MODEL DEVELOPMENT

System Modeling Language (SysML) activity diagrams were developed, beginning with activities to be performed within the Space Navigator environment. These activities were then allocated to the human and agent. Once these activities were allocated to a human and agent, other inherent activities were defined to facilitate communication of system state as control is passed between the human and agent (Bindewald, Miller, & Peterson, 2014; Goodman, Miller, & Rusnock, 2015). A simplified diagram is shown in Figure 13.3. Activity diagram representing the actors and actions in the static model. which divides the activities among three primary sections (i.e., "swim lanes"). Each section represents the activities of the environment, the human operator, or the agent. The environment nodes in Space Navigator are shown in the center swim lane. The environment starts the model, generates ships, alters no-fly zone and bonus locations, operates the timer, and halts the activity at game completion.

The human's attention and actions during game play are facilitated through a loop shown in the left swim lane of Figure 13.3. The human continuously repeats two high-level functions; determining which ship to select and drawing a trajectory for a ship. This loop is completed for ships that have no drawn trajectories and for those that have undesirable trajectories. A possible player strategy would be to work to their capacity as they try to earn the highest score, leveraging the agent to draw paths they do not have time to draw. This behavior is depicted through the path in the human swim lane of Figure 13.3, which includes identifying background items, identifying ships without routes, selecting a ship, and drawing a route. However, the human could decide to permit the agent to draw some or all initial paths, freeing capacity to attend to other tasks within the game, such as monitoring the environment. Thus, a task node, indicated by the first decision node in the human swim lane, is used to simulate a human's decision to either initiate ship selection or monitor the environment. Simultaneously, the agent is selecting ships and drawing trajectories as shown in the right swim lane of Figure 13.3. The agent is constantly monitoring all ships on screen and drawing a route for any ship that does not have a route once the trigger delay time has passed. As a path is drawn, the environment is updated.

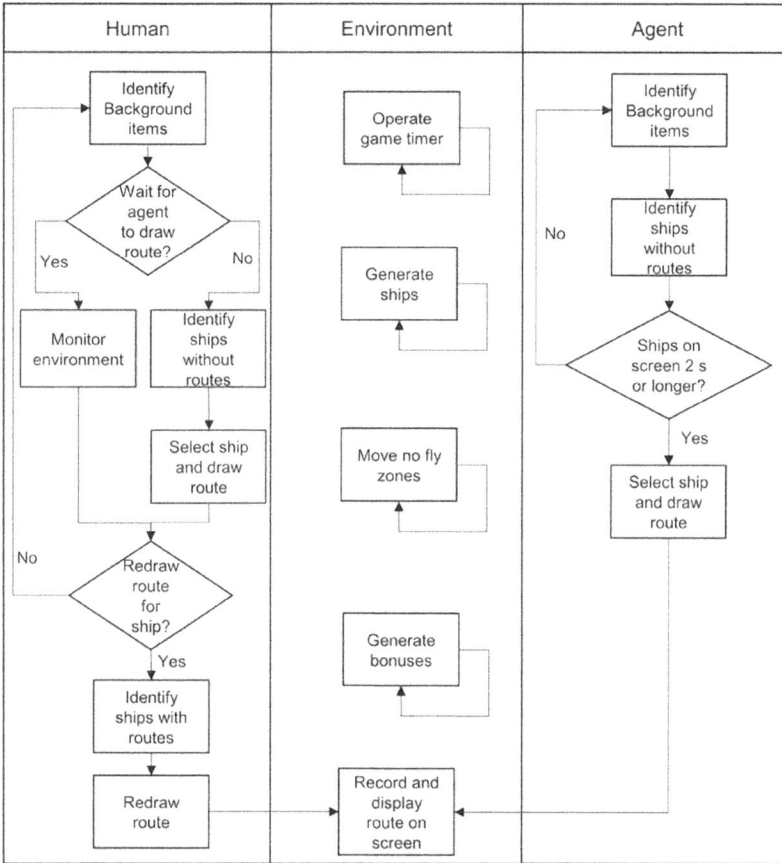

FIGURE 13.3 Activity diagram representing the actors and actions in the static model. Vertical swim lanes are used to designate actions performed by the specified actor.

13.2.3 BASELINE HITL EXPERIMENT

The baseline HITL experiment included a group of 36 volunteers to play the game with and without the aid of the autonomous agent under relatively high workload conditions (i.e., event rate of 2.0 s). In this experiment, the agent responded slightly faster than the average human response. The average intra-trajectory draw time for the average human, without the agent's assistance, was 2.6 s. This experiment has been discussed previously (Goodman, Miller, Rusnock, & Bindewald, 2016). As this experiment is not the focus of the current research, the reader is referred to this reference for further details. Data from this experiment provided dynamic model task time distributions and other behavioral and performance data for the human both in the presence and in the absence of the agent. Additionally, the baseline experiment provided insight into the reliance of the human operator upon the agent for a single task time, as shown in Figure 13.4. The probability the human permitted the agent to draw a trajectory varied as a function of the number of ships on screen. The

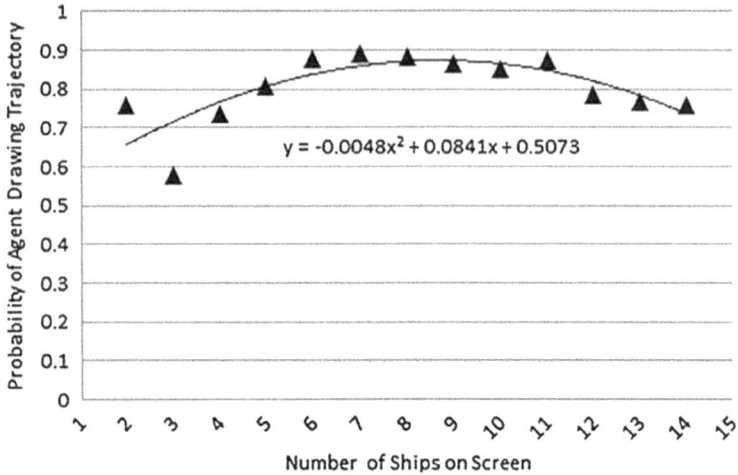

FIGURE 13.4   Graph displaying the probability of the agent drawing a route as a function of the number of ships on the screen. The third-order regression line, with equation, was used in calculating the reliance algorithm in the IMPRINT model.

human drew a higher percentage of trajectories when there were few ships on screen, likely corresponding to lower workload conditions. The human also drew a higher percentage of trajectories when a larger number of ships were on screen, likely corresponding to conditions where the agent was unable to draw trajectories that would not result in collisions.

13.2.4   DYNAMIC MODEL DEVELOPMENT

To model timing in the context of a human-agent team, the Improved Performance Research Integrated Tool (IMPRINT), a discrete-event simulation environment was applied (Mitchell, 2000). The system was modeled through a task network, a high-level depiction of which is shown in Figure 13.3. Human workload is assessed in terms of visual, auditory, cognitive and psychomotor (VACP) difficulty for each task (Bierbaum, Szabo, & Aldrich, 1989; Mitchell, 2000). Task times were drawn from distributions of performance times. The frequency of the tasks, as well as the time necessary to perform each task, results from a stochastic process. The stochastic process permits the modeler to represent the variability within the system. Various system allocations can then be modeled by allocating specific tasks to the human or agent.

This model employed the relationship between the number of ships on screen and the human's reliance on the agent, shown in Figure 13.4, to represent the likelihood a human will decide to draw a trajectory. Humans can draw a route for a ship that does not have a trajectory, or they can "redraw" a trajectory for a ship that has an existing, but presumably poor, trajectory. Following the draw route node, it is assumed that the human focuses again on determining a ship to route. When the simulated human draws a trajectory, the simulated environment is updated and the agent is precluded from drawing a trajectory for the corresponding ship.

After the simulated human or agent has designated a trajectory for a ship, a new entity is created in the simulation model representing the ship traversing its path. The model assumes a ship continues along its trajectory for a length of time drawn from a time distribution representing time-on-screen and is removed from the simulation after the time elapses. There are three possible end results for a ship: collision, destination reached, and off-screen traversal. Ships arrive at these end results according to probabilities associated with the number of ships on screen and the operator (human or agent) that drew the route. This model was validated against gameplay data from the baseline experiment for an agent delay time of 2.0 s.

13.2.5 SIMULATION EXPERIMENT

Considering that the simulation was built upon an experiment with a fixed agent delay time, it was necessary to include assumptions within the model regarding the changes in human behavior as a result of variable agent delay times. To incorporate this parameter, it was assumed that if the agent drew the line as soon as the ship appeared, the human would never have time to initiate a route. However, in the case that the agent requires an infinite amount of time before drawing the route, the human cannot rely upon the agent to draw any route. As noted earlier, based upon the baseline experiment the humans' average ship selection cycle (i.e., the time between initiating routes for separate ships) was determined to be 2.6 s. Although this time was not normally distributed as it was truncated near zero, variability was estimated by calculating the standard deviation, which was near 3 s. Using three standard deviations above the mean of 2.6 s, it was assumed that the agent would not be able to initiate a route at a delay time of 11.6 s because the human would have already drawn the initial route. Assuming the probability of a human drawing a route when the agent delay time was zero and that the probability of a human drawing a route when the agent delay time was 11.6 s was the same as it would be without an agent, a linear equation relating delay time to probability of agent draws was created. This linear model was used to shift the regression line, shown in Figure 13.4, vertically as a function of agent delay time. For every 1.0 s change in agent delay time, the baseline probability value of agent draws was incremented or decremented by the slope of the regression line (i.e., 0.1058), within the bounds that the probability must be between 0 and 1.

A series of simulations were conducted, altering the agent delay time in each simulation. Six conditions were evaluated for the agent delay time, including the mean time for a participant to select a ship (2.6 s), plus one, two, and three standard deviations (i.e., 5.6, 8.6, and 11.6 s), as well as an agent delay time near zero (e.g., 0.1 s) and the 2.0 s delay from the baseline experiment. The six scenarios were each simulated 100 times, each scenario applying the same random seed sequence. The average scores, VACP workload, and trajectories drawn were calculated for each scenario.

13.2.6 VALIDATION HITL EXPERIMENT

A HITL study was designed to test the hypothesis that agent delay time is related to human behavior and system performance.

13.2.6.1 Participants

The experiment involved 20 volunteer participants (4 female, 16 male) with an average age of 26.5 years (range of 21–38). The participants reported average use of tablet computers between 1 and 2 times a week and gaming consoles between 1 and 3 times per month. The participants were screened for color vision deficiencies using the Ishihara 14 plate color test (Ishihara, 1962).

13.2.6.2 Experimental Design and Procedure

A within-subjects design was applied in which each participant completed two phases of training and 15, 5-min experimental trials. The first phase of training consisted of two or more fully manual (no automated agent) trials. Participants were given the opportunity to play the game fully manually as many times as they wished, at home or in the laboratory, but were required to complete at least two 5 min, in laboratory sessions before starting the second phase of training. The second phase was used to familiarize the participant with the automated agent. This phase was conducted in the laboratory prior to the experiment and contained two 5-min trials with agent delay times of 2.0 and 6.0 s. Each participant then completed five experimental blocks, each containing three, 5-min trials at one of the delay times. The participant was informed of the delay time before beginning the trial and the delay time was consistent throughout each block. The delay times of 0.1, 2.6, 5.6, 8.6, and 11.6 s were assigned to participants through a Graeco-Latin Square Design. Ships were spawned on screen at a fixed rate of one ship every 2.0 s. Ships began movement along a random trajectory immediately after appearing on screen. A new Bonus appeared every 30 s and no-fly zones moved to a new random location every 60 s. Score and behavioral metrics were collected for each five-minute trial but NASA-TLX values were collected only at the end of the third 5-min trial within each experimental block.

13.2.6.3 Apparatus

Participants completed the experiment in the confines of a laboratory and were permitted breaks between blocks. Player data collection was performed on one of two Microsoft Surface Pro 3 tablet computers running the Windows 8 operating system. Workload information was collected through NASA-TLX (Hart & Staveland, 1988), with workload values on each scale collected using a paper questionnaire after each experimental block. Space navigator recorded the time of each event, including the presentation of each ship, the presentation or movement of each no-fly zone or bonus, the selection of a ship for trajectory generation, trajectory completion, as well as ship arrivals, collisions, bonus pickups, and no-fly zone entries and exits. The game also recorded the score at the end of each trial.

13.2.7 Simulation Extension Experiment

Agent delay time was varied in the first simulation experiment and the verification HITL experiments. However, based on earlier research (Rouse, 1977), this event rate (e.g., the rate at which ships appeared within the space navigator environment) was also expected to affect human and team performance. Therefore, an additional simulation experiment was conducted to explore the interaction of agent delay time and event rate. The first simulation experiment and HITL experiments utilized a high event rate with a new ship appearing every 2.0 s, faster than the average human response time. The simulation extension experiment, applied the IMPRINT model to explore the preferred agent delay time at slower event rates of 3.0 and 4.0 s by increasing the event rate and simulating a range of agent delay times to identify the agent delay which provided the highest score.

13.2.8 DATA ANALYSIS

The VACP workload values from the simulation experiment were calculated by computing a time-weighted average from the 100 simulation runs for each delay time condition. Average NASA-TLX values from the validation HITL experiment were calculated across the individual NASA-TLX scores for each scale for each delay time condition. The resulting workload data from the two measures were normalized to a standard normal distribution, to permit comparison of the relative modeled objective workload and experimentally-reported subjective workload scores. The raw scores for the remaining dependent variables, including score, the number of trajectories drawn by the participants (e.g., human draws), the number of trajectories drawn by the agent (e.g., agent draws) the number of trajectories redrawn by the human (e.g., redraws), were subjected to analysis. Although the number of redraws was analyzed, this number was further divided into the number of trajectories drawn by the human and then redrawn by the human (e.g., human redraws) and the number of trajectories drawn by the agent and redrawn by the human (e.g., agent redraws) for the HITL experiment. Primary data analysis was performed using Analysis of Variance (ANOVA). Greenhouse-Geisser correction was applied to the repeated measures ANOVA results when the experimental results did not meet the assumption of sphericity as indicated by Mauchley's test. Tukey pairwise comparisons were applied to compare condition means.

13.3 RESULTS

Both the simulation and HITL experiment results indicated an inverse relationship between performance and workload as a function of agent delay time, as shown in Figure 13.5. A single factor ANOVA indicated that score decreased with increasing agent delay time for the simulation, $F(5,594) = 43.78$, MSE $= 832,236$, $p < 0.001$, $\eta_p^2 = 0.261$. A two-factor, repeated-measures ANOVA indicated a similar effect for the HITL experiment, $F(4,64) = 11.92$, MSE $= 1,493,073$, $p < 0.001$, $\eta_p^2 = 0.581$. Tukey pairwise comparisons for the simulation indicated that there were four significantly different groups of scores; the groups including the 0.1 and 2.0 s delay times, the 2.0 and 2.6 s delay time, the 2.6 and 5.6 s delay times and the 8.6 and 11.6 s delay times. For the HITL experiment, the score was significantly greater for the 0.1 and 2.6 s agent delay times than for the 5.6, 8.6, and 11.6 s agent delay times. Mauchly's test indicated that the assumption of sphericity was violated for the delay by trial interaction ($X^2(9) = 58.09$, $p < 0.012$), therefore the degrees of freedom were corrected using Greenhouse-Geisser estimates of sphericity ($\varepsilon = 0.125$). The

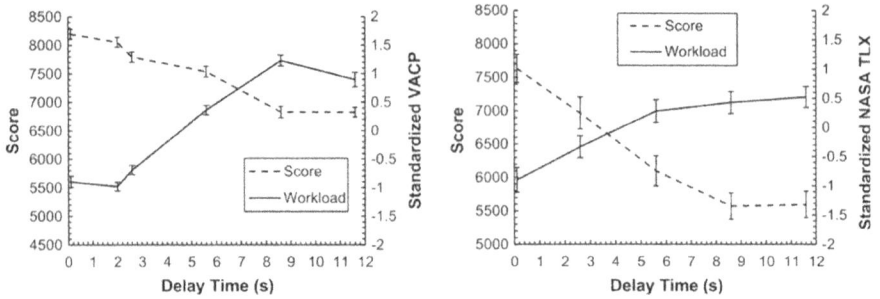

FIGURE 13.5 Graphs depicting score and standardized workload values as a function of agent delay time for the simulation (a) and HITL experiment (b). Error bars indicate standard error.

two-factor, repeated-measures ANOVA results indicated that score was not affected by the trial order but was affected by the interaction of delay time and trial order, $F(3.94,63) = 10.73$, MSE $= 3,213,122$, $p < 0.0001$, $\eta_p^2 = 1.000$. Further analysis of this effect indicated that the score was higher for the first trial than the following two trials when the delay time was 5.6 s or less but the score was lower for the first trial than the following trials when the delay time was 11.6 s.

Workload increased as a function of agent delay time for the simulation as indicated by a single factor ANOVA $F(5,142) = 104.7$, MSE $= 0.22$, $p < 0.0001$, $\eta_p^2 = 0.758$. Mauchly's test indicated that the assumption of sphericity was violated for workload in the HITL experiment ($X^2(9) = 37.07$, $p < 0.001$), therefore degrees of freedom were corrected using the Greenhouse-Geisser estimates of sphericity ($\varepsilon = 0.621$). The single factor, repeated measures ANOVA indicated that workload was significant as a function of delay time in the HITL experiment, $F(2.5,47.2) = 17.288$, MSE $= 0.678$, $p < 0.0001$, $\eta_p^2 = 0.476$. Workload is shown as a function of agent delay time in Figure 13.5. As shown workload for the simulation and HITL experiment, increases during the intermediate delay time conditions, but asymptotes for longer agent delay times. The simulation results, however, indicate a floor in workload for agent delay times between 0.1 and 2 s, which is not indicated by the HITL experimental results. This trend is supported by the post-hoc tests as three groups of means were observed for the simulation data, including a first group for delay times equal to or less than 2.6 s, a second group for delay times of 5.6 s, and a third group for delay times equal to or greater than 8.6 s while the means for the HITL experimental increased for each change in agent delay time up to the 5.6 s but did not statistically increase for longer delay times. While the general trends are similar, the transition times between asymptotic and increasing workload values differ between the model and experimental results. The results overall are as expected. Workload does not increase for short delay times, where the agent reacts faster than the human is capable; workload increases for agent delay times where the human and agent each contribute to initial trajectories; and workload asymptotes for lengthy delay times, where the agent does not contribute meaningfully to the team.

The number of agent draws decreased with increases in agent delay time for the simulation, $F(5,594) = 35,784$, MSE $= 2,679,998$, $p < 0.001$, $\eta_p^2 = 0.998$, as indicated

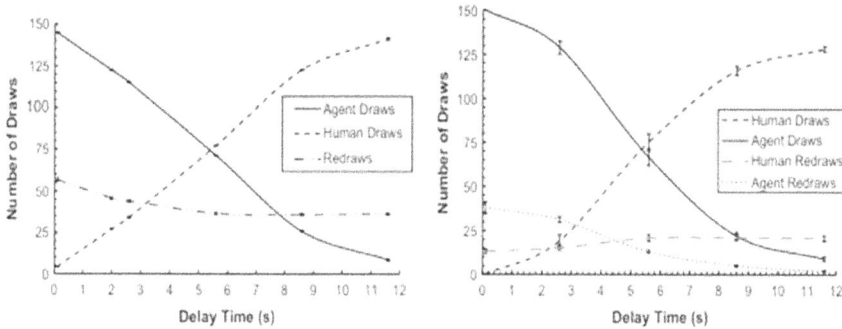

FIGURE 13.6 Graphs depicting agent draws, as well as human draws and redraws, as a function of agent delay time for the simulation (a) and HITL experiment (b). Error bars indicate standard error.

by a single factor ANOVA and for the HITL experiment $F(4,64)=205$, $MSE=968$, $p<0.001$, $\eta_p^2=0.928$ as indicated by a two-factor, repeated-measures ANOVA. The number of human redraws also decreased with increases in agent delay time for the simulation, $F(5,594)=174$, $MSE=39.4$, $p<0.001$, $\eta_p^2=0.625$, and the HITL experiment, $F(4,64)=7.63$, $MSE=50.1$, $p<0.001$, $\eta_p^2=0.323$. The number of human draws increased with increases in agent delay time for both the simulation, $F(5,594)=31,975$, $MSE=10$, $p<0.001$, $\eta_p^2=0.998$, and the HITL experiment, $F(4,64)=168.3$, $MSE=977$, $p<0.001$, $\eta_p^2=0.913$. The trend for each of these relationships is shown in Figure 13.6. The effects of trial and the interaction of delay time and trial on the number of agent draws, human redraws, and human draws were not statistically significant in the HITL experiment.

Post-hoc analysis for the simulation indicated that the number of agent and human draws was significantly different between each pair of agent delay times. These results differed slightly from the HITL experiment where the human draws were not statistically different for the two longest agent delay times. Therefore, when the agent creates routes at approximately the same speed or faster than the human (e.g., has a delay time less than the average human draw time of 2.6 s), the human only initiates routes between 2% and 20% of the time. When the agent delay time is one to three standard deviations slower than the average time required for a human to draw a trajectory, the number of human-initiated routes increases to between 50% and 95% of all routes drawn. Furthermore, the model correctly anticipated that the largest shift in performance would occur when the trigger time was adjusted from 5.6 to 8.6 s. This led to the score decreasing by 10%, as neither the human nor the agent was able to assign high-quality trajectories to all ships in a timely fashion.

Notice that the human never drew initial routes for ships at the 0.1 s delay time. However, a sequence could occur where the agent drew the first route for a ship followed by the human drawing a second and third route for the same ship. This sequence would be counted as 1 Agent Draw, 1 Agent Redraw, and 1 Human redraw as the agent drew the initial route, the human redrew over an agent's route, and the human redrew over their own route. Therefore, the experimental results in Figure 13.6

show the effect of not only agent redraws, but human redraws. The effect of agent delay time on simulated human redraws generated three different groupings; a delay time of 0.1 s produced the most redraws, delay times of 2.0 and 2.6 s produced fewer redraws and delay times of 5.6, 8.6, and 11.6 s produced the fewest redraws. The HITL experiment results created two groupings with 0.1 and 2.6 s agent delay times resulting in statistically more human redraws than the 5.6, 8.6, and 11.6 s agent delay times. Post-hoc analysis indicated the presence of two groupings with the number of agents redraws greater when the agent delay time was 0.1 or 2.6 s than for longer agent delay times. Therefore, as the agent draws more routes, the human devotes more effort to redrawing trajectories originally formed by the agent. The data shows a relatively steady decrease in agent redraws, whereas the human redraws stay relatively the same as the delay time increases.

Although the performance from the simulation and average performance and behavior in the HITL experiment were similar, the participants did not all behave similarly. Clustering participants by the total number of initial trajectories drawn indicated a subgroup of eight of the participants (indicated as G2) who were more aggressive in drawing trajectories than their counterparts (indicated as G1). A three-factor ANOVA including group, delay time and the interaction of these factors indicated that the effect of group was significantly significant $F(1,265) = 194.1$, $MSE = 263$, $p < 0.0001$, $\eta_p^2 = 1.00$ and the interaction of agent delay time with group was statistically significant, $F(4,265) = 30.2$, $MSE = 263$, $p < 0.0001$, $\eta_p^2 = 1.00$.

The number of initial draws performed by G2 was higher than performed by G1 for all but the 0.1 s delay time, as shown in Figure 13.7. On average the eight individuals in G2 drew 30 more trajectories for the 2.6 s agent delay time and 70 more trajectories for the 5.6 s agent delay time condition than individuals in G1, indicating their preference for drawing initial trajectories, as opposed to acquiescing to this function to the agent. Perhaps surprisingly, neither the effect of group, $F(1,265) = 0.02$, $MSE = 2,810,132$, $p < 0.892$, $\eta_p^2 = 0.052$, or the interaction of group with agent delay,

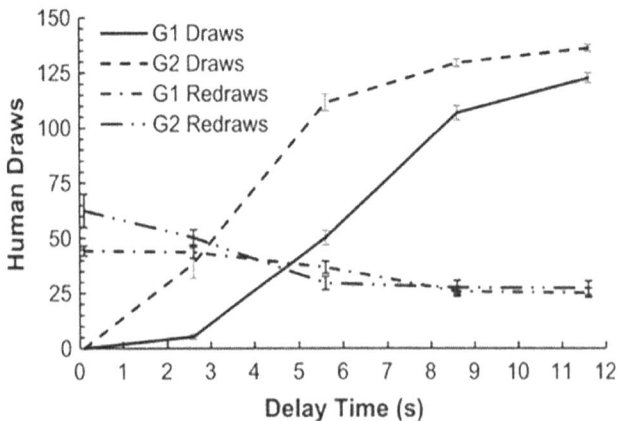

FIGURE 13.7   Human draws and total redraws as a function of delay time for each participant subgroup. Error bars indicate standard error.

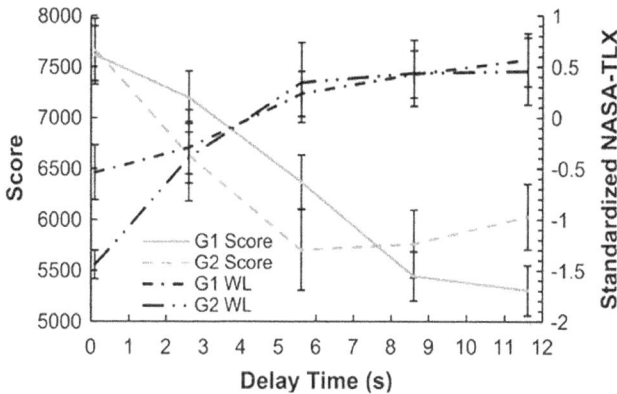

FIGURE 13.8 Score and workload as a function of delay time for each participant subgroup. Error bars indicate standard error.

$F(4,265) = 1.75$, $p < 0.973$, $\eta_p^2 = 0.177$, had a significant effect on participants' score, as shown in Figure 13.8. While the score for G1 increased as a function of the inverse of delay time, the score for G2 does not increase for delay times less than 5.6 s. Therefore, it would appear that G2 does not benefit from the presence of the agent in the same way as G1. The number of redraws and workload are also shown in Figures 13.7 and 13.8 respectively. Each of these variables was statistically different by the group at the 0.1 s delay time, but not at any other delay time, with G1 redrawing more frequently and reporting a lower workload at the 0.1 s agent delay time than G2.

13.4 DISCUSSION

As hypothesized, the timing of the agent had a significant impact on both the performance of the human-agent team and the behavior of the human actors within this team. Therefore, it would appear that the timing of an agent can be manipulated to modulate human engagement within a human-agent team. Thus, the role assumed by the human team member is influenced by the timing of the autonomous agent. As illustrated by the results at the 0.1 and 2.6 s delay times, the human may assume a supervisory role if the time required for an agent to complete the primary task is equal to or faster than the average rate of human performance.

However, if the agent requires more time than the human to initiate an action, humans are more likely to dedicate significant resources to completing what is now a primary task to maximize team performance. This supposition was illustrated by the fact that the participants drew the majority of trajectories when the agent's delay time was 5.6 s or longer. This result is supported by both the simulation and experimental results and is in general agreement with results obtained from earlier queuing theory-based models for decision systems (Rouse, 1977).

This finding is particularly significant in adaptive automation systems. Traditionally, these systems modulate human task load to affect human perception of workload or engagement by changing the tasks that are automated (Feigh et al.,

2012; Scerbo, 1996). This method of modulating the human tasks, however, adds complexity to the interface as it becomes necessary to communicate the change in task responsibility. The current research illustrates that modulating the timing of an agent for a significant task within the interface modulates task load, affecting human perceived workload. This change in the behavior of the agent may provide an alternate, continuous, and perhaps more natural adaptation within an automation system.

All human participants in the experiment were not equally likely to adapt their roles based on the timing of the agent. Rather, a subgroup of eight of the twenty participants was much more likely to create initial trajectories than the remaining participants, particularly at agent delay times near the average human response time as determined from the baseline HITL experiment. This proclivity for assuming the role that the agent could fulfill did not have a statistical effect on the performance given the relatively short duration of the experimental task. However, it is unclear if this behavior would continue or result in a relative difference in performance in a longer duration task.

It should be noted that in the baseline and validation HITL experiments, unlike much of the previous research and many applications (Molloy & Parasuraman, 1996; Riley, 1997), the supervisory role was not a vigilance task. Instead, the human was actively redrawing significant numbers of the agent's routes, on average redrawing eight trajectories per minute. However, this redraw rate was not significantly higher than the redraw rate for other conditions. Therefore, the participants perceived a lower workload as the agent delay time decreased, and the participants had more time to dedicate to updating their situation awareness and redrawing routes. This change apparently resulted in improved decisions as the overall team score improved by nearly 40% as the agent delay time decreased. Thus, improved team performance could be attributed to the combination of the agent having superior performance and speed than the human. Experiments with the agent performing alone indicate that the average score increased from 6,180 to 2,092 to 3,638 as the agent delay time decreases from 11.6 to 2.6 to 0.1 s. As these scores are substantially lower than the team performance, the increase in team score over the human's score is not solely due to the faster response of the agent alone. However, because the average score for the agent acting alone increases as the agent's delay time decreases, improving the timeliness of the agent can have a positive effect on performance within the current task environment.

The expectation of human-performance simulations, such as the IMPRINT simulations applied in this research, is to provide outputs that are general estimates of human and agent performance within the context of the application. For this research, IMPRINT represented the general effect of variable delay time on performance, behavior, and workload. The redraws and score were not as accurate for human and agent draws, as the simulation consistently projected higher values than observed in the validation HITL experiments for redraws and score. Nonetheless, the simulation conveyed similar results and trends as the validation HITL experiment. As the simulations can predict score within the environment, this model can be applied to optimize score across ship spawn rates to determine the optimum score as a function of various spawn rate and delay time combinations. Extending the use of the simulation model in this way indicates that while an agent delay time of 0.1 s provides the highest score at a 2.0 s spawn rate, longer agent delay times are desired for slower ship spawn rates. Specifically, the model indicates that when a new ship is spawned every 3.0 s, agent delay time has a smaller effect on score than when a ship is spawned every 2.0 s. However, the 2.6 s agent delay time results in the highest score. When the time between ship spawns is increased to 4.0 s, an agent delay time of 8.6 s results in the highest score. While the direction of these results is consistent with altering the agent delay time to maintain human engagement as spawn rate (i.e., task load imposed upon the team) decreases, the model does not account for changes in human motivation that may occur under less demanding conditions. Therefore, the development of recommendations for agent delay time or other logic to support agent timing will likely require additional research and are likely to be task-dependent. For instance, the simulation and HITL experiment indicated that the human and agent contributed nearly an equal number of paths when the agent delay time was in the 4–5 s range in this game with a ship spawn rate of 2 s. This time appears to provide the human enough time to perform what they believe to be a satisfactory scan to determine impending collisions, redraw routes to avoid collisions, as well as determine and draw routes for half of the ships under this particular condition. These results will vary as the subtasks and the subtask difficulty varies. For example, fewer ships need to be scanned and evaluated as the number of ships on screen decreases. Reducing the number of ships on screen will also reduce the number of redraws required as the probability of collision is reduced. Therefore, this one change to the task environment provides multiple impacts on the time required for the human to perform the task and more dramatic changes in the task environment are likely to have a larger impact on the desired timing of the agent.

13.5 CONCLUSION

This research indicates that as human-agent teaming conceptually matures, timing of the agent needs to be thoroughly evaluated, as it significantly influences human-agent team performance when the human and the agent are interdependent and share the goals. Additionally, the timing of the agent significantly influences the role assumed by the human within the team. Specifically, the results indicate that in an environment in which human performance is overly constrained by the available time to successfully complete all tasks, the employment of even a very simple agent, which responds faster than the human, can have a significant, positive influence on team performance. Experimental results indicate that some, but not all, individuals readily change their behavior in response to the timeliness of the agent. While clear performance differences between the groups of individuals who readily adapt their behavior to the agents' timing and those who do not were not illustrated, the general trends in score indicate that team performance increased as a function of decreasing delay time for the individuals who more readily adapted their behavior based upon changes in agent timing (e.g., G2). This trend was not consistent for the less adaptive group across all agent delay times. Although it is difficult to generalize these results to less time-constrained environments, modeling results indicate that it will likely be appropriate to slow the artificial agent's timing in less demanding environments, permitting the human to assume additional responsibility within the team. Developing a further understanding of the role of agent timing within dynamic environments may prove advantageous for the design of future human-agent teams.

ACKNOWLEDGMENT AND DISCLOSURE

The authors gratefully acknowledge the financial support of the Air Force Office of Scientific Research, Computational Cognition and Machine Intelligence Program. Additionally, we extend our thanks to Jayson Boubin for his contributions to model development and Dr. Gilbert Peterson for assisting agent development.

The views expressed in this document are those of the authors and do not reflect the official policy or position of the United States Air Force, the United States Department of Defense, or the United States Government.

REFERENCES

Arthur, W., Edwards, B. D., Bell, S. T., Villado, A. J., Station, C., & Bennett, W. (2005). Team task analysis: Identifying tasks and jobs that are team based. *Human Factors*, 47(3), 654–669.

Bice, G. W., Skoog, M. A., & Howard, J. D. (1990). Aircraft ground collision avoidance and autorecovery systems device. *United States: United States Patent*, 4(924), 401.

Bierbaum, C. R., Szabo, S. M., & Aldrich, T. B. (1989). Task analysis of the UH-60 mission and decision rules for developing; a UH-60 workload prediction model. Volume I: Summary report (AD-A210 763). Alexandria, VA.

Bindewald, J. M., Miller, M. E., & Peterson, G. L. (2014). A function-to-task process model for adaptive automation system design. *Journal of Human Computer Studies*, 72(12), 822–834. http://dx.doi.org/10.1016/ j.ijhcs.2014.07.004.

Brookings, J. B., Wilson, G. F., & Swain, C. R. (1996). Psychophysiological responses to changes in workload during simulated air traffic control. *Biological Psychology*, 42(3), 361–377. http://dx.doi.org/ 10.1016/0301–0511(95)05167-8.

Bruemmer, D. J., Marble, J. L., & Dudenhoeffer, D. D. (2002). Mutual initiative in human-machine teams. In *Proceedings of the IEEE 7th Conference on Human Factors and Power Plants* (pp. 22–30). http://dx. doi.org/10.1109/HFPP.2002.1042863.

Dixon, S. R., & Wickens, C. D. (2003). Control of multiple-UAVs: A workload analysis. In *12th International Symposium on Aviation Psychology*, Dayton, OH.

Feigh, K. M., Dorneich, M. C., & Hayes, C. C. (2012). Toward a characterization of adaptive systems: A framework for researchers and system designers. *Human Factors*, 54(6), 1008–1024. http://dx.doi.org/ 10.1177/0018720812443983.

Fitts, P. M., & Peterson, J. R. (1964). The information capacity of the human motor system in controlling the amplitude of movement. *Journal of Experimental Psychology*, 67(2), 103–112. http://dx.doi.org/ 10.1037/h0045689.

Goodman, T., Miller, M. E., & Rusnock, C. F. (2015). Incorporating automation: Using modeling and simulation to enable task reallocation. In *Proceedings of the 2015 Winter Simulation Conference*, Huntington Beach, CA.

Goodman, T., Miller, M. E., Rusnock, C. F., & Bindewald, J. (2016). Timing within human-agent interaction and its effects on team performance and human behavior. In *CogSIMA 2016* (pp. 35–41). San Diego, CA: IEEE.

Hancock, P. A., & Chignell, M. H. (1988). Mental workload dynamics in adaptive interface design. *IEEE Transactions on Systems, Man and Cybernetics*, 18(4), 647–658. http://dx.doi.org/10.1109/21.17382.

Hart, S. G., & Staveland, L. E. (1988). Development of NASA-TLX (task load index): Results of empirical and theoretical research. *Advances in Psychology*, 52(C), 139–183.

Hilburn, B., Molloy, R., Wong, D., & Parasuraman, R. (1993). Operator versus computer control of adaptive automation. In *International Symposium on Aviation Psychology*, Columbus, OH (pp. 161–166).

Hoc, J.-M. (2001). Towards a cognitive approach to human–machine cooperation in dynamic situations. *International Journal of Human Computer Studies*, 54(4), 509–540. http://dx.doi.org/10.1006/ ijhc.2000.0454.

Hyman, R. (1953). Stimulus information as a determinant of reaction time. *Journal of Experimental Psychology*, 45(3), 188–196. http://dx. doi.org/10.1037/h0056940.

Ishihara, S. (1962). *Ishihara Pseudo Isochromatic Test Chart Book*. Tokyo Japan: Paragon.

Licklider, J. C. R. (1960). Man-computer symbiosis. *IRE Transactions on Human Factors in Electronics*, HFE-1(1), 4–11. http://dx.doi.org/ 10.1109/THFE2.1960.4503259.

Miller, M. E., Lourette, R. W., Fellegara, P. C., Bussi, C. A., Tekek, M. J., Hunter, M. E., & Kerr, D. R. (2001). User interface for electronic image viewing apparatus. *United States: United States Patent*, 6(310), 648.

Mitchell, D. K. (2000). *Mental Workload and ARL Workload Modeling Tools*. Aberdeen Proving Ground, MD.

Molloy, R., & Parasuraman, R. (1996). Monitoring an automated system for a single failure: Vigilance and task complexity effects. *Human Factors*, 38(2), 311–322. http://dx.doi.org/10.1518/001872096779048093.

Parasuraman, R., & Riley, V. (1997). Humans and automation: Use, misuse, disuse, abuse. *Human Factors*, 39(2), 230–253.

Parasuraman, R., Sheridan, T. B., & Wickens, C. D. (2000). A model for types and levels of human interaction with automation. *IEEE Transactions on Systems, Man, and Cybernetics. Part A, Systems and Humans: A Publication of the IEEE Systems, Man, and Cybernetics Society,* 30(3), 286–297. http://dx.doi.org/10.1109/3468.844354.

Riley, V. (1997). Operator reliance on automation: Theory and data. In R. Parasuraman & M. Mouloua (Eds.), *Automation and human performance: Theory and applications* (pp. 19–35). Boca Raton, FL: Lawrence Erlbaum Associates, Inc.

Rouse, W. B. (1977). Human-computer interaction in multitask situations. *IEEE Transactions on Systems, Man & Cybernetics*, 7(5), 384–392. http://dx.doi.org/10.1109/ TSMC.1977.4309727.

Rouse, W. B. (1981). Human-computer interaction in the control of dynamic systems. *ACM Computing Surveys*, 13(1), 71–99. http://dx. doi.org/10.1145/356835.356839.

Rouse, W. B., Edwards, S. L., & Hammer, J. M. (1993). Modeling the dynamics of mental workload and human performance in complex systems. *IEEE Transactions on Systems, Man, and Cybernetics*, 23(6). http://dx.doi.org/10.1109/21.257761.

Rump, S., Steiner, M., & Douglas, B. (1996). Method for controlling the triggering sensitivity of a vehicle automatic braking process to match driver behavior. *United States: United States Patent*, 5(535), 123.

Scerbo, M. W. (1996). Theoretical perspectives on adaptive automation. In R. Parasuramn & M. Mouloua (Eds.), *Automation and Human Performance: Theory and Applications* (pp. 37–63). Mahwah, NJ: Lawrence Earlbaum Associates.

Smeaton, J., & De Moura, F. R. S. (1751). An engine for raising water by fire: Being an improvement on Savery's construction, to render it capable of working itself. *Philosophical Transactions of the Royal Society A*, 64, 445–460. http://dx.doi.org/10.1098/rstl.1763.0053.

Warm, J. S., Parasuraman, R., & Matthews, G. (2008). Vigilance requires hard mental work and is stressful. *Human Factors,* 50(3), 433–441. http://dx.doi.org/10.1518/001872008X312152.

Wiener, E. L., & Curry, R. E. (1980). Flight-deck automation: Promises and problems. *Ergonomics*, 23(10), 995–1011. http://dx.doi.org/10.1017/CBO9781107415324.004.

14 Spatialized Audio Improves Call Sign Recognition during Multi-Aircraft Control[*]

Sungbin Kim, Michael E. Miller,
Christina F. Rusnock, and John J. Elshaw
US Air Force Institute of Technology

CONTENTS

[*] Reprinted with permission: Kim, S., Miller, M. E., Rusnock, C. F., & Elshaw, J. J. (2018). Spatialized audio improves call sign recognition during multi-aircraft control. *Applied Ergonomics*, *70*, 51–58.

14.1 INTRODUCTION

Many domains, including military, aviation, and emergency response, rely upon the transmission of large volumes of auditory communication to permit coordination and interaction among teams of individuals. This communication often involves the transmission of information critical to the performance of a subset of the individuals, as well as noncritical information, which does not directly affect the performance of these individuals. This noncritical information can provide important contextual information and improve situation awareness, thus elimination of this noncritical information is not desirable. In these domains, individuals often utilize multiple radio channels and communication aids (such as operator call signs) to filter critical information from noncritical information.

This research is motivated by a potential application in Unmanned Aerial Systems (UASs). Current military UASs require two or more people to operate a single Unmanned Aerial Vehicle (UAV) (Calhoun and Draper, 2015). However, a concept referred to as Multi-Aircraft Control (MAC) has been proposed, allowing one UAS operator to control multiple UAVs, (Franke et al., 2005). Operation of military UAVs can be divided into five phases: launch, transit to mission area, mission, transit to base, and recovery. Previous research has shown relatively low workload during the transit phases, making these phases amenable to MAC (Colombi et al., 2012). Aside from manpower reductions, such a concept is appealing because it could alleviate mission operators from the tedious responsibility of long transits, potentially improving operator effectiveness during critical mission segments. However, a number of human factors concerns regarding MAC have yet to be addressed.

The current research recognizes that the UAS operator is exposed to a large amount of auditory information. In the MAC concept, the amount of information one operator must process will increase proportionally to the number of vehicles. Additionally, transit operators would transition aircraft, each with a unique call sign, to and from locations where other operators assume control. Therefore, the call signs of the aircraft the transit operator controls will be ever changing. It has been hypothesized that due to the ever changing call signs controlled by the transit operators, accuracy and response time will be degraded as the number of call signs increases (Amaddio et al., 2015). We concur with this hypothesis.

Significant research has been conducted to understand the utility of spatialized auditory displays in domains where operators must monitor multiple radio channels simultaneously (Haas et al., 1997; MacDonald et al., 2002; McAnally and Martin, 2007; McAnally et al., 2002). This research into auditory displays employing multichannel radios illustrated that spatialized audio aids user comprehension of simultaneously spoken messages when these messages are presented at multiple locations on the azimuthal plane. However, in this system implementation, the operators were expected to listen to all radio channels, even while these channels were simultaneously presenting information, to determine if any of the messages contain their call sign. Changes in speaker spectral composition or intensity also aid comprehension of simultaneous speakers from multichannel radios (Brungart et al., 2002). Unfortunately, simple manipulations, such as sound intensity, also reduce the operator's ability to comprehend information from the quieter radio channel.

Spatialized auditory displays have also been used to encode spatial information, such as direction or distance (Cengarle, 2012; Haas and Stachowiak, 2007; Maza et al., 2009; Philbrick and Colton, 2014; Simpson et al., 2005; Trouvin and Schlick, 2004). In these applications, the user's visual system is often cued with sound to reduce the time necessary for visual search.

The target systems for the current research will employ multichannel radios and could employ spatialized audio to aid the separation of information on separate channels as employed by Haas et al. (1997) and others. Alternately, the UAV's spatial location with respect to the operator or operator's area of control could be encoded with location in a spatialized auditory display, similar to the systems evaluated by Cengarle (2012) and others. However, neither of these manipulations is likely to aid the operator in remembering, recognizing, or responding to multiple, critical UAV call signs. The current research sought to utilize a spatialized auditory display system in a different way. Specifically, the system was assumed to employ a voice recognition system capable of automatically recognizing call signs on each of the multiple radio channels and to determine whether each call sign was critical, i.e., corresponded to a UAV under the control of the operator. As shown in the left side of Figure 14.1, the voice recognition system then parsed critical from noncritical radio calls and encoded relevance through the spatial location.

The objective of this research was to investigate whether an operator would benefit from a system that automatically classifies call signs based on relevance and then encodes relevance using spatial location. Specifically, the system presented call signs the operator is responsible for monitoring (critical information) in a different spatial location than call signs of aircraft controlled by other operators (noncritical information). We hypothesize that operators will rely upon the spatial location cue to simplify the task structure necessary to distinguish critical from noncritical information. This simplification is expected to decrease task times and operator workload compared to the traditional system, as shown in Figure 14.1. Specifically, we hypothesize that the operator will rely upon the spatial information rather than their memory or memory aids to determine the relevance of auditory information. This benefit is expected to be particularly impactful during high workload conditions. Therefore, we expected this influence to be greater as the number of call signs increased and it might also depend upon information criticality.

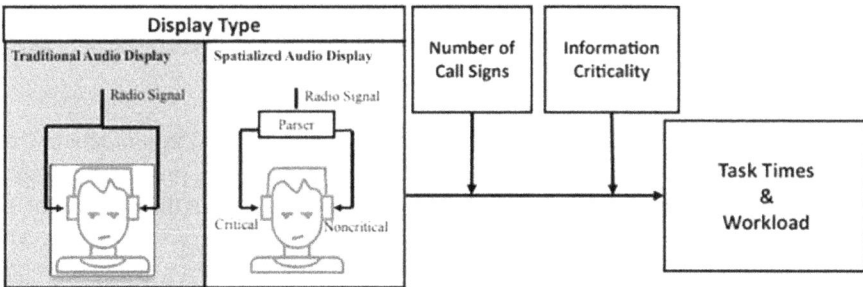

FIGURE 14.1 Graphical depiction of the experimental conditions and their hypothesized relationship with task time and workload.

14.2 METHOD

Our method consisted of three phases. First, an experiment was designed. Second, a model was constructed to simulate human performance during the experiment. This model assumed certain human behavioral changes in response to the number of critical call signs and the auditory display system, permitting response time and workload to be estimated for each experimental condition. The results of the model were used to establish clear hypotheses for changes in human performance and workload during the human-in-the-loop experiment. Finally, the experiment was conducted to test the hypothesized changes in response time and workload.

14.2.1 HUMAN-IN-THE-LOOP EXPERIMENT

14.2.1.1 Experimental Design

The human-in-the-loop experiment employed a two-by-two-by-two, within-subjects, experimental design including the independent variables of number of critical call signs (3 or 7), auditory display system (traditional or spatialized audio) and call sign relevance (critical or noncritical). The traditional audio system employed monaural, diotic sound. The order of the number of call signs and display system type was counterbalanced. However, the participants always completed both call sign conditions for one audio display system type before experiencing the other. Call sign relevance was randomized with equal numbers of critical and noncritical call signs within each block, where each block was formed from a combination of display system type and number of call signs.

When using the spatialized audio display, critical call signs were presented to the observer's right ear as most individuals show a left hemisphere (Jung et al., 2003) and right-ear advantage for speech (Studdert-Kennedy and Shankweiler, 1970). Therefore, most listeners are more likely to remember information from their right ear better than from their left (Kimura, 1967).

14.2.1.2 Participants

Twenty-four individuals (2 females, 22 males) volunteered from among the military and civilian workforce on Wright-Patterson Air Force Base. The ages of the participants ranged from 22 to 39 with a mean of 29 and a standard deviation of 4.4 years. The educational background of these participants is representative of U.S. Air Force UAS operators. All subjects were fluent in English had obtained or were near completion of a bachelor's degree, and self-reported having no known hearing deficiency.

14.2.1.3 Experimental Procedure

At the start of the experiment, each participant was exposed to 1 min of spatialized audio, during which time they were told which ear the signal was present. Next, they completed nine trials indicating whether the voice was present in their left, right, or both ears and were required to successfully indicate 8 of 9 correct responses. The participants were then instructed on and given two, 1-min practice sessions for the full task, including practicing with each auditory display system with five call signs during each practice. After a 2-min break, the participant began the experiment following the assigned order of conditions.

Each participant completed four conditions, each of which included 60 pre-recorded radio calls with equal numbers of critical and noncritical call signs. Participants were not informed of the ratio of critical to noncritical call signs. Prior to each experimental condition, participants were given a list of the UAV call signs they were responsible for monitoring and were instructed how to respond to critical and noncritical call signs. The participants listened to each radio call, categorized the radio call into one of the two categories (critical or noncritical) and responded as quickly and accurately as possible. If a call sign was noncritical, the operator was required to type '0' on a keypad and press the 'Enter' key. Alternately, if the call sign was critical, the operator needed to recall an alpha-numeric code from the radio message, employ the code to determine the location on a grid, type the number from the grid on the keypad, and press the 'Enter' key.

Before each experimental condition, the participants were informed of the type of audio display and the number of call signs for the subsequent condition. Instructions for the spatial audio display included the additional instruction that critical call signs will be presented to the right ear and noncritical call signs will be presented to the left ear.

After completing each experimental condition, participants completed a NASA-TLX questionnaire (Hart and Staveland, 1988). When all experimental conditions were complete, the participants ranked the importance of the six NASA-TLX scales (i.e., mental/physical/temporal demand, performance, effort, and frustration) to determine relative weights. Each participant then completed a SWORD workload questionnaire (Stanton et al., 2010) and a brief survey concerning their subjective impression of the conditions they had experienced.

14.2.1.4 Apparatus

The experiment was conducted in a 2×2 m cubicle in a quiet laboratory using the Multi-Modal Chat (MMC) Monitor Client Program (Finomore et al., 2010). The program logged all entered data and the time between the start of the audio presentation and activation of the 'Enter' key. Bose QC15 noise canceling headphones and a laptop were used to present the auditory information. Participants entered numbers into a ten-digit keypad.

The experiment employed a modified version of the Coordinate Response Measure (CRM) task. Initial stimuli included the standard speech corpus with eight call signs, four colors, and numbers between one and eight recorded by eight talkers (Bolia et al., 2000). The four colors and eight numbers corresponded to columns and rows in a matrix containing a required response for critical call signs. As some experimental conditions required seven critical call signs and additional noncritical call signs, the original CRM corpus was augmented to include a total of 19 different call signs. Voices of an additional nine individuals were recorded, each recording a different randomly selected subset of the 19 call signs paired with the four colors and eight numbers. In this modified task, a critical call sign list and the number grid containing the response information for critical call signs were printed on paper and arranged in the workspace prior to each experimental condition. The call sign list and number grid were unique for each experimental condition. Each participant heard only one message at a time, instead of using the simultaneous presentation of multiple radio calls, as is common in the standard CRM task.

14.2.2 Model Description

The model employed the Improved Performance Research Integration Tool (IMPRINT) (Mitchell, 2000) to predict the impact of changes in task structure on task completion times and perceived workload. This tool has been applied to perform a number of early system assessments, including human versus system function allocation, mental workload estimation, and prediction and assessment of human performance (Allender, 2000). IMPRINT models require the determination of a task structure, time distributions for each task, and workload values for those tasks.

The model captured the task structure shown in Figure 14.2. As displayed in this figure, the task structure for the traditional audio system requires an operator to

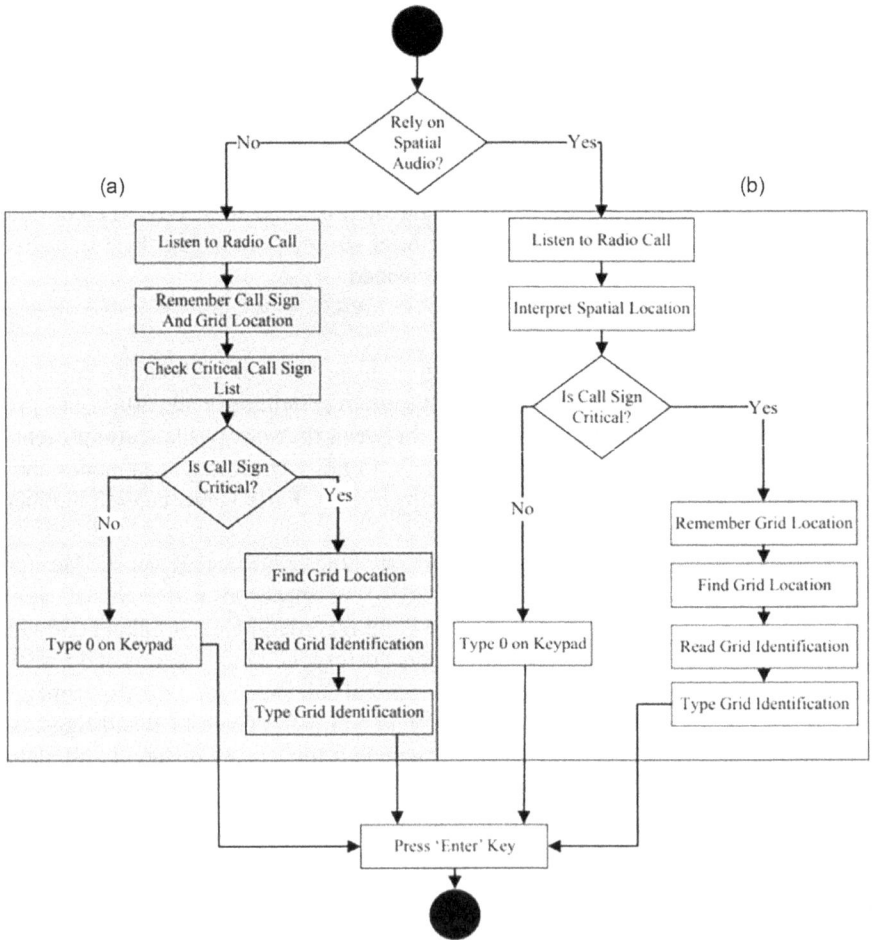

FIGURE 14.2 Activity diagram depicting the anticipated task structure for binaurally-presented, monophonic audio (a) and spatial audio (b).

perform the activities indicated in the shaded portion of the flow chart. For either audio system, users listened for an auditory message. For the traditional system, if an incoming auditory message contained a critical call sign, the user, stored the call sign and coordinates in working memory, checked the call sign list to determine if the call sign was on the list, determined a number at the intersection of the grid coordinates, entered this number into a computer, and then pressed enter. If the message did not contain a critical call sign, the user stored the call sign and coordinates into working memory, checked the call sign list to verify the call sign was not on the list, then entered zero into the computer and pressed enter.

The spatial audio cue should permit the user to streamline the process. The streamlined process is depicted in the unshaded portion of Figure 14.2. With the presence of the spatial cue, we predict that the operator will shed the more complex cognitive and physical tasks associated with call sign recognition and relevance determination. These tasks will be replaced with the simpler perceptual task of location interpretation. Thus, the model assumes that the spatial audio cue will simplify the task structure by eliminating tasks associated with referencing memory aids and determining information relevance when responding to multiple critical call signs. The simplified task structure should decrease response time and workload as hypothesized. The hypothesized improvements in task times and workload will not be observed if participants do not rely on the spatial audio cue.

Task time distributions for the IMPRINT model were based on the results of an earlier experiment (Amaddio, 2015). To estimate workload, each task in the network was assigned a workload value using the Visual, Auditory, Cognitive, Psychomotor (VACP) method (Bierbaum et al., 1989). VACP is based on multiple resource theory (Wickens, 2002) and provides activity-based demand values for each of seven resource channels: visual, auditory, cognitive, fine psychomotor, gross psychomotor, speech, and tactile. The VACP values were populated based upon input from a group of four subject matter experts who were trained in the use of this method. These individuals were asked to provide independent estimates using the VACP demand charts. After independent estimates were provided by each individual, a consensus-building discussion was conducted to reconcile disagreements.

The model allowed the simulation of the full factorial, three-factor design. Simulations were conducted, repeating each experimental condition with 25 model runs, each of which included 30 call sign instructions. It should be noted that the task time distributions were calculated across individuals and therefore, the between-run variability was expected to be less than the between-subject variability in the human-in-the-loop experiment.

14.2.3 Data Analysis

14.2.3.1 Model Data Analysis

We collected VACP workload values for each task. These values were multiplied by each task time to compute a time-weighted average VACP value across the entire time of the trial. The time-weighted average workload and average time to respond to a call sign were subjected to appropriate repeated-measures ANOVAs.

14.2.3.2   HITL Data Analysis

Response times were subjected to repeated-measures, three-factor ANOVA. Each participant's NASA-TLX ratings were multiplied by the appropriate weights for each of the workload dimensions and summed to determine a composite NASA-TLX score (Hart, 2006). The SWORD ratings were normalized to determine the relative workload for each condition and subject (Stanton et al., 2010). Workload data were subjected to repeated-measures, two-factor ANOVA. Accuracy was recorded as a binary response, with accurate responses requiring accurate entry of the numerical information. Chi-square or Fisher Exact Probability tests were conducted on the accuracy data, depending on the expected frequency (Siegel, 1956).

14.3   RESULTS

14.3.1   MODEL RESULTS

The three-factor, repeated-measures ANOVA indicated a significant three-way interaction of display, number of call signs, and information criticality on response time. ($F(1,24) = 15.19$, MSE $= 0.003$, $p < 0.001$, $\eta_p^2 = 0.388$). To explore this interaction, separate two-way ANOVAs were conducted for critical and noncritical conditions. As shown in Figure 14.3a and b, the model predicted a decrease in response time for both critical ($F(1,24) = 20.09$, MSE $= 0.002$, $p < 0.001$, $\eta_p^2 = 0.456$) and noncritical ($F(1,24) = 137.12$, MSE $= 0.002$, $p < 0.001$, $\eta_p^2 = 0.851$) call signs for the spatial audio system as compared to the traditional audio system. This decrease is substantial for

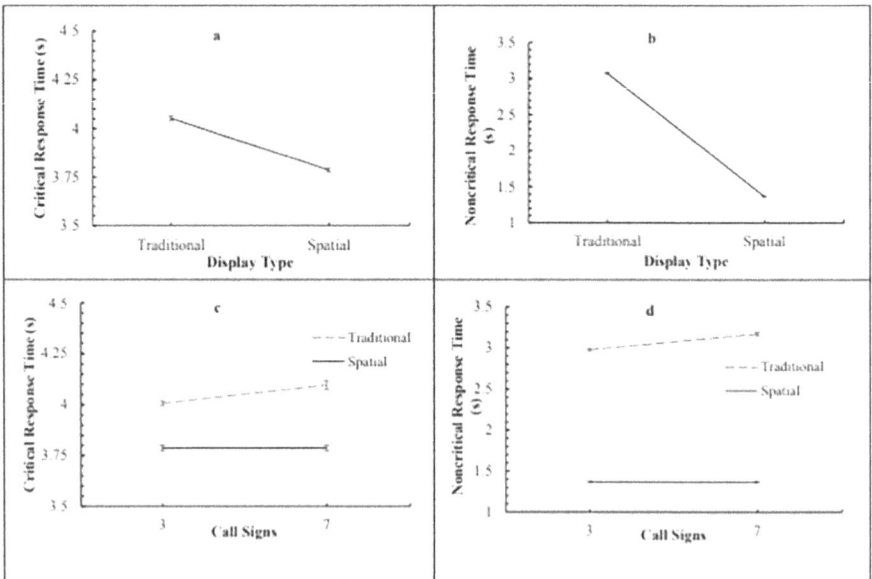

FIGURE 14.3   Model response times for critical (a, c) and noncritical (b, d) call signs for each display type and number of call signs. Error bars indicate standard error of the mean.

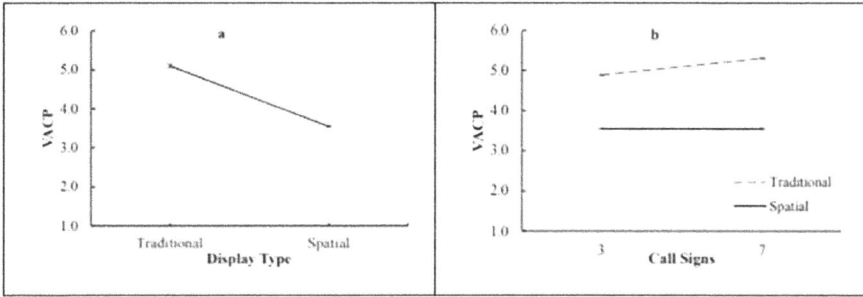

FIGURE 14.4 Model workload values for each audio display type (a) and the interaction of audio display type with number of call signs (b). Error bars indicate standard error of the mean.

the noncritical call sign condition as the task of determining whether a call sign is relevant or not is the longest duration task in this condition. Additionally, Figure 14.3c and d show that the model predicted an interaction between call sign and audio system for both critical $(F(1,24)=20.09$, MSE$=0.002$, $p<0.001$, $\eta_p^2=0.456$) and noncritical $(F(1,24)=137.12$, MSE$=0.002$, $p<0.001$, $\eta_p^2=0.851$) call signs. In this interaction, response time increased as the number of call signs increased for the traditional audio system. However, this increase was not present for the spatial audio system.

Similar to response time, the model predicted a decrease in perceived workload as the audio system changes from traditional to spatial as shown in Figure 14.4a. An ANOVA supports the main effect of audio system as significant $(F(1,24)=9{,}250.8$, MSE$=0.007$, $p<0.001$, $\eta_p^2=0.990$). A significant interaction is also supported for workload $(F(1,24)=171.7$, MSE$=0.007$, $p<0.001$, $\eta_p^2=0.641$) as shown in Figure 14.4b. This interaction indicated a smaller change in predicted workload for the spatial audio system when the number of call signs was expanded from 3 to 7 than for the traditional audio system.

14.3.2 HUMAN-IN-THE-LOOP EXPERIMENT

Response times measured in the human-in-the-loop experiment are depicted in Figure 14.5. A three-way repeated-measures ANOVA indicated all effects and all interactions were significant, including the three-way interaction of information criticality, audio system and call sign $(F(1,23)=53.05$, MSE$=0.013$, $p<0.001$, $\eta_p^2=0.698$). To decompose this interaction, separate two-way repeated-measures ANOVAs were conducted for the critical and noncritical call signs. The effects of critical call signs are shown in Figure 14.5a and c. The repeated measures ANOVA indicated a significant effect of audio system $F(1,23)=16.24$, MSE$=0.263$, $p<0.001$, $\eta_p^2=0.414$. The participants responded slightly more rapidly in the presence of spatial audio than traditional audio. There was no support for an effect of either number of call signs $F(1,23)=0.30$, MSE$=0.53$, $p=0.589$, $\eta_p^2=0.013$ or the interaction of audio system and number of call signs $F(1,23)=0.20$, MSE$=0.015$, $p=0.656$, $\eta_p^2=0.009$ on response time for critical call signs.

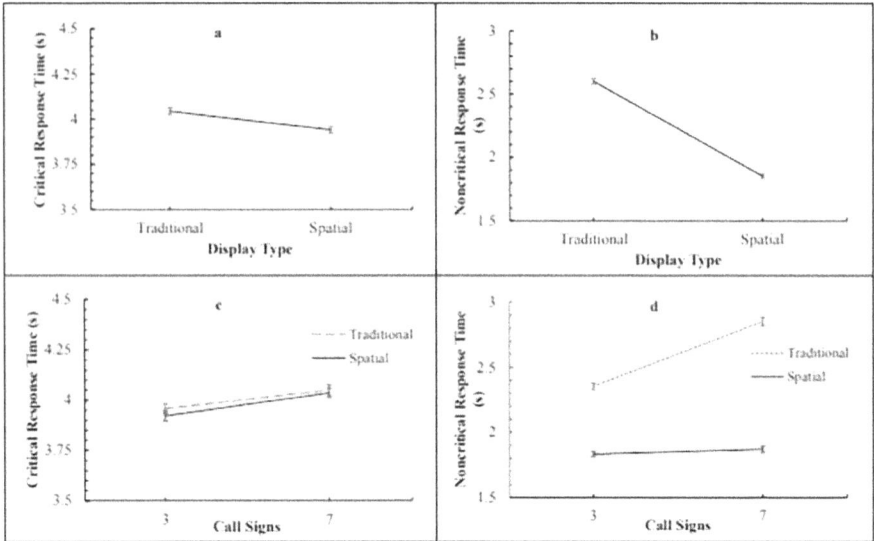

FIGURE 14.5   Response time for critical (a and c) and noncritical (b and d) call signs as a function of display type and number of call signs. Error bars indicate standard error of the mean.

For the noncritical call signs, the effect of audio system $(F(1,23)=51.36,$ MSE$=0.086, p<0.001, \eta_p^2=0.691)$, the number of call signs $(F(1,23)=156.94,$ MSE $0.086, p<0.001, \eta_p^2=0.872)$, and the interaction of audio system and the number of call signs $(F(1,23)=56.81,$ MSE$=0.022, p<0.001, \eta_p^2=0.712)$ were all significant. As shown in Figure 14.5b, response time decreased significantly for the spatial audio display as compared to the traditional display when noncritical call signs were presented. Furthermore, while the response time increased for the traditional audio display as the number of call signs increased, response time did not increase significantly for the spatial audio display with an increase in the number of call signs.

Figure 14.6 shows model NASA-TLX and SWORD workload values as a function of audio system type and the interaction of the audio system with the number of call signs. The repeated measures ANOVA supported the hypothesis that the participants' NASA-TLX ratings varied as a function of audio system $(F(1, 23)=19.2,$ MSE$=23.35, p<0.001, \eta_p^2=0.455)$, number of call signs $(F(1, 23)=16.2,$ MSE$=66.95, p<0.001, \eta_p^2=0.413)$ and the interaction of audio system with the number of call signs $(F(1, 23)=11.98,$ MSE$=27.34, p=0.002, \eta_p^2=0.342)$. Similarly, the SWORD ratings varied as a function of audio system $(F(1, 23)=57.30,$ MSE$=0.002,$ $p<0.001, \eta_p^2=0.714)$, number of call signs $(F(1, 23)=32.95,$ MSE$=0.006, p<0.001,$ $\eta_p^2=0.589)$ and the interaction of audio system with the number of call signs $(F(1, 23)=19.54,$ MSE$=0.002, p<0.001, \eta_p^2=0.459)$. Generally, the data displayed for both NASA-TLX and SWORD workload ratings in Figure 14.6 were in agreement with the trends we predicted based on the model with the exception that NASA-TLX values did not increase as the number of call signs increased from 3 to 7 for the spatial audio display.

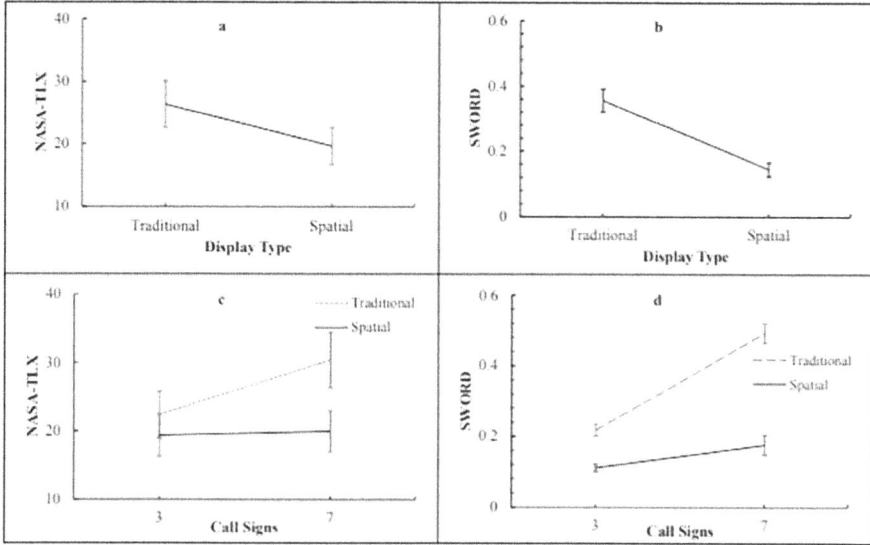

FIGURE 14.6 Response time for critical (a and c) and noncritical (b and d) call signs as a function of display type and number of call signs. Error bars indicate standard error of the mean.

Response accuracy was high across all conditions, ranging from just over 96%–100% for each condition. The statistical tests indicate that response accuracy was not related to any of the experimental manipulations. During the post-experiment questionnaire, all participants indicated that, given the choice, they would use the spatial audio system over the traditional audio system. Comments indicated that the spatial auditory system decreased perceived workload, stress, and response time.

14.4 DISCUSSION

The current research explored the application of a spatial audio display to aid operators in determining information relevancy within a UAS MAC scenario. This system differs from previous multichannel spatial audio displays in which it is assumed that different radio communication channels are presented at different spatial locations (Haas et al., 1997; MacDonald et al., 2002; McAnally and Martin, 2007; McAnally et al., 2002). In these previous displays, each radio channel can present both critical and noncritical information. While these displays aid the operator in segregating the information on one channel from another, the operator must attend to all radio communication information to parse critical from noncritical information.

The system explored in the current research employed a parser to automatically determine critical from noncritical information. This information, regardless of radio channel, was presented in different spatial locations. Our research provided evidence that a spatial audio system in which critical and noncritical call signs originate in different locations improved performance and reduced workload in UAS MAC scenarios. Specifically, both modeling and experimental results indicated that the spatial audio system reduced UAS operator workload and response times while listening for critical call signs from a pool of critical and noncritical call signs. In fact, workload and the time necessary for operators to decide to ignore noncritical call signs were found to be largely independent of the number of call signs when the number of critical call signs varied from 3 to 7. Therefore, such a system could be expected to provide a significant benefit over a traditional audio display when the operator is required to monitor a large number of critical call signs.

We theorized that this system would change the task structure for the operator, replacing working memory demand and physical activity necessary to confirm the presence or absence of a call sign on a call sign list with the simple perceptual task of sound localization. The results support this theory for noncritical call signs. Specifically, while employing the spatial audio display, the experimentally obtained task times for noncritical call signs did not increase with an increase in call signs as predicted by the model. Additionally both the experimental results and model indicated an increase in response time as the number of call signs increased for the traditional audio display. The participants did not fully adopt this change in task structure for critical call signs. While the model predicted similar trends in task times for the critical call signs as observed for the noncritical call signs, the experimental results indicated that response time increased as the number of call signs increased for critical call signs. As response time increased with call signs, it appeared that the participants verified the call sign against the call sign list as this is the only task in which performance time varied as a function of the number of call signs. Therefore the results indicated that while the operators employed the spatial audio cue to filter noncritical call signs, they continued to verify at least a portion of the critical call signs on the critical call sign list. This interpretation is consistent with post-experiment survey results, as 11 out of 24 (45.8%) participants indicated that they could easily and quickly disregard irrelevant information when using the spatial audio display in an open-ended question. Similar comments were not made with regard to the critical call signs.

This apparent natural bias has implications for system design. As it is unlikely that any automated system will be 100% reliable, this natural human bias implies that the system should be calibrated toward exhibiting false alarms (i.e., classifying noncritical information as critical, type I error) rather than misses (i.e., classifying critical information as noncritical, type II error). Such a system bias would ensure that the human operator would be more likely to detect and react appropriately to system errors.

This recommendation is counter to the recommendation made with respect to the automation of visual tasks. In this domain, it has been recommended that systems should be biased toward misses rather than false alarms (Meyer et al., 2014; Rice and McCarley, 2011). Researchers within the visual domain have found false alarms to be more impactful on human performance than misses, perhaps due to the saliency of the false alarms (Rice and McCarley, 2011). These researchers have noted that the saliency of the false alarms leads to greater operator disruption and reduces operator reliance on automation.

Similarly, our research would predict that false alarms within auditory displays are likely to be recognized by the human operator due to their bias toward verifying the system classification of critical information. Therefore, it is possible that biasing the system toward false alarms will increase the saliency of automation errors, reducing trust and automation reliance. However, biasing the system toward misses is likely to increase the number of automation errors that both the system and operator will fail to detect given the human bias observed in this research. Certainly, the impact of operator trust and reliance requires further research for the envisioned spatial audio display.

The application of spatial audio reduced workload and the benefit increased with increasing numbers of call signs. Previous investigations of spatial audio displays have not demonstrated similar changes in operator workload (MacDonald et al., 2002). However, unlike previous system implementations, the assumed implementation permits the user to disregard noncritical information under high workload conditions. It is believed that this attribute of the current system facilitated a decrease in workload not observed in previous spatial audio display implementations. MacDonald et al. (2002), however, employed a more realistic aircraft control task as a secondary task in their research. It is possible that the workload advantages of spatial audio are insignificant with respect to the workload imposed by aircraft control and the observed changes in operator workload would not be observed in the presence of a complex secondary task. Further, the spatial audio cues could be less salient in the presence of masking audio signals (May and Walker, 2017). Therefore, it is important to investigate the potential benefit of the current spatial audio system within a more realistic environment.

Similar systems might be applied in manned aircraft, military, and emergency response communications. For example, while military aircraft pilots only fly one aircraft, they often monitor the radio for calls from several individuals within a workgroup (Haas and Stachowiak, 2007). Conversations with manned aircraft pilots indicate that they typically adjust the sound level of radio frequencies to permit them to monitor channels their workgroup employs with a higher priority than channels that are less likely to contain communications for their workgroup. This strategy for handling voice communications using multichannel radios has been discussed in the ergonomics literature (Brungart et al., 2002). Similarly, Air Traffic Controllers adjust the sound level of radio frequencies for aircraft within their area of responsibility differently from the sound level of radio frequencies used by aircraft outside their area of responsibility. The use of spatial audio display as tested in the current research could be applied to aid operators in these domains to differentiate critical from noncritical information through the use of spatial audio displays.

14.5   CONCLUSION

The current research demonstrated reduced operator response time and workload for a spatial audio display capable of presenting critical and noncritical information in spatially separated locations. This system was shown to be particularly advantageous when the operator was responsible for monitoring a radio for multiple call signs. Such a system may provide an operator advantage in MAC systems. Additionally, the research suggests that participants relied upon this spatial cue to a greater extent when attempting to disregard noncritical call signs than when confirming critical call signs. This spatial audio display concept would likely require an automated speech recognition system to facilitate call sign identification. As a result, further research is required to determine the impact of recognition errors on the utility and reliability of such a system.

14.5.1 FUNDING

This work was supported by the Air Force Office of Scientific Research (F4FGA06085J001), Computational Cognition and Machine Intelligence Program.

14.5.2 ACKNOWLEDGMENT

The authors would like to acknowledge Colonel Anthony Tvaryannas for motivating this research and Dr. Victor Finnomore for assistance in creating the experimental research method.

The views in this article are those of the authors and do not necessarily reflect the official policy or position of the Department of the Air Force, Department of Defense, nor the U.S. Government.

REFERENCES

Allender, L., 2000. Modeling human performance: impacting system design, performance and cost. In: *Proceedings of the Military, Government and Aerospace Simulation Symposium: 2000 Advanced Simulation Technologies Conference*, pp. 139–144.

Amaddio, K.M., 2015. *The Cognition of Multiaircraft Control (MAC): Cognitive Ability Predictors, Working Memory, Interference and Attention Control in Radio Communication*. Air Force Institute of Technology.

Amaddio, K.M., Miller, M.E., Elshaw, J., Finnomore, V., 2015. The cognition of multiaircraft control (MAC): proactive interference and working memory capacity. In: *Proceedings of the International Symposium on Aviation Psychology*. Dayton, OH.

Bierbaum, C.R., Szabo, S.M., Aldrich, T.B., 1989. Task Analysis of the UH-60 Mission and Decision Rules for Developing; a UH-60 Workload Prediction Model, vol. I Summary Report (AD-A210 763). Alexandria, VA

Bolia, R.S., Nelson, W.T., Ericson, M.A., Simpson, B.D., 2000. A speech corpus for multi-talker communications research. *J. Acoust. Soc. Am.* 107 (2), 1065–1066. https:// doi. org/10.1121/1.428288.

Brungart, D.S., Ericson, M.A., Simpson, B.D., 2002. Design considerations for improving the effectiveness of multitalker. In: International Conference on Auditory Display, pp. 1–6 (Kyoto, Japan).

Calhoun, G.L., Draper, M.H., 2015. Display and control concepts for multi-UAV applications. In: Valavanis, K.P., Vachtsevanos, G.J. (Eds.), *Handbook of Unmanned Aerial Vehicles*. Springer Netherlands, Dordrecht, pp. 2443–2473. https://doi. org/10.1007/978-90-481-9707-1.

Cengarle, G., 2012. *3D Audio Technologies: Applications to Sound Capture, Post-production and Listener Perception*. Pompeu Febra University.

Colombi, J.M., Miller, M.E., Schneider, M., McGrogan, M.J., Long, C.D.S., Plaga, J., 2012. Predictive mental workload modeling for semiautonomous system design: implications for systems of systems. *Syst. Eng.* 15 (4), 448–460. https://doi.org/10.1002/sys. 21210.

Finomore, V., Popik, D., Castle, C., Dallman, R., 2010. Effects of a network-centric multimodal communication tool on a communication monitoring task. *Proc. Hum. Factors Ergon. Soc. Annu. Meet.* 54 (25), 2125–2129. https://doi.org/10.1177/154193121005402501.

Franke, J.L., Zaychik, V., Spura, T.M., Alve, E.E., 2005. Inverting the operator/vehicle ratio: approaches to next generation UAV command and control. In: *Proceedings of the Association for Unmanned Vehicle Systems*, pp. 1–11.

Haas, E.C., Gainer, C., Wightman, D., Couch, M., Shilling, R., 1997. Enhancing system safety with 3-D audio displays. In: *Proceedings of the Human Factors and Ergonomics Society*, pp. 868–872. https://doi.org/10.1177/107118139704100230.

Haas, E.C., Stachowiak, C., 2007. Multimodal displays to enhance human robot interaction on-the-move. In: *Proceedings of the 2007 Workshop on Performance Metrics for Intelligent Systems - PerMIS '07*. ACM Press, New York, pp. 135–140. https://doi.org/10.1145/1660877.1660895.

Hart, S.G., 2006. NASA-TLX: 20 years later. In: *Proceedings of the Human Factors and Ergonomics Society Annual Meeting*, pp. 904–908.

Hart, S.G., Staveland, L.E., 1988. Development of NASA-TLX (task load index): results of empirical and theoretical research. *Adv. Psychol.* 52 (C), 139–183.

Jung, P., Baumgärtner, U., Bauermann, T., Magerl, W., Gawehn, J., Stoeter, P., Treede, R.D., 2003. Asymmetry in the human primary somatosensory cortex and handedness. *Neuroimage* 19 (3), 913–923. https://doi.org/10.1016/S1053-8119(03)00164-2.

Kimura, D., 1967. Functional asymmetry of the brain in dichotic listening. *Cortex* 3 (2), 163–178. https://doi.org/ https://doi.org/10.1016/S0010-9452(67)80010-8.

MacDonald, J.A., Balakrishnan, J.D., Orosz, M.D., Karplus, W.J., 2002. Intelligibility of speech in a virtual 3-D environment. *Hum. Factors* 44 (2), 272–286.

May, K.R., Walker, B.N., 2017. The effects of distractor sounds presented through bone conduction headphones on the localization of critical environmental sounds. *Appl. Ergon.* 61, 144–158. https://doi.org/10.1016/j.apergo.2017.01.009.

Maza, I., Caballero, F., Molina, R., Peña, N., Ollero, A., 2009. Multimodal interface technologies for UAV ground control stations. *J. Intell. Rob. Syst.* 57 (1–4), 371–391. https://doi.org/10.1007/s10846-009-9351-9.

McAnally, K., Martin, R.L., 2007. Spatial audio displays improve the detection of target messages in a continuous monitoring task. *Hum. Factors* 49 (4), 688–695. https://doi.org/10.1518/001872007X215764.

McAnally, K.I., Bolia, R.S., Martin, R.L., Eberle, G., Brungart, D.S., 2002. Segregation of multiple talkers in the vertical plane: implications for the design of a multiple talker display. In: *Proceedings of the Human Factors and Ergonomics Society*, pp. 588–591.

Meyer, J., Wiczorek, R., Günzler, T., 2014. Measures of reliance and compliance in aided visual scanning. *Hum. Factors J. Hum. Factors Ergon. Soc.* 56 (5), 840–849. https://doi.org/10.1177/0018720813512865.

Mitchell, D.K., 2000. *Mental Workload and ARL Workload Modeling Tools*. Aberdeen Proving Ground, Harford County.

Philbrick, R.M., Colton, M.B., 2014. Effects of haptic and 3D audio feedback on operator performance and workload for quadrotor UAVs in indoor environments. *J. Robot. Mech.* 26 (5), 1–12.

Rice, S., McCarley, J.S., 2011. Effects of response bias and judgment framing on operator use of an automated aid in a target detection task. *J. Exp. Psychol. Appl.* 17 (4), 320–331. https://doi.org/10.1037/a0024243.

Siegel, S., 1956. *Nonparametric Statistics for Behavioral Sciences*. McGraw-Hill, New York.

Simpson, B.D., Bolia, R.S., Draper, M.H., 2005. Spatial audio display concepts supporting situation awareness for operators of unmanned aerial vehicles. In: Vincenzi, D.A., Mouloua, M., Hancock, P.A. (Eds.), *Human Performance, Situation Awareness, and Automation: Current Research and Trends HPSAA II, Volumes I and II*. Psychology Press, San Francisco, CA, pp. 61–65.

Stanton, N.A., Salmon, P.M., Walker, G.H., Baber, C., Jenkins, D.P., 2010. *Human Factors Methods*. Ashgate Publishing, Farnham.

Studdert -Kennedy, M., Shankweiler, D.P., 1970. Hemispheric specialization for speech perception. *J. Acoust. Soc. Am.* 48 (2), 579–594. https://doi.org/10.1121/1. 1912174.

Trouvin, B., Schlick, C., 2004. A study of audio and visual context switch indicators in a multirobot navigation task. In: *Proceedings of the 2004 IEEE International Conference on Systems*, pp. 2821–2868.

Wickens, C.D., 2002. Multiple resources and performance prediction. *Theor. Issues Ergon. Sci.* 3, 159–177.

Section 5

Electromagnetics

15 Method for Generating a Figure of Merit for the Selection of Antennas Under Volumetric Constraint*

Daniel W. Stambovsky
US Air Force Institute of Technology

Ronald J. Marhefka
The Ohio State University

Andrew J. Terzuoli
US Air Force Institute of Technology

CONTENTS

* Reprinted with permission: Stambovsky, D. W., Marhefka, R. J., & Terzuoli, A. J. (2020). Method for Generating a Figure of Merit for the Selection of Antennas Under Volumetric Constraint. *IEEE Transactions on Antennas and Propagation*, 69(4), 1952–1958.

DOI: 10.1201/9781003220978-20

15.1 INTRODUCTION

Generating an apples-to-apples comparison of possible antennas with an eye to maximizing performance for applications with volumetric restrictions can be challenging. Most previous work has focused on the theoretical limits of antenna performance parameters in a given volume [1], limits on Q-factor [2], or limits on other fundamental properties, such as radiation efficiency [3, 4]. While these provide important theoretical bounds to antenna performance, there remains a practical difficulty in selecting a design or off-the-shelf (OTS) antenna, which most effectively utilizes limited available space. There have been some attempts to build a single metric, such as electrical volume [5], but they have limitations in the amount and type of the information that they provide restricting their usefulness. The closest effort to this work is the recent outstanding publication by Kirov [6] who notes the insufficiency of standard antenna comparison techniques and generates four criteria that are the functions of gain, bandwidth, and volume to test antenna suitability. Some results from this work will be used in the following; however, it should be noted that the method in [6] is only applied to circularly polarized patch antennas. A novel method, a parameter called the A-B parameter, is proposed, which differs both in its treatment of volumetric concerns as well as its structure. It consists of two polar points constructed from electrical and physical properties commonly provided by antenna designers and provides a comparative graphical representation of antenna volume and electrical performance, independent of radiation mechanism. This allows a more extensive comparison of relevant electrical and geometric parameters than the single-parameter construction, limitations of which are discussed in Section III. Before continuing to the method, expounded below two caveats must be addressed: first, the A-B parameter only indirectly reflects the physics of radiating systems. Unlike parameters based on the physics of antennas, such as the Chu–Harrington limit referenced before [2], the A-B parameter pertains to operational considerations such as available space and acceptable gain or bandwidth requirements for a specific application. Second, it is important to understand that any metric is a tool that requires a proper application. Before using the A-B parameter (or other metric or method) to select an antenna, the engineer must have a general idea of the desired uses and restrictions of the application. Does the application require an extremely large bandwidth? Is high gain or broad geographic coverage required? In general, higher performance in one area will come at the cost of lower performance elsewhere, so understanding of system requirements is a presupposition to the use of the A-B parameter for antenna selection.

15.2 ANTENNA A-B PARAMETER

15.2.1 CONSTRUCTION AND PLOTTING

1. **A-B Construction**: As indicated, the A-B parameter consists of two polar coordinates

$$A = a_1 e^{a_2 \pi i}$$

$$B = b_1 e^{b_2 \pi i} \tag{15.1}$$

where a_1 is the gain, a_2 is the fractional bandwidth

$$BW_f = \frac{f_{high} - f_{low}}{f_{center}} = \frac{2(f_{high} - f_{low})}{f_{high} + f_{low}} \tag{15.2}$$

b_1 is the longest linear dimension (LLD)[†] in units of wavelength, especially the center wavelength of the lowest band, and b_2 is the profile height of the antenna with a propagation direction[‡] of z and linear dimensions x, y, and z measured in wavelengths (again the center wavelength of the lowest operational band)

$$b_2 = \frac{z}{x + y + z} \tag{15.3}$$

2. **Notes on Gain:** Gain, of course, varies across the bandwidth both because of the frequency-dependent nature of resonance and impedance matching and the shifting relative size of antenna aperture to wavelength across the band. For a simplified 2-D aperture, for instance, the directivity (and thus upper bound of the gain) is proportional to the effective area over the wavelength $D = \left(4\pi\, A / \lambda^2\right)$, which means that the gain for any single antenna will not be constant as operational frequencies change. It is necessary, then, to select a gain value consistently across all antennas being compared with this method so that a proper apples-to-apples comparison can be made with clear, explicit knowledge of how the metric pertains to antenna performance. Example methodologies include average gain across the band

$$\bar{G} = \frac{1}{f_{high} - f_{low}} \int_{f_{low}}^{f_{high}} G(f)\, d\,f \tag{15.4}$$

peak gain, or some other method. In addition, antenna performance may be measured in a number of ways, such as directivity, IEEE gain, or realized gain, and the particular measure used to generate the A-B parameter will depend on which metric best fits the problem at hand as well as the available information on the antennas being compared.

$$1 - \left|\frac{Z_{in} - Z_O}{Z_{in} - Z_O}\right|^2 = \frac{1}{2}. \tag{15.5}$$

3. **Notes on Bandwidth:** To paraphrase Aristotle, bandwidth is said in many ways. Fractional bandwidth is the simplest to utilize in the A-B parameter since it varies from 0 to 2, which maps easily onto the radial angle. This works very well for narrow band antennas (e.g., patches, dipoles, and most

---

[†] Defined as
[‡] If profile height is not the limiting dimension for a specific design, z can be the limiting dimension or the LLD.

resonant antennas), as f increases; however, $f_{low} f_{high}$ and (15.2) approaches 2 asymptotically, and thus, in antennas with a large bandwidth, the fractional bandwidth increases nonlinearly making fractional bandwidth a poor measure. See the discussion on the application of the A-B parameter to ultra-wideband (UWB) antennas in the following. An additional permutation to the stated bandwidths of antennas is the method of high- and low-frequency selections for the bandwidth calculation. Usually, these are taken to be the half-power impedance points (impedance bandwidth).

However, this is not always the case, and at times, a VSWR bandwidth is given, often 2:1 which is a power reduction of only 0.85 rather than 0.5. Less commonly, a pattern, or gain, bandwidth is used, identifying the frequencies where gain drops by 3 dB. Regardless of how the limits of bandwidth are used, the A-B parameter will allow direct comparison only if the engineer is careful to be consistent across all antennas being evaluated.

4. **Notes on Physical Parameters:** Profile height has been selected for b_2 as being "low profile" is a common desired antenna property, but it need not be the direction used if there are other, more pressing spatial limitations. Here again, we see how the antenna selection must be made with preexisting knowledge of application and design space if the volumetric limiting factor for antenna selection is in a direction other than the primary direction of radiation that can be used as the numerator of b_2 instead of z. Alternatively, for omnidirectional antennas or in cases where there is no single direction of concern, the LLD can be used to replace z. An added consideration is the nature of the geometric restriction, it may be that only one part of the antenna (say the portion above a vehicle skin) is subject to this restriction. In such a case, it is more useful to build the B coordinates from the parts of the antenna subject to this restriction, being careful to prevent 0 or infinite values. In a case where there is no geometry in the limiting region, the comparison becomes one of electrical performance alone, limiting the usefulness of the A-B parameter. Once again, the good sense of the engineer must come into play to ensure comparisons are being made appropriately. For applications such as reflector antennas, the A-B parameter may be utilized effectively, as shown in Figure 15.1. Depending on the needs of the designer, the standard method given in (15.3) can be used where b_2 becomes a measure of [height/($2\times$diameter)]. If the limiting factor is dish diameter, however, b_2 reduces to [$1/(1-($height/diameter$))$]. Of course, the placement of the secondary reflector (or feed for antennas designed in such a way) is related to the focal point of the dish, and the (*f/d*) ratio is key to such antenna design [7]. Regardless the A-B parameter method of comparison holds true for this antenna type, as long as the antennas being weighed for use have comparable bandwidth measures (see Section 15.2.3). Note as well that, in a practical environment, other considerations may come into play for antenna selection, for example, if a reflect array design is lighter than a parabolic reflector and so requires less supporting structure, this difference may play into the A-B construction. Ultimately, the nature of the volumetric

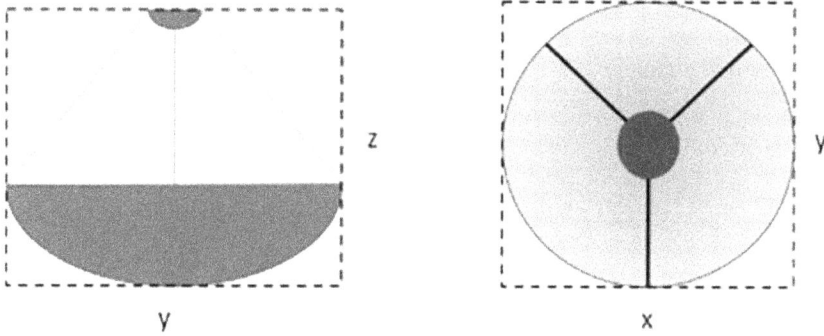

FIGURE 15.1   Example reflector dish antenna showing the physical measures that would go into the b_2 calculation.

TABLE 15.1
Antenna Identifiers

Antenna Identifier	Antenna Type
α	Crescent Moon Pair Patch [9]
β	CP Crossed Dipole Over Seivenpiper Reflector [10]
γ	Dual-Polarized Stepped Slot Antenna Array [11]
δ	Compact Open Slot [12]
ε	Pasternack PE9887-1 I Standard Gain Horn [13]
η	WB Multiple-Microstrip Dipole [14]

limitations under which the designer is operating will determine if and what support equipment plays part in the B point.

Similar practical considerations are to be taken into account when applying the A-B parameter to wire antennas, for example, a Yagi-Uda antenna, such as the design elucidated in [8], has a stated set of dimensions of $1,365 \times 215 \times 218$ mm with the longest obviously being in the direction of propagation, so probably taking the z spot in the B point.

The decision to add any necessary mounting brackets to the geometry of the antenna would again depend on the use case and the specific geometric restrictions placed on the engineer. A whip monopole or wire dipole that relies on extremely thin wires has an LLD perpendicular to the primary direction of propagation. Here again, though, as long as the radius of the wire is not 0, the A-B parameter should function well, especially since these are not highly directional antennas so using the LLD as z as indicated above is probably the most useful. The A-B points for the antennas listed in Table 15.1 can be seen in Table 15.2. Note that, for this table, the gain values are average gain and that geometric measurements include the ground plane.

TABLE 15.2
Antennas A-B Values

Antenna Identifier	a_1	a_2	b_1	b_2
α	8.07	0.756	1.394	0.098
β	5.08	0.397	1.342	0.038
γ	13.52	0.469	1.351	0.052
δ	3.96	1.216	0.615	0.020
ε	11.54	1.399	6.222	0.336
η	8.84	0.483	0.783	0.105

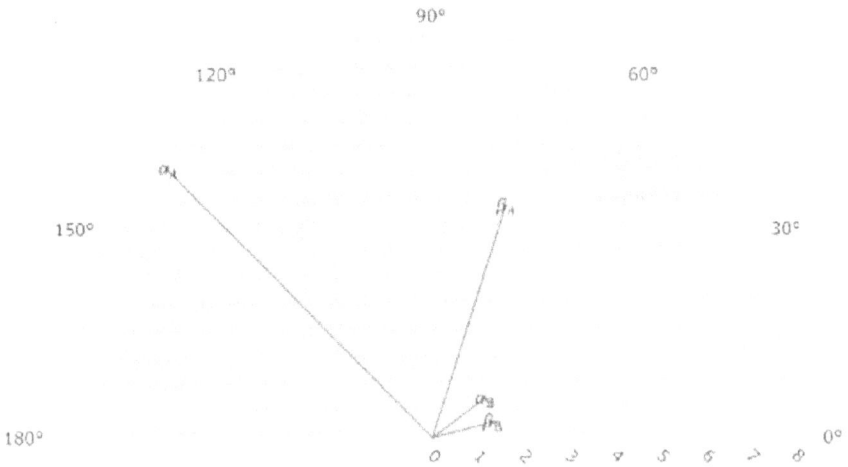

FIGURE 15.2 Plot of A-B parameter for crescent pair patch antenna [9] labeled α and CP dipole with AMC reflector (β) from Table 15.1.

5. **Plotting:** The A-B parameter comes into its own when plotted on a polar graph. Using the origin as a third point, the A and B lines are easily comparable among antennas. Thus, for example, the A-B parameter for antenna α from Table 15.1 can be compared on a plot with antenna β, as shown in Figure 15.2. It can clearly and immediately be seen that α has superior electrical performance both in bandwidth and gain. When it comes to geometric considerations, however, the design focus of β as a low-profile solution comes into its own. The benefit of the A-B parameter for antenna selection is the ability to see hardware tradeoffs graphically and use that as a tool to choose the correct design.

Plotting additional antennas should help drive this home, as shown in Figure 15.3. With all six plotted, performance parameters can be compared: clearly, the horn (ε) performed exceptionally far and away from the largest bandwidth and having the

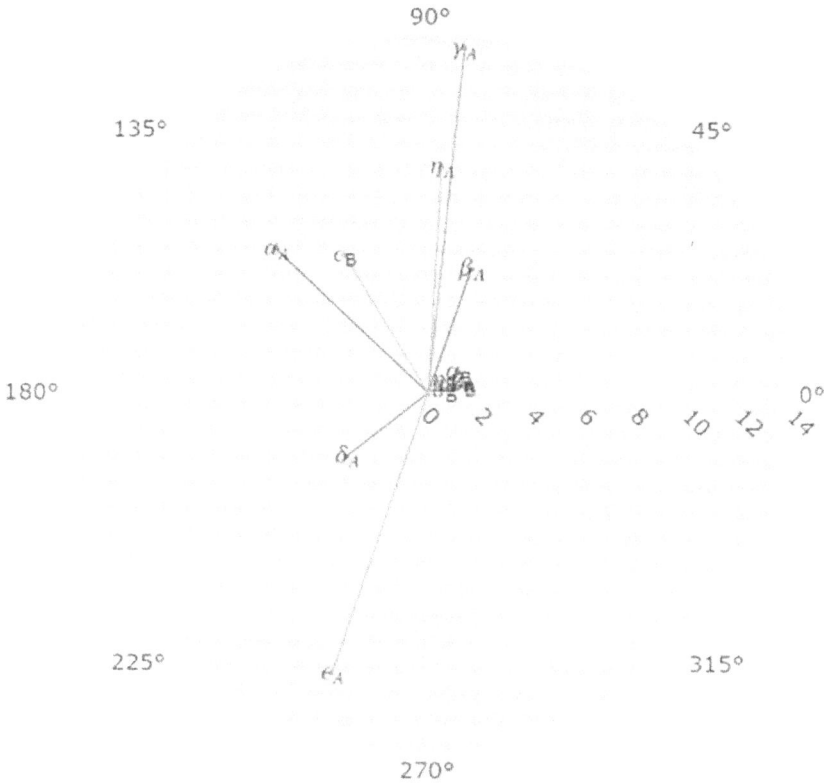

FIGURE 15.3 Plot of A-B parameter for all six antennas from Table 15.1.

second highest gain value. However, its geometric size is predictably far larger than the other plotted antennas, resulting in an $\in_B$ point at 120.8o. To compare with the B-points of other antennas, we can expand the inner part of the graph, as shown in Figure 15.4. Here, we can, for example, appreciate how effectively antenna δ utilizes its space and complete the analysis for the specific antenna use case.

Although the four-part A-B parameter lends itself to polar plotting, it can be plotted as a bar graph, as shown in Figure 15.5. This construction allows the parameters to be plotted in dB scale ($10 \times \log_{10}$) helping to make the appropriate comparisons. The dB metric is, of course, a comparative measure, and in these cases, the LLD is being compared with one wavelength (again center frequency of the lowest band), and the profile (b_2) is compared against the nonphysical case of a 1-D line. So for, an ideal whip monopole using LLD as z and, with no measurable width, the profile would be z

$$b_{1-\text{monopole}} = \frac{z}{z + x + y} = \frac{z}{z + 0 + 0} = 1. \tag{15.6}$$

It should be noted that a perfect cube would plot to $-4.77\,\text{dB}$ and values below that indicate that: (i) the z-direction is the direction of propagation that is not the LLD

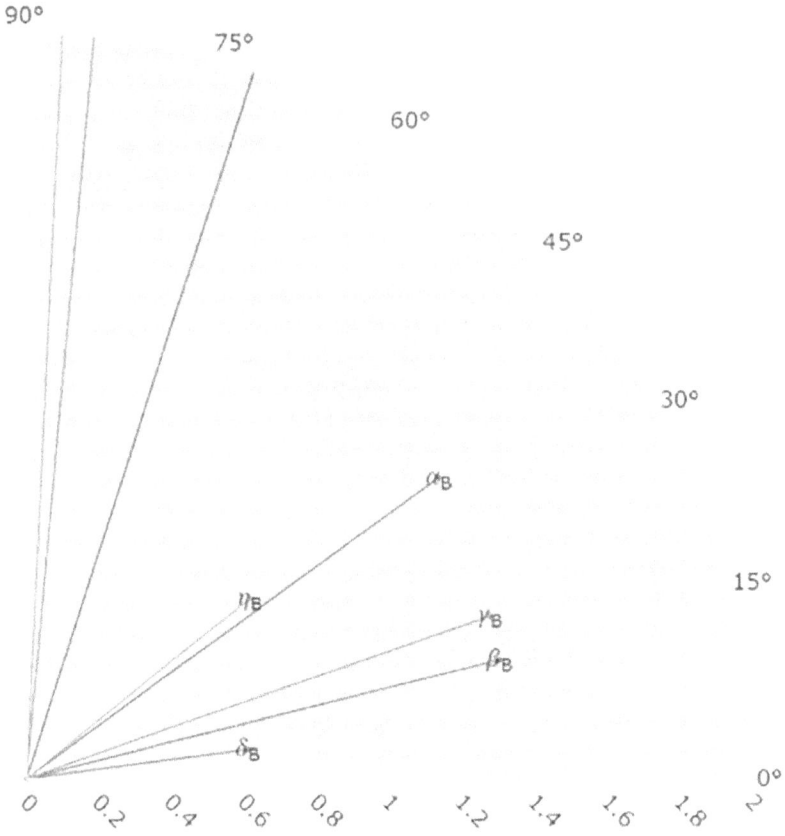

FIGURE 15.4   Plot of A-B parameter values from Table 15.1. This is a zoomed-in view at the cluster of B points in the first quadrant close to the origin (the exception being the horn with its dramatically larger geometry).

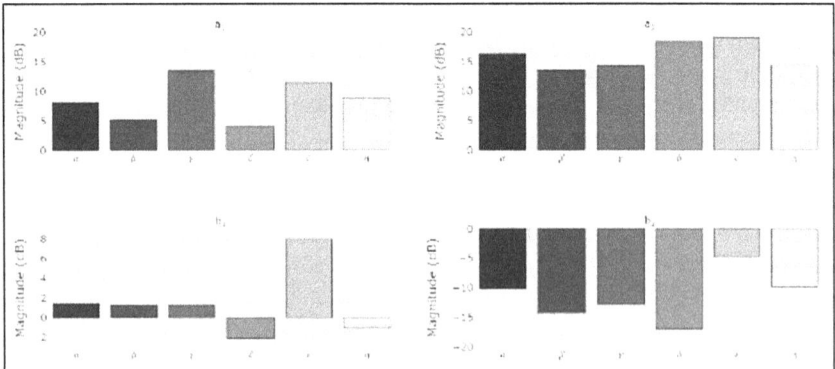

FIGURE 15.5   Bar plot representation of the A-B parameter for the six antennas listed in Table 15.1, broken into component values for simplified comparison.

and (ii) the larger the magnitude of this value, the lower profile the antenna is. Thus, antenna δ that has a b_2 value of $-16.99\,\text{dB}$ is structurally similar to a patch antenna.

15.2.2 MULTIBAND ANTENNAS

Multiband antennas add complexity to the parameterization of antenna performance; the simplest way to construct an A-B parameter for a multiband antenna is simply to add the percent bandwidths and use a gain weighted by the proportion of the bandwidth at that gain. For example, if an antenna had two bands, a lower with 10% fractional bandwidth and 8 dB of gain and an upper with 5% fractional bandwidth and 9 dB of gain, the total bandwidth would be 15%, and the weighted gain would be

$$\left[8\left(\frac{10}{15}\right) + 9\left(\frac{5}{15}\right) \right] = 8.33. \tag{15.7}$$

Or more generally, for an antenna with n bands, each with a fractional bandwidth b_n and gain g_n

$$G_{\text{weight}} = \sum_n \frac{g_n b_n}{\sum_n b_n}. \tag{15.8}$$

Note that this does not take into account the distance between bands, which is a weakness of this method.

15.2.3 ULTRAWIDEBAND ANTENNAS

Ultrawideband (UWB) antennas are becoming increasingly important in a wide variety of applications, but they add a wrinkle to the A-B parameter.

Due to the limitations of the percent bandwidth calculations over large bands (greater than 2:1), a ratio bandwidth is used

$$B\,W_r = \frac{f_h}{f_l} \tag{15.9}$$

and is expressed as $B\,W_r : 1$. The ratio bandwidth is related to the fractional bandwidth defined above

$$B\,W_f = \frac{B\,W_r - 1}{B\,W_f + 1}. \tag{15.10}$$

The asymptotic relationship between the two becomes apparent when the two are plotted in Figure 15.6. Since the ratio bandwidth is used for wideband applications, the A-B parameter (if it is to be useful in a broad range of applications) must be able to adapt to ratio bandwidth. Ratio bandwidth, however, is not numerically limited to between 0 and 2, so, in order to compare antennas, in this class, an alternate method is needed to calculate a_2. Again, the engineer selecting the antenna comes to the table

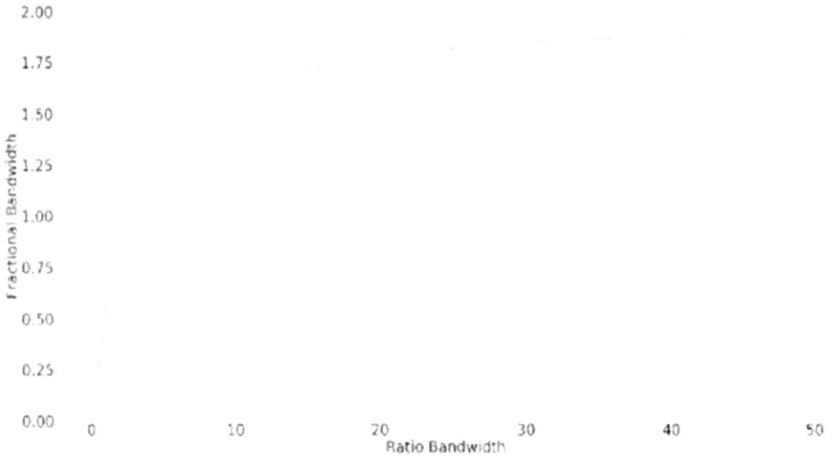

FIGURE 15.6   Plot of fractional bandwidth as a function of ratio bandwidth. Generally, fractional bandwidth is utilized on the lower end where it behaves linearly.

with preexisting knowledge of the type of system that he or she is looking for and choose the appropriate method for comparison. For a set of antennas to be evaluated whose bandwidths are given as fractional bandwidth, first, the order of magnitude of the largest given B W_r value is taken as O_{max}. Thus, for example, if two UWB antennas are being compared, with 80:1 and 10:1 stated bandwidths respectively, $O_{max} = 10$. Alternately, for two antennas 110:1 and 150:1, $O_{max} = 100$. From O_{max}, reduced bandwidth values are calculated from the given X numbers
BW$_r$

$$B\ W_{red} = \frac{B\ W_r}{O_{max} \times 10} \tag{15.11}$$

This ensures that B W_{red} is a value between 0 and 1, allowing it to be used as a percentage. For the abovementioned examples, B W_{red} for the first set of equations would be 0.8 and 0.1 and, for the second set, would be 0.11 and 0.15, respectively. These values are, in turn, used to make the a_1 term of the A-B parameter by setting them as a percent value of 2

$$a_1 = 2B\ W_{red} \tag{15.12}$$

Now, the radial value of the A point varies between 0 and 2, mapping properly onto the polar plot.

15.2.4   COMPARISON WITH KIROV'S RESULTS

As noted earlier, Kirov's recent article [6] also proposes a reframing of antenna comparison and selection and offers a series of proposed criteria to enable it. Here, the authors will generate and plot A-B parameters for the top five antennas, as listed in

TABLE 15.3
Kirov Antennas

Antenna Identifier	Antenna Type
$K1$	Stacked Microstrip Broadband Antenna [15]
$K2$	X-band Bidirectional [16]
$K3$	Couple-fed Bow Tic Microstrip Antenna [17]
$K4$	Stacked Annular-Ring Microstrip Antenna [18]
$K5$	Wideband Shorted Patch [19]

TABLE 15.4
Kirov A-B Values

Antenna Identifier	a_1	a_2	b_1	b_2
$K1$	5.7900	0.0740	0.4303	0.0003
$K2$	3.8700	0.0968	0.7238	3.852×10^{-5}
$K3$	6.3100	0.0740	0.4662	0.002
$K4$	7.8300	0.0637	0.7052	6.7751×10^{-5}
$K5$	3.8000	0.0118	0.3484	0.0001

[6, Table 4], not as criticism, but to provide a comparison with that work. It will be at the discretion of the reader to determine which method best suits his or her particular problem space. The antennas listed in [6] are all circularly polarized patch antennas, and the top five antennas listed in [6, Table 4] are seen in Table 15.3.

Using the average gain and bandwidth values listed [6] along with geometric numbers from each respective paper, A-B parameters can be computed for each of the five antennas and listed in Table 15.4. Kirov A-B Values.

As expected for the variations of patch antennas, the designs are relatively low bandwidth and low profile (even in stacked configurations), properties reflected in the a_2 and b_2 values, respectively. The resultant A-B parameters are then plotted in Figures 15.7–15.9 using the method seen in Section 15.2.

While Kirov's four criteria in sum provide more gain information than the A-B parameter, which is based on a single gain value (average gain for the calculations in this article), it provides less volumetric information. In addition, since the A point is composed of electrical performance information, while the B point pertains to volumetric measures, it is easy to compare both characteristics of the antennas being compared. Nonetheless, both methods have advantages, and the engineer can select the methodology best suited to a specific design problem.

15.3 ALTERNATE VOLUMETRIC FIGURE OF MERIT

An additional metric can be calculated to reduce the A-B parameter to a single number at the cost of some informational content of the parameter. This metric is similar enough to an efficiency that it will be called "volumetric efficiency" here and denoted as V_{Eff}.

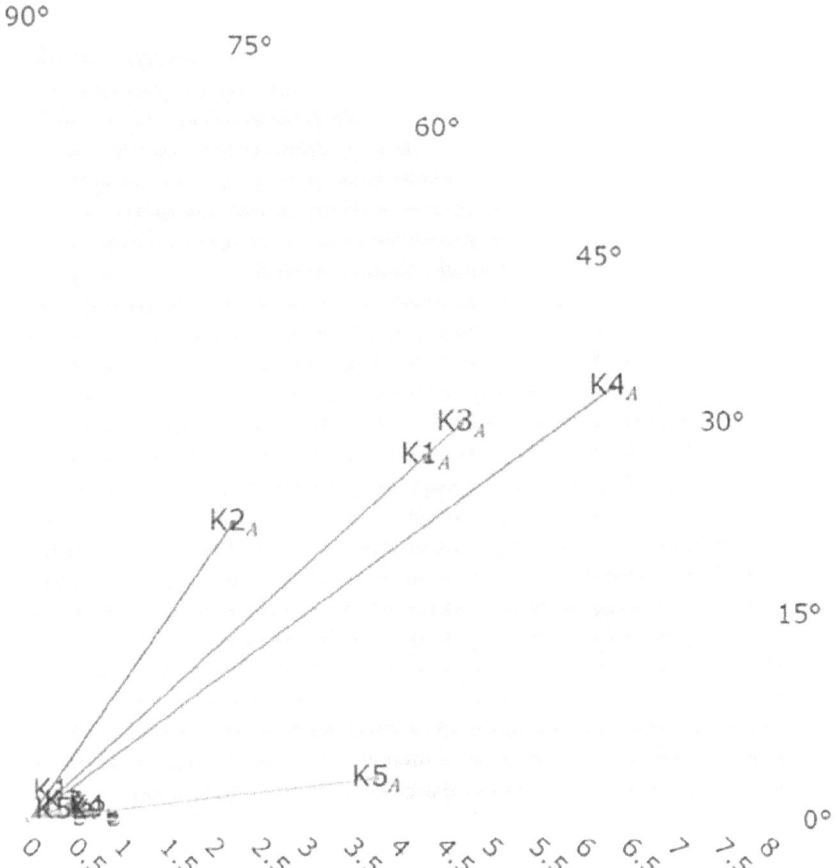

FIGURE 15.7   Plot of the A-B parameter for the five antennas from Kirov's article [6] listed in Table 15.3.

This volumetric efficiency is a dimensionless quantity calculated from the A-B parameter by taking one minus the area subtended by point B over that subtended by point A and is calculated as follows:

$$V_{\text{Eff}} = 1 - \frac{2b_2 b_1^2}{a_2 a_1^2}. \tag{15.13}$$

Although these may be compared to our six exemplar designs, they are, with the exception of the horn, ostensibly low profile, so the volumetric efficiency is close to one with little variation. Table 15.5. Volumetric Efficiency bears this out. For completeness, the V_{Eff} values for the antennas from the Kirov article [6], as listed in Table 15.3. Kirov Antennas, are calculated in Table 15.4. Kirov A-B Values. While this method may assist in antenna selection, the information lost in constructing V_{Eff} may be detrimental. For example, $\delta>$'s large bandwidth has come at the expense of

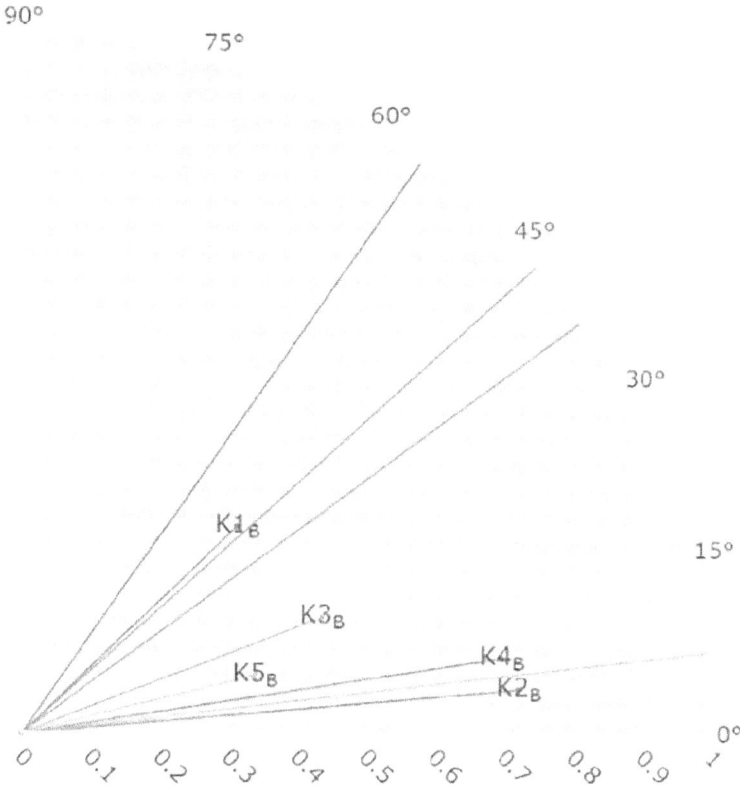

FIGURE 15.8 Plot of the A-B parameter for the five antennas from Kirov's article [6] listed in Table 15.3. This is a zoomed-in view at the cluster of B points in the first quadrant close to the origin.

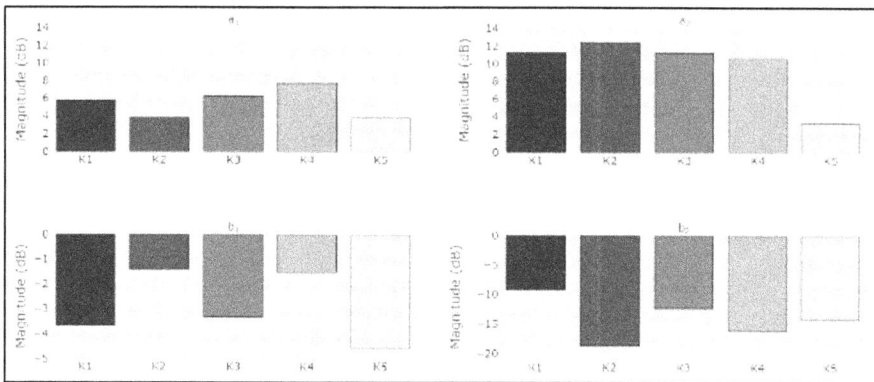

FIGURE 15.9 Bar plot representation of the A-B parameter for the five antennas from Kirov's article [6] listed in Table 15.3. Note that the scales are different on the b_1 and b_2 plots.

TABLE 15.5
Volumetric Efficiency

Antenna Identifier	V_{Eff}
α	0.9923
β	0.9867
γ	0.9978
Δ	0.9992
E	0.603
H	0.9966

TABLE 15.6
Volumetric Efficiency of Kirov Analyzed Antennas

Antenna Identifier	V_{Eff}
$K1$	0.9943
$K2$	0.9968
$K3$	0.9973
$K4$	0.9980
$K5$	0.9827

high-gain values that other antennas achieve, and it may be that δ 's increased gain, reflected in V_{Eff}, is not worth the cost.

It is clear that the A-B parameter will not replace all other metrics in the antenna selection process, and it does have some limitations. While it provides a gain measure, it does not give any hint of radiation geometry or how that geometry shifts across the bandwidth of the antenna. In addition, if there are antenna systems that cannot physically scale linearly with changes in operating band A-B parameter, comparisons may not yield actionable information. Finally, care must be taken that all metrics used are comparable: some authors publish directionality versus realized gain, and bandwidth values are not always selected to the same standard and may, for example, be measured in terms of return loss, 3-dB gain points, or axial ratio among others. With a little care, however, appropriate comparables may be selected.

15.4 CONCLUSION

The A-B parameter provides a powerful tool in antenna selection, especially where geometric concerns are prominent. Ultimately, however, antenna analysis and design are simply complex and difficult, and attempts to simplify aspects of the process (such as the A-B parameter) will leave out important elements, radiation pattern being the most glaring. While there are no physical bounds to using the A-B parameter for any antenna with nonzero geometric dimensions (again this metric is not directly based on the limitations of radiating physics), the wavelength agnostic nature of comparisons that this figure of merit engenders can lead to unfeasible conclusions. For example, the patch in [9] may simply not be feasible to construct at HF scales, and there are limits to the practical size of horn antennas. Thus, the A-B parameter is not a catchall, but, if thoughtfully used, it can provide a useful method to assist in making selection decisions in geometrically restricted design environments.

REFERENCES

1. R. F. Harrington, "Effect of antenna size on gain, bandwidth, and efficiency," *J. Res. Nat. Bureau Standards*, vol. 64D, no. 1, pp. 1–12, Jan./Feb. 1960.
2. L. J. Chu, "Physical limitations of omni–directional antennas," *J. Appl. Phys.*, vol. 19, no. 12, pp. 1163–1175, Dec. 1948.
3. M. Shahpari and D. V. Thiel, "Fundamental limitations for antenna radiation efficiency," *IEEE Trans. Antennas Propag.*, vol. 66, no. 8, pp. 3894–3901, Aug. 2018.
4. M. Shahpari, D. V. Thiel, and A. Lewis, "An investigation into the gustafsson limit for small planar antennas using optimization," *IEEE Trans. Antennas Propag.*, vol. 62, no. 2, pp. 950–955, Feb. 2014.
5. H. Arai, "Electrical antenna volume for a scale of downsizing," in *Proc. IEEE Antennas Propag. Soc. Int. Symp.*, Ann Arbor, MI, USA, Jun. 1993, pp. 1854–1857.
6. G. S. Kirov, "Evaluation of the frequency bandwidth and gain properties of antennas: Characteristics of circularly polarized microstrip antennas," *IEEE Antennas Propag. Mag.*, vol. 62, no. 3, pp. 74–82, Jun. 2020.
7. C. A. Balanis, *Antenna Theory: Analysis and Design*, 4th ed. Hoboken, NJ: Wiley, 2016.
8. *Airmax 900 MHz Yagi Antenna Datasheet*, MIMO High-Gain Antenna Model, Ubiquiti Networks, Inc, 2016.
9. J. Guo, Y. Zou, and C. Liu, "Compact broadband crescent moon-shape patch-pair antenna," *IEEE Antennas Wireless Propag. Lett.*, vol. 10, pp. 435–437, 2011.
10. D. Feng, H. Zhai, L. Xi, S. Yang, K. Zhang, and D. Yang, "A broadband low-profile circular-polarized antenna on an AMC reflector," *IEEE Antennas Wireless Propag. Lett.*, vol. 16, pp. 2840–2843, 2017.
11. R. Lian, Z. Wang, Y. Yin, J. Wu, and X. Song, "Design of a lowprofile dual-polarized stepped slot antenna array for base station," *IEEE Antennas Wireless Propag. Lett.*, vol. 15, pp. 362–365, 2016.
12. W. Liu, Y. Yin, W. Xu, and S. Zuo, "Compact open-slot antenna with bandwidth enhancement," *IEEE Antennas Wireless Propag. Lett.*, vol. 10, pp. 850–853, 2011.
13. *Waveguide antennas technical data sheet*, document PE9887-11, 2011.
14. Z. Zhou, Z. Wei, Z. Tang, and Y. Yin, "Design and analysis of a wideband multiple-microstrip dipole antenna with high isolation," *IEEE Antennas Wireless Propag. Lett.*, vol. 18, pp. 722–726, 2019.
15. H.-C. Lien, H.-C. Tsai, Y.-C. Lee, W.-F. Lee, and W.-F. Lee, "A circular polarization microstrip stacked structure broadband antenna," *PIERS Online*, vol. 4, no. 2, pp. 259–262, 2008.
16. J. Shen, C. Lu, W. Cao, J. Yang, and M. Li, "A novel bidirectional antenna with broadband circularly polarized radiation in x-band," *IEEE Antennas Wireless Propag. Lett.*, vol. 13, pp. 7–10, 2014.
17. H.-C. Lien, Y.-C. Lee, and H.-C. Tsai, "Couple-fed circular polarization bow tie microstrip antenna," *PIERS Online*, vol. 3, no. 2, pp. 220–224, 2007.

18. K. Ding, T. Yu, and Q. Zhang, "A compact stacked circularly polarized annular-ring microstrip antenna for GPS applications," *Prog. Electromagn. Res. Lett.*, vol. 40, pp. 171–179, 2013.

19. X. Tang, H. Wong, Y. Long, Q. Xue, and K. L. Lau, "Circularly polarized shorted patch antenna on high permittivity substrate with wideband," *IEEE Trans. Antennas Propag.*, vol. 60, no. 3, pp. 1588–1592, Mar. 2012.

16 First Principle Scintillation Characterization of Natural and Artificial Disturbances on V/W Band Signals in the Ionosphere Using the Multiple Phase Screen Technique*

Andrew J. Knisely, and Andrew J. Terzuoli
US Air Force Institute of Technology

CONTENTS

16.1 INTRODUCTION

A disturbance to the Ionosphere can cause a change in the distribution of the electron content. A common natural occurrence that can contribute to this alteration are geomagnetic storms. Geomagnetic storms result from a compression of the magnetosphere due to the arrival of a solar wind discontinuity. The electric fields, currents,

---

* Reprinted with permission: Knisely, A. J., & Terzuoli, A. J. First Principle Scintillation Characterization of Natural and Artificial Disturbances on V/W Band Signals in the Ionosphere Using the Multiple Phase Screen Technique. In *2020 XXXIIIrd General Assembly and Scientific Symposium of the International Union of Radio Science* (pp. 1–4). IEEE.

DOI: 10.1201/9781003220978-21

and particle precipitation increase as a large amount of energy are deposited into the Ionosphere. The auroral E-region electron densities increase in conjunction with ion and electron temperatures at high latitudes. At mid-latitudes, the wind speeds increase and drive the F region plasma toward higher altitudes which can result in ionization enhancements. For big storms, the enhanced neutral winds and composition changes penetrate toward the equatorial region [1].

High Altitude Nuclear Explosions (HANES) contribute an extraordinary amount of energy to the Ionosphere. Up to three-fourths of the energy yielded by the nuclear explosion may be expended in ionizing the atmosphere, resulting in changes that are characteristic of the burst and the debris altitude. Nuclear explosions at high altitudes may affect a considerable portion of the Ionosphere in a manner similar to solar activity; however, the mechanisms of the interactions with the atmosphere are quite different. Due to the complexities of these interactions, descriptions of the typical changes to be expected from a nuclear explosion are often not applicable or very meaningful without explicit statements of the conditions [2].

During a nuclear explosion, the air inside the fireball is at a temperature of many thousands of degrees. Electron density and collision frequency are high in addition to the absorption of electromagnetic waves. The regions around the fireball are ionized in varying degrees by the initial thermal and nuclear radiations and by the delayed particles from the radioactive debris. As the detonation altitude increases, the radiation can escape at greater distances and the electron density will reach values at which electromagnetic signal propagation can be affected. In the D region of the Ionosphere, the most persistent absorption of electromagnetic waves will take place. In the E and F regions, the frequency of particle collisions is low, and refraction is the predominant effect [2].

The Ionosphere behavior, in general, can be described as a plasma. The local electric field (E-field) governs the movement of the plasma, and the particle is also a source of the electric field. Electrons typically respond much more rapidly to an E-field and ion-motion can be neglected. A group of electrons in the plasma will move in response to the wave field. This movement alters the local concentration of the electric charge, thereby making their own contribution to the electric field [3].

Computationally modeling the fluctuations of electron density in the Ionosphere caused by natural and artificial disturbances typically involves the use of phase screens. A multiple phase screen (MPS) model compresses the thick Ionosphere layer into a series of thin screens that sample the irregularities on a finite grid. This method avoids the computational limitations of discretizing a large layer while calculating the E-field at high frequencies. A metric to measure the effects of randomized Ionosphere irregularity fluctuations is the scintillation index. The scintillation index is the normalized variance of intensity fluctuations defined in formula (16.1). The variable "I" is the intensity calculated from the absolute square of the electric field, i.e.

$$S_4 = \sqrt{\frac{\langle I^2 \rangle - \langle I \rangle^2}{\langle I \rangle^2}} \qquad (16.1)$$

TABLE 16.1

Phase Screen Parameters

Parameter	Value
L: Grid Length [km]	10 L_0
L_0: Outer Scale [km]	3
I_1: Inner Scale [m]	150
N: Spatial Samples	4,096

The following section describes the process of developing the MPS model and the test cases for examining first principle signals transmitted in V/W frequency bands.

16.2 METHODOLOGY

The multiple phase screen (MPS) model begins with defining the parameters of each phase screen. The phase screen converts the fluctuations of electron density into a phase variance and power spectral density in the spatial frequency domain. The grid length of the phase screen is dictated by the outer scale (largest) of the irregularities while the sample index is dictated by the inner scale (smallest) of the irregularities. The outer and inner scale sizes are chosen on the order of kilometers and meters, respectively, as often found in the literature (see [4]). Table 16.1 summarizes these phase screen parameters.

A total of five phase screens are used in this MPS model. The region between each phase screen is considered free space. The propagation through free space is accomplished using the split-step Fourier method, also known as the spectral method. The spectral method solves the paraxial approximation of the scalar Helmholtz equation to relax the phase sampling requirements and ease the computational expense of high-frequency wave propagation. This is also known as the parabolic wave equation (16.2), given by

$$\frac{\partial \overline{U}(x,z)}{\partial z} = \frac{1}{2\underline{y}} \frac{\partial^2 \overline{U}(x,z)}{\partial x^2} \tag{16.2}$$

where "U" is the amplitude and γ is the propagation constant of the wavefront consisting of the attenuation parameter set equal to zero and the propagation constant equal to $2\pi/\lambda$. The formulation for the parabolic wave equation is written as an interior problem with natural boundary conditions is

$$BC \quad \begin{cases} \underline{U}(0,z)=0 \\ \underline{U}(L,z)=0 \end{cases} \quad 0 \leq z \leq Z_{max}$$

$$IC \quad \underline{U}(x,0)=\underline{\phi}(x) \quad 0 \leq x \leq L$$

where "$\phi(x)$" is the initial input function, "L" is the grid length of the wavefront samples, and "z" is the distance along the propagation path with respect to altitude.

The spectral method technique relies on transforming from spatial position to spatial frequency domains and using trigonometric global basis functions to approximate the continuous PDE defined in (16.2). The derivative is approximated by the sequence of Fourier modes in the following manner:

$$\frac{d^k U_N}{dx^k}(x_j) = \sum_{|n| \le N} (in)^k \hat{a}_n e^{\left[\frac{2\pi i}{L}\right] n(x_j)} \tag{16.3}$$

The Fast Fourier transform is taken to acquire the Fourier U coefficients, $\hat{a}_n$ via

$$\hat{a}_n = fft(U) = \frac{1}{N} \sum_{j=0}^{N-1} U(x_j) e^{inx_j} \tag{16.4}$$

The second-order spatial derivative is formulated into a square diagonal matrix where each dimensional length is equal to the length of the input vector. The sequence of Fourier modes is arranged to be evenly uniform across the spatial×grid. Applying this process, take the Fourier transform on both sides of the parabolic wave equation:

$$\int_{-\infty}^{\infty} \left[\frac{\partial^2 \overline{U}(x,z)}{\partial x^2}\right] e^{-\beta x} dx = 2\underline{y} \int_{-\infty}^{\infty} \left[\frac{\partial \overline{U}(x,z)}{\partial z}\right] e^{-\beta x} dx \tag{16.5}$$

This forms an ODE in the spatial frequency domain, i.e.

$$-K^2 \widetilde{\underline{U}}(K,z) = 2\underline{y} \frac{\partial \widetilde{U}(K,z)}{\partial z} \tag{16.6}$$

Integrating both sides of the equation yields

$$\int_{z_1}^{z_2} -\frac{K^2}{2\underline{y}} dz = \int_{z_1}^{z_2} \frac{1}{\widetilde{\underline{U}}(K,z)} d\widetilde{\underline{U}}(K,z) \tag{16.7}$$

The solution is then inverse Fourier transformed back to the spatial position domain, i.e.

$$\widetilde{\underline{U}}(K,z_2) = \widetilde{\underline{U}}(K,z_1) e^{-\left(\frac{K^2}{2\underline{y}}\right)\Delta z} \tag{16.8}$$

$$\widetilde{\underline{U}}(x,z_2) = ifft\left(\widetilde{\underline{U}}(x,z_1)\right) \tag{16.9}$$

The number of phase screens is dictated by adequate sampling of this free space when using the spectral method. The minimum frequency examined in this chapter includes 378 MHz. By using this frequency and a distance of 380 km between each screen, the free space sampling requirement can be verified by

$$\Delta z < \frac{2L\Delta x}{\lambda}$$

$$\rightarrow 380,000 < \frac{2(30,000)\left(\dfrac{30,000}{4,096}\right)}{(3\times10^8 / 378\times10^6)}$$

$$\rightarrow 380,000 < 553,710$$

Figure 16.1 illustrates the MPS model along with a superimposed electron density distribution whose density variances are assigned to a respective phase screen.

The test cases collected for the natural disturbance analysis for V/W band signals are shown in Figure 16.2. The electron density data collection was acquired from NeQuick, an online database. NeQuick is a three-dimensional and time-dependent Ionosphere electron density model based on an empirical climatological representation of the Ionosphere [5].

Figure 16.3 shows a sequence of an electron density distribution data set acquired after a computationally modeled nuclear detonation. The data is acquired from a plasma physics-based software.

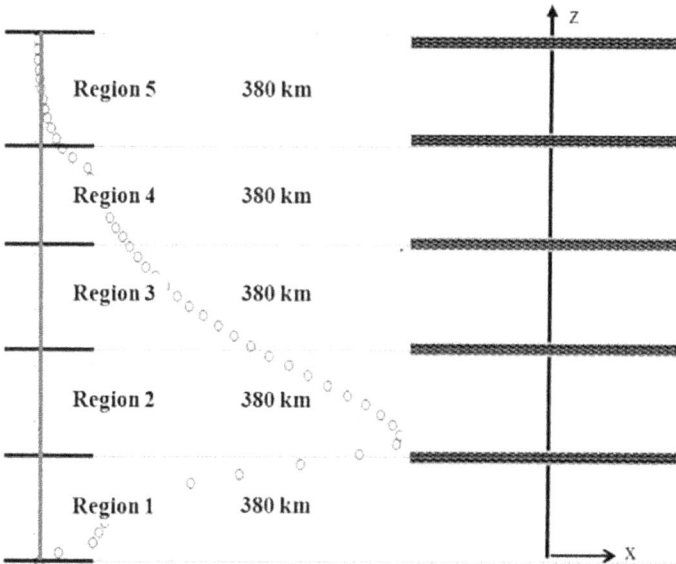

FIGURE 16.1   MPS layout.

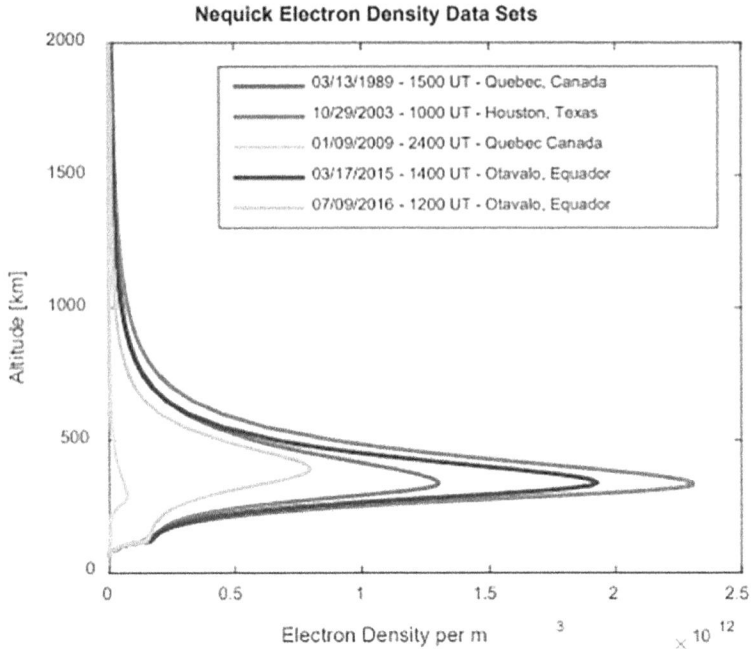

FIGURE 16.2   Electron density distributions.

The MPS simulations in this chapter are conducted at the frequencies of 378 MHz, 40 GHz, 72 GHz, 84 GHz, and 96 GHz. The following section shows the resulting scintillation calculations from these MPS simulations. Each signal generated from the transmitter has an initial amplitude of unity and a phase of zero. The E-field is considered to be a uniform plane wave prior to phase screen incidence.

16.3   RESULTS

Figures 16.4 and 16.5 show a significant difference between the low-frequency and high-frequency scintillation behavior of the signal. At 378 MHz, the scintillation saturates close to unity during the more severe geomagnetic storms. At 84 GHz, the scintillation is much less insignificant, only on the order of 10^{-3}.

The HANE results show that V/W band signals are greatly impacted by the excitation of the electron content in the Ionosphere after a nuclear detonation. The significant scintillation occurs in the initial 45 s of the blast and begins to settle after 1 minute. The electron density content is high enough to yield a large phase variance when converting the electron density to the power spectrum. The larger power spectrum results in a greater magnitude output from the phase screen realization, thus interfering greatly with the intensity of the E-field. It should be noted that the model is considered the first principle as many of the equations used to represent phases of the nuclear detonation are based on generalized descriptions involving nuclear and plasma physics phenomenon.

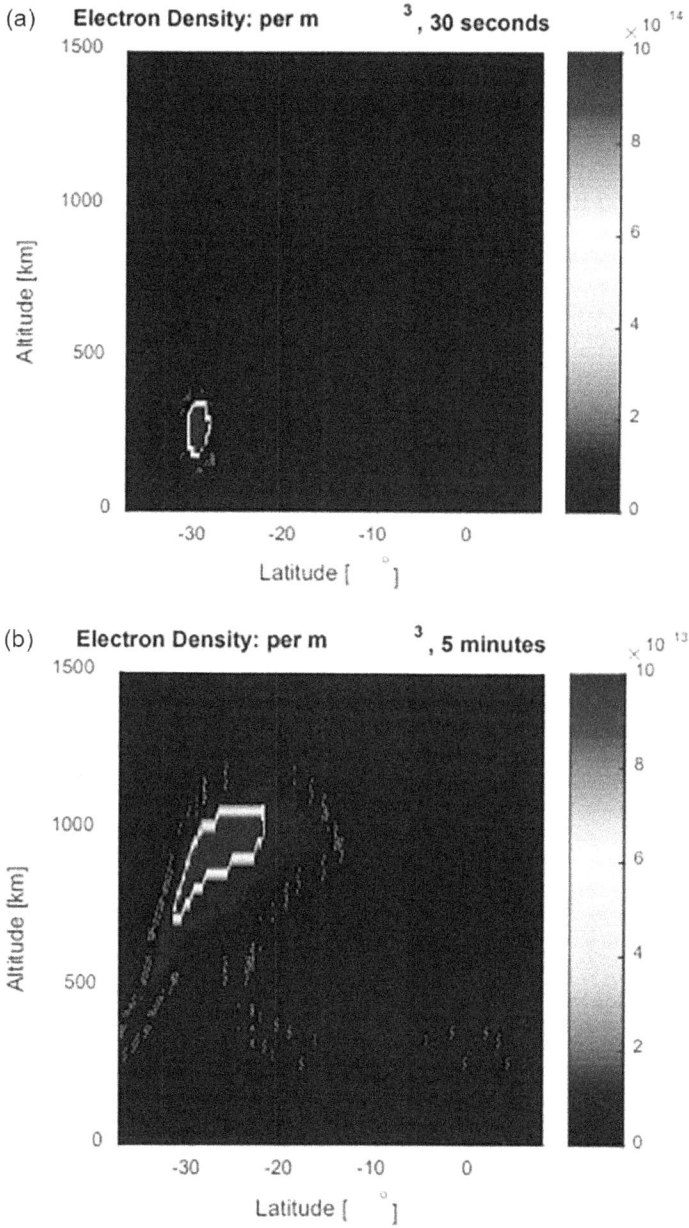

FIGURE 16.3 Nuclear detonation electron density distribution: 30 s after (a) and 5 min after (b).

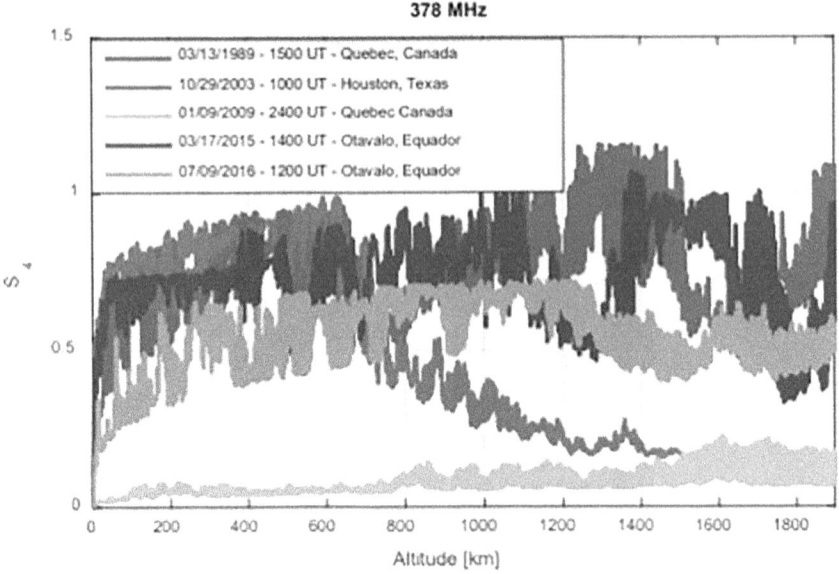

FIGURE 16.4 Scintillation curve for 378 MHz signal (Part A).

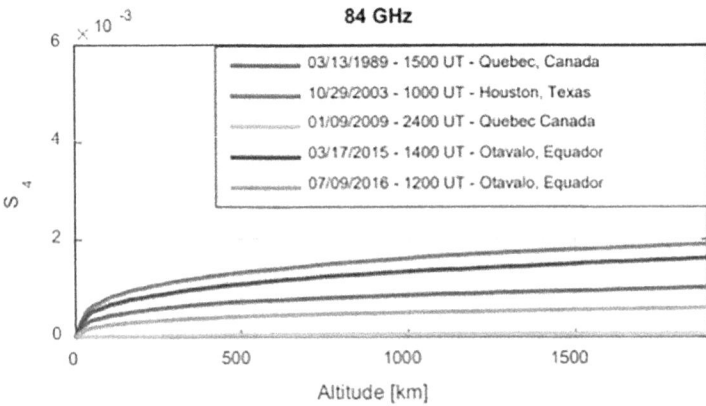

FIGURE 16.5 Scintillation curve for 378 MHz signal (Part B).

16.4 CONCLUSION

The results demonstrate that the environmental factors in the Ionosphere must be considered in extreme cases of electron density fluctuations caused by Ionosphere disturbances. Even though the geomagnetic storms yielded little impact on a V/W band signal, a HANE event has a significant impact. Future research will demonstrate the

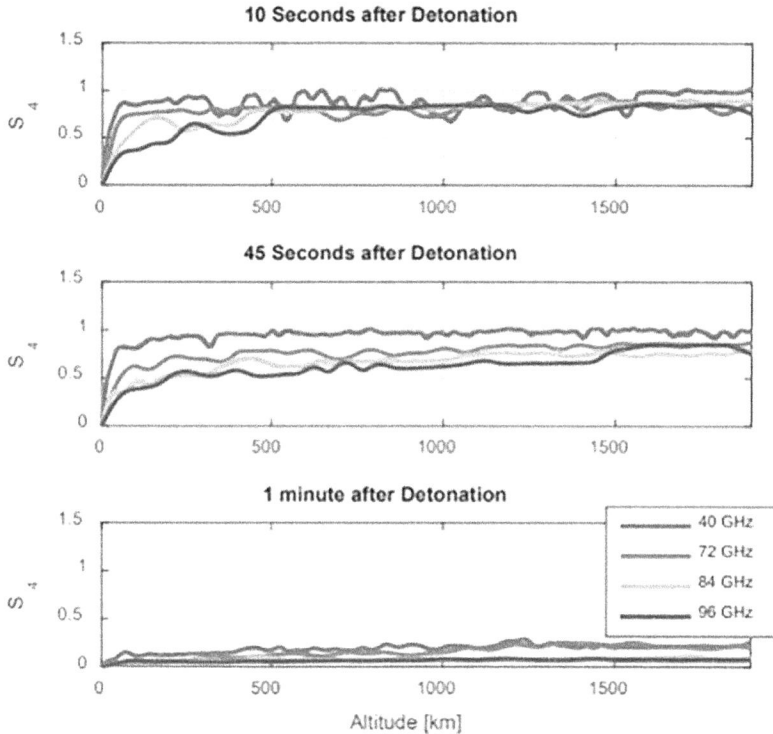

FIGURE 16.6 Scintillation curve time sequence after nuclear detonation.

temporal impact of such events on a wideband signal that incorporates the multiple phase screen techniques to simulate ionic effects. The correlation between phase change and time delay must be understood for HANE events. The significance of these results will lead to a model that can assist with the design of assured communication systems.

16.5 ACKNOWLEDGMENTS

We would like to thank our research assistants for their continued dedication and support on this research effort.

REFERENCES

1. R.W. Schunk, *Ionospheres: Physics, Plasma Physics, and Chemistry*, Cambridge University Press, Cambridge, UK, 2000.
2. Samuel Glasstone, and Philip J. Dolan, "The Effects of Nuclear Weapons", United States Department of Defense and the Energy Research and Development Administration, Washington, DC, 1977.
3. R.0. Dendy, *Plasma Dynamics*, Oxford Science Publications, Clarendon Press, 1990.

4. Dennis L Knepp, "Propagation of Wide Bandwidth Signals Through Strongly Turbulent Ionized Media", electronic file available at: https://apps.dtic.mil/dtic/tr/fulltext/u2/a131355.pdf, 1982.

5. Nava, B., P. Coisson and S.M. Radicella, "A New Version Of The NeQuick Ionosphere Electron Density Model", *Journal of Atmospheric and Solar-Terrestrial Physics*, doi:10.1016/j.jastp.2008.01.015.

17 First Principle Computational EMI Model of V and W Wideband Signal Temporal Delay Induced by a HANE in the Ionosphere*

Andrew J. Knisely and Andrew J. Terzuoli
US Air Force Institute of Technology

CONTENTS

17.1 INTRODUCTION

The SATCOM infrastructure supports many everyday electronic devices including long-range terrestrial networks, global positioning systems, and remote sensing data collections. The quality and performance of SATCOM are dependent on the environmental sources of ambiguity. The operable frequency of the communication system will also affect the interaction between the signal and the physical nature of the environment.

---

* Reprinted with permission: Knisely, A. J., & Terzuoli, A. J. (2020, September). First Principle Computational EMI Model of V and W Wideband Signal Temporal Delay Induced By A HANE in the Ionosphere. In *2020 International Symposium on Electromagnetic Compatibility-EMC EUROPE* (pp. 1–5). IEEE.

DOI: 10.1201/9781003220978-22

The Ionosphere is an atmospheric layer of free ions and electrons between the altitudes of 60 and 2,000 km. The terrestrial Ionosphere at all latitudes has the tendency to separate into stratified layers: D, E, and F. The electron content structure of the Ionosphere during day time hours can be described as steadily increasing from 100 to 300 km (E and F layers), where it begins to decline and decrease at altitudes greater than 300 km. During night hours, the electron content in the E layer (200 km) drops off, giving an electron content profile. A group of electrons in the Ionosphere plasma will move in response to an incident wave field. This movement alters the local concentration of the electric charge, thereby making their own contribution to the electric field as a signal propagates through this region [1].

A disturbance to the Ionosphere can cause a change in the distribution of the electron content. A common natural occurrence that can contribute to this alteration are geomagnetic storms. Geomagnetic storms result from a compression of the magnetosphere due to the arrival of solar wind discontinuity. The electric fields, currents, and particle precipitation increase as a large amount of energy is deposited into the Ionosphere. The auroral E-region electron densities increase in conjunction with ion and electron temperatures at high latitudes. At mid-latitudes, wind speeds increase and drive the F-region plasma toward higher altitudes, which can result in ionization enhancements. For big storms, the enhanced neutral winds and composition changes will penetrate toward the equatorial region [2].

In this chapter, the steady-state Ionosphere is modeled using the Chapman function profile in the magnetic EastWest direction. This provides an ideal Gaussian distribution with a peak electron density. More information regarding the formulation of the Ionosphere using the Chapman profile can be found in [3]. The HANE burst effect is added to this background electron density to provide the disturbed Ionosphere profile. The following section describes the phase process of the burst effect.

17.2   HANE EFFECTS IN THE IONOSPHERE

High-altitude nuclear explosions contribute an extraordinary amount of energy to the Ionosphere. Up to three-fourths of the energy yielded by the nuclear explosion may be expended in ionizing the atmosphere, resulting in changes that are characteristic of the burst and the debris altitude. Nuclear explosions at high altitudes may affect a considerable portion of the Ionosphere in a manner similar to solar activity; however, the mechanisms of the interactions with the atmosphere are quite different. Due to the complexities of these interactions, descriptions of the typical changes to be expected from a nuclear explosion are often not applicable or very meaningful without explicit statements of the conditions [4].

During a nuclear explosion, the air inside the fireball is at a temperature of many thousands of degrees. Electron density and collision frequency are high in addition to the absorption of electromagnetic waves. The regions around the fireball are ionized in varying degrees by the initial thermal and nuclear radiations and by the delayed particles from the radioactive debris. As the detonation altitude increases, the radiation can escape at greater distances and the electron density will reach values at which electromagnetic signal propagation can be affected. In the D region of the Ionosphere, the most persistent absorption of electromagnetic waves will take place.

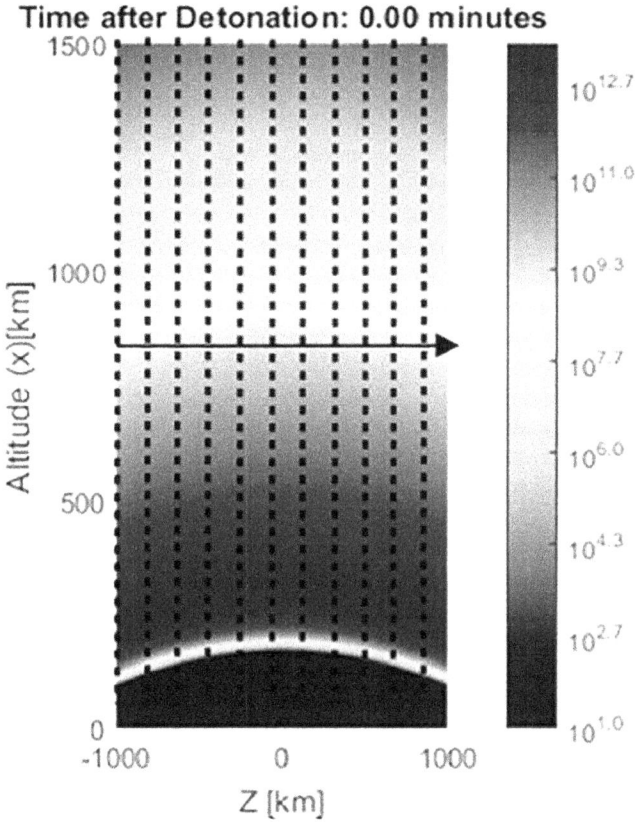

FIGURE 17.1 Electron Density Distribution: 0.0 min with indication of phase screen placement (black dashed lines).

In the E and F regions, the frequency of particle collisions is low, and refraction is the predominant effect [4].

Figures 17.1–17.3 illustrates the sequence of the electron density profile on the Chapman layer as a 1 megaton HANE burst occurs at an altitude of 100 km. Note that in Figure 17.1, the black dashed lines indicate the placement of 11 phase screens and the arrow indicates the direction of the wave propagation. Phase screens are used to compress the electron density fluctuation layer into thin screens of samples described by a power spectral density and are auto-correlated with complex Gaussian realizations. Phase screens are a crucial component of the wideband signal model described in the following section.

17.3 WIDEBAND SIGNAL MODEL

The wideband signal applies for a dispersive (frequency selective) communication channel. A transmit signal is modulated using a complex envelope on a carrier frequency. The parabolic wave equation provides the realizations of the signal's

Time after Detonation: 0.50 minutes

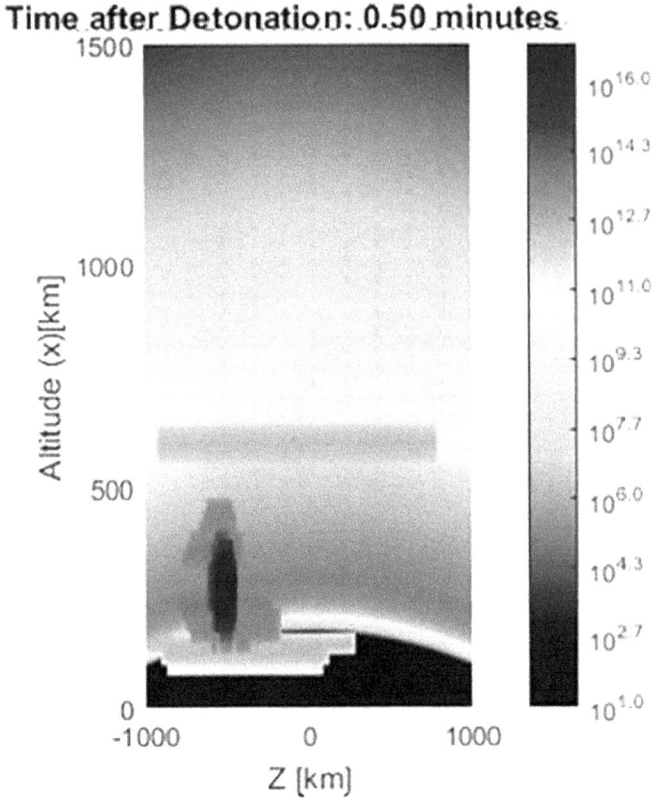

FIGURE 17.2 Electron density distribution: 0.5 min.

amplitude and phase as it propagates spatially in the frequency domain. The signal is Fourier transformed inversely into the time domain at the receiver. More detailed information regarding this methodology can be found in [5]. The following equation represents the modulation of the parabolic wave equation realization at the receiver:

$$v\left(x, z_r, \tau\right) = \frac{1}{2\pi} \int_{-Bw/2}^{Bw/2} M\left(f\right) U\left(x, z_r, f\right) e^{1i(2\pi f)} df \qquad (17.1)$$

$M(f)$ is the amplitude of the modulation waveform at a particular frequency f, $U(x, z, f)$ is the PWE realization and Bw is the frequency bandwidth. The modulation is implemented as a band limited triangular pulse spectrum, represented in the frequency domain as (17.22), where T_c is the pulse duration:

$$M(f) = \begin{cases} \frac{T_c}{2}\left[\mathrm{sinc}\left(2\pi f_c T_c\big/4\right)\right] & \text{if } |f| \le \dfrac{Bw}{2} \\[2mm] \text{else} & 0 \end{cases} \qquad (17.2)$$

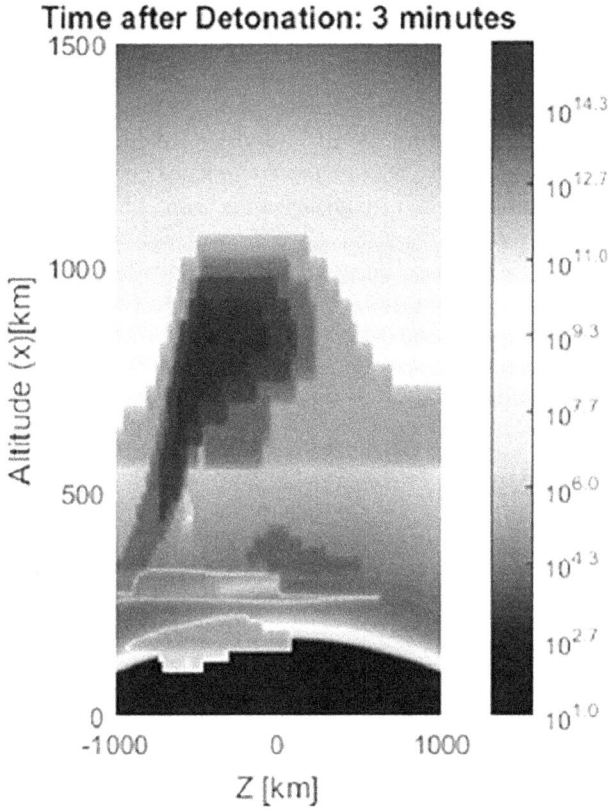

FIGURE 17.3   Electron density distribution: 3 min.

A wave incident on a lens or a phase screen will exhibit time delay, corresponding to a variation of phase. The region in between each phase screen is considered free space. The wideband signal is subject to a divergent Gaussian lens coupled with a Kolmogorov phase screen by multiplication of a phasor on the incident E-field. The Gaussian lens represents the deterministic loud that causes divergence of the E-field away from the receiver when passing through high ionic content. The phase screen provides the stochastic effect of the electron density fluctuations. The Gaussian lens phase is given below:

$$\phi = 100 \left(\frac{f_c}{f} \right) e^{(-[x-15000])/(12000)^2} \tag{17.3}$$

The phasor formulation of the incident E-field is input into the parabolic wave equation. The integral form of the parabolic wave equation is solved computationally using the split-step Fourier method:

$$\int_{-\infty}^{\infty} \left[\frac{\partial^2 \overline{U}(x,z)}{\partial x^2} \right] e^{-\beta x} dx = 2\gamma \int_{-\infty}^{\infty} \left[\frac{\partial \overline{U}(x,z)}{\partial z} \right] e^{-\beta x} dx \tag{17.4}$$

$U(x, z)$ is the complex E-field, β is the propagation constant, and γ is the propagation parameter. Integrating the modulated parabolic wave equation realization at the receiver over the bandwidth yields the time delay of the signal. The time delay is averaged over the temporal grid and is measured in units called chips, the inverse pulse duration. The time delay jitter is the variance in the measured time delay across the grid. Each metric provides a useful means to quantify the temporal performance of the signal given a transmission frequency and a propagation distance.

17.4   RESULTS

Below in Figure 17.4, the sequence of a modulated E-field propagated at 100 MHz is shown after a nuclear detonation.

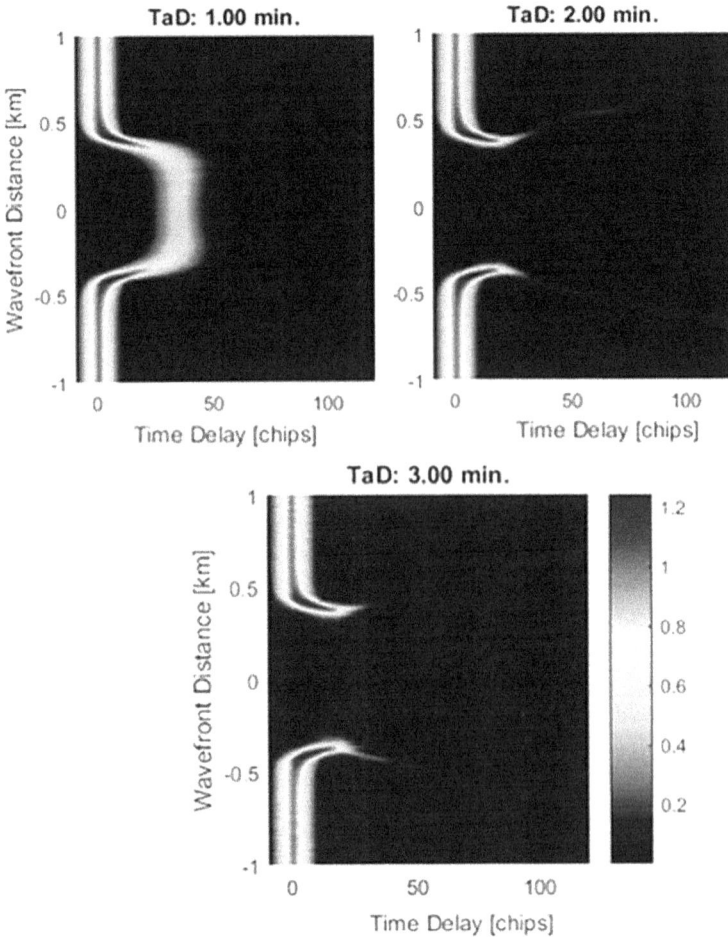

FIGURE 17.4   Temporal delay sequence for received 100 MHz wideband signal at times after detonation (TaD).

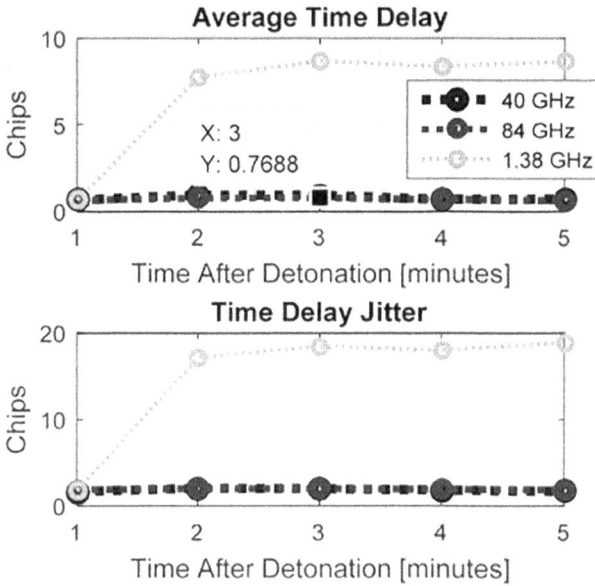

FIGURE 17.5 Average time delay and jitter occurring over the time after detonation.

The delay is well defined by the faint band that forms and spreads outwardly. The low-frequency signal is unable to negotiate the severe disturbance and electron density fluctuations caused by the HANE, subjecting to significant diffraction. The diffraction makes it difficult to quantify the time delay in this case as much of the signal undergoes deconstructive interference. In Figure 17.5, the 1.38 GHz case demonstrates that well after the initial HANE burst, the time delay saturates at around eight chips for a considerable amount of time.

In the V/W band, the time delay is considerably less (fluctuating at around 0.8 chips) for the 40 and 84 GHz cases. The time jitter also appears to decline significantly as well when transmission is at a higher frequency. The shorter time delay indicates that the phase of the E-field does not deviate as significantly from the initial phase compared to the case of low-frequency wave propagation. Higher ionic content will generally cause a low-frequency wave to have a larger phase variation and thus a longer time delay at the receiver.

The nature of these results demonstrates that the disturbed Ionosphere behaves as a frequency selective channel. This notion supports the significance of studying wideband signals not only to examine physical behavior but to ascertain the best possible quality of signal transmission. The sequence in Figure 17.6 provides a global perspective of the time delayed signal transmitted at 1.38 GHz. As altitude is increased above the location of the initial burst at 2.00 min, the signal becomes less obscured by the high shift in the ionic content modeled by phase screens. Eventually, the delay is predominantly due to the background Chapman function profile modeled in conjunction with the Gaussian deterministic cloud lens.

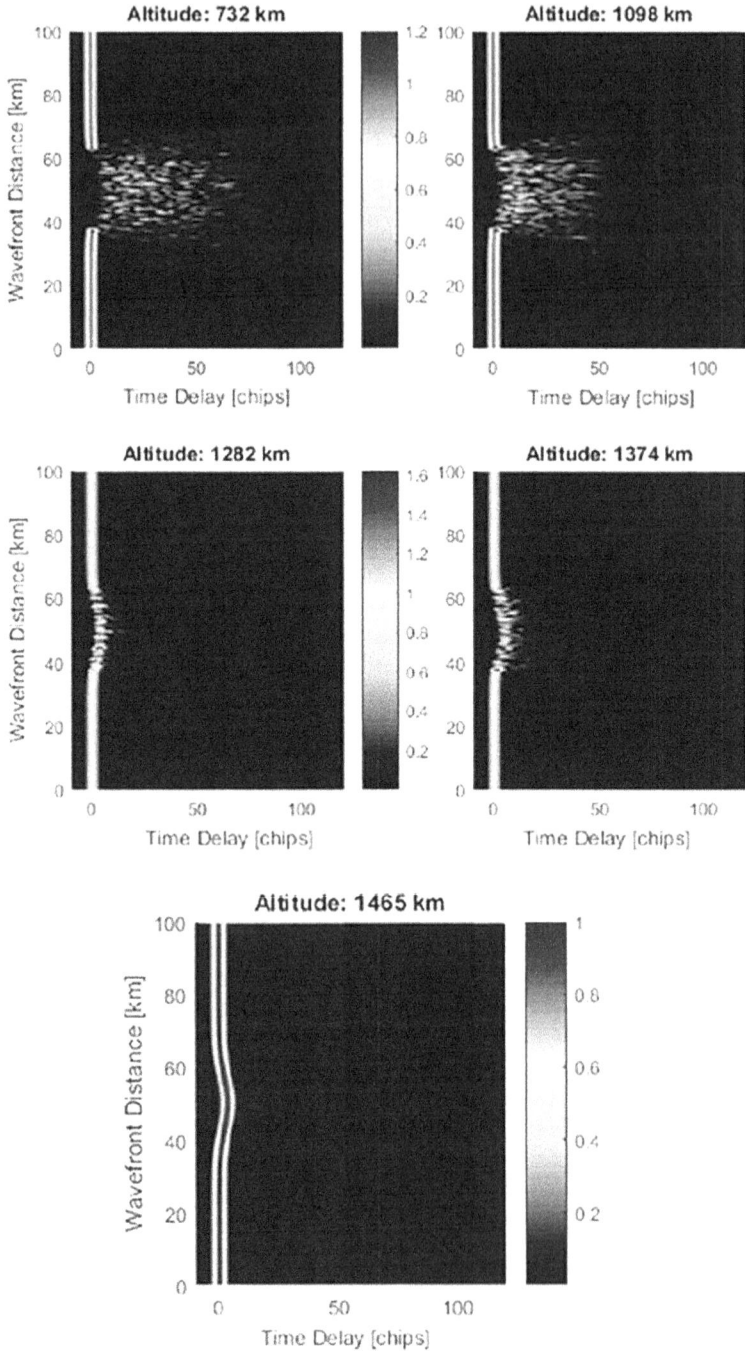

FIGURE 17.6   Sequence of the time delayed signal at 1.38 GHz.

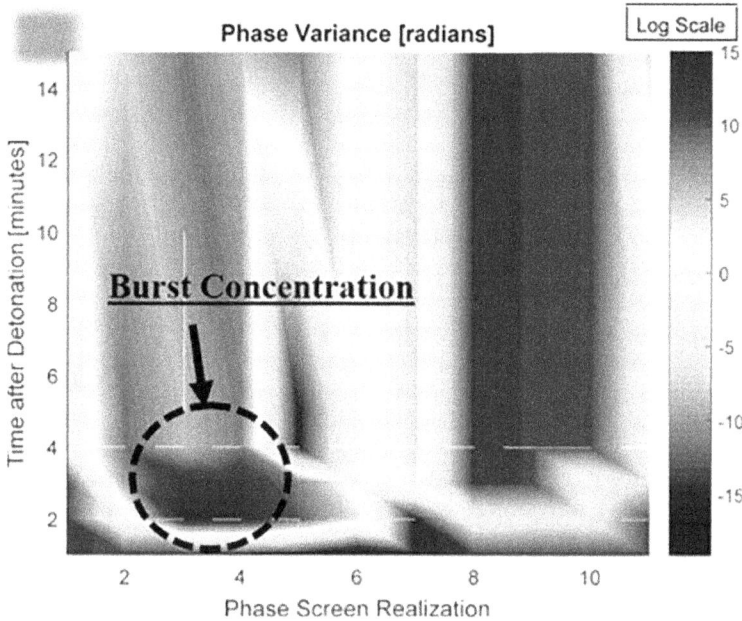

FIGURE 17.7 Phase variances as a function of time after nuclear detonation, 732 km altitude.

As mentioned in the previous section, the phase screen samples the irregular slab layer that is the Ionosphere. The phase variation of this medium is calculated from the variance in the electron density fluctuations. Local regions of high ionic content from the HANE burst will cause a large persistent shift in the phase variance as the energy from the HANE is expended. In Figures 17.7 and 17.8, the distribution of the electron density fluctuations among the 11 phase screens is shown as a logarithmic magnitude of the phase variance. In Figure 17.7, the phase variance decreases for the initial phase screens and increases for the later phase screens as the plume in the electron density propagates upward over time. This is a slow process that causes the time delay to saturate for a lengthy duration after the initial burst. This EMI effect has been known to cause radio blackouts, especially in terrestrial communications that operate in lower bands. In Figure 17.8, the receiver plane altitude is increased above the burst cloud, demonstrating a general decrease in phase variance which ultimately contributes to a shorter time delay. At an altitude above 1,400 km, the phase variance is too small to contribute significant ionic effects from the burst.

17.5 CONCLUSION

The first principle results acquired from the computational model demonstrate the promising characteristics of using V/W band transmission to counteract the EMI caused by a HANE on a wideband signal. The low time delay jitter in the V/W band signals provides ease of characterizing the impact on communication systems. It is

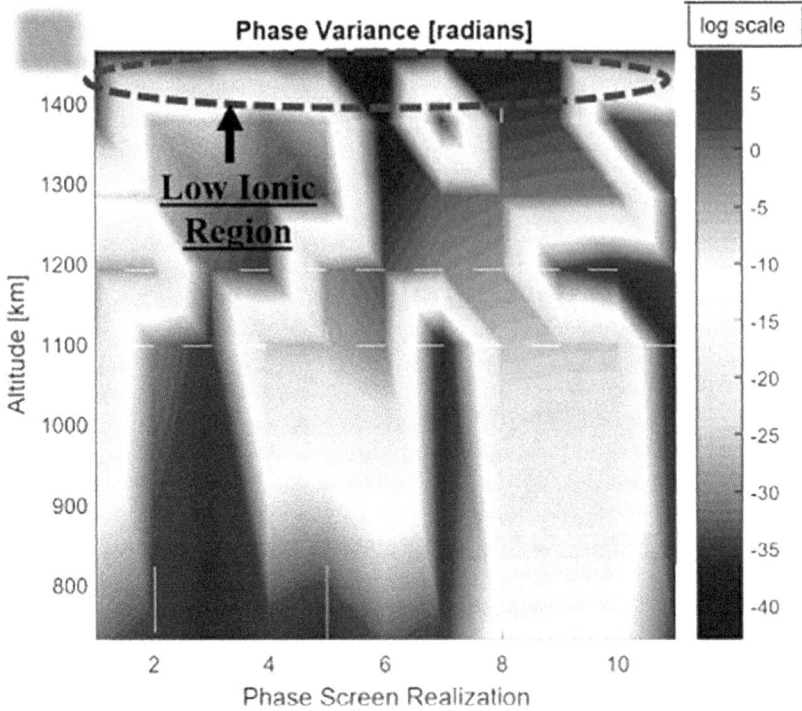

FIGURE 17.8 Phase variances as a function of altitude at 2.00 min after nuclear detonation.

important to note that this model assumes a lateral propagation that does not cross through the Troposphere.

Particulate matter in the Troposphere can cause natural absorption and extinction to occur at higher frequencies, where the wavelength starts to match the particle size. Scattering effects from this particulate matter must also be considered. The dimensionality and geometry of the problem formulation must also be evaluated. The model presented in this chapter is a one-dimensional propagation path that is in direct line-of-sight to the receiver. Modeling of more complex transmission angles involves solving a three-dimensional parabolic wave equation and to assume isotropic irregularities spanning in two dimensions in the phase screens.

17.6 FUTURE WORK

Future work will examine the tropospheric interaction of the HANEs and the impact on a wideband signal to determine the adequate power levels to counteract signal extinction and other forms of atmospheric turbulence. Other environmental changes to the model may include the occurrence of multiple HANE events occurring at various altitudes to examine fallout in signal performance in relation to the communication systems proximity near the burst. The rate at which the time delay declines as frequency increases must also be further evaluated. The research presented in this paper is an initial step toward laying the groundwork for designing EMC assured SATCOM systems that are operable in a wide array of environments by utilizing the uncontested V/W bands.

ACKNOWLEDGMENT

We would like to thank our research assistants for their continued dedication and support on this research effort.

REFERENCES

1. R.O. Dendy, "*Plasma Dynamics*", Oxford Science Publications, Clarendon Press, 1990.
2. R.W. Schunk, "*Ionospheres: Physics, Plasma Physics, and Chemistry*", Cambridge University Press, Cambridge, 2000.
3. Charles S. Carrano, Keith M. Groves, Ronald G. Caton, Charles L. Rino, and Paul R. Straus, "Multiple Phase Screen Modeling of Ionospheric Scintillation Along Radio Occultation Raypaths", *Radio Science*, Vol. 46, RS0D07, doi: 10.1029/2010RS004591, 2011.
4. Samuel Glasstone, and Philip J. Dolan, "The Effects of Nuclear Weapons", United States Department of Defense and the Energy Research and Development Administration, Washington, DC, 1977.
5. Dennis L Knepp, "Propagation of Wide Bandwidth Signals Through Strongly Turbulent Ionized Media", electronic file available at: https://apps.dtic.mil/dtic/tr/fulltext/u2/a131355.pdf, 1982.

18 Wideband SATCOM Model

Evaluation of Numerical Accuracy and Efficiency*

Andrew J. Knisely and Andrew J. Terzuoli
Institute of Electrical & Electronics Engineers (IEEE)

CONTENTS

18.1 INTRODUCTION

18.1.1 IONOSPHERE SCINTILLATION MODEL

Satellites are often used in long-range remote sensing applications to avoid hazardous regions, measure a process without disturbance, and to probe large volumes economically and efficiently [1]. Longer wavelength microwave radiation can penetrate through cloud cover, haze, dust, and all but the heaviest rainfall [1]. The longer

---

* Reprinted with permission: Knisely, A. J., & Terzuoli, A. J. (2020). WIDEBAND SATCOM MODEL: EVALUATION OF NUMERICAL ACCURACY AND EFFICIENCY. *ISPRS Annals of Photogrammetry, Remote Sensing & Spatial Information Sciences, 5*(3).

DOI: 10.1201/9781003220978-23

379

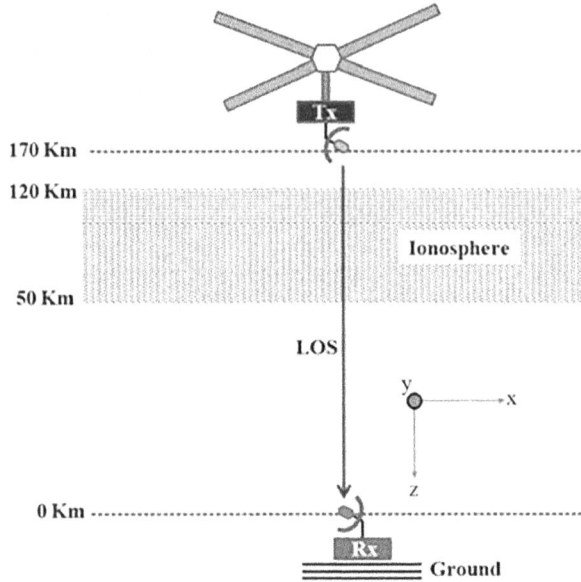

FIGURE 18.1   SATCOM data link to ground receiver.

wavelengths are not susceptible to atmospheric scattering which affects shorter optical wavelengths. It is for this reason that the UHF band is used for the model described in this chapter. The particular process explored in this research is the post data collection of a passive remote sensing satellite. The data acquired from the receiver is processed and transmitted down to an earth-based receiver or ground receiver station as illustrated in Figure 18.1.

Ionosphere scintillation is of major concern to a radio frequency (RF) signal, as it can cause the incident electric field (E-field) to delay in time and phase. The amplitude of the waveform also fluctuates due to constructive and deconstructive interference as the random striations of the Ionosphere interact with the E-field.

18.1.2 Parabolic Wave Equation

The parabolic wave equation (free space form) is given by the following Eq. (18.1):

$$\left(\frac{\partial^2}{\partial x^2} - 2\underline{\gamma}\frac{\partial}{\partial z}\right)\underline{\bar{U}}(x,z) = 0 \tag{18.1}$$

where $\underline{\bar{U}}$ represents the complex waveform in the frequency domain whose field vector points in the x direction and propagates in the z-direction. This equation is derived from the scalar Helmholtz equation by assuming a slow-varying envelope on the wavefront propagation given by the substitution of (18.2):

$$\underline{\bar{E}} = \underline{U}(x,z)e^{-\underline{\gamma}_z z} \tag{18.2}$$

The slow-varying envelope assumptions relax the phase sampling requirements that would otherwise be significant in full-wave simulations of wavefront propagation. This is especially necessary for the SATCOM problem as shown in Figure 18.1, where the propagation distance is long relative to the wavelength of the E-field.

18.1.3 GAUSSIAN LENS AND PHASE SCREENS

The ionic content and scintillation characteristics can be modeled with a divergent Gaussian lens and phase screen, respectively. The divergent lens is representative of high ionic content which contributes a positive phase to the E-field. The phase screen discussed in this research utilizes Kolmogorov theory by taking a slab layer of ionospheric irregularities and compressing it down to a thin layer. Mathematically, the effect is represented as a phasor convolved into the phasor of the incoming waveform. The screen contains a number of grid points that sample the Ionosphere's irregularities ranging within a pre-defined inner and outer scale. The incoming EM wave is constructed to have a total spatial width that is dependent on spatial frequency. The specific realizations of the random phase are obtained by sampling a distribution of phase shifts that have statistical properties determined by the statistics of the electron density irregularities [2]. These statistics are specified by the spatial power spectral density of the irregularities. Relating the power spectral density of the phase and the electron density begins with the two-dimensional autocorrelation of phase and electron density fluctuations. This relationship is translated to the one-dimensional phase power spectral density of spatial dependence by integrating the autocorrelation function over the spatial wave propagation distance.

The numerical phase screen PSD is presented in the following form:

$$S_\phi(K_x) = \frac{\sigma_\phi^2 (2\pi)^{-1/2} l_i (l_i/L_o)^{(m-1)/2} K_{m/2}\left(l_i\sqrt{K_x^2 + 1/L_o^2}\right)}{K_{(m-1)/2}(l_i/L_o)\left(l_i\sqrt{K_x^2 + 1/L_o^2}\right)^{m/2}} \qquad (18.3)$$

where,

$$\sigma_\phi^2 = \sigma_{N_e}^2 (2\pi l_i L_o)^{1/2} r_e^2 \lambda^2 \Delta z \frac{K_{(m-1)/2}(l_i/L_o)}{K_{(m-2)/2}(l_i/L_o)} \qquad (18.4)$$

Variable parameters are defined as:

σ_ϕ^2 is the phase variance, $\sigma_{N_e}^2$ is the electron density variance

L_o is the outer scale of the plasma, l_i is the inner scale of the plasma

K_x is the spatial frequency of the phase screen grid points corresponding to the x direction

m is the slope of the power spectral density, K_n is the Bessel function of order n

Proper implementation of a phase screen requires that the following conditions are satisfied:

Sample Grid: $L > 5 L_o$

Sample index: $\Delta x \leq \frac{l_i}{3}$

where L is the length of the spatial grid and Δx is the spatial resolution of the samples along the grid.

Other conditions exist for the behavior at the boundaries of the phase screen. These effects are minimized by establishing continuity at the boundaries of the screen and the numerical algorithms that solve the parabolic wave equation.

18.1.4   NUMERICAL TECHNIQUES

The numerical techniques used for solving the parabolic wave equation (PWE) are the finite difference method (FDM) and the spectral method (SM). Each method has its respective advantages and disadvantages. FDM uses localized interpolants to discretize the continuum. FDM approximations are pointwise and constrained to a uniform grid. SM uses global trigonometric interpolants that are very accurate when the input function is smooth (continuously differentiable) and uniform. The computational cost of implementing each algorithm is dependent on the input data size and the methodology for solving the PWE to acquire solutions. FDM will require matrix inversion, a computationally expensive operation. However, SM's accuracy can only be improved if more points are used to sample a progressively non-uniform input function. Thus, a trade-off will occur where it becomes impractical to use SM over the localized approximants of FDM in terms of computational cost.

18.2   METHODOLOGY

18.2.1   PARABOLIC WAVE EQUATION ANALYTIC SOLUTION

The PWE analytic solution is acquired by using the classic separation of variables technique on the original partial differential equation (PDE). This effectively converts the PDE into two ordinary differential equations (ODEs). The first ODE in terms of z is first order and can be solved by direct integration. The second ODE with respect to x is in the Sturm-Liouville form. It has a general solution that can be solved by substituting the boundary conditions into the ODE and simplifying. The result is a series solution with a sum of coefficients. The ODE solutions in terms of X and Z are recombined to form the final solution. The solution is refined further as it is determined that the coefficients are equivalent to Fourier integration of the initial condition $\phi(x)$. The final equation is represented in (185).

$$\bar{U}(x,z) = \sum_{n=0}^{\infty} \left[\frac{2}{L} \int_0^L \phi(x) \sin\left(\frac{n\pi}{L}x\right) dx \right] \sin\left(\frac{n\pi}{L}x\right) e^{-\frac{1}{2\gamma}(\lambda^2)z} \qquad (18.5)$$

where λ is the wavelength and γ is the propagation parameter.

L is the length of the wavefront grid.

It should be noted that the summation series is truncated at 1,500 when programmed in a computer, as this value provides a solution that does not vary significantly. It demonstrates the impracticality of using the analytic solution due to the duration of computation time required to solve the series for a solution at a particular distance in the interior of the domain. Instead, this analytic solution will be used to determine the accuracy of the numerical methods applied to solve the PWE at the receiver distance downrange from the satellite. More information on the separation of variable techniques applied to PDEs can be found in [3].

18.2.2 FINITE DIFFERENCE METHOD

The FDM uses local pointwise approximations defined on a finite, uniform grid. The particular formulation of the PWE is implemented by using the Crank-Nicolson scheme. The Crank-Nicolson scheme is unconditionally stable, meaning that any aspect ratio (spatial steps of z and x) can be selected to solve the problem while maintaining error bounds that do not grow. This is the nature of implicit numerical schemes. The Crank-Nicolson formulation of the PWE is given by (18.6):

$$\frac{U_j^{n+1} - U_j^n}{(\Delta z)} = \frac{1}{2}\left[\frac{1}{2\bar{\gamma}_R}\right]\left[\frac{U_{j+1}^{n+1} - 2U_j^{n+1} + U_{j-1}^{n+1}}{(\Delta x)^2} + \frac{U_{j+1}^n - 2U_j^n + U_{j-1}^n}{(\Delta x)^2}\right] \quad (18.6)$$

The Crank-Nicolson scheme must be constructed in a matrix formulation that can be implemented in a computer program.

The initial step is to isolate the forward propagating variables to the left-hand side of the equation:

$$\underline{U}_j^{n+1} - \alpha\left[\underline{U}_{j+1}^{n+1} - 2\underline{U}_j^{n+1} + \underline{U}_{j-1}^{n+1}\right] = \underline{U}_j^n + \alpha\left[\underline{U}_{j+1}^n - 2\underline{U}_j^n + \underline{U}_{j-1}^n\right]$$

Combine like terms and simplify:

$$\left[-\alpha\underline{U}_{j+1}^{n+1} + (1+2\alpha)\underline{U}_j^{n+1} - \alpha\underline{U}_{j-1}^{n+1}\right] = \left[-\alpha\underline{U}_{j+1}^{n+1} + (1-2\alpha)\underline{U}_j^{n+1} + \alpha\underline{U}_{j-1}^{n+1}\right]$$

Computer implementation of the equation in the form above will require a tri-diagonal matrix that maintains the spatial orientation of the "j" and "n" terms.

$$\begin{bmatrix} 1+2\alpha & -\alpha & 0 & . & . & . & & & 0 \\ -\alpha & 1+2\alpha & -\alpha & 0 & . & . & . & & 0 \\ 0 & -\alpha & 1+2\alpha & -\alpha & & & . & & \\ . & 0 & -\alpha & . & & . & & & . \\ . & . & 0 & & . & & . & & \\ . & . & & . & & . & & . & \\ . & . & & & . & & . & 0 & \\ & & & & & . & & -\alpha & \\ 0 & . & . & . & & 0 & -\alpha & 1+2\alpha \end{bmatrix}\begin{bmatrix} \underline{U}_1^{n+1} \\ \underline{U}_2^{n+1} \\ . \\ . \\ . \\ . \\ . \\ . \\ \underline{U}_N^{n+1} \end{bmatrix}$$

$$
=
\begin{bmatrix}
1-2\alpha & \alpha & 0 & \cdot & \cdot & \cdot & & 0 \\
\alpha & 1-2\alpha & \alpha & 0 & \cdot & \cdot & \cdot & 0 \\
0 & \alpha & 1-2\alpha & \alpha & \cdot & \cdot & & \\
\cdot & 0 & \alpha & \cdot & \cdot & & & \\
\cdot & & 0 & \cdot & \cdot & & & \\
\cdot & \cdot & & \cdot & \cdot & \cdot & & \\
\cdot & \cdot & & & \cdot & \cdot & \cdot & \\
& & & & & \cdot & \cdot & 0 \\
& & & & & & \cdot & \alpha \\
0 & \cdot & \cdot & \cdot & & 0 & \alpha & 1-2\alpha
\end{bmatrix}
\begin{bmatrix}
\underline{U}_{-1}^{n} \\
\underline{U}_{-2}^{n} \\
\cdot \\
\cdot \\
\cdot \\
\cdot \\
\cdot \\
\cdot \\
\cdot \\
\underline{U}_{-N}^{n}
\end{bmatrix}
$$

Let the tri-diagonal matrix on the left side of the equation be "A", and the right side tri-diagonal matrix be "B" and "I" is an identity matrix:

$$\underline{U}^{n+1}[I+A]=[I-B]\underline{U}_j^n$$

$$\underline{U}_j^{n+1}=[I+A]^{-1}[I-B]\underline{U}_j^n \tag{18.7}$$

18.2.3 SPECTRAL METHOD

The SM improves the convergence of a solution by using more grid points to approximate the derivative of a function. The SM is known to be accurate to $O(n\log(n))$, where n is the order of the grid size used to sample points [4]. The objective is to formulate a periodic domain using a trigonometric basis on an equi-spaced grid. As the number of samples increases, the error should decrease provided the solution is infinitely differentiable and smooth. For a function with p - derivatives the vth spectral derivative typically has accuracy $O(h^{p-v})$.

The derivative is approximated by the sequence of Fourier modes in the following manner:

$$\frac{d^k U_N}{dx^k}(x_j)=\sum_{|n|\le N}(in)^k \hat{a}_n e^{\left[\frac{2\pi i}{L}\right]n(x_j)} \tag{18.8}$$

where k is the order of the derivative, n is the Fourier mode, L is the length of the spatial domain, and $\hat{a}_n$ is the Fourier coefficient after taking the fast Fourier transform.

An appropriate finite or discrete representation of the solution must be selected by using an interpolating function between the values $U(x_j)$ at some suitable points (or nodes) x_j:

$$U_N(x)=\sum_{n=0}^{N}\hat{a}_n \varphi_n(x) \tag{18.9}$$

The Fast Fourier transform is taken to acquire the Fourier U coefficients, $\hat{a}_n$:

$$\hat{a}_n = fft(U) = \frac{1}{N}\sum_{j=0}^{N-1}U(x_j)e^{inx_j} \tag{18.10}$$

The second order spatial derivative is approximated with the square of the Fourier mode sequence and formulated into a square diagonal matrix where each dimensional length is equal to the length of the input vector:

$$U_{xx} = -n^2\hat{a}_n \tag{18.11}$$

The sequence of Fourier modes is arranged to be evenly uniform across the spatial x grid:

$$n = 1i*\left[-N\!\!\big/\!_2 + 1, 0, N\!\!\big/\!_2 - 1\right] \tag{18.12}$$

The spatial frequency to spatial position conversion integration transform is:

$$\text{Position}: \ f(x) = \frac{1}{2\pi}\int_{-\infty}^{\infty} f(K)e^{-j\beta x}\,dK \tag{18.13}$$

The spatial position to spatial frequency conversion integration transform is:

$$\text{Frequency}: \ F(K) = \int_{-\infty}^{\infty} f(x)e^{-j\beta x}\,dx \tag{18.14}$$

Applying this process to the PWE, take the spatial Fourier transform on both sides:

$$\int_{-\infty}^{\infty}\left[\frac{\partial^2\bar{U}(x,z)}{\partial x^2}\right]e^{-\beta x}\,dx = 2\gamma\int_{-\infty}^{\infty}\left[\frac{\partial\bar{U}(x,z)}{\partial z}\right]e^{-\beta x}\,dx \tag{18.15}$$

This forms an Ordinary Differential Equation in the spatial frequency domain:

$$-K^2\tilde{\underline{U}}(K,z) = 2\gamma\frac{\partial\tilde{\underline{U}}(K,z)}{\partial z} \tag{18.16}$$

The solution frequency domain solution is inversed Fourier transformed back to the spatial position domain:

$$\tilde{\underline{U}}(K,z_2) = \tilde{\underline{U}}(K,z_1)e^{-\left(\frac{K^2}{2\gamma}\right)\Delta z}$$

$$\tilde{\underline{U}}(x,z_2) = ifft\left(\tilde{\underline{U}}(x,z_1)\right) \tag{18.17}$$

18.2.4 WIDEBAND MODULATED SIGNAL

The wideband signal applies for a dispersive (frequency selective) communication channel. A transmit signal is modulated using a complex envelope on a carrier frequency. The PWE provides the realizations of the signal's amplitude and phase as it propagates spatially in the frequency domain. The signal is Fourier transformed inversely into the time domain at the receiver. More information regarding this methodology can be found in [2]. The following equation represents this process:

$$v(x, z_r, \tau) = \frac{1}{2\pi} \int_{-Bw/2}^{Bw/2} M(f) U(x, z_r, f) e^{1i(2\pi f)} df \qquad (18.18)$$

where $M(f)$ is the amplitude of the modulation waveform at a particular frequency f, $U(x, z_r, f)$ is the PWE realization and Bw is the frequency bandwidth.

The frequency range is specified in (18.19):

$$f = -\frac{Bw}{2} + f_c : \Delta f : \frac{Bw}{2} + f_c \qquad (18.19)$$

where f_c is the carrier frequency and Δf is the frequency index

The modulation is implemented as a band-limited triangular pulse spectrum, shown in Figure 18.2, represented in the frequency domain as (18.20):

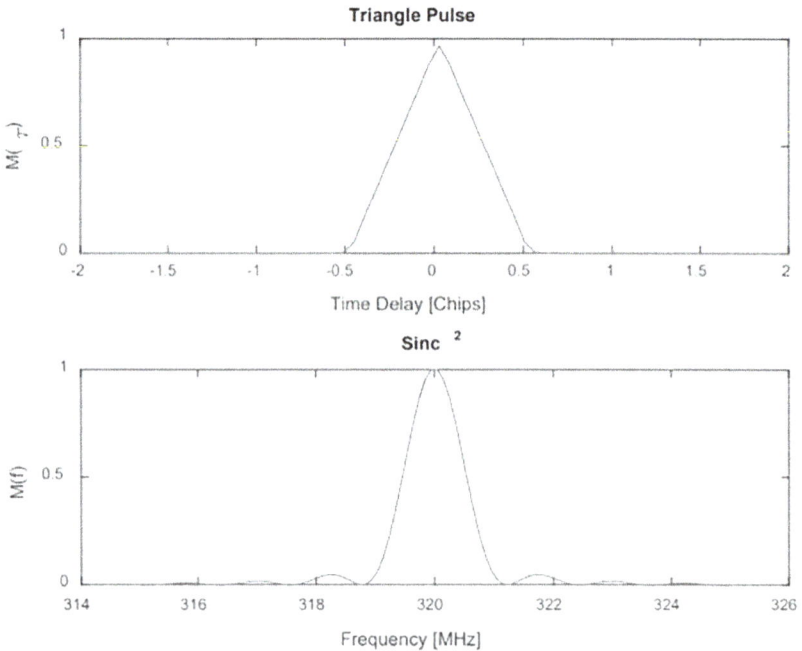

FIGURE 18.2   Triangular modulation scheme.

$$M(f) = \begin{cases} \dfrac{T_c}{2}\left[\text{sinc}\left(2\pi f_c T_c\middle/4\right)\right]^2 & \text{if } |f| \le \dfrac{Bw}{2} \\ \text{else} & 0 \end{cases} \qquad (18.20)$$

where T_c is the pulse duration.

A wave incident on a lens or a phase screen will exhibit time delay, corresponding to a variation of phase. The phase and time delay are related by the following relationship:

$$\tau_d = \frac{\phi}{2\pi f_c} \qquad (18.21)$$

where τ_d and ϕ are the time delay and phase, respectively. The following relationships establish the link between time and frequency sampling rates:

$$\tau = T\middle/M, \quad Bw = \frac{1}{\Delta\tau} \qquad (18.22)$$

from which the frequency index can be assigned:

$$\Delta f = 1\middle/T \qquad (18.23)$$

where $\Delta\tau$ is the time delay index, T is the total time duration, Bw is the delayed bandwidth, and M is the number of samples taken over the time duration. The time delay vector is:

$$\tau_d = -\frac{T}{2} : \Delta\tau : \frac{T}{2} \qquad (18.24)$$

The wideband signal is subject to a divergent Gaussian lens or a Kolmogorov phase screen by multiplication of a phasor. The Gaussian lens phasor is given below:

$$\phi = 2.5\left(\frac{f_c}{f}\right)e^{\left(-[x-0.5L]\right)/(150)^2} \qquad (18.25)$$

$$U_{\text{lens}} = U_{\text{inc}}e^{1i\phi} \qquad (18.26)$$

The phase screen phasor involves establishing an inner and outer scale of the irregularity located within the Ionosphere. The phase variance ordinarily depends on the variance of the electron density for each respective altitude that it is collected. In this chapter, the phase variance is pre-defined to examine a significant variance in the spectral density. A high variance in density can occur in the case of Ionosphere disturbances such as geomagnetic storms.

$$S_\phi(K_x) = \frac{\left(\sigma_\phi^2\right)\left[(2\pi)^{-1/2}l_i\left(l_i/L_o\right)\right]^{((m-1)/2)}\left[K_{m/2}\left(l_i\sqrt{K_x^2 + \left(1/L_o\right)^2}\right)\right]}{K_{(m-1)/2}\left(l_i/L_o\right)\left(l_i\sqrt{K_x^2 + \left(1/L_o\right)^2}\right)^{m/2}} \qquad (18.27)$$

Table 18.1
Simulation Input Parameters

Parameter	Value
Spatial grid length	5,100 m
Outer scale	1,000 m
Inner scale	100 m
Frequency	320 MHz
Grid samples	2^{12}
Phase screen variance	10
Pulse duration	6.88×10^{-9} s

σ_ϕ^2 is the phase variance, L_0 is the outer scale of the plasma, l_i is the inner scale of the plasma, K_x is the spatial frequency of the phase screen grid points corresponding to the x direction, m is the slope of the power spectral density, K_n is the Bessel function of order n

$$\text{PSD} = \frac{\sum_{R=1}^{N}\left[\sqrt{\frac{1}{2}}\left(\text{normrnd}\,(0,1)+i\left[\text{normrnd}\,(0,1)\right]\right)\left(\frac{S_\phi L}{2\pi}\right)^{1/2}\right]_R}{N} \tag{18.28}$$

$$\phi_{\text{phz}} = \text{real}\left(\text{ifft}\,(\text{PSD})\right) \tag{18.29}$$

In Table 18.1, the input parameters for the Gaussian lens and phase screen simulations in the wideband model are provided.

18.3   RESULTS

The following results show that the hypothesis is correct regarding the numerical accuracy and computational performance of the FDM and SM algorithms.

18.3.1   AMPLITUDE AND PHASE OF RECEIVED FIELD

In Figure 18.3, the amplitude plot of the received waveform shows the edge diffraction caused by the Gaussian lens. Note the fluctuations in the magnitude toward the exterior of the wavefront. The diffraction is also apparent in the phase.

In Figure 18.4, the magnitude and phase of the E-field fluctuate in a non-deterministic manner as approximated by the Kolmogorov phase screen's representation of the power spectral density in the frequency domain.

The SM average error comparison to the analytic solution declines by twelve orders of magnitude while the FDM average error declines by three orders of magnitude when comparing the phase screen error response and the Gaussian lens error response. The phase screen causes issues for the SM algorithm at the left and right-hand boundaries. The global trigonometric interpolants of the SM algorithm are not

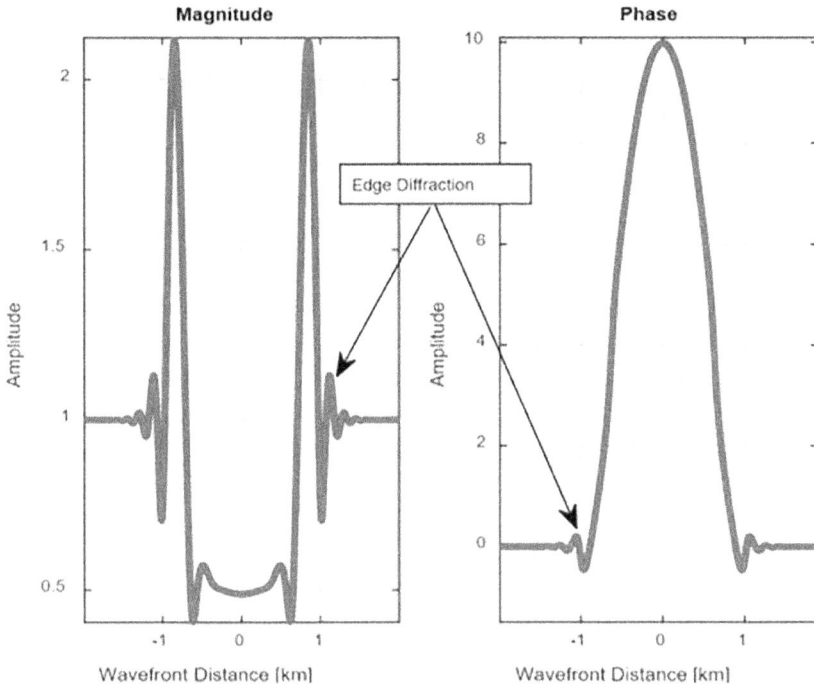

FIGURE 18.3 Divergent Gaussian lens interaction on incident E-field, the received E-Field at 170 km.

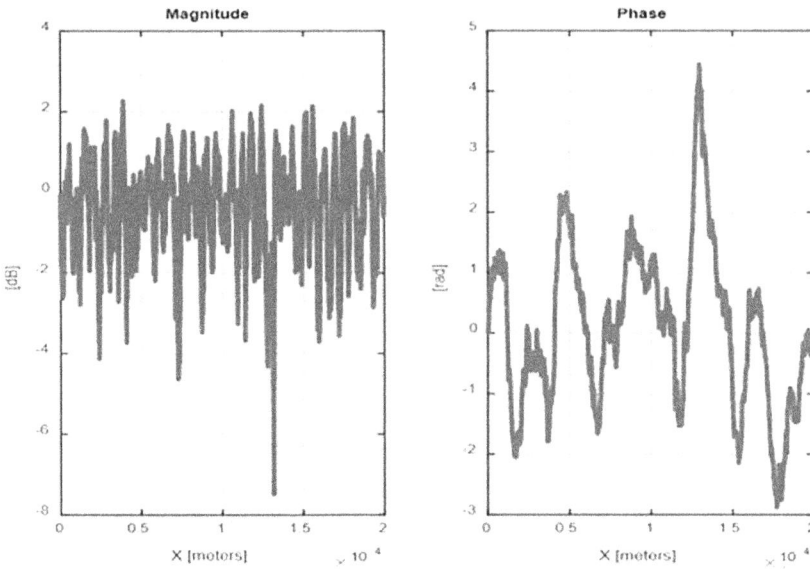

FIGURE 18.4 Kolmogorov phase screen interaction on incident E-field, the received E-Field at 170 km.

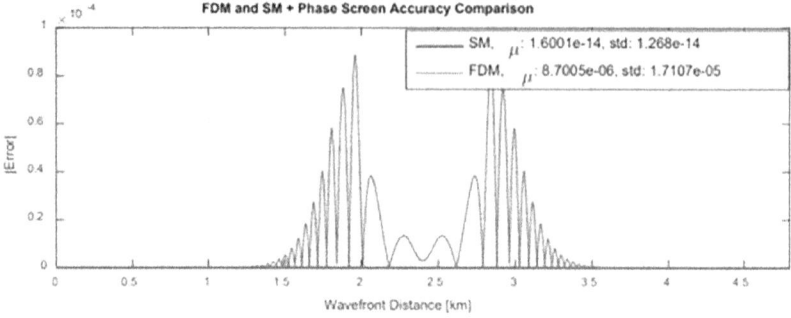

FIGURE 18.5 Gaussian lens FDM and SM error comparison.

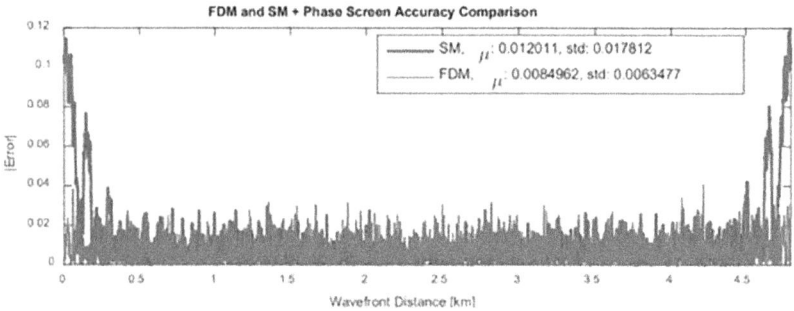

FIGURE 18.6 Phase screen FDM and SM error comparison.

as effective in computing the electric field for non-uniform input functions observed with phase screens due to an absence of smoothness and periodicity. This is also reflected in the variance and standard deviation of the error as shown in Figures 18.5 and 18.6 for a distance of 70,000 m.

18.3.2 TIME DELAY

The time delay occurs as a result of the phase delay in the E-field's propagation. Figure 18.7 shows the global time delay in terms of chips. The color bar represents the amplitude of the electric field as seen on the map. The wavefront of the E-field spans the vertical axis. The secondary (delayed) wavefront that protrudes outward from the center of the initially received wavefront is caused by the edge diffraction from the Gaussian lens as noted in Figure 18.3.

In Figure 18.8, the error between the analytic global time delay plot and the corresponding numerical method is shown for the Gaussian lens problem. The solution from FDM is delayed by an average of approximately 0.0006 chips compared to the analytic solution. The SM algorithm does not have an average delay differential from the analytic solution. Generally, the magnitude of the overall error is smaller than in the FDM algorithm by two full orders of magnitude. The calculation of the

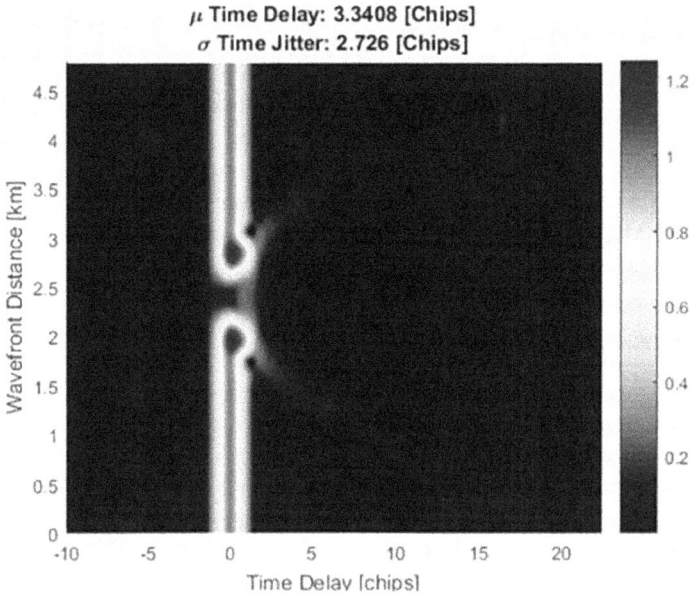

FIGURE 18.7 Global time delay, analytic solution plot.

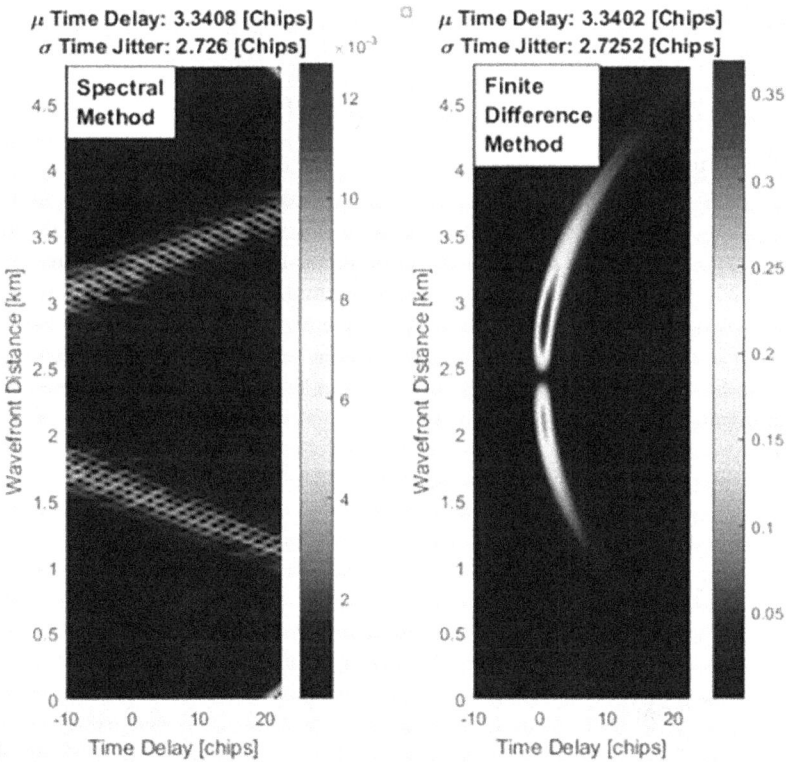

FIGURE 18.8 Global error difference (Gaussian lens interaction).

time jitter appears to be an exact match between the SM and analytic solutions as the FDM algorithm differs by 0.0008. Potential floating point error could contribute to the inaccuracy that causes the FDM algorithm to have an artificially added delay compared to the analytic solution. It can be inferred that the floating point error accumulation grows with the propagation distance of the E-field.

A well-defined band occurs on the right side of Figure 18.8 as a result of the FDM algorithm solution mismatch with the analytic solution. The left side shows no such band as the SM algorithm generally agrees with the analytic solution and has minimal inaccuracies.

18.3.3   COMPUTATIONAL EFFICIENCY

The FDM algorithm is significantly outperformed by the SM algorithm as shown in Table 18.2. The computer used in these simulations is an Intel(R) CORE I7(TM), 6GB RAM, 64-bit operating system.

The time was calculated in the PWE solution over a propagation distance, followed by the time duration of calculating the wideband frequency responses.

18.4   CONCLUSION

The FDM algorithm is greatly outperformed in computational efficiency by the SM algorithm. The inversion of a matrix required in the FDM algorithm is a significant disadvantage. As expected, the accuracy declines in SM and FDM when the input function encounters a phase screen due to the non-uniformity of the function. Despite the decline in accuracy, SM still closely resembles the analytic algorithm in the received amplitude and phase solutions. The subtle differences in accuracy contributed to the differences observed in the calculated time delays, as FDM did not resemble the analytic solution as closely as SM. Parallel computing should be examined to alleviate the time issues in the FDM algorithm so that the wideband frequency responses can be calculated simultaneously. Spline interpolation may also help achieve faster computation times by reducing the samples along the wavefront required by the algorithm's input function. Future work will investigate implementing Spectral Element Methods (SEM) to improve the accuracy of the local interpolants along the E-field wavefront grid. SEM will provide a more robust approximation of non-uniform input functions encountered when modeling scintillation with phase screens.

Table 18.2
Computation Times of FDM and SM

	FDM (s)	SM (s)
PWE Algorithm	1.3333	0.0141
Wideband Algorithm	37208.59	0.784629

ACKNOWLEDGMENTS

We would like to thank our research assistants, Cameron Brown, Ryan Kinkade, and Shawn McTaggart for their input, conversation, endless support, and encouragement in this research effort.

REFERENCES

1. Canada Centre for Remote Sensing (CCRS), "Fundamentals of Remote Sensing", Natural Resources Canada, electronic file available at https://www.nrcan.gc.ca/sites/www.nrcan.gc.ca/files/earthsciences/pdf/resource/tutorfundam/pdf/fundamentals_e.pdf, 2019.
2. Dennis L. Knepp, "Propagation of Wide Bandwidth Signals Through Strongly Turbulent Ionized Media", electronic file available at: https://apps.dtic.mil/dtic/tr/fulltext/u2/a131355.pdf, 1982.
3. Farlow, S.J. *Partial Differential Equations for Scientists and Engineers*", John Wiley & Sons, Inc. Canada, 1982.
4. Kyle A. Novak, *Numerical Methods for Scientific Computing: The Definitive Manual for Math Geeks*", Lexington, Kentucky, 2017.

Section 6

Materials Technologies in Aerospace Applications

19 Understanding Flow Characteristics in Metal Additive Manufacturing[*]

Carl R. Hartsfield, Travis E. Shelton,
Gregory R. Cobb, Ryan A. Kemnitz,
and Joseph Weber
US Air Force Institute of Technology

CONTENTS

19.1 INTRODUCTION

Additive manufacturing (AM) has become a profound fabrication technique with remarkable flexibility. The localized processing of AM offers the ability to print small complex passages (diameter < 0.25 mm) within structures. This microscale manufacturability has implications for the design and fabrication of flow devices. Devices such as injectors, nozzles, and heat exchangers are capable of being printed

[*] Reprinted with permission: Hartsfield, C. R., Shelton, T. E., Cobb, G. R., Kemnitz, R. A., & Weber, J. (2021). Understanding Flow Characteristics in Metal Additive Manufacturing. *Journal of Aerospace Engineering*, *34*(6), 04021082.

DOI: 10.1201/9781003220978-25

using AM (Petrovic et al. 2011; Anilli et al. 2018; Townsend et al. 2016; Anderson et al. 2020; Waters 2018; Steitz et al. 2013; Durkee et al. 2020; Tommila et al. 2021) while reducing part and labor complexity. SpaceX has experimented with printing thrust chambers and flown printed oxidizer valve housings to reduce lead time and expense (Howell 2014). Large flow structures like this would be less sensitive to the tens of micrometer-scale roughness expected from AM (Tommila et al. 2021), simply by virtue of their large size, supporting flow for 500 kN of thrust and up. However, small-scale flow devices from this process alleviate constraints that conventional manufacturing can pose. Finer features with more complex geometry have been realized through AM. In some cases, the process has shown significant cost reductions relative to traditional manufacturing. One study found that the cost of printed rocket nozzles was approximately 25% of the traditionally manufactured nozzles' cost (Tommila 2017) due to savings in labor.

Inconel 718 (IN 718, American Special Metals Corporation, Miami) is a material capable of being processed by AM. It is a Ni-Fe superalloy that has excellent corrosion resistance (Thomas et al. 2006) and microstructure stability (Sabelkin et al. 2019) that make it an excellent material for pipe flow applications. Many studies have reported on the basic material properties; however, no overarching analysis and design projections of the fundamental flow regimes have been fully reported. Manufacturing properties, such as surface roughness, have been studied, and the results have shown high average surface roughness ($R_a \geq 7\,\mu m$) to be inherent to the AM process (Shelton et al. 2020). Printed metal, specifically using laser powder bed fusion (LPBF), lacks the smooth, consistent surface produced by many traditional, subtractive techniques. Conventionally, Moody's (1944) diagrams were produced to forecast pipe flow pressure losses based on empirical data. To conclude this, an effective sand-grain roughness is assigned to the flow path surface based on empirical data. Drawn-Over-Mandrel (DOM) tubing is a high standard for smoothness in flow applications and commonly has a sand-grain surface roughness of 1.5 μm (Bergman et al. 2011). Sand-grain roughness effectively assumes a surface made up of many identical diameters of tangent spheres, in which the roughness dimension is the diameter of those spheres (Adams et al. 2012). Printed metal has a finish far from a uniform sand-grain roughness, with larger protrusions projecting from a relatively smooth base.

Understanding the underlying effect of this modern manufacturing technique and the associated surface roughness requires studying with a perspective toward applications. Adams et al. (2012) developed a simple method of estimating equivalent sand-grain roughness based on the arithmetic average surface roughness, R_a, which is the average of the absolute deflection from the mean profile of a surface. The result was a constant (5.683) applied to the R_a based on calculations derived from many vector paths taken across an array of equally sized spheres arranged in a hexagonal pattern. This arrangement corresponds to the ideal sand-grain configuration. According to this method, the printing process yields an estimated sand-grain roughness nearing 40 μm; this roughness is in the range of commercial steel tubing (46 μm, presumably hot rolled, welded seam tubing or similar) values commonly used for the calculation of friction factors and pressure losses in flow devices (Bergman et al. 2011). However, previous research by Tommila et al. (2021) indicates that these values may not be accurate in reflecting the actual effective roughness that should be used in design predictions. Their work on additively manufactured rocket nozzles found that loss predictions using this methodology were well short of the experimental losses experienced, although some of these losses were due to compressibility effects, likely including shocks within the nozzle (Tommila et al. 2021). Some difference was expected, as the Adams et al. (2012) formula was based on average roughness for calculating the sand-grain texture.

High surface roughness coupled with small fluid passages leads to extremely high-pressure losses with low flow rates. This research was performed to experimentally determine the equivalent sand-grain roughness and associated friction factors when evaluating AM parts, specifically those manufactured of IN 718 using LPBF. In this study, the fundamental flow characteristics within AM are vital when designing with this newer manufacturing technique. Empirical data must be formed to understand all the impacts on fluid mechanics that AM imposes on the design process. From this research, it is expected that empirical flow data will more accurately represent equivalent or effective sand-grain roughness for AM parts and scalability.

19.2 MATERIALS AND METHODOLOGY

19.2.1 FLUID FLOW ANALYSIS CALCULATIONS

This work focuses on the roughness impacts of small printed passages. The analysis approach can be put into perspective with the Moody diagram and associated equations (Bergman et al. 2011; White 2006). Pressure drop is the result of friction during fluid flow in a length of a pipe. The pressure drop in a pipe can be derived by the Darcy-Weisbach equation (assuming incompressible flow, constant cross-section, and constant properties), as defined in Eq. (19.1)

$$\Delta p = -\int_{p_1}^{p_2} dp = f \frac{\rho u_m^2}{2D_H} \int_{x_1}^{x_2} dx = f \frac{\rho u_m^2}{2D_H} L \tag{19.1}$$

where L = length of pipe; D_H = diameter; ρ = density; u_m = velocity; and f = friction factor. From this, a large pressure drop would be expected from the following variables: a long pipe (L), small diameter (D_H), high density (ρ), high velocity (u_m), or high friction factor (f). Making mathematical substitutions, a suitable derivation for experimental measurements is designated in Eq. (19.2)

$$\Delta p = f \frac{\rho}{2D_H} \left(\frac{4\dot{m}}{\rho \pi D_H^2} \right)^2 L = f \frac{8\dot{m}^2}{\rho \pi^2 D_H^5} L \tag{19.2}$$

The mass flow rate can be measured or controlled, while the friction factor, f, must be calculated.

Experimental determination of the friction factor, f, generally uses measured pressure drop and mass flow to calculate a friction factor. In 1944, Moody published a diagram (Moody 1944) showing the general trends between the friction factor and the Reynolds number. The chart shows the dependence of the Darcy friction factor on the flow Reynolds number (Re_D) and the relative roughness of the inside surface of the pipe (ε/D). A low Reynolds number leads to laminar flow, in which the Darcy friction factor is independent of the roughness and calculated by Eq. (19.3)

$$f = \frac{64}{Re_D} \qquad (19.3)$$

A transitional flow ($2{,}000 < Re_D < 10{,}000$) is significantly less affected by roughness than higher Reynolds number flows. A higher Reynolds number leads to turbulent flow, where the Darcy friction factor is significantly impacted by the relative roughness. It also leads to a region dominated by roughness, referred to as fully rough, which is characterized by a near absence of the friction factor dependence on the Reynolds number (Bergman et al. 2011). Moody's work was based on the experimental data, while other researchers, including Colebrook and Haaland, produced analytic functions to fit the Moody data. Haaland's formula [Eq. (19.4)] (White 2006) is explicit if the Reynolds number and relative roughness are known

$$\frac{1}{\sqrt{f}} \approx -1.8 \log_{10}\left[\frac{6.9}{Re_D} + \left(\frac{\varepsilon/D}{3.7} \right)^{1.11} \right] \qquad (19.4)$$

This makes Haaland's formula useful for the prediction of a friction factor in most design problems. On the other hand, the friction factor is determined implicitly using Colebrook's formula [Eq. (19.5)] (White 2006). Thus, if the Reynolds number and relative roughness are known, an iterative process is required to find the appropriate friction factor. While Colebrook's formula is more time-intensive, this formulation is slightly more accurate to the experimental data

$$\frac{1}{\sqrt{f}} \approx -2.0 \log_{10}\left[\frac{\varepsilon/D}{3.7} + \frac{2.51}{Re_D \sqrt{f}} \right] \qquad (19.5)$$

The increased accuracy of the Colebrook formulation is desirable for this research and was utilized in this study. The Colebrook formula can be rearranged to deliver an expression for relative roughness explicitly in terms of the Reynolds number and friction factor [Eq. (19.6)]

$$\frac{\varepsilon}{D} = 3.7\left(10^{-(1/2.0\sqrt{f})} - \frac{2.51}{Re_D \sqrt{f}} \right) \qquad (19.6)$$

Defining an appropriate Reynolds number is necessary for defining the flow characteristics. For internal flows, the convention is to use the Reynolds number based on hydraulic diameter, Re_D, as calculated in Eq. (19.7)

$$Re_D = \frac{\rho u_m D_H}{\mu} = \frac{4\dot{m}}{\rho \pi D_H} \qquad (19.7)$$

where D_H = hydraulic diameter, which is the effective flow diameter seen by the fluid. For experimentation, the mass flow was determined and controlled while the hydraulic diameter was calculated as the viscosity of the fluid, and the temperature was known. The hydraulic diameter, D_H, was calculated according to Eq. (19.8)

$$D_H = \frac{4 A_c}{P} \tag{19.8}$$

where both the cross-sectional area, A_c, and the perimeter, P, are affected by significant roughness.

A representation of sand-grain roughness is shown in Figure 19.1. First, the presence of significant roughness generally reduces the cross-sectional area. Geometrically, the area is reduced by the scalloping of the flow path edges, while the perimeter is also significantly increased. The most simplistic calculations have the perimeter increased by a factor of $\pi/2$, regardless of the size of the roughness. However, a significant contribution to the perimeter calculation is the very narrow gap between two adjacent and tangent spheres. When this void fills with fluid, there will not be significant flow through it as a result of viscous effects. To remove the gap from the calculation of the perimeter and, as a result, make the perimeter a function of the sand-grain roughness size, a limiting minimum gap, δ, was assumed to be $2\delta = 2\,\mu m$ (fillet area between circles). It was assumed that any gap smaller than this would not contribute to the flow area, especially for very small protrusions, such as that in DOM tubing ($\varepsilon = 1.5\,\mu m$). This assumption makes the perimeter very close to the smooth circle perimeter, while at higher roughness, the perimeter becomes much larger.

From these assumptions, the cross-sectional area can be calculated

$$A_c = \frac{\pi}{4} D_{nom}^2 - \left(\frac{\pi}{\theta} \right) \left(\frac{\pi - 2}{2} \right) R_{SG}^2 \tag{19.9}$$

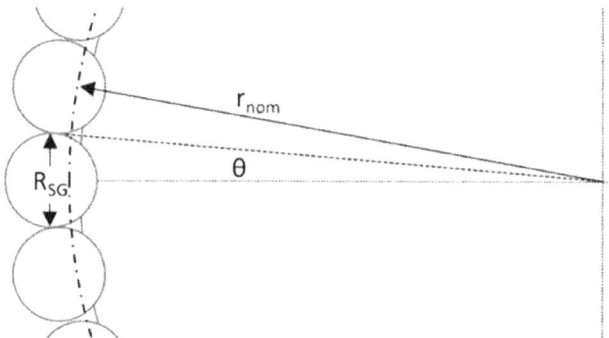

FIGURE 19.1 (Color) Internal tube geometry showing sand-grain roughness impact (θ is a measure of the angular extent of a sand-grain particle, while r_{nom} is the nominal radius of the tube or pipe).

and the perimeter of the tube

$$\phi = \tan^{-1}\left(\frac{\sqrt{R_{SG}\delta - \delta^2}}{\frac{R_{SG}}{2} - \delta}\right)$$

(19.10)

$$P_A = 2\left(\frac{2\pi}{\theta}\right)\left[\frac{R_{SG}}{2}\left(\frac{\pi}{2} - \theta - \phi\right) + \delta\right]$$

The sand-grain roughness yields a significant impact on the hydraulic diameter. Twenty micrometers of sand-grain roughness increases the calculated perimeter [Eq. (19.10)] and decreases the cross-sectional area [Eq. (19.9)], resulting in a calculated hydraulic diameter [Eq. (19.8)] that is only 83%–84% of the observed or measurable diameter for the smallest tubes assessed, while 1.5 µm of roughness (DOM tubing) yields a hydraulic diameter very close to the observed diameter. This reduction in hydraulic diameter works with the wall roughness to significantly affect the pressure drop in a small tube.

19.2.2   EXPERIMENTAL SAMPLES

Thirty samples were processed on a GE Concept Laser (General Electric Additive, Cincinnati) M2 Cusing LPBF metal printer. All samples were synthesized from IN 718 powder (purchased from Powder Alloy Corporation, Loveland, Ohio) on Inconel build plates in a single print. Each printed sample had a 127 mm long flow section of a constant diameter. Figure 19.2 shows the cross-sectional flow path of a printed sample. The fluid enters this section through a smooth contoured funnel (6 mm long) entrance at the input feed. External features of the printed tubes include a cylindrical base, printed 12.7 mm in diameter and then machined to a 1/4″ (6.35 mm) National Pipe Thread to seal into a Swagelok T fitting (Swagelok, Solon, Ohio). Just above the cylindrical base is a hexagonal section printed on the exterior to provide easy installation into the fittings.

Figure 19.3 represents the 30 samples' locations on the substrate during the print process. As seen, they are in a grid pattern inclined 15° counterclockwise. The samples were semi-randomly distributed to minimize potential inside diameter effects with respect to position. Samples were printed in sets of five, each with six different inside diameters: 0.381 mm (0.015″), 0.762 mm (0.030″), 1.143 mm (0.045″), 1.524 mm (0.060″), 2.540 mm (0.100″), and 3.175 mm (0.125″). The range of inside diameters was selected to extend as low as the processing capability allowed and up

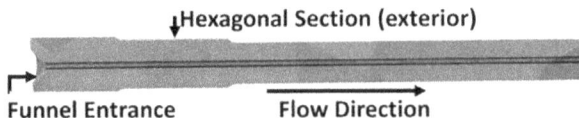

↓Hexagonal Section (exterior)

Funnel Entrance   Flow Direction

FIGURE 19.2   Representative geometry of a printed sample.

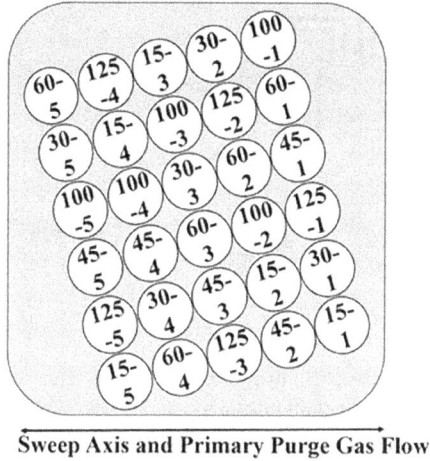

Sweep Axis and Primary Purge Gas Flow

FIGURE 19.3 Distribution of printed tubes on a build plate.

to a maximum that would lead to velocities in the water feed tubing at less than half of the flow velocity in the printed samples to ensure that most of the gas feed pressure would reflect pressure drop in the printed sample. At the maximum, the pressure drop per length for the largest tubes would be at least five times higher than that of the inlet tubing. Individual part numbers shown in Figure 19.3 consist of the nominal inside diameter in thousandths of an inch, followed by an index (1–5) of parts within that nominal size. The 381 μm (0.015″) ID tubes were printed and tested; however, it was found that their flow rates were extremely low. The pressure sensor recording was saturated after approximately 10 min of recording, and well over one-half million data points were recorded with less than 1 g of fluid exiting the system. The low total mass of fluid through the tube, coupled with an accuracy of 0.1 g for the total mass measurement, leads to very high uncertainty in these samples' flow rate calculations. The flow rate uncertainty generated high uncertainty in both the friction factor and roughness calculations. As a result, those results are not presented in this study.

19.2.3 EXPERIMENT SETUP

The experimental setup consists of a pressurized inert gas source linked to a water tank through a regulator (Figure 19.4). The water tank is partially filled with ice before it is filled with water to maintain an approximately 0°C temperature. This setup allows the flow through the printed tubes to be a known state of density and viscosity. The water tank is pressurized from the top and has an outflow feed from the bottom. The outflow goes through another valve to a Swagelok T fitting. The specimens are on the direct path, and a pressure sensor is on the cross path. The pressure sensor monitors the gauge pressure loss across the printed specimens. The printed tube flows into an open reservoir. The mass change in the reservoir and the time of positive pressure are both measured to calculate the mean mass flow for the experiment.

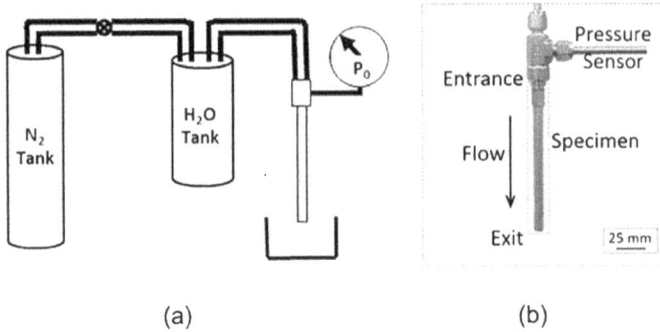

(a)   (b)

FIGURE 19.4   (Color) (a) Schematic of the experimental setup with the pressurized tank forcing water through the experimental nozzle as seen in (b).

After each run, pressure readings (sampled at a rate of approximately 1 kHz), the total mass of water that passed through the system, the elapsed time, and the printed sample geometry were used to determine where the data converged using the following procedure:

1. Calculate the mean pressure drop during the test:

$$\overline{\Delta p} = \frac{1}{\Delta t} \int_{t_1}^{t_2} p\, dt \tag{19.11}$$

2. Calculate a mean mass flow rate during the test from the total mass and elapsed time:

$$\overline{\dot{m}} = \Delta m / \Delta t \tag{19.12}$$

3. Estimate an effective sand-grain roughness and calculate the effective cross-sectional area, perimeter, and hydraulic diameter using Eqs. (19.13)–(19.15) (reprinted in this study):

$$A_c = \frac{\pi}{4} D_{nom}^2 - \left(\frac{\pi}{\theta}\right)\left(\frac{\pi-2}{2}\right) R_{SG}^2 \tag{19.13}$$

$$P_A = 2\left(\frac{2\pi}{\theta}\right)\left[\frac{R_{SG}}{2}\left(\frac{\pi}{2} - \theta - \phi\right) + \delta\right] \tag{19.14}$$

$$D_H = \frac{4 A_c}{P} \tag{19.15}$$

4. Calculate the effective friction factor during the test from the pressure drop and mass flow:

$$f = \frac{\rho \pi^2 D_H^5}{8 L \dot{m}^2} \Delta p \tag{19.16}$$

5. Calculate the Reynolds number from Eq. (19.17) (reprinted in this study)

$$\text{Re}_D = \frac{4\dot{m}}{p\pi D_H} \tag{19.17}$$

6. Calculate effective relative roughness from Eq. (19.18) (reprinted in this study)

$$\frac{\varepsilon}{D} = 3.7\left(10^{-(1/2.0\sqrt{f})} - \frac{2.51}{\text{Re}_D \sqrt{f}}\right) \tag{19.18}$$

7. Calculate a new effective sand-grain roughness from the relative roughness and tube hydraulic diameter.
8. Return to Step 3 with an updated sand-grain roughness estimate. Note that it proved necessary to artificially damp the oscillations in this process by only changing the Step 3 estimate by about 20% of the difference between that previous estimate and the Step 7 calculation. This led to convergence, defined as less than 1 nm of change in the sand-grain roughness value, in less than 10 iterations in all evaluated cases.

19.3 RESULTS AND DISCUSSION

19.3.1 PREDICTED ROUGHNESS

A fully developed flow is important for this study. In turbulent flow, Bergman et al. (2011) indicate that a flow becomes turbulent above a length/diameter (L/D) ratio of 10. All samples in this study were the same length, resulting in the smallest sample having an L/D ratio of 333 and the largest diameter L/D as low as 40. The highest L/D (smallest diameter) tubes were expected to experience only laminar flow, while all other L/Ds would only experience turbulent flow. Table 19.1 shows the experimental

TABLE 19.1

Predicted Characteristics of Test Samples According to Geometry

Nominal Diameter (mm)	(L/D) max	Nominal Relative Roughness	Predicted Re_D (min)	Predicted Re_D (max)	Roughness- Re_D Product (min)	Roughness-Re_D Product (max)
0.381	333	0.105	600	2,700	63	284
0.762	167	0.052	2,000	9,300	105	488
1.143	111	0.035	3,400	19,300	119	676
1.524	83	0.026	5,600	32,400	146	843
2.540	50	0.016	13,600	78,000	211	1,209
3.175	40	0.013	20,400	112,000	255	1,456

sample geometries manufactured and the associated predicted flow characteristics, including the expected minimum and maximum Reynolds numbers. Both were based on expected performance with a nominal 40 μm of equivalent sand-grain roughness. Calculations for minimum and maximum predicted Reynolds numbers were based on assumed pressure drops of 34 kPa (5 psi) and 1 MPa (150 psi), respectively. This assumed that all feed pressure was dropped in the printed tube samples and reflected minimum and maximum feed pressures that are at the edge of what would be achievable with the hardware available. The minimum pressure of 5 psi, is absolutely the minimum readable value on the pressure regulator used. The water tank was rated for 1.45 MPa (210 psi) maximum pressure, and for safety purposes, no feed pressures were allowed in excess of about two-thirds of the maximum allowable pressure (0.97 MPa). The potentially achievable pressure range allows an approximately 5× variation in the Reynolds number, which generally covers the areas of curvature in the Moody diagram friction factor curves.

According to White (2006), the product of the relative roughness and Reynolds number is an important analysis tool for correlating roughness to the friction factor. When this product is less than 10 (very smooth surfaces and/or very low Reynolds numbers), roughness is not important to the flow or pressure loss. This is the case for most relatively smooth walls at laminar (<2,000) Reynolds numbers. In comparison, if the product is more than 1,000 (very rough surfaces or very large Reynolds numbers), the roughness dominates, and the friction factor is largely independent of the Reynolds number. As shown in Table 19.1, several of these test samples were expected to approach the fully rough state within the given pressure and mass flow limitations, especially at higher feed pressures, higher mass flow rates, and Reynolds numbers. Observed roughness in excess of the preliminary estimates will only push further into the fully rough zone.

19.3.2 MEASURED GEOMETRIES

Figure 19.5a–f presents representative optical images of the as-printed diameter for each sample size. Measurements were made on each sample as a circular overlay of the actual image. As seen in Figure 19.5, the as-printed passages were highly circular.

Figure 19.6 illustrates the relative error of the as-printed to the as-designed diameters for the five test articles in each size manufactured. As the nominal and actual diameter points show, on average, the samples are relatively close (<3% error) to the intended diameter. However, the absolute errors (between design diameter and actual diameter), plotted in gray, show variations within an average diameter (~6% error). These variations are more prevalent at smaller diameters. The average diameter error could result in a variation of 11.5% (or greater) in the expected flow area for the smaller printed cross-sections. These errors do not show a significant dependence on size and are apparently caused by the resolution of the print process. The resolution is determined by the minimum layer height for the M2 cusing printer (20 μm), the mean powder diameter (approximately 35 μm), and the laser beam width (50 μm).

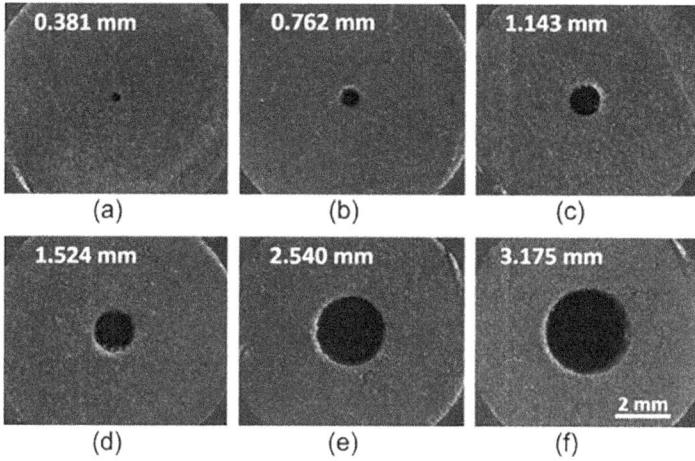

FIGURE 19.5 Optical images normal to the channel exit. These channels were measured with optical tools.

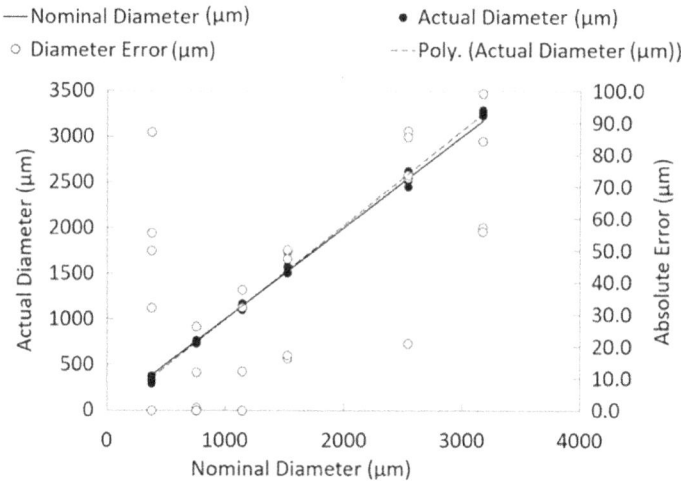

FIGURE 19.6 Experimentally measured sample diameters in relation to theoretically designed diameters.

19.3.3 ANALYSIS OF THE EFFECTIVE ROUGHNESS

Figure 19.7 shows the flow testing results of the Darcy friction factor with respect to the Reynolds number for a subset of samples. The experimental results follow analytic curves based on the mean relative roughness calculated. Empirically, these curves exhibit the same trends observed in Moody diagrams generated from flow testing traditionally manufactured tubes. The friction factor increases as a result of increasing the e/D ratios. Also, within each e/D ratio, the friction factor decreases

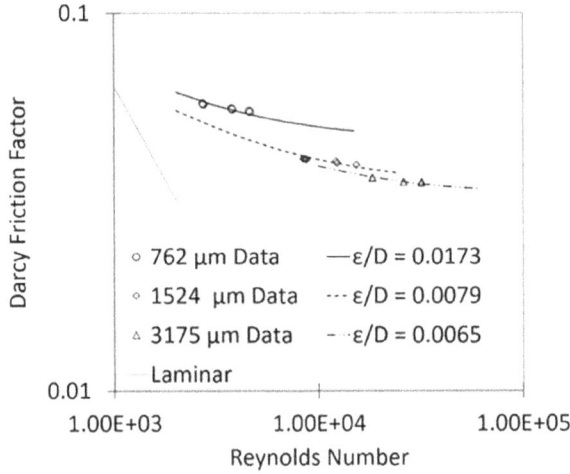

FIGURE 19.7 Additively manufactured experimental pipe flow results on Moody diagram lines.

TABLE 19.2
Summary of Test Samples and Flow Test Results

Nominal Diameter (μm)	Mean Printed Diameter (μm)	COV for Printed Diameter (%)	Mean Sand-Grain Roughness (μm)	COV for Sand-Grain Roughness (%)
762	756	1.97	16.04	22.22
1,143	1,144	2.22	17.23	48.12
1,524	1,547	2.37	18.64	32.27
2,540	2,565	2.99	17.54	19.32
3,175	3,258	0.87	22.24	22.95

as the Reynolds number increases. Similar results were seen for each set of samples and are, therefore, not presented. One item to note from the data is the variation in roughness between sample parts. A disproportional relative roughness was seen between geometric flow sizes. If the absolute roughness were similar in all samples, the relative roughness would decrease proportionally with the increasing diameter. This decrease would cause a shift from a higher friction factor at small diameters to a lower friction factor at larger diameters. Experimentally, this is not the case, as the 3,175-μm sample demonstrated only a slight decrease in the friction factor from the 1,524-μm sample, even though it is more than twice the diameter. This trend was present in every size sample, with variations exhibited in both diameter and roughness (Table 19.2) at each nominal tube size.

Table 19.2 was constructed from experimental flow data, in which the average and standard deviation of the five samples with respect to the as-printed diameter and absolute sand-grain roughness are shown. The mean sand-grain roughness varied in range from 16.04 to 22.24 μm. The author's previous research (Shelton et al. 2020) indicated that the mean sand-grain roughness would be approximately 40 μm. However, these results are at least 40% less than predicted. Two observations were made that would account for this difference. First, the flow testing effective roughness was considerably lower than predicted. In this study, the sand-grain roughness was about 40% of the estimate calculated at a comparable diameter, based on average roughness measurements from the previous research. There are two possible explanations for this rise: either the average surface roughness value used was not the right metric for the printed surface, or the characteristics of the roughness are such that the Adams, Grant, and Watson correlation [Eq. (19.1)] is not applicable. The Adams, Grant, and Watson correlation (Adams et al. 2012) tied sand-grain roughness to the average roughness measured with a profilometer. Table 19.4 shows several roughness statistics measured for individual tubes of different diameters, as compared to the observed roughness derived from the friction factor based on pressure drop and mass flow data. There is no correlation between the observed roughness based on the friction factor and any of the standard roughness statistics, even when comparing data from the same part. Second, the variation in roughness within the five replicates of a given sample size is high. It will be shown that this is largely driven by the location of the part on the build plate. An individual sample's roughness is likely affected by asymmetries within the build chamber affecting the purge flow of argon gas and, to a lesser degree, the loss of focus as the laser beam is scanned across the build area. The absence of any correlation between effective sand-grain roughness, as measured, and any of the observable roughness characteristics for the sectioned tubes that were measured indicates that the high variability in roughness even between samples of the same nominal diameter is not the reason the observed roughness is much lower than the Adams, Grant, and Watson correlation [Eq. (19.1)] would predict (Adams et al. 2012).

19.3.4 Measured Average Roughness

The surface roughness was verified for multiple samples using a Zeiss laser scanning microscope (LSM) 700. Five of the test samples were sectioned lengthwise along the flow path using electrode discharge machining (EDM) and probed with the LSM to determine the interior surface's roughness, as seen in Figure 19.8. The effective sand-grain roughness for these samples was closest to the mean value for their respective sets. The average roughness was measured at four locations axially along the tube and averaged to get a value for the sample. Visually, this characterization technique shows random protrusions from the metal powder adhered to the inner surface wall. They tend to be randomly distributed and occasionally form dense clusters. Overall, the flow passage is well defined. This data is summarized in Table 19.4. As shown, the average roughness ranges from 7.48 to 14.28 μm, while the effective sand-grain roughness ranges from 15.87 to 21.24 μm. Visually and qualitatively, the apparent roughness of the samples tracked with the average roughness measured with the LSM. The 45-2 and 100-5 samples are visually smoother and about 1/2–2/3 the average roughness value of the other samples.

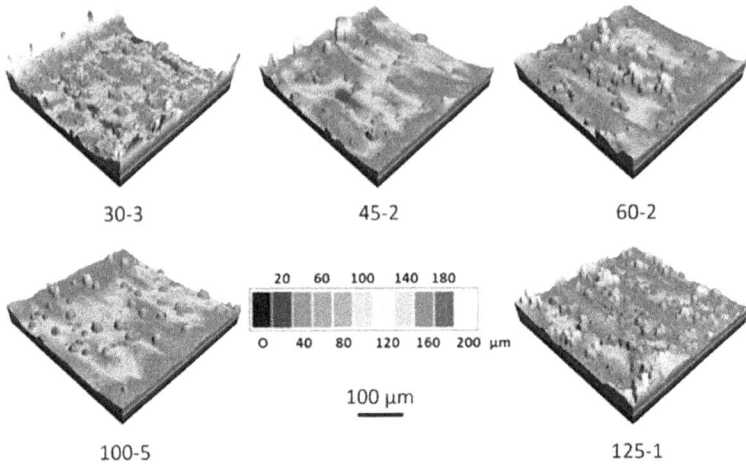

FIGURE 19.8 (Color) Surface topography of the interior surface; particles distributed across the surface are present.

TABLE 19.3

Laser Scanning Microscope Roughness for Specific Tubes with Effective Sand-Grain Roughness

Tube ID	Average Roughness (μm)	Roughness Standard Deviation (μm)	Effective Sand-Grain Roughness (μm)	Sand-Grain Roughness Standard Deviation (μm)
30-3	14.28	2.03	15.87	1.11
45-2	8.05	1.08	16.47	0.86
60-2	13.53	2.53	20.02	0.69
100-5	7.48	0.28	16.84	0.46
125-1	12.33	1.67	21.24	0.62

Note: Flat surface (exterior) values based on the average of all tubes. Interior values are based on a single sample of each size, which was selected because they were closest to the mean effective sand-grain roughness for that tube size.

The quantitative roughness results are tabulated in Table 19.3. As seen, the average roughness (R_a) is in the range of 7.48–14.28 μm, without showing a clear trend between tube IDs. The average roughness of the samples also does not strongly correlate with the effective sand-grain roughness (ε) calculated from the pressure drop in the tubes. Samples were also measured on different tangents of the exterior hexagonal section to measure a roughness independent of any surface curvature effects. Between the two result methods, there is no real trend relating to the average roughness and the effective sand-grain roughness. These results indicate the average roughness has not captured the important surface effects on fluid flow from the additive process.

TABLE 19.4

Measured Roughness Parameters

Nominal Size (mm)	Effective Sand Grain (µm)	RSA (µm)	RSQ (µm)	RSZ (µm)	RSV (µm)
0.762	15.87	16.65	22.12	182.7	56.29
1.143	16.47	10.89	15.32	118.8	29.60
1.524	20.02	11.15	13.85	76.91	27.82
2.540	16.84	8.79	10.99	56.19	25.21
3.175	21.24	12.32	15.98	101.5	57.60

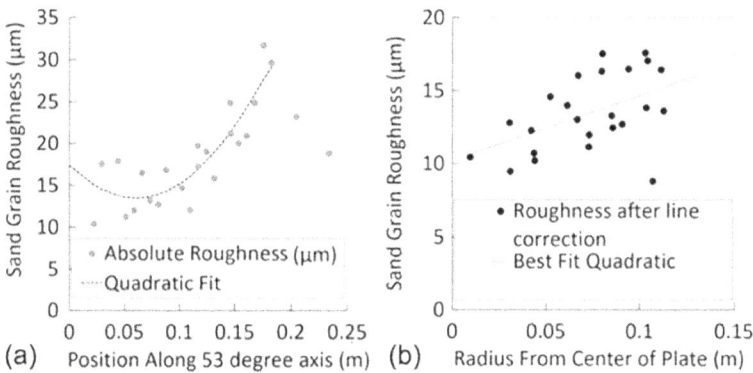

FIGURE 19.9 (a) Variation sand-grain roughness for individual tubes with position; and (b) roughness variation with radius after subtraction of axis correction.

There are several other standard roughness statistics listed in Table 19.4, which also show variations. These variations differed mostly in a scalar multiple from the measured average roughness, R_{SA}, and, therefore, did not correspond well to the measured effective sand-grain roughness.

Based on the experimental measurements, there is little indication of a relationship between the effective sand grain and these statistical roughness values. In all of these cases, there is no relatively simple, monotonic functional relationship (i.e., linear or quadratic) to be drawn between the LSM measured roughness statistics and the effective sand-grain roughness.

19.3.5 BUILD LOCATION

Each build location was evaluated to understand its influence on creating spread in the data. Figure 19.9 represents the roughness of the samples with respect to the build location. The roughness was a function of distance from a normal line at 37° from the vertical axis. The normal line passed through the lowest part in the array (part number 15-5), and its slope intersects the lower leftmost part in the print pattern. As shown in Figure 19.9, there is a clear dependence on print position for effective roughness. One potential effect from this is the flow of inert (argon) gas across the build area (from right to left) to clear contaminants resulting from the laser melting and sintering process. The front edge (near 0.00 m) is near a smooth vertical wall, and the rising roughness is likely due to lower gas flow in the region. The back edge (near 0.25 m) has similar issues of flow, which is complicated by a presence of a cavity along the wall to allow a germanium lens for infrared imaging of the print process. This cavity could create a recirculation zone that adversely affects the argon flow across the print surface, leading to increased flow turbulence and roughness near this zone, slightly away from the wall (near the 0.2 m point). Even without the asymmetry of the window cavity, the parabolic distribution of roughness would be an expected feature due to corner flow effects on the print surface.

After removing the dependence from the gas flow on the position along a linear axis from the data, a linear trend was revealed (Figure 19.9b) with respect to the sample location from the radius of the center. Because the laser is steered about the build area by moving a mirror that is centered over the build plate at the top of the build chamber, the roughness depends on the radius from the nadir point below that mirror. As the beam moves further from the nadir point, the path length is longer, and the diffraction-limited beam size increases. In addition, any pointing angle error is magnified by the increased distance. Figure 19.9b shows the distribution of points with radius from that center point. In this case, the best fit quadratic correlation was a straight line.

The flow path diameter was then studied to understand its contribution to the effective sand-grain roughness. Geometric sample diameters were evaluated to determine the laser processing influence on roughness. Due to the laser scan speed being held constant, the laser would take longer to make a full scan around a larger fluid channel relative to a smaller channel. This concept could potentially lead to different thermal gradients among samples during the additive process. Figure 19.10 shows the variation of roughness after correcting for position along the 53° point from the horizontal axis and the radius from the center of the build plate. Statistically, no trend becomes apparent due to the channel size. Empirically, there is a potential

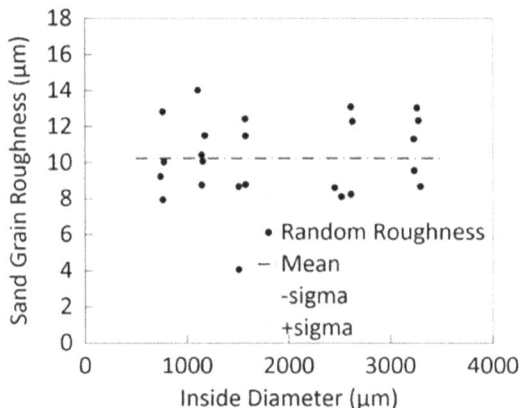

FIGURE 19.10 Effective sand-grain roughness after correction for axis and radius.

dependence on the diameter, with larger diameters being slightly rougher. However, the statistical assessment showed that it was difficult to separate the distributions at different diameters. The distribution of roughness at the largest diameter could not be confidently separated from that of the smallest diameter, so any dependence on diameter is considered negligible.

Analysis from all contributions resulted in a mean linear axis dependency of 5.16 µm, a mean radius dependence of 3.41 µm, and a mean contribution from the remaining roughness of 10.23 µm. The sum of these three contributions is 18.80 µm. For a more conservative approach in design, a root-mean-square (RMS) value is desired for the entire build plate and results in 7.16 µm (mean 5.41 µm) for the axis dependence and 4.63 µm (mean 4.34 µm) for the radius contribution. The remaining RMS roughness for the samples, after correcting for axis and radial contributions, was 10.47 µm, resulting in a sum of RMS values of 22.26 µm. As mentioned, this is a conservative estimate of roughness and would consistently overestimate pressure losses. A more accurate estimate of the sum of the means across the entire plate would be 19.98 µm. However, this estimate has a much higher risk of underestimating the pressure drop. The constants for the correction quadratics and offsets can be found in the following equations. All Cartesian positions are measured relative to the front left corner of the build plate as it sits in the printer.

The position on the axis is

$$L_{axis} = \left(x - 0.05 \text{ m}\right)\cos(53°) = (y - 0.015 \text{ m})\sin(53°) \tag{19.19}$$

The dependence due to a position along the axis is

$$\Delta\varepsilon_{axis} = 1073L_{axis}^2 - 129.42L_{axis} \quad (\text{in µm}) \quad \text{for } L_{axis} < 0.182 \text{ m} \tag{19.20}$$

$$\Delta\varepsilon_{axis} = 1073(0.364 - L_{axis})^2 - 129.42(0.364 - L_{axis}) \text{ (in µm) for } L_{axis} > 0.182 \text{ m} \tag{19.21}$$

The radius from the center of the plate is

$$R = \sqrt{(x - 0.125 \text{ m})^2 + (y - 0.125 \text{ m})^2} \tag{19.22}$$

The dependence on the radius is

$$\Delta\varepsilon_{axis} = 44.46R \quad (R \text{ in m}, \Delta\varepsilon \text{ in µm}) \tag{19.23}$$

where L_{axis} = length along the inclined axis (m); x and y = positions on the build plate measured from the front, left corner of the build plate (m), with x in the left-right axis and y in the front-back axis; $\Delta\varepsilon_{axis}$ = incremental change in sand-grain roughness associated with that axial position (µm); R = radius from the center of the plate (m); and $\Delta\varepsilon_{radius}$ = incremental change in sand-grain roughness associated with the radius from the center of the plate (µm).

A final dependence study was conducted on the baseline roughness to the powder dimensions (without location). From processing, the powder at the surface was

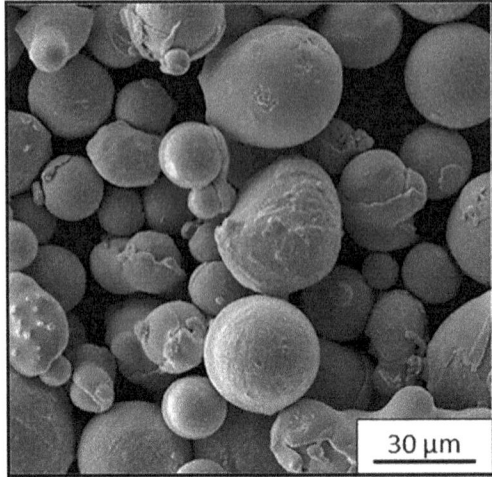

FIGURE 19.11   Scanning electron microscope image of Inconel powder.

potentially not fully melted but sintered to the bulk sample while maintaining some of the basic packed powder shapes. Powder size characteristics, as measured by the manufacturer, were determined by sieving powder through different mesh sizes. An electron microscope image of a typical powder sample is shown in Figure 19.11. Material certification data from the powder supplier, Powder Alloy Corporation, based on powder sieving through different size mesh screens in accordance with ASTM B214-16 standard (ASTM 2021), showed that the powder batch used for printing had 0.0% content greater than a #270 mesh size (53 μm particle size), 2.0% content greater than a #325 mesh size (44 μm particle size), and no content smaller than 15 μm (PAC 2019). Assuming a Gaussian distribution of powder such that the percentages round to the precision given on the specification (<0.05% powder over 53 μm, 1.95%–2.05% of powder between 44 and 53 μm, and <0.05% of powder less than 15 μm), an estimation of the mean powder size range and the coefficient of variation of the powder size can be calculated. The mean powder diameter would be between 32.4 and 35.1 μm, with the associated coefficients of variation of 17.3% and 12.4%, respectively. These values result in less than 0.01% of the powder exceeding 53 μm in diameter but allow up to 0.05% below 15 μm. If the fraction below 15 μm is required not to exceed that above 53 μm, the range tightens considerably: 34.02–35.09 μm diameter with coefficients of variation between 12.4% and 14.4%. Using the middle of this smaller range as the nominal mean for the powder and calculating the associated standard deviation results in a mean of 34.56 μm and standard deviation of 4.63 μm, respectively, and a 13.4% coefficient of variation. This correlates to an effective sand-grain roughness contribution due to the powder size being 29.6% of the mean powder size, while the standard deviation would be 49.2% of the powder standard deviation. There is a certain physical sense to this as the mean size of protrusions caused by the packed powder, which is at least somewhat melted together, would be less than the radius of the powder. While the variation would be reduced in magnitude, it might not be reduced as much as fully melted or nonpacked powder. Overall, this is a much better match to the observed data than the Adams et al. (2012) correlation based on the average roughness for this surface.

19.4   CONCLUSIONS

Flow designs incorporating AM allow for unprecedented opportunities. Fundamental flow characteristics for additively manufactured tubes have not been fully character- ized and, for that reason, were experimentally tested in this work. Testing showed the additive samples' pipe flow behaves similarly to conventional pipe flow. The relative friction factor increased with an increasing ε/D ratio and behaved consistently to the Colebrook formulation for friction factor variation with the Reynolds number. This observation fundamentally holds true for the mechanics; however, the proportional- ity of how this ratio decreased relative to pipe sizes is more inconsistent. Each printed pipe has independent artifacts, with an inherent effective sand-grain roughness, par- tially dependent on the position within the print location where it was manufactured. Even with variation, the effective sand-grain roughness was consistently on the order of half the estimated mean powder size (46.4%–64.4%, with an average of 53.1%) for the IN 718 powder used in this study. This should provide a reasonable estimate of effective sand-grain surface roughness for those wishing to implement additively manufactured fluid paths into practical devices. It was also clear from laser scanning microscopy that traditional roughness statistics are not fully representative of the surface qualities resulting from LPBF AM, and using those roughness estimates may dramatically overpredict pressure losses.

19.5   DATA AVAILABILITY STATEMENT

Some or all data, models, or code that support the findings of this study are available from the corresponding author upon reasonable request.

ACKNOWLEDGMENTS

The authors would like to thank the administration, laboratory technicians, and machine shop of the Air Force Institute of Technology and the Air Force Research Laboratory for their support throughout this research. This research was primarily funded by sponsors who have requested anonymity. Additionally, this research was supported in part by an appointment of the Student Research Participation Program at the US Air Force Institute of Technology administered with the Oak Ridge Institute for Science and Education through an interagency agreement between the US Department of Energy and USAFIT.

NOTATION

The following symbols are used in this chapter:
  A_c = cross-sectional area (m²);
  COV = coefficient of variation (dimensionless);

D = diameter (m);
D_H = hydraulic diameter (m);
f = Darcy friction factor (dimensionless);
ID = inside diameter (m);
L = tube length (m);
L_{axis} = length along axis (m);
m = mass (kg);
P = perimeter (m);
p = static pressure (Pa);
R_a = average roughness (μm);
R_{SG} = radius of sand-grain particles (μm);
Re_D = Reynolds number, based on diameter (dimensionless);
r = radius (m);
T = temperature (K);
t = time (s);
u_m = mean velocity (m s);
δ = limiting minimum gap width (μm);
ε = absolute sand-grain roughness (μm);
μ = dynamic viscosity (Pa·s);
π = 3.14159265...; and
ρ = density (kg m³).

REFERENCES

Adams, T., C. Grant, and H. Watson. 2012. "A simple algorithm to relate measured surface roughness to equivalent sand grain roughness." Int. J. Mech. Eng. Mechatron. 1 (1): 66–71. https://doi.org/10.11159/ijmem.2012.008.

Anderson, W. S., S. D. Heister, B. Kan, and C. Hartsfield. 2020. "Experimental study of a hypergolically ignited liquid bipropellant rotating detonation rocket engine." J. Propul. Power 36 (6): 851–861. https:// doi.org/10.2514/1.B37666.

Anilli, M., A. G. Demir, and B. Previtali. 2018. "Additive manufacturing of laser cutting nozzles by SLM: Processing, finishing and functional characterization." Rapid Prototyping J. 24 (3): 562–583. https://doi.org/10.1108/RPJ-05-2017-0106.

ASTM. 2021. "Standard test method for sieve analysis of metal powders." ASTM B214-16. West Conshohocke, PA: ASTM.

Bergman, T. L., A. S. Lavine, F. P. Incropera, and D. P. Dewitt. 2011. *Fundamentals of Heat and Mass Transfer*. 7th ed. Hoboken, NJ: Wiley.

Durkee, M., C. McCain, A. Quinton, K. Stewart, L. Utley, and K. P. Rouser. 2020. "AIAA 2020-1122: Design and test of a small-scale, additively manufactured, liquid-cooled rocket nozzle." In *Proc., AIAA SciTech 2020 Forum*. Orlando, FL: American Institute of Aeronautics and Astronautics. https://doi.org/10.2514/6.2020-1122.

Howell, E. 2014. "SpaceX taking 3D printing to the final frontier." Accessed April 19, 2021. https://www.space.com/26899-spacex-3d -printing-rocket-engines.html.

Moody, L. 1944. "Friction factors for pipe flow." Trans. Am. Soc. Mech. Eng. 66 (8): 671–678.

PAC (Powder Alloy Corporation). 2019. *PAC Material Characteristics Report: IN718 Powder*. Loveland, OH: PAC.

Petrovic, V., J. Vicente Haro Gonzalez, O. Jordá Ferrando, J. Delgado Gordillo, J. Ramo´n Blasco Puchades, and L. Portolés Grinan. 2011.˜ "Additive layered manufacturing: Sectors of industrial application shown through case studies." Int. J. Prod. Res. 49 (4): 1061–1079. https://doi.org/10.1080/00207540903479786.

Sabelkin, V., G. Cobb, T. Shelton, M. Hartsfield, D. Newell, R. O'Hara, and R. Kemnitz. 2019. "Mitigation of anisotropic fatigue in nickel alloy 718 manufactured via selective laser melting." Mater. Des. 182 (Nov): 108095. https://doi.org/10.1016/j.matdes.2019.108095.

Shelton, T. E., D. J. Stelzer, C. R. Hartsfield, G. R. Cobb, R. P. O'Hara, and C. D. Tommila. 2020. "Understanding surface roughness of additively manufactured nickel superalloy for space applications." Rapid Prototyping J. 26 (3): 557–565. https://doi.org/10.1108/RPJ-02-2019-0049.

Steitz, D. E., K. K. Martin, and E. Dick. 2013. "NASA, industry test additively manufactured rocket engine injector." Accessed April 16, 2021. https://www.nasa.gov/press/2013/july/nasa-industry-test-additively-manufactured-rocket-engine-injector-0/#.YHn5ESWSmUk.

Thomas, A., M. El-Wahabi, J. Cabrera, and J. Prado. 2006. "High temperature deformation of Inconel 718." J. Mater. Process. Technol. 177 (1–3): 469–472. https://doi.org/10.1016/j.jmatprotec.2006.04.072.

Tommila, C. 2017. "Performance losses in additively manufactured low thrust nozzles." M.S. thesis, Dept. of Aeronautics and Astronautics, Air Force Institute of Technology.

Tommila, C., C. Hartsfield, J. Redmond, J. Komives, and T. Shelton. 2021. "Performance impacts of metal additive manufacturing of very small nozzles." J. Aerosp. Eng. 34 (2): 04020115. https://doi.org/10.1061 /(ASCE)AS.1943-5525.0001229.

Townsend, A., N. Senin, L. Blunt, R. K. Leach, and J. S. Taylor. 2016. "Surface texture metrology for metal additive manufacturing: A review." Precis. Eng. 46 (Oct): 34–47. https://doi.org/10.1016/j.precisioneng.2016.06.001.

Waters, M. C. 2018. "Analysis of additively manufactured injectors for rotating detonation engines." M.S. thesis, Dept. of Aeronautics and Astronautics, Air Force Institute of Technology.

White, F. M. 2006. *Viscous Fluid Flow*. New York: McGraw Hill.

20 Fatigue Life Modeling and Experimentation of Additively Manufactured Components with Respect to Defect Size and Location[*]

Daniel G. Miller, Ryan A. Kemnitz, and Ramana V. Grandhi
US Air Force Institute of Technology

Luke C. Sheridan
US Air Force Research Laboratory

CONTENTS

[*] Reprinted with permission: Miller, D., Kemnitz, R., Grandhi, R. V., & Sheridan, L. C. (2022). Fatigue Life Modeling and Experimental Validation of Additively Manufactured Turbine Blade with Respect to Defect Size and Location. In *AIAA SCITECH 2022 Forum* (p. 0671).

DOI: 10.1201/9781003220978-26

20.1 NOMENCLATURE

$\Delta\sigma$ = Applied cyclic stress range
$\Delta\sigma_0$ = Applied cyclic stress range for the defect-free fatigue limit
$\Delta\sigma_N$ = Applied cyclic stress range for the defect-free fatigue life of N cycles
N = Fatigue life measured in the number of cycles
a = Defect length
a_0 = Critical defect length
a_c = Fracture defect length
a_N = Initial crack length to last N cycles
$a_{0,N}$ = Critical defect length for N cycles fatigue life
ΔK = Stress Intensity Factor (SIF)
ΔK_{TH} = Threshold SIF for crack growth
ΔK_{1C} = Fracture SIF
Y = defect shape factor
R = Applied stress load ratio

20.2 INTRODUCTION

The additive manufacturing (AM) process incorporates a number of defects into any generated structure from small voids to large cracks. These defects are a source of crack growth and fatigue life degradation for any AM component. Several studies have looked at how the defect size can be used to predict the fatigue life of AM components [1–4] based on the Kitagawa-Takahashi (KT) fatigue life model and the El-Haddad fatigue life model [5, 6]. The current techniques assume that the worst-case defect will occur at the highest stress concentration point. This creates a conservative fatigue life estimate that restricts the component's geometry design space and inhibits the development of AM components for limited life applications. This research augments the El-Haddad model with finite element analysis (FEA) to explore the effects of defect location in conjunction with defect size to predict where and when critical fatigue failure could occur. The experimental component studies fatigue bars in axial loading and a turbine blade design subjected to vibrational loading fatigue tests. The experimental fatigue life data is compared to the model predicted life to baseline the model quality as an accurate representation of the relationship between defect size, location, and fatigue life. The measured defect locations are evaluated against the model predicted failure locations to assess the accuracy of the augmented model to predict where a fatal defect can form for a given design life and maximum defect size.

Both the KT and original El-Haddad models are analytical models to define the boundary between the infinite and finite fatigue life based on the cyclic stress range and the crack size found on a part [5]. The El-Haddad model was recently modified to extend predictive capabilities to the finite life region based on the size of defect and the applied cyclic stress range [4].

This chapter develops the criteria to predict where a component will fail at using the finite fatigue life El-Haddad model in conjunction with FEA. The critical failure location criteria are developed using a simulated defect distribution across a specimen model with the design choices of maximum defect size and design life. Critical failure location predictions are then verified through hardware testing.

20.3 BACKGROUND

20.3.1 AM DEFECTS

The LoF defect type is of particular interest to fatigue life as a crack initiation point due to the size and shape. LoF defects tend to form flat separations that cover a relatively large area (Figure 20.1). This creates a large stress concentration point that propagates crack growth. Void defects tend to be much smaller and spherical in nature (Figure 20.1). The difference in size and lack of any sharp edges typically minimizes the importance of void defects in fatigue life studies. However, a small defect in a high-stress location can be more damaging to fatigue life than a large defect in a low-stress region.

20.3.2 FATIGUE LIFE MODELING WITH DEFECTS

The original El-Haddad model [6] is a modification of the KT diagram and applies a small crack correction to estimate the fatigue limit variation in the presence of various defect sizes. The KT fatigue life model is made up of two parts (Figure 20.2). Line 1 is based on the defect-free stress fatigue limit of the material. Any cyclic stress higher than the fatigue limit will result in fatigue crack growth leading to failure, any lesser stress is assumed to have an infinite life. Line 2 is based in linear elastic fracture mechanics (LEFM) and applies the threshold stress intensity factor (SIF) to split infinite and finite fatigue life. Re-arranging Eq. 20.1, and solving for $\Delta\sigma$ over a range of crack sizes, a, defines the boundary between the crack growing to failure or not based on the material crack growth properties [7]. Applying the Murakami root area defect characterization, when the flaw is near the surface the shape factor, Y, is assumed to be 0.65, and 0.5 for internal flaws [8]. Equation 20.2 defines the boundary for when a defect can be considered a surface flaw vs. internal flaw [9]. When the ratio of radius of the defect, r, divided by the distance from the defect center to surface, h, is greater than 0.8, the defect in question may be treated as a surface flaw.

$$\Delta K = Y * \Delta\sigma * \sqrt{\pi * a} \tag{20.1}$$

$$r/h \leq 0.8 \tag{20.2}$$

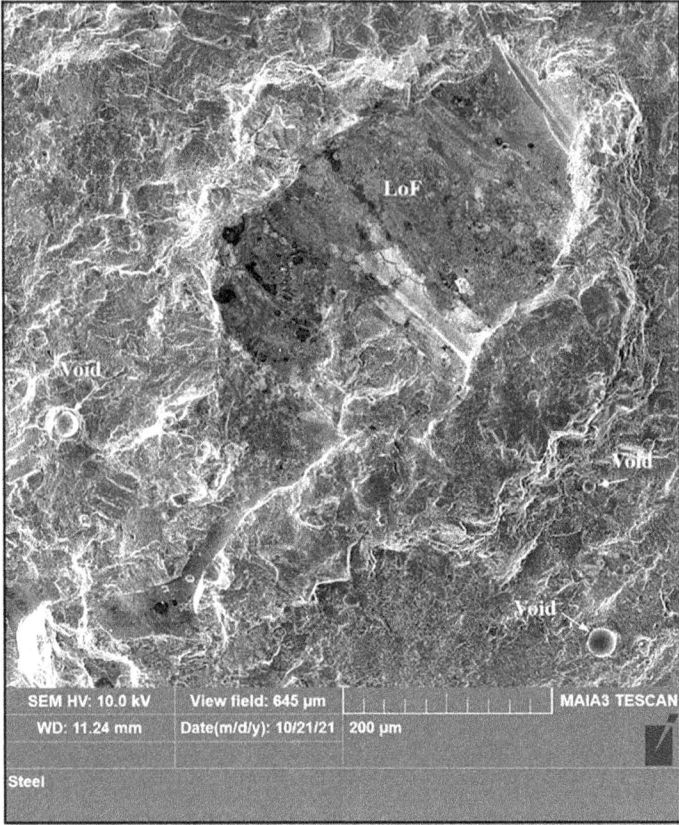

FIGURE 20.1   Relative size differences between a LoF defect and void defects.

The El-Haddad model combines the defect-free fatigue limit with the LEFM line to generate line 3 (Figure 20.2 & Eq. 20.3) bounding the infinite life fatigue limit. The critical crack length, a_0, is the defining point where LEFM becomes the dominant crack growth process and is calculated by applying the material SIF threshold, ΔK_{TH}, and defect-free fatigue limit, $\Delta\sigma_0$, to Eq. 20.1, resulting in Eq. 20.4.

$$\Delta\sigma = \Delta\sigma_0 * \sqrt{\frac{a_0}{a + a_0}} \tag{20.3}$$

When the defect being analyzed is significantly smaller than the critical crack size, the material stress fatigue limit dictates if a defect will grow into a crack and cause failure. As the initial defect size increases, the model becomes dominant by LEFM to determine the defect sizes that could grow to failure.

$$\Delta K_{TH} = Y * \Delta\sigma_0 * \sqrt{\pi * a_0} \tag{20.4}$$

The El-Haddad fatigue limit model was previously converted to model finite fatigue life (Eq. 20.5) [4]. The new model applies the Basquin fit of the material S-N curve

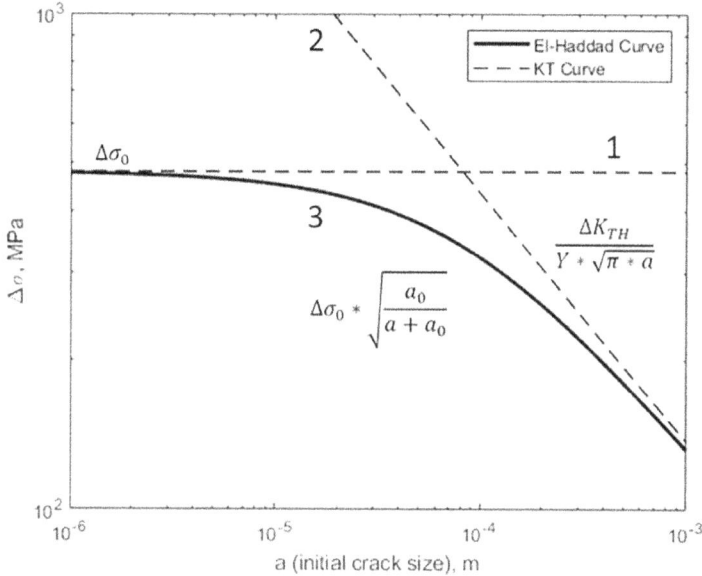

FIGURE 20.2 KT model and El-Haddad model for infinite fatigue life.

to generate a defect-free applied stress range, $\Delta\sigma_N$, based on the desired life, N, (Eq. 20.6) where A and b are material constants, and R is the load ratio (Eq. 20.7).

$$\Delta\sigma = \Delta\sigma_N * \sqrt{\frac{a_{0,N}}{a+a_{0,N}}} \tag{20.5}$$

$$\Delta\sigma_N = A*(1-R)*N^b \tag{20.6}$$

$$R = \frac{\sigma_{min}}{\sigma_{max}} \tag{20.7}$$

The critical crack length, $a_{0,N}$, is the initial defect size that will grow to failure in the desired life according to the integrated Paris Law (Eq. 20.8). In Eq. 20.8, C_0 and n are material constants, and Y is the shape factor assumed for the initial defect. The final crack length at failure, a_c, is found by solving Equation 9. ΔK_{1C} is the material failure SIF, $\Delta\sigma_N$ is from equation 6, and Y_f is the shape factor at failure.

$$a_{0,N} = \left[\left(a_c^{\left(1-\frac{n}{2}\right)}\right) - N*\left(1-\frac{n}{2}\right)*C_0*\left(Y*\Delta\sigma_N*\sqrt{\pi}\right)^n\right]^{\frac{1}{1-\frac{n}{2}}} \tag{20.8}$$

$$\Delta K_{1C} = Y_f * \Delta\sigma_N * \sqrt{\pi*a_c} \tag{20.9}$$

Applying the outputs from Eqs. 20.6 and 20.8 into Eq. 20.5 across a range of finite fatigue life values and crack sizes yields Figure 20.3. View A is in the orientation of a traditional S-N curve. As the cyclic stress decreases, the fatigue life increases. The top bound in view A is the Basquin fit (Eq. 20.6) using the material properties from Table 20.1. View B shows the traditional El-Haddad curves calculated at various design life levels. As the initial crack size increases, the allowable applied stress range to maintain a constant life curve decreases. The bend in the curve between material fatigue limit and LEFM shifts to the right as desired life decreases. This shift means that as the life requirements decrease, larger defects are allowed to exist before LEFM becomes the dominant crack growth mechanism. If the defects are kept below $a_{0,N}$, they have a negligible impact to the fatigue life when compared to the material fatigue limit. When the crack size gets above $a_{0,N}$, LEFM and the material crack growth properties become important factors in determining fatigue life.

There is a defect location dependence in Figure 20.3 based on the stress profile of the component being tested and the volume of the component that experiences the maximum applied stress. For a specimen such as fatigue bars subject to uniform uniaxial loading, there is a substantial volume of material experiencing the maximum applied stress. This enables the assumption that a large defect will form

---

TABLE 20.1

El-Haddad Model Material Properties for the Basquin Equation (Eq. 20.6) and Paris Law (Eq. 20.8)

A	b	C_0	n
4,623.4	−0.1558	1.25e-14	4

The properties were previously fit to the El-Haddad model using AM fatigue bars that were manufactured and processed to the same standards applied in this research [4].

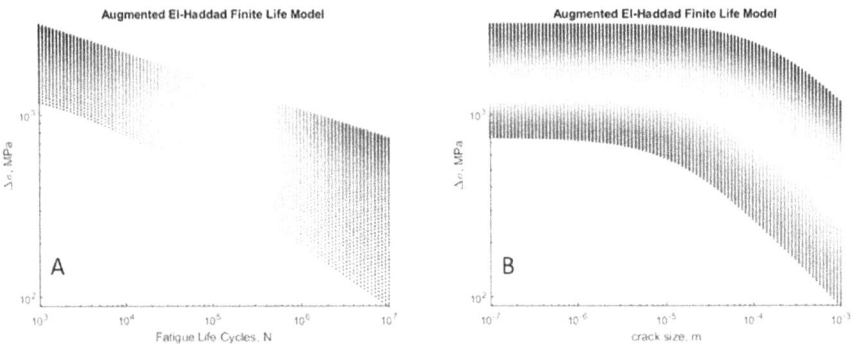

FIGURE 20.3   Finite Life surface relating initial crack size, cyclic stress at the fracture initiation point, and final fatigue life. (a) Life/$\Delta\sigma$ relationship. The Basquin equation defines the upper limit of the model. (b) Crack size/$\Delta\sigma$ relationship. Traditional El-Haddad curves build from Eq. 20.5 for different design lives, N.

in the maximum stress region and initiate failure. The defect location information is simplified down to surface or interior defect and only impacts the applied shape factor [10]. In other cases, such as bending loads, the volume of material subject to the maximum applied stress is very small. In three-point bending tests, failure has been documented in AM components due to defects located away from the maximum stress location [11]. The augmented El-Haddad model (Figure 20.3) is a surface defined by three values: the defect size, the applied stress, and the life associated with growing the defect to failure. In many real-world situations, the applied stress varies throughout the geometry and is determined by the component geometry and the applied boundary conditions. Knowledge of two values enables the prediction of the third. For simple structures, the defect location applied stress may be directly calculated, but for more complicated systems, a finite element model (FEM) creates a map to relate location to stress for an assumed load case.

20.4 SIMULATION

Simulations using the augmented El-Haddad model utilize a generic turbine blade design subject to a vibrational load state to generate a mapping of stress to location. An FEM is created to estimate the stress profile across the turbine blade when the base is fixed and the blade is subjected to a vibrational load to induce the second bending mode. The turbine blade (Figure 20.4a) is meshed with 223,500 quadratic hexahedral elements and processed in Abaqus 6.14. When the grip is fully constrained on the top and bottom surfaces and the model is subjected to a dynamic frequency analysis, the second bending mode and the associated normalized stress map are computed (Figure 20.4b) After completion of the FEM simulation, the grip is removed leaving just the blade material composed of 47,500 elements with 210,600 nodes.

20.4.1 DEFECT GENERATION

The defect sizes and locations needed for the augmented El-Haddad model are initially simulated using a fitted defect distribution (Figure 20.5) for nickel-based superalloy 718 [12]. The distribution mean fit equation was developed for the same material across a range of print parameters encompassing those applied to this research. The integers on the Y-axis correspond to the number of nines in the cumulative distribution function (CDF) probability (ex: $y = 3$, $F = 0.999$). Defect sizes of assumed

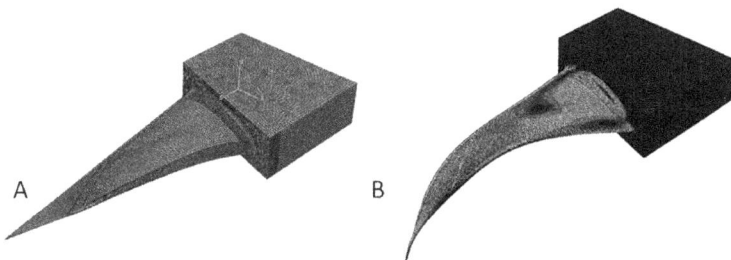

A   B

FIGURE 20.4   (a) Mesh of the generic turbine blade. (b) Second bending mode.

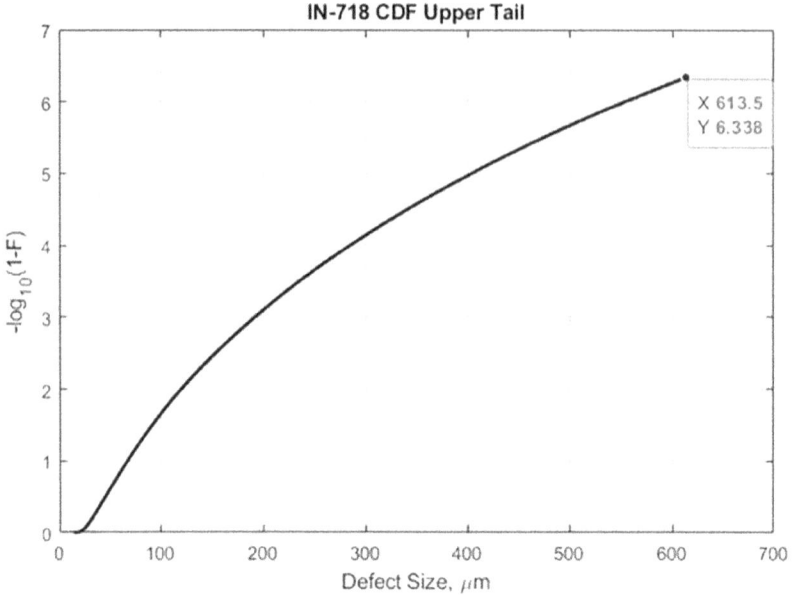

FIGURE 20.5 Defect CDF for nickel-based superalloy 718.

spherical shape are pulled at random from the population until the volume of defects reaches 0.1% of the total volume of the blade. The procedure generates 155,000 to 160,000 defects that are randomly assigned by a uniform distribution to FEM nodes within the turbine blade. Generated defects range from 14 μm up to 960 μm.

20.4.2 STRESS MAP

The FEM frequency analysis calculates the natural frequency, mode shape, and relative stress and strain profile for the first five modes. The second bending mode for the constrained turbine blade is 1,379 Hz and the shape is seen in Figure 20.4b. The maximum stress for this load condition is located on the surface of the blade at the root. Only 2.5% of the blade volume is subject to stress levels within 50% of the maximum stress. From Eq. 20.1, if the stress level is cut in half, the defect size must increase by a factor of four to maintain the same SIF. The Abaqus default stress and strain profiles are normalized to a maximum displacement of 1 unit, by scaling the stress values that generated Figure 20.4b to the stress range calculated by Eq. 20.6 for a desired defect-free life, an estimate of the applied stress is generated for every point within the turbine blade.

Coupling the developed stress map with the locations of the defects randomly simulated across the turbine blade creates the second parameter needed for the augmented El-Haddad model: the applied stress at the defect locations. For these simulations, the maximum applied stress is scaled to 1,075 MPa, which corresponds to a defect-free design life of 10^6 cycles. Applying the defect size and stress at location for every defect in one case to the augmented El-Haddad model creates Figure 20.6. Every defect has an associated fatigue life based on the maximum stress

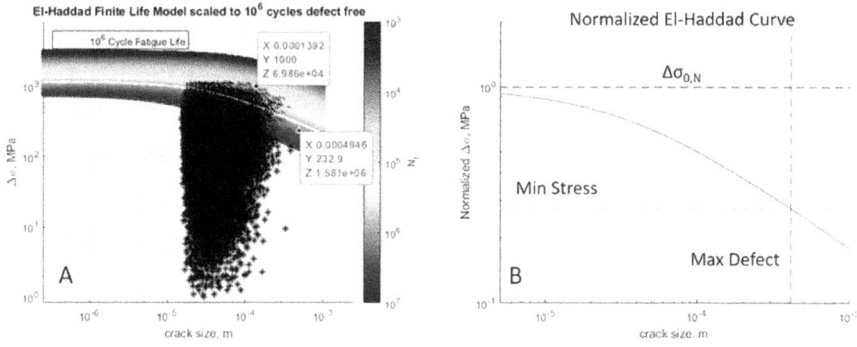

FIGURE 20.6 (a) Finite life model with 159,000 defects randomly distributed throughout the component. Sizes range from 0.015 to 0.41 mm. (b) Minimum stress level that could cause failure for a design life and maximum defect size.

at its location that is calculated without any influence from the other defects. The white line in Figure 20.6a represents the 10^6 fatigue life curve. Any defect above the curve is predicted to cause premature failure. Defects below the curve are predicted to fail after the design life. In this case, the largest defect generated is 495 μm diameter at a location that only experiences 21.7% of the maximum stress. The largest defect has been randomly placed in a location that does not experience enough stress to grow the crack within the design life. The first failure point is predicted at a 139 μm diameter defect at a location experiencing 93% of the maximum stress. Only 4% of the 1,064 cases ran predict the largest generated defect to cause the first failure, while 30% of the cases fail due to a small defect at the maximum stress location.

20.4.3 MODEL OUTPUTS

There are many outputs that can be generated from the model. The simplest outputs involve picking two of the values and solving for the third as was done in Figure 20.6a where the defect size and stress are defined, and the life is predicted. The model works just as easy to pick a desired life and defect size to determine the maximum allowed stress at the defect location. This level of output only requires knowledge of the material properties to build the augmented El-Haddad model.

The next level of outputs makes inferences on parameter limits. Using the applied defect distribution (Figure 20.5) with a desired probability of occurrence to generate a maximum defect size and applying a design life, N, a minimum stress level is identified by the intersection of the El-Haddad curve for N cycles and the maximum defect expected (Figure 20.6b). This value of minimum stress defines the stress level where defects will not interfere with the material fatigue life. Likewise, by intersecting the maximum defect size with $\Delta\sigma_{0,N}$ (Figure 20.6b), the worst-case life assessment is made. This is the life that will be achieved if the largest defect does coincide with the maximum stress location.

The final level of outputs requires knowledge of the geometry and load case. By plotting every location on the geometry that experiences the minimum stress from

Design Life: 10^6
Max Defect: 400µm

Design Life: 10^6
Max Defect: 200µm

Design Life: 10^5
Max Defect: 400µm

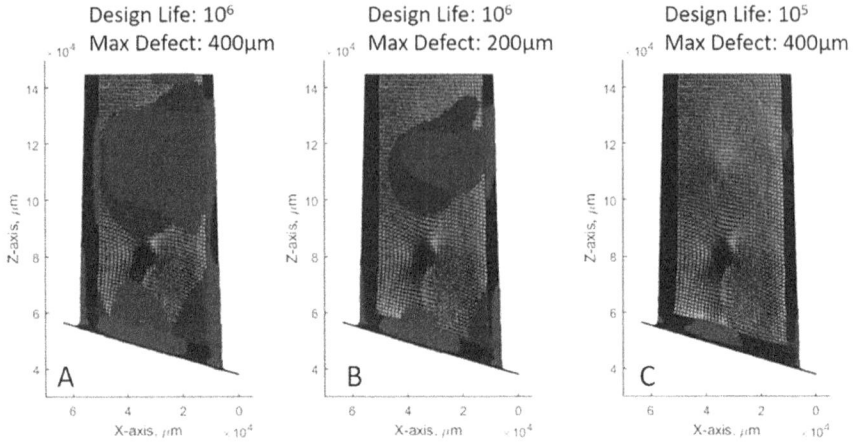

FIGURE 20.7   Turbine blade geometry denoting the critical failure locations. (a) 10^6 cycle design life, assumed maximum defect of 400 µm. (b) 10^6 cycle design life, assumed maximum defect of 200 µm. (c) 10^5 cycle design life, assumed maximum defect of 400 µm.

Figure 20.6b or higher, critical failure locations are predicted. Both shorter design lives and smaller maximum defects reduce the size of the critical failure locations. In the examples here (Figure 20.7), when the CDF probability is desired at 0.99999, the maximum defect corresponds very near to 400 µm. Similarly, when the CDF probability is only desired at 0.999, the maximum defect is approximately 200 µm. From Figure 20.7a and b, the reduction in maximum defect size of interest moves the intersection of the maximum defect and El-Haddad curve up and to the left, allowing higher stresses to be considered non-critical and reducing the critical failure location volume. Figure 20.7a–c keeps the same maximum defect size, but decreases the design life from 10^6 to 10^5 cycles. The reduction in design life shifts the El-Haddad curve up and to the right, minimizing the effects of LEFM on the fatigue life for the same defect sizes. These critical failure maps inform where a component needs to be inspected for a given load case and expected defect population.

20.5   HARDWARE DEVELOPMENT

To verify the functionality of the augmented El-Haddad model and the accuracy of the outputs, ten turbine blades and eighteen cylindrical specimens were printed from nickel-based superalloy 718 (Figure 20.8). Post-print, the cylindrical specimen was machined to fatigue bars according to ASTM E466-15 [13]. To bypass the uncertainties associated with using standard material properties in the FEM, three of the fatigue bars were subject to monotonic tensile testing to measure the Young's Modulus for the print to improve the accuracy of the FEM. The density and Young's Modulus have been directly measured for this build as 8.14 e^{-3} g/mm^3 and 159 GPa, respectively. The ten turbine blades and the remaining 15 fatigue bars were subject to fatigue testing. These components have been studied to collect fatigue life data with respect to failure-inducing defect sizes and locations.

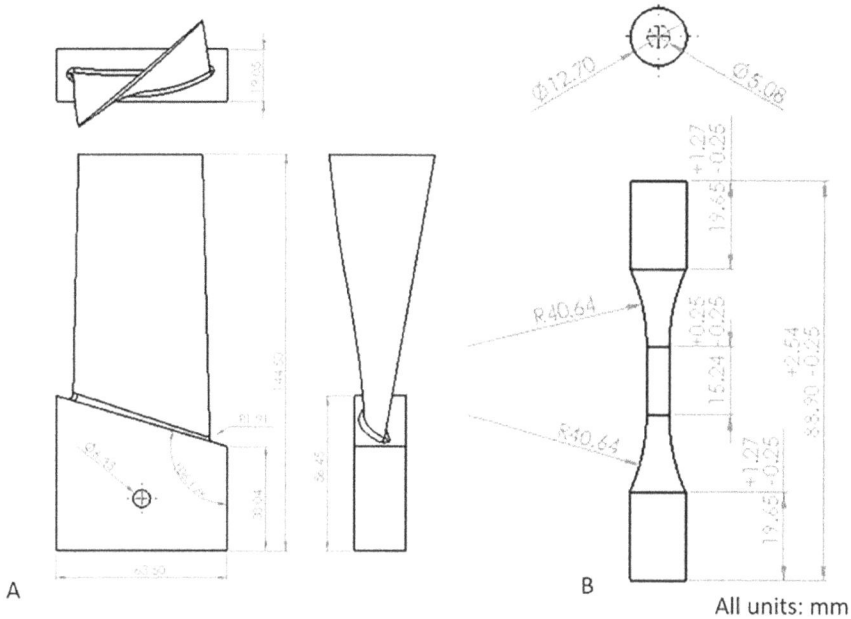

FIGURE 20.8 (a) Dimensions for the turbine blades. (b) Dimensions for the fatigue bars [13].

20.5.1 PROCESSING

The parts were printed on an M2 Cusing laser powder bed fusion (LPBF) printer with three regional parameter settings: core, skin, and contour (Table 20.2). The contour scan is a single thin line around the edge of the surface to ensure a good melt along the geometry surface. The skin scan is the small region from the surface ~1.5 mm into the geometry. This is a tight scan pattern near the surface to create a high-density region near the surface of the component with fewer defects. Finally, the core scan is a high-power setting that scans every other build layer on the interior of the part following an island scan pattern. The core parameters sacrifice print quality to improve print speed.

TABLE 20.2
M2 C using Printer Scan Settings

	Contour	Skin	Core
Power (W)	120	180	320
Layer height (µm)	40	40	80
Scan speed (mm/s)	280	800	700
Spot size (µm)	50	130	180
Trace spacing (µm)	N/A	105	130
Offset (µm)	90	95	130

Post-print, the parts were processed by a stress relief heat treatment, machining, age hardening, and a final polish process. The stress relief heat treatment put the whole build plate into an oven at 1,000 °C for 1 h and then left the oven to cool back to room temperature.

The specimens were then removed from the build plate. The cylindrical specimen was machined down according to ASTM E466-15 [13] to match the specification in Figure 20.8b. The turbine blades were machined along the grip to ensure parallel surfaces on the front and back and enlarge the grip hole to 19 mm (0.75 in) for a secure mount to the test fixture. The blades were rough and polished with sandpaper to remove the oxidation layer before aging. The sanding process reduced the surface roughness to a value of $S_a = 4.35 \pm 1.26$ μm. The surface roughness measurements are in accordance with ISO 25178 [14].

The aging process was performed according to AMS2774-G [15]. The nickel-based superalloy 718 components were heated to 718°C and held for 8 h, cooled to 621°C and held for another 8 h, and finally air cooled. Post aging, the turbine blades and fatigue bars were subject to final polishing. The turbine blades were polished using a pneumatic dremel with polishing stones and finished with a very fine polishing pad. The final surface roughness for the turbine blades is $S_a = 3.19 \pm 0.81$ μm.

20.5.2 Geometry Deviations

Due to AM printing processes, post-processing thermal environments, and various polishing steps involved to make each turbine blade ready for testing, some level of structural deviation is expected between the turbine blade computer-aided design (CAD) file and the final blade geometries. Using Advanced Topology Optimalogy System (ATOS) scans to generate a point cloud of each turbine blade surface, a measured surface geometry is built for each blade. Three scans were taken for each turbine blade and averaged together to create a geometrically accurate surface for each blade. The scanned surfaces have a maximum variation of 76 μm from the individual scans to the mean blade surface, and an average variation of 25 ±5.5 μm across all of the blade specimen. Finally, the turbine blade FEM (Figure 20.4a) is morphed to align with each unique turbine blade surface using FEMorph to create an FEM for each turbine blade that matches the final geometry for every specimen. When comparing the scanned blade surfaces to the CAD file (Figure 20.9a), there is an average variation of 315 ±9 μm with a maximum variation of 750 μm. In every case, the printed geometry tilts slightly forward of the CAD file so that the largest variation is at the blade tip (Figure 20.9b).

Using the unique FEM for each turbine blade and the measured material properties from the monotonic testing, a frequency analysis is run for every turbine blade. Where the CAD-based FEM predicts the second bending at 1,379 Hz, the geometrically variant turbine blade FEMs predict the second bending at 1,364 ±2.5 Hz. From each unique turbine blade FEM, individual stress maps are generated from the second bending mode.

20.6 TESTING

Two different fatigue tests were run to verify the augmented El-Haddad model. The first is an axial tension-compression test applied to the fatigue bar specimen

Blade 01:
Accuracy from CAD Blade: 0.01279 in

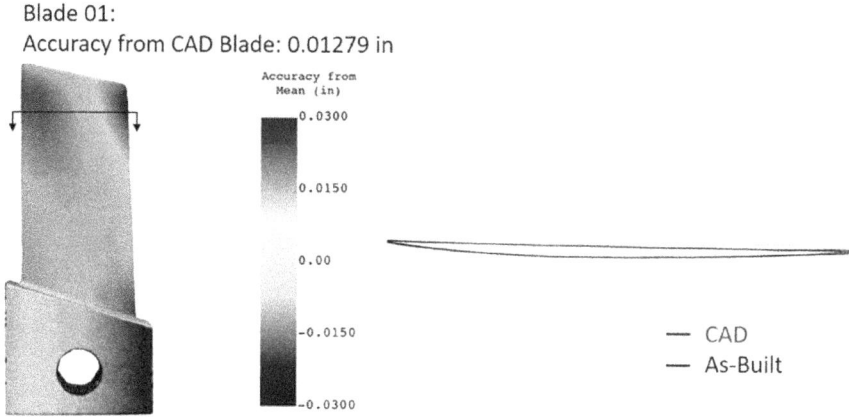

FIGURE 20.9 (a) Heat map of the variations between the CAD file and blade 01. (b) Cross-section (top down view) of variation between the scanned surface geometry and the CAD file. The blade edges are slightly shorter and the blade surface tilts away on the right edge.

executed on an 810 MTS Load Frame with a 100 kN Load Cell. The second fatigue test is a vibrationally induced bending test with the turbine blades performed on an Unholtz-Dickie 20K Electrodynamic Shaker Table. The testing sets the maximum stress level seen by each sample, the fatigue life to failure, and by studying the fracture surface, the failure defect size and location. Combining the defect location with the stress map for the part yields the stress experienced at the defect. The failure defect size, defect location stress, and fatigue life are all the data needed to determine if the augmented El-Haddad model is an accurate representation of finite fatigue life with respect to defects and if it can accurately predict where failure will initiate from.

20.6.1 Fatigue Bar Testing

The fatigue bar testing is performed at three different stress levels and runs at 20 Hz until failure. Table 20.3 depicts the test results. Columns 2 and 3 are the data generated directly from the fatigue tests. To convert from σ_{max} to $\Delta\sigma$ for the augmented El-Haddad model, the maximum stress is multiplied by $(1-R)$, where $R = -1$. Column 4 is generated by studying the fracture surfaces under a Scanning Electron Microscope (SEM) to identify the initial failure defect sizes. On every fatigue bar, the failure defect is identified as an LoF defect on or near the specimen surface. With failure defects ranging in size from 236 μm up to 656 μm, every fatigue bar failed to a large defect outside the 99.95 percentile of the assumed defect distribution depicted in Figure 20.5. Defects 400 μm or larger have the probability of occurrence once per fatigue bar based on the assumed defect distribution and total fatigue bar volume. Factoring in that the region of maximum stress is 28% of the total volume, it is expected that four to five of the fifteen fatigue bars should fail due to defects ≥400 μm. In this data set, a total of seven defects ≥400 μm are visible on the fracture surfaces. Fatigue Bars 07, 08, and 10 all contain defects over 400 μm that did not initiate the fatal crack. Fatigue Bar 08 (Figure 20.10a) contains two defects ≥400 μm

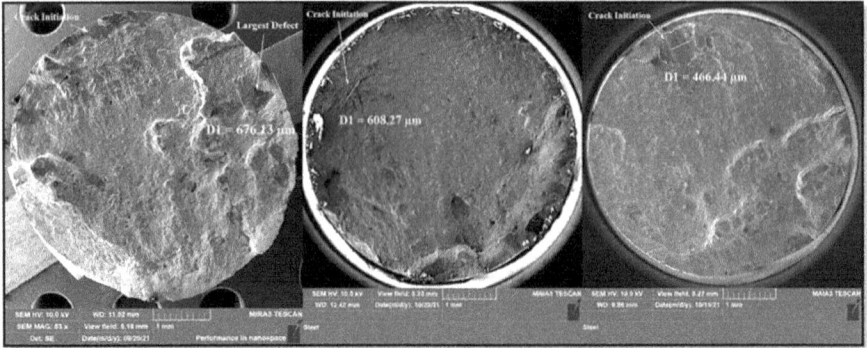

FIGURE 20.10   Fatigue bar fracture surfaces that potentially start from internal defects. (a) Fatigue Bar 08, while a larger internal defect initiated one crack growth region, the dominant region grew from a surface defect. (b) Fatigue Bar 09. (c) Fatigue Bar 10.

and grew from the smaller defect probably because the larger defect does not touch the surface and is assessed as an internal defect according to Eq. 20.2. Fatigue Bar 10 (Figure 20.10c) also contains two of these large defects and grew the fatal crack from the larger defect. Fatigue Bar 07 (Figure 20.11c) grew from a much smaller defect (248 μm). While the larger defect propagated a crack growth region, it is not clear why the smaller defect region grew faster and dominant the total crack growth.

The initiating defects for Fatigue Bars 09 and 10 (Figure 20.10b and c) do not touch the edge of their bars. Applying Eq. 20.2 to the fatal defects yields values of 0.22 and 0.70 respectively, concluding that these defects may safely be classified as internal defects.

Fatigue Bars 02, 05, 07, 08, and 13 all appear to have multiple regions of independent crack growth (Figures 20.11 and 20.10a). In every case of multiple crack growth regions, the failure defect is taken as the defect that initiated the largest crack growth region.

Column 5 is the stress experienced at each fracture surface. Measuring the diameter of the fatigue bar at the fracture surface allows the calculation of the stress experienced by the failure location. Figure 20.12 depicts the location of every fracture surface with respect to the fatigue bar dimensions. In every case, the fracture surface is at or near the maximum applied stress. The largest deviation is Fatigue Bar 04, which is far enough into the flare to experience stress at 93% of the maximum applied stress. Fatigue Bar 04 is an example of an extremely large defect at a lower stress location that dictates the final failure.

The application of columns 4 and 5 of Table 20.3 to the augmented El-Haddad model enables a predictive fatigue life that may be compared to the experimental fatigue life from column 3 to assess the quality of the model to capture the relationship between defect size, applied stress, and fatigue life.

20.6.2   TURBINE BLADE TESTING

The turbine blade testing is a more involved process to control the applied stress. A Polytec OFV 500 Laser Vibrometer measures the peak blade velocity at one point

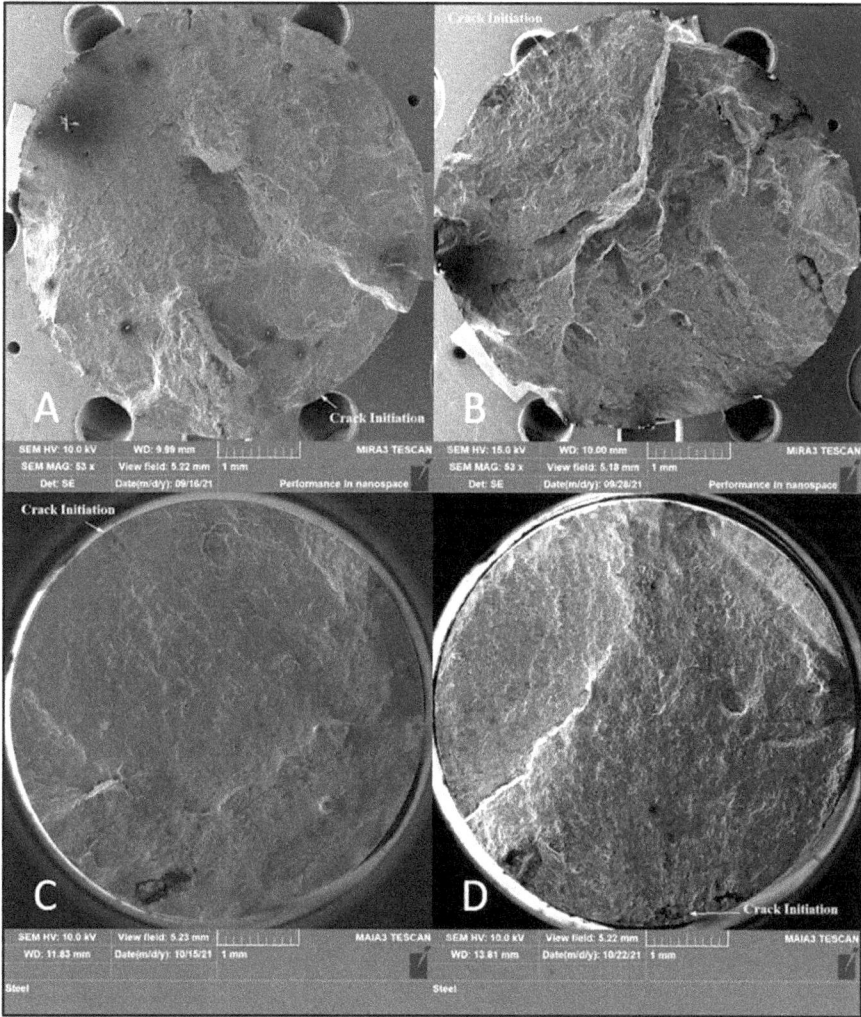

FIGURE 20.11 Fatigue bar fracture surfaces with multiple crack initiation points. Identified initiation points belong to the dominant crack growth region for each specimen. (a) Fatigue Bar 02: contains four different crack growth regions. (b) Fatigue Bar 05: contains two different crack growth regions. (c) Fatigue Bar 07: contains two different crack growth regions. (d) Fatigue Bar 13: contains two different crack growth regions.

on each blade during testing. Two strain gauges, for redundancy, are applied along the length of the blade to measure the axial strain (Figure 20.13). Application of a strain gauge at the point of maximum stress is not feasible due to the geometry of the turbine blade. Instead filtering techniques identify two gauge locations where the FEM strain value has a range of less than $1\ e^{-4}$ mm/mm across a 2 mm radius and that the average strain within the 2 mm radius is within 1% of the maximum strain in the region. The filtering criteria ensure that the strain measurements represent

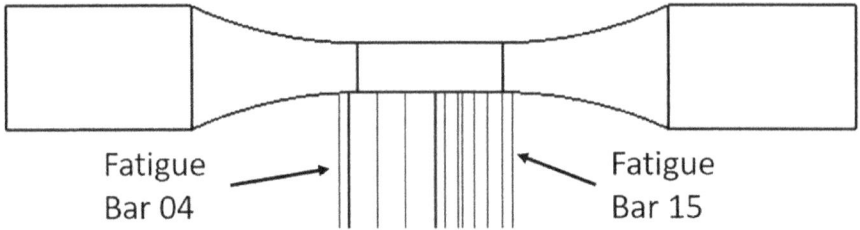

FIGURE 20.12 Failure locations for every fatigue bar test. While the majority fail within the neck of the specimen, a small group fail in the flare with Fatigue Bars 04 and 15 being the farthest out on each end.

TABLE 20.3
Fatigue Bars Subject to Axial Fatigue Testing and Fractography

Fatigue Bar	Applied σ_{Max} (MPa)	Fatigue Life (N)	Defect Size (μm)	Defect σ (MPa)
01	345	230,416	236	341
02	517	22,356	262	517
03	221	464,646	314	218
04	345	75,059	538	321
05	517	23,888	448	516
06	221	553,711	425	219
07	345	118,499	248	345
08	517	20,368	402	514
09	221	1,075,687	269 (I)	220
10	345	95,388	656 (I)	345
12	517	23,776	370	514
13	221	541,165	261	217
15	345	134,758	355	335
16	517	23,250	330	517
17	221	5,883,002	351	220

The (I) next to Defect Size denotes that the defect is an internal defect.

the expected peak strain at their locations. The applied strain gauges are 1.6 mm across ensuring that any minor variation in gauge placement does not influence the accuracy of relating the measurement values to the FEMs. The primary strain gauge is co-located with the maximum stress associated with the second bending mode (Figure 20.4b). Using the linear relationship between the measured peak velocity and the strain gauge measurements [16], the stress and strain at the gauge locations are controlled by setting the peak measured velocity in a closed feedback loop in the 20K electrodynamic shaker table. From each turbine blade's unique FEM, the relationship between maximum stress/strain and stress/strain at the gauge locations is calculated. While the exact ratio changes for each turbine blade FEM, on average, the primary gauge measures 44.3% of the maximum strain with a standard deviation of 2%.

By extension, the same ratio applies to the maximum stress. Applying the transforms of laser vibrometer peak velocity to strain gauge measurement, strain to stress at the strain gauge location, and measured stress to peak stress fully defines the system to control the maximum stress from the peak velocity.

The point of maximum stress for every turbine blade is along the root where the blade meets the grip (Figure 20.13). The maximum stress location varies between each turbine blade due to small geometric variations. The ten-turbine blade FEMs predict three primary regions where the maximum stress appears: left, right, and center of blade on the convex side near the root. Due to the unique geometry of each turbine blade FEM, even blades that trend to similar peak stress locations have some regional variation. Five of the ten turbine blades have their maximum stress point along the center. Turbine Blades 03 and 08 have their maximum stress points on the right edge of the blade and Turbine Blades 01, 05, and 07 all have their maximum points on the left edge of the blade. While the global maximum stress point varies due to the geometric variations between the turbine blades, the three identified regions are local maximum stress points for every blade (Figure 20.4b).

The turbine blade fatigue testing applies the stress step function test (Eq. 20.10) [16,17]. The desired fatigue life is selected for the test, N_t, specified in Table 20.4. The sample is cycled at an initial peak stress level until N_t is reached, then the stress is stepped up by an incremental value. For the testing here, the incremental value is

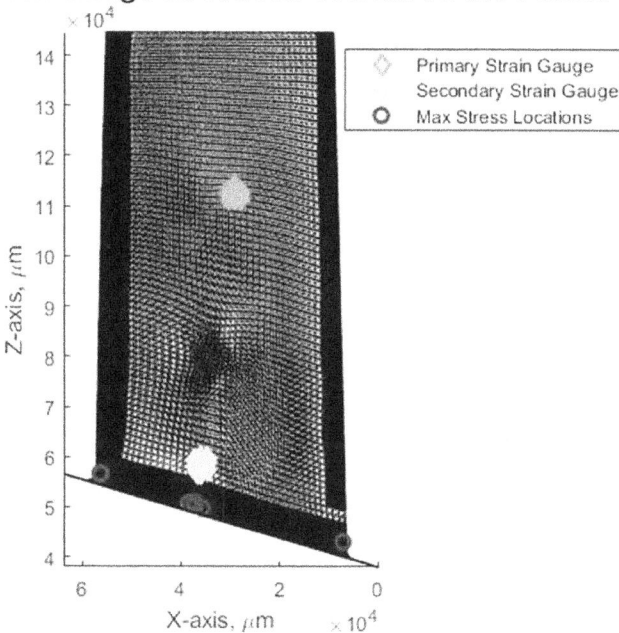

FIGURE 20.13 Locations of the primary and secondary strain gauges. Gauges attached to the same locations for every turbine blade. The position of the maximum stress point for each unique FEM.

TABLE 20.4

Turbine Blades Subject to Vibration Induced Bending Fatigue Tests and Fractography

Blade ID	Applied σ_{Max} (MPa)	Fatigue Life (N)	Defect Size (µm)	Defect Type	Defect σ (MPa)
01	569	10^6	19	Notch (Surface)	531
02	437	5×10^5	41	LoF (Internal)	419
03	653	10^6	195	LoF (Surface)	534
04	310	10^7	23	Notch (Surface)	296
05	717	10^5	44	LoF (Internal)	635
06	350	5×10^6	207	LoF (Surface)	328
07	431	5×10^6	60	LoF (Surface)	324
08	469	10^7	-	-	-
09	503	5×10^5	116	LoF (Surface)	499
10	717	10^5	33	LoF (Surface)	589

34.5 MPa (5 ksi). The fatigue stress level, σ_a, is the linear interpolation between the last stress level that reaches N_t cycles, σ_{pr}, and the stress level that causes failure, σ_f based on the percentage of completion for the final step, $\dfrac{N_f}{N_t}$, where N_f is the number of cycles ran in the final step.

$$\sigma_a = \sigma_{pr} + \frac{N_f}{N_t} * (\sigma_f - \sigma_{pr}) \tag{20.10}$$

The results of the fatigue testing for each turbine blade are found in Table 20.4. The fatigue testing of the turbine blades reached the limits of the vibration table without growing any of the cracks to full separation. Applying a Fluorescent Penetrate Inspection (FPI) to each blade surface revealed the location and width of each crack (Figure 20.14). With the exceptions of Blades 01 and 08, the cracks initiated in locations near the center of the blade root. In Blade 03, the crack initiated at the left edge of the blade, far from the Blade 03 global peak stress. In Blade 08, the crack initiated along the right edge in close proximity to its global peak stress. After growing the cracks, the blades were separated to expose the full fracture surface by applying notches as the crack tips. During the separation process, the failure initiation point for Blade 08 was mistakenly destroyed.

Fractography on the blade crack growth surfaces provides the critical defect size (Table 20.4, column 4) and location that appears to initiated crack growth for each turbine blade. Unlike the fatigue bars where LoF defects caused every failure, two of the turbine blades (Blades 01 and 04) failed due to notches on the surface associated with surface roughness. The other blade failure defects are LoF defects. The failure defects for Blades 02 and 05 do not touch the surface, applying equation 2 yields values of 0.15 and 0.32 respectively, classifying those defects as internal. Applying the measured defect locations back to the FEM for each blade generates the ratio between maximum stress and defect location stress to generate column 5. The calculated stress at the defect locations ranges from 75% to 99% of the maximum stress.

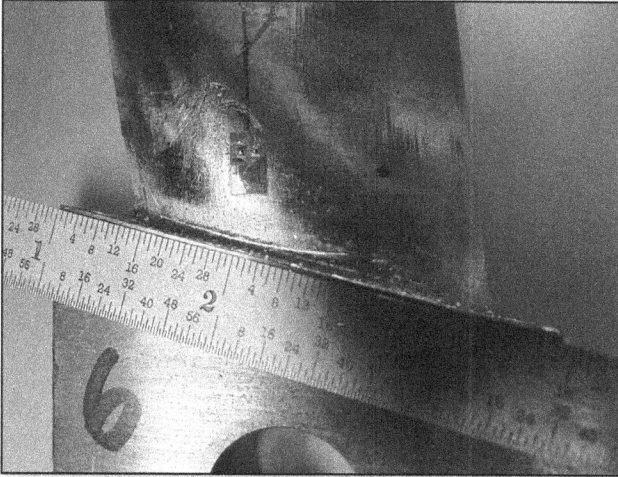

FIGURE 20.14 FPI of Blade 06. Measurements indicate that the crack front is approximately 27 mm long and started near the center of the blade root.

20.7 RESULTS

Applying the measured defect sizes and the defect stresses from Tables 20.3 and 20.4 into the El-Haddad model predicts life values. The predicted life for each specimen is compared to the measured fatigue lives to generate Figure 20.15. The predicted versus measured fatigue life plot informs the quality of the augmented El-Haddad model against the experimental data sets. Applying the measured defect locations for each turbine blade to their unique FEMs with the experimental fatigue life evaluates the ability of the augmented El-Haddad model to predict critical failure locations.

20.7.1 FATIGUE LIFE EVALUATION

The fatigue bar data trends very well with the augmented El-Haddad model that was trained from prior testing [4]. Only three of the fifteen specimens (Fatigue Bars 09, 10, and 17) tested more than twice as long as the model predicted (Figure 20.15). Fatigue Bars 09 and 10 failed due to an internal defect instead of a surface defect. The model assumes a surface defect that has a higher shape factor, Y, and is therefore conservative when an internal defect becomes the root cause of error. When the model is adjusted to account for an internal defect both data points fall within the $2\times$ bands. The final fatigue bar specimen that that model under-predicts the fatigue life is Fatigue Bar 17, which lasted 16.4 times longer than the model prediction. It is unclear at this time why Fatigue Bar 17 lasted so much longer than the rest of the set. The remaining fourteen fatigue bars have a mean ratio of measured life over predicted life of 1.54 ± 0.59. With the material properties built into the El-Haddad model, the fatigue bar data demonstrates a slightly conservative trend in the model prediction while keeping very accurate results.

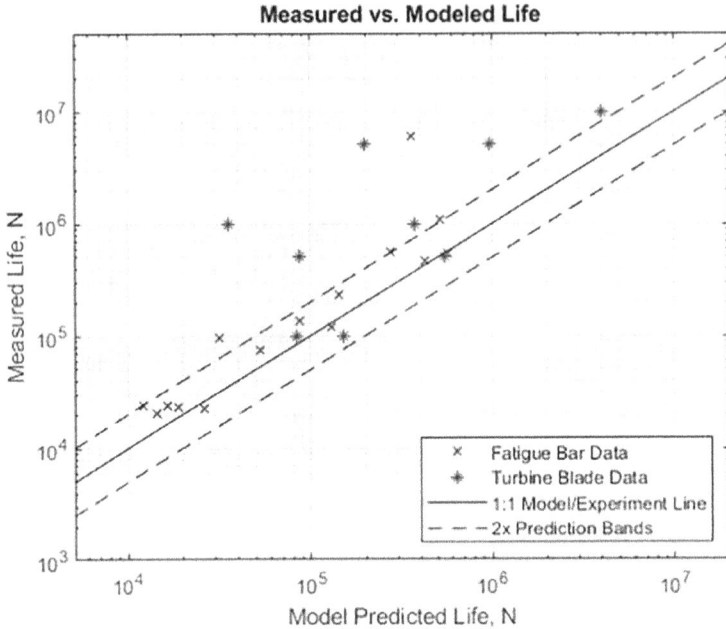

FIGURE 20.15 Measured fatigue life of the axially loaded fatigue bars and the bending loaded turbine blades versus the augmented El-Haddad model predictions based on the measured fatal defect size and location.

The turbine blade data demonstrates the same trend as the fatigue bars with an increase in the scatter. Evaluating all nine turbine blade data points has a mean ratio of 8.08 ± 10.86 for the measured life over the predicted life.

Turbine Blade 03 has the largest variation between the model prediction and the experimental data at a ratio of 28.4. This specimen is one of the two unique cases where the crack grew from the edge of the blade instead of the center. The edge growth means that the crack crosses the blade at a very early stage and that it is only attached to the main body on one side. The critical defect associated with Blade 03 is also smoother than any of the defects seen on the other fracture surfaces implying that there was additional friction between the fracture surfaces not seen in any of the other specimens. The smoother surface makes identification of the initial size harder to measure. As a result, anything that might be part of the original defect is factored into the reported defect size in Table 20.4 creating an over-estimation of the defect size being processed through the augmented El-Haddad model. There is also a strong possibility that the extra degree of freedom imparted by having half of the crack surface free of any applied loads or constraints absorbs some of the applied energy and further inhibits the crack growth.

Removing Turbine Blade 03 from the set due to the unique nature of failure, the remaining eight blades have a mean of 5.54 ± 8.27. In either case, there is insufficient data to reject the hypothesis that both the fatigue bars and turbine blades use the same trend line.

The other data point of note is Turbine Blade 06 which measures 25.5 times longer than the El-Haddad model predicts for the measured defect size and location. No cause has been attributed at this time to Blade 06 for why the specimen survived as long as it did.

Figure 20.16 depicts all of the test specimens against the El-Haddad model. There is a clear separation in failure defect sizes between the fatigue bars and turbine blades despite being printed on the same build plate. Every fatigue bar contains sufficiently large defects to have the fatigue life dominated by LEFM. In contrast, the turbine blades failed due to smaller defects and fall on both sides of the El-Haddad critical crack length, $a_{0,N}$, making the material fatigue limit a more dominant mechanism in the failure.

20.7.2 CRITICAL FAILURE EVALUATION

Evaluating the critical failure locations looks at a single design life and a maximum defect size in conjunction with the augmented El-Haddad model and FEM. Figure 20.17 outlines the steps required to generate the critical failure locations on any component. This research built and tested a number of specimen and verified that the initial finite life El-Haddad model creates an accurate representation of the relationship between defect size, location, and desired life. For the expected maximum defect size, the best practice is to generate a defect distribution to inform on the maximum defect size of interest. In this section, maximum defect sizes are identified based on the measured failure defects in the turbine blades to explore the accuracy of

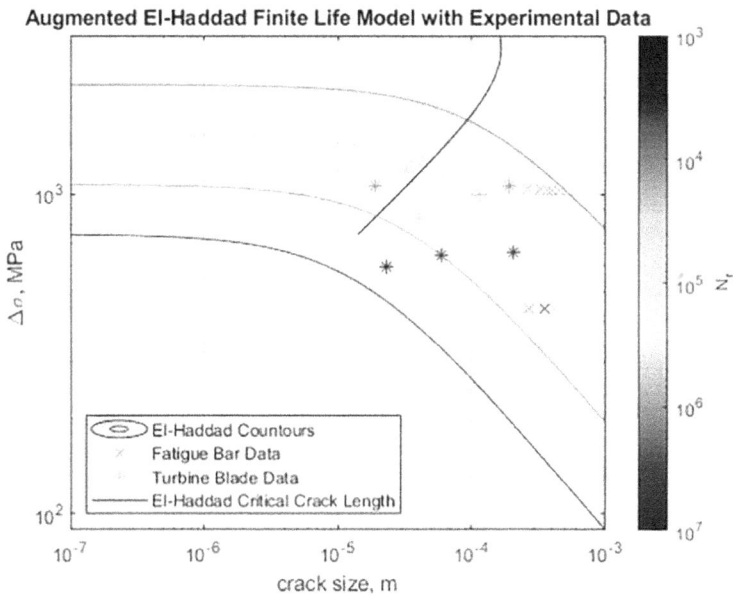

FIGURE 20.16 Fatigue bar and turbine blade experimental data on the augmented El-Haddad model. The black line depicts the El-Haddad critical crack length, $a_{0,N}$. Any defects larger than $a_{0,N}$ are dominant by LEFM. Any defects smaller than $a_{0,N}$ are fatigue limit dominated.

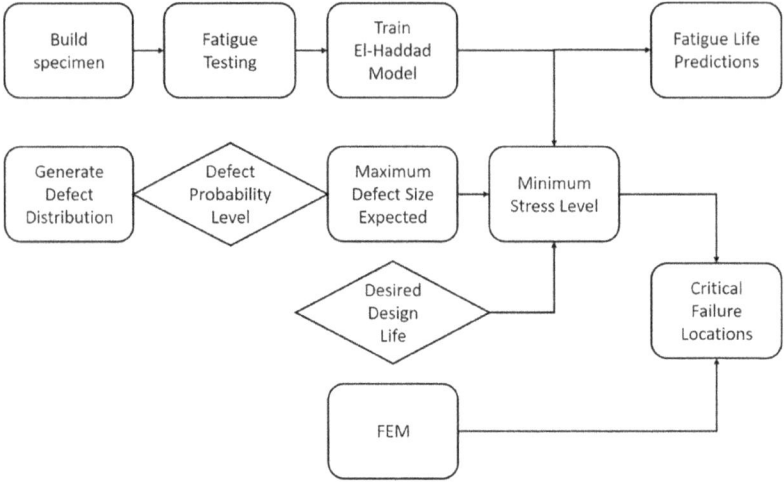

FIGURE 20.17 Flow chart to generate predictions of fatigue life and critical failure locations.

the identified critical failure locations. The desired design life for each turbine blade is the tested fatigue life value (Table 20.4). Finally, the minimum stress level for each choice of design life and defect size is combined with the blade unique FEMs to identify every location that could initiate failure. In every case here, the measured failure locations fall within the predicted failure locations.

For Turbine Blades 01 and 03 with a fatigue life of 10^6 cycles, the largest failure defect is measured at 195 µm on Turbine Blade 03. Figure 20.18 looks closely at Turbine Blade 03 with respect to the choice of maximum defect size when determining the critical failure locations. Figure 20.18a shows the predicted failure areas where a defect 50 µm has the potential to grow into a fatal crack. The measured location for the 195 µm defect is just outside of the identified region at the left edge.

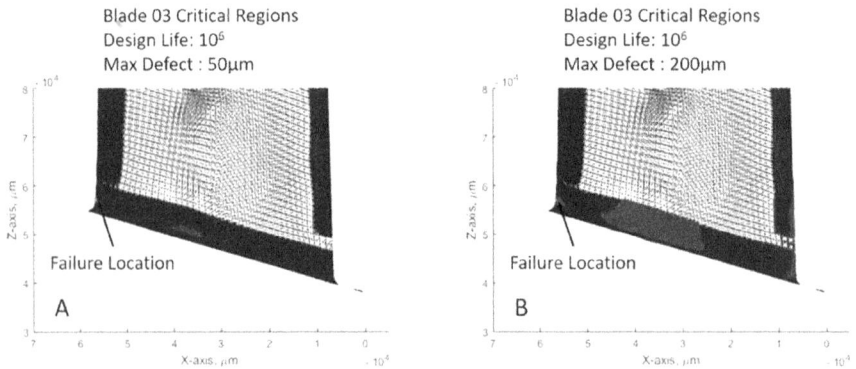

FIGURE 20.18 Critical failure locations for turbine blade 03 assuming (a) A defect ≤ 50 µm causes failure and (b) A defect ≤ 200 µm causes failure.

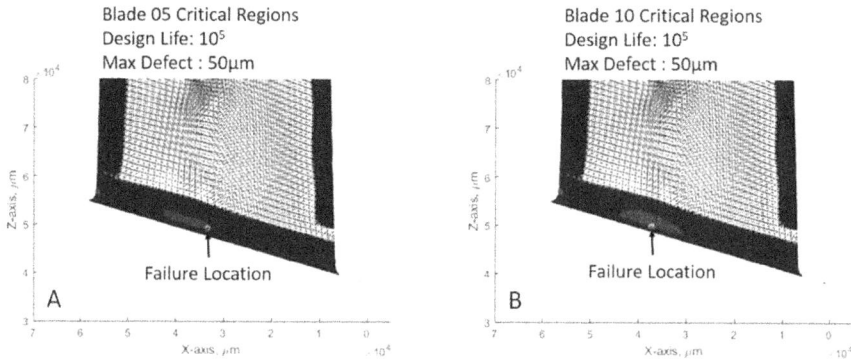

FIGURE 20.19 Critical failure locations for (a) Turbine blade 05 and (b) Turbine blade 10. In both cases the identified failure defect falls within the bounds predicted by the augmented El-Haddad model.

When the evaluation maximum defect size is increased to 200 μm (Figure 20.18b), the predicted area of crack growth expands to include the identified failure location.

The predicted critical failure locations for Turbine Blades 05 and 10 (Figure 20.19) are depicted for the case of 105 cycles fatigue life and a failure-inducing defect of 50 μm. Both turbine blades fail due to defects smaller than 50 μm, and in both cases, the measured defect location falls within the predicted failure location along the center of the blade.

20.8 CONCLUSION

This study walks through combining the finite fatigue life El-Haddad model with an FEM to emphasize the importance of the defect location in addition to the defect size. The augmented El-Haddad model demonstrates the ability to accurately predict the fatigue life for both axial and vibrationally loaded specimens. Here the values for defect size and stress are measured from the fracture surface to minimize the uncertainties in the defect that causes failure; however, the model does not require that to generate a fatigue life prediction. Any technique to generate a defect size and location, when coupled with an FEM will generate the parameters required to make a fatigue life prediction. Other techniques that could be used to make a fatigue life prediction are: surface penetrating scans (e.g., X-ray scans), defect estimations from process monitoring during a print (e.g., in-situ data collections), or stochastic evaluations from a known or assumed defect population.

The model is also shown to accurately predict where failure can start for a design life and defect size. By linking the augmented El-Haddad model with an FEM load case, maximum expected defect, and desired component life, the model predicts sections of the component where failure can start at. From an inspection view, the capability to minimize the search region for root causes of failure saves time and money. By developing a better understanding of potential defect sizes and location based on AM printer settings and materials, this feature of the augmented El-Haddad model will become even more accurate.

REFERENCES

1. Beretta, S., and Romano, S., "A comparison of fatigue strength sensitivity to defects for materials manufactured by AM or traditional processes," *International Journal of Fatigue*, Vol. 94, 2017, pp. 178–191. https://doi.org/https://doi.org/10.1016/j.ijfatigue.2016.06.020, URL https://www.sciencedirect.com/science/article/pii/S0142112316301645, fatigue and Fracture Behavior of Additive Manufactured Parts.

2. Beretta, S., Gargourimotlagh, M., Foletti, S., du Plessis, A., and Riccio, M., "Fatigue strength assessment of "as built" AlSi10Mg manufactured by SLM with different build orientations," *International Journal of Fatigue*, Vol. 139, 2020, p. 105737. https://doi.org/https://doi.org/10.1016/j.ijfatigue.2020.105737, URL https://www.sciencedirect.com/science/article/pii/ S0142112320302681.

3. Sheridan, L., "Primary Processing Parameters and Their Influence on Porosity and Fatigue Life of Additively manufactured Alloy 718," Dissertation, Wright State University, Dayton, OH, 2020.

4. Sheridan, L., "A modified El-Haddad model for versatile defect tolerant design," *International Journal of Fatigue*, Vol. 145, 2021, p. 106062. https://doi.org/https://doi.org/10.1016/j.ijfatigue.2020.106062, URL http://www.sciencedirect.com/science/article/pii/S0142112320305946.

5. Ciavarella, M., and Monno, F., "On the possible generalizations of the Kitagawa–Takahashi diagram and of the El Haddad equation to finite life," *International Journal of Fatigue*, Vol. 28, No. 12, 2006, pp. 1826–1837. https://doi.org/https://doi.org/10.1016/j.ijfatigue.2005.12.001, URL https://www.sciencedirect.com/science/article/pii/S0142112306000120.

6. El-Haddad, M., Smith, K., and Topper, T., "Fatigue Crack Propagation of Short Cracks," *Journal of Engineering Materials and Technology*, Vol. 101, 1979, pp. 42–46.

7. Aigner, R., Pusterhofer, S., Pomberger, S., Leitner, M., and Stoschka, M., "A probabilistic Kitagawa-Takahashi diagram for fatigue strength assessment of cast aluminium alloys," *Materials Science and Engineering: A*, Vol. 745, 2019, pp. 326–334. https://doi.org/https://doi.org/10.1016/j.msea.2018.12.108, URL https://www.sciencedirect.com/science/article/pii/ S0921509318317994.

8. Murakami, Y., "Analysis of stress intensity factors of modes I, II and III for inclined surface cracks of arbitrary shape," *Engineering Fracture Mechanics*, Vol. 22, No. 1, 1985, pp. 101–114. https://doi.org/https://doi.org/10.1016/0013-7944(85)90163-8, URL https://www.sciencedirect.com/science/article/pii/0013794485901638.

9. Romano, S., Brückner-Foit, A., Brandão, A., Gumpinger, J., Ghidini, T., and Beretta, S., "Fatigue properties of AlSi10Mg obtained by additive manufacturing: Defect-based modelling and prediction of fatigue strength," *Engineering Fracture Mechanics*, Vol. 187, 2018, pp. 165–189.

10. Romano, S., Miccoli, S., and Beretta, S., "A new FE post-processor for probabilistic fatigue assessment in the presence of defects and its application to AM parts," *International Journal of Fatigue*, Vol. 125, 2019, pp. 324–341. https://doi.org/https://doi.org/10.1016/j.ijfatigue.2019.04.008, URL http://www.sciencedirect.com/science/article/pii/S014211231930132X.

11. Du Plessis, A., and Beretta, S., "Killer notches: The effect of as-built surface roughness on fatigue failure in AlSi10Mg produced by laser powder bed fusion," *Additive Manufacturing*, Vol. 35, 2020, p. 101424.

12. Sheridan, L., Gockel, J. E., and Scott-Emuakpor, O. E., "Primary processing parameters, porosity production, and fatigue prediction for additively manufactured alloy 718," *Journal of Materials Engineering and Performance*, Vol. 28, No. 9, 2019, pp. 5387–5397.

13. E466-15, A., *Standard Practice for Conducting Force Controlled Constant Amplitude Axial Fatigue Tests of Metallic Materials*, ASTM International, West Conshohocken, PA, 2015.
14. 25178-2, I., *Geometrical Product Specifications (GPS) - Surface Texture: Areal - Part 2: Terms, Definitions and Surface Texture Parameters*, International Organization for Standardization, Geneva, Switzerland, 2012.
15. 2774-G, A., *Heat Treatment Nickel Alloy and Cobalt Alloy Parts*, SAE Technical Standards Board, 2020.
16. Scott-Emuakpor, O., Shen, M.-H. H., George, T., Cross, C. J., and Calcaterra, J., "Development of an improved high cycle fatigue criterion," *Journal of Engineering for Gas Turbines and Power*, Vol. 129, 2007, pp. 162–169. https://doi.org/https://doi.org/10.1115/1.2360599.
17. Maxwell, D. C., and Nicholas, T., "A rapid method for generation of a Haigh diagram for high cycle fatigue," *Fatigue and Fracture Mechanics: 29th Volume*, ASTM International, 1999.

21 Analysis and Simulation of Hypervelocity Gouging Impacts for a High-Speed Sled Test*

John D. Cinnamon and Anthony N. Palazotto
US Air Force Institute of Technology

CONTENTS

21.1 INTRODUCTION

The Holloman High-Speed Test Track (HHSTT) is a U.S. Air Force test facility that conducts hypervelocity impact testing. The HHSTT performs this testing using rocket-propelled sleds that ride on a 15.5 km steel rail track. The rocket sleds are held on the rails by steel "shoes/slippers." Figure 21.1 illustrates the sled and shoe geometry. Current testing capability allows for impact testing in the 3 km/s range, with near-term goals extending the sled velocity to greater than 3.5 km/s. At these velocities, the shoes impact the rail, due to the motion experienced by the test sled, with both a vertical and downrange component. The severity of the impact varies, but in certain cases can result in catastrophic failure of the test sled. Inspections of the

---

* Reprinted with permission: Cinnamon, J. D., & Palazotto, A. N. (2009). Analysis and simulation of hypervelocity gouging impacts for a high speed sled test. *International journal of impact engineering*, *36*(2), 254–262.

DOI: 10.1201/9781003220978-27

445

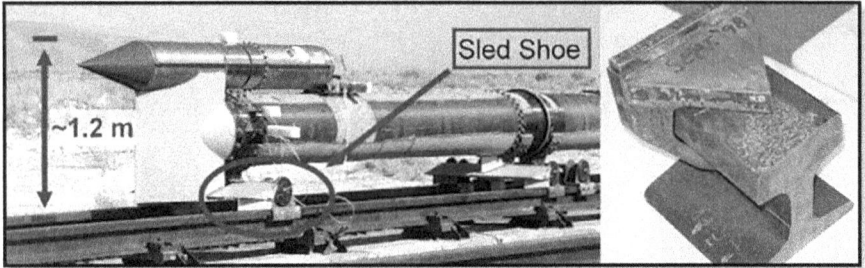

FIGURE 21.1 HHSTT sled/shoe geometry and shoe/rail section with gouge.

rails after test events have revealed damage in the form of gouges [1]. All high-speed (above 1.5 km/s) test runs result in some form of gouging – only the frequency and depth of the gouges vary throughout a series of tests. Gouges associated with high-energy impacts are characterized by a teardrop-shaped shallow depression in the material (see Figure 21.1 and [1–6] for more detail). These gouges typically are the result of a high-pressure core developing at the shoe/rail material interface and high viscoplastic deformation that leads to material mixing (i.e., shoe material becoming embedded within the rail, and vice versa) [1–6].

Simulations of the hypervelocity gouging process have previously been conducted [2–6] using the hydrocode CTH [7,8]. A plane-strain model of the shoe/rail interaction was developed as an accurate, and computationally necessary, representation of the actual impact geometry [2–6]. These previous models were limited by the material constitutive relationships present within the CTH code. Therefore, an effort was undertaken to perform experimentation on the specific materials (VascoMax 300 for the shoes, 1080 steel for the rails) to arrive at accurate Johnson–Cook and Zerilli– Armstrong constitutive models [9,10]. The constitutive model parameters were adjusted to more accurately match split Hopkinson bar (SHB) and validated against Taylor impact test results. The experimentation was performed on the actual materials at the HHSTT. The 1080 steel samples came from rail sections and the VascoMax 300 samples were machined from material used at the HHSTT to fabricate the sled shoes. Preliminary characterization experiments were limited to SHB and Taylor impact tests, with strain rates in the 10^3/s range. The simulations of the SHB and Taylor impact tests using the material flow models accurately replicated both of these experimental tests (in most cases less than a 2% error in predicting material deformation, at worst a 9% error) [10,11]. However, the high-energy impact environment of gouging typically generates strain rates well in excess of those levels.

Therefore, flyer plate impact experiments were subsequently conducted to extend the material constitutive behavior into the 10^4–10^5/s strain-rate regime [12–16]. The CTH code was used to modify the flow models deduced from the SHB and Taylor impact tests to make them capable of matching the measured stress waves in the flyer plates. These new models were then used within CTH to very successfully simulate previously conducted laboratory hypervelocity gouging experiments – that is, material deformation was exactly predicted by CTH within the experimental measurement error bars. The gouges generated in CTH matched other experimentally observed characteristics, such as material microstructure changes due to thermoplastic material deformation, extremely well.

These observations concerning the characteristics of gouging and the associated thermal events that are evident when the post-gouge material is metallurgically examined were previously reported [17–19]. These actual gouges from hypervelocity gouging impacts provide a validation scenario for the simulations of these high-energy impact events. With the new validated material constitutive models, the HHSTT sled/rail gouging phenomenon was accurately modeled within the CTH code and preliminary observations were made regarding the conditions, which lead to gouging. The results of those simulations match experimentally observed gouge characteristics [14–16].

In this chapter, we extend our previous work. The refined (via the use of SHB, Taylor impact, and flyer plate impact tests) and validated (via comparison to a laboratory hypervelocity gouging impact experiment) Zerilli–Armstrong material flow models of VascoMax 300 and 1080 steel, implemented within the hydrocode CTH, are applied to the HHSTT gouging phenomenon. The goal is to make recommendations regarding test parameter modifications that will mitigate the occurrence of hypervelocity gouging on the HHSTT rails. Conditions leading to wearing damage to the HHSTT rails are differentiated from those that precipitate hypervelocity gouging damage. Conclusions are drawn from the simulations regarding the mitigation of hypervelocity gouging at the HHSTT.

21.2 OVERVIEW OF VALIDATED MATERIAL FLOW MODELS

The typical fashion in which material deformation is considered within computational codes is through constitutive models. In dynamic deformation, a simple description of the yield stress is insufficient. More complex material models are necessary – which modify the flow stress σ, by accounting for variations in strain, strain rates, and temperature. One of the most widely implemented constitutive models is the Zerilli–Armstrong (Z–A) model [20].

$$\sigma = A + (c_1 + c_2\sqrt{\varepsilon})\,e^{(-c_3 T + c_4 T \ln \dot{\varepsilon})} + c_5 \varepsilon^n \tag{21.1}$$

where $\dot{\varepsilon}$ is the equivalent plastic strain rate, and A, c_1, c_2, c_3, c_4, c_5, and n are experimentally determined constants. The Zerilli– Armstrong is a materials science-oriented constitutive model approach. For instance, $c_1 = 0$ for a face-centered cubic (FCC) material lattice, while $c_2 = 0$ for a body-centered cubic (BCC) lattice.

The existing constants for the Z–A material strength model for both VascoMax 300 and 1080 steel were derived from low to midrange strain-rate (approx. 10^3/s) experimental tests (simple quasistatic tests and SHB tests) [9–11]. These relatively simple tests can provide information on stress and strain at given temperatures for strain rates in the $1–10^3$/s range. The calibrated Z–A constants derived from these tests were validated for use within CTH by simulating a midrange strain-rate Taylor impact test [10]. Successful replication of the Taylor impact test results was obtained from CTH simulations using these material models. It should be noted here that earlier work also focused on using the well-known Johnson–Cook constitutive model

TABLE 21.1

Zerilli–Armstrong Model Constants for VascoMax 300 and 1080 Steel

Constant	Steel	VascoMax 300
Z–A: A (GPa)	0.825	1.42
Z–A: c_1 (GPa)	4.0	4.0
Z–A: c_2 (GPa)	0	0
Z–A: c_3 (eV^1)	160.0	79.0
Z–A: c_4 (eV^1)	12.0	3.0
Z–A: c_5 (GPa)	0.266	0.266
Z–A: n	0.289	0.289

[21], due to the relative simplicity in generating the experimentally derived constants. The Zerilli–Armstrong model was later adopted for this work, due to its ability to better match higher strain-rate material behavior – a critical feature for simulating gouging.

The high-energy impacts at the HHSTT generate material strain rates on the order of 10^4–10^6/s [4–6]. In order to generate data to extend the previously developed flow models, a higher strain-rate uniaxial test is required – the flyer plate impact experiment [12–16]. This flyer plate experiment was able to generate strain rates of 10^4–10^5/s. By using CTH to refine the calibration of the Z–A material flow models to accurately replicate experimentally measured stress waves within the flyer plates, an accurate high strain-rate material flow model was developed for both VascoMax 300 and 1080 steel. In order to validate these newer strength models, a scaled hypervelocity impact experiment was devised and conducted in order to provide experimental observations to compare against CTH simulations. The CTH model was very successful in replicating the experimentally observed impact phenomenon [12–16]. Table 21.1 summarizes the Zerilli–Armstrong constants derived from these previous efforts.

In the shoe/rail impact regions undergoing deformations that have strain rates above 10^5/s, the equation-of-state (EOS) of the material dominates the solution of the conservation equations. CTH, fortunately, has a significant embedded tabular EOS database, based on experimental hypervelocity impacts, for both 1080 steel and VascoMax 300. In a typical simulation of either the scaled hypervelocity gouge test, or the HHSTT sled simulation, there are regions of the impact experiencing strain rates from 1/s to 10^6/s. Therefore, a validated material constitutive model (which can cover strain rates at 10^5/s and below), in conjunction with an accurate EOS (which covers higher strain rates), is necessary to compute the entire impact event.

21.3 CTH MODELING OF THE HHSTT GOUGE SCENARIO

With the validated Zerilli–Armstrong flow stress model, and the built-in EOS tables for these two materials, a CTH model can be created and used to simulate the HHSTT gouging scenario with high confidence in the accuracy of the material models. A full-scale model (i.e. the actual physical sled dimensions) of the sled shoe/rail

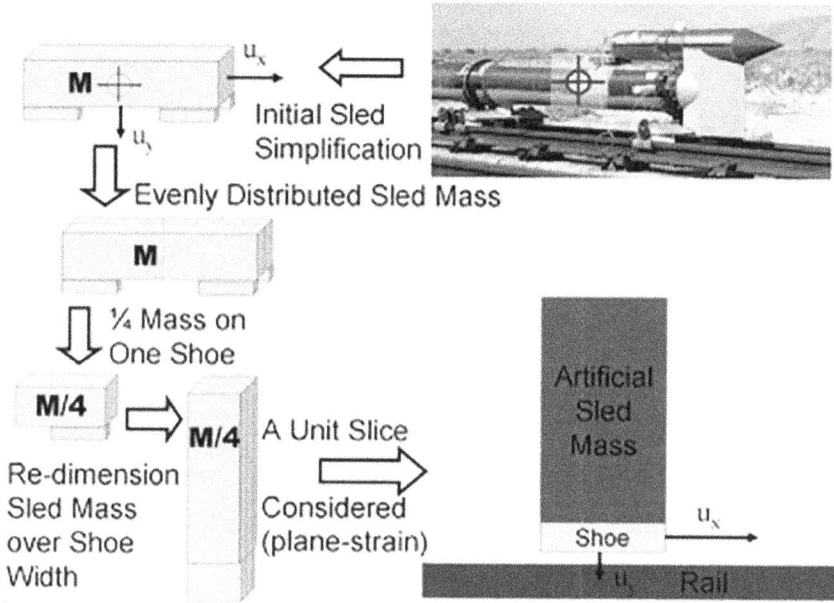

FIGURE 21.2 Plane-strain model of HHSTT sled shoe/rail [6].

gouging problem was previously developed [4–6], but used a nonspecific material strength model for both the sled shoes and rails. By modifying that previous effort with these new validated material models, a further investigation was made into the gouging phenomenon. The model of the sled/rail interaction is a 2-D plane-strain implementation, as illustrated in Figure 21.2 [6]. A more detailed discussion of the 2-D modeling effort, including validation of the CTH, results in this plane-strain simulation (for the short time of impact we are considering in this work, approx. 20–40 μs), is available in the literature [12–15].

As previous studies found, there are two primary mechanisms that seem to be required to replicate the observed physical gouging phenomenon in CTH simulations. The first mechanism is providing the moving shoe with a sufficiently large vertical impact velocity component to deform the rail and initiate material mixing and gouging. This vertical impact creates a "hump" of rail material out in front of the shoe that is a recognized precursor to gouging, at least in simulations [19].

The second kind of mechanism that creates gouging is simulating shoe contact with a rail discontinuity, rail roughness, or an asperity [4–6]. A collision of the shoe with these kinds of rail conditions causes gouging in the simulations at the horizontal (downrange) velocities being considered – i.e. 3 km/s. These types of rail discontinuities represent actual conditions possible (probable) at the HHSTT.

Both of these types of gouge-inducing impact mechanisms were evaluated within CTH simulations to ascertain the conditions that lead to gouging (vice wear) with the goal of determining which HHSTT parameters could be modified to reduce the probability of future gouging incidents. Wear is defined here as mechanical damage to

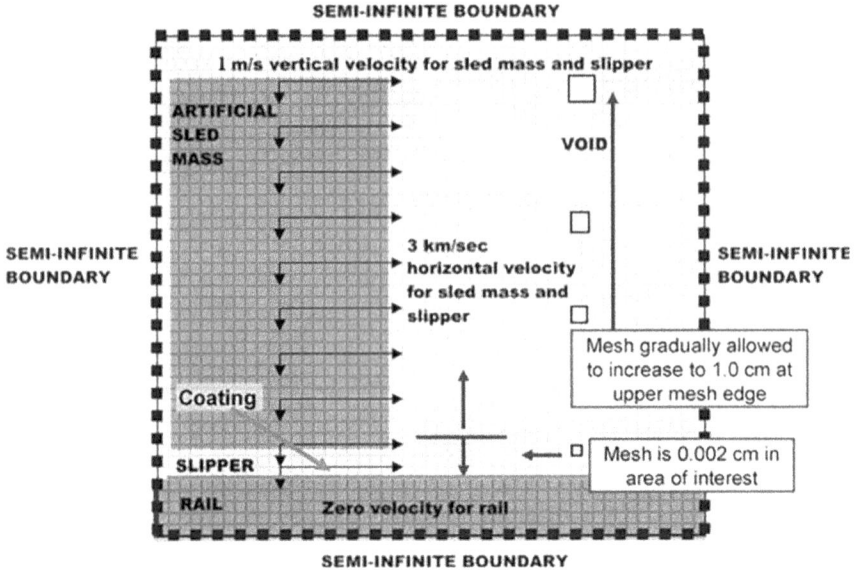

FIGURE 21.3 CTH model of HHSTT impact scenario.

the shoe or rail in which material mixing and large-scale deformations do not occur. At the HHSTT, these wear impacts are characterized by rail "scrapes" and damage to the rail coating.

The shoe/rail impact scenario was modeled using a mesh-converged value for CTH cell size reported previously [4–6]. In addition to the exhaustive mesh-convergence study conducted previously [4–6], the mesh-convergence cell size was re-verified by the authors when the new material flow model parameters were input into CTH. This model also included the coatings currently used at the HHSTT (iron oxide and epoxy) to mitigate gouge initiation. CTH possesses both of these materials in its current database – both for a Z–A strength model and a tabular EOS implementation. It should be noted here that the material flow models for the coatings were validated in CTH simulations of the Taylor impact test and the scaled hypervelocity impact test previously reported [10–16]. Figure 21.3 illustrates the model constructed in CTH [19].

21.3.1 Vertical Impact

The vertical impact case considers a sled oscillating during its high-speed run and impacting the rail at a prescribed vertical velocity while at a downrange velocity of 3 km/s. By varying that vertical impact speed in the impact sled simulations, it became clear that lower than a critical vertical velocity threshold, the impact would only lead to a wearing-type result. Above that critical velocity threshold, hypervelocity gouging would occur. In other words, a threshold vertical kinetic energy exists, above which gouging is created. Gouging includes dynamic phenomena like material "jetting" and mixing – in which shoe material remains within the rail after impact and vice versa. Impacts that are simulated below this identified threshold vertical kinetic energy cause a glancing impact that can be described as "scraping" or "wear." There is permanent material deformation and removal, but it does not develop into gouging. This concept of a threshold is analogous to a threshold penetration energy described in previous work [16].

The critical vertical impact velocity determined from extensive CTH simulations was approximately 50 m/s. This far exceeds the HHSTT estimated vertical impact velocities of 1–2 m/s. That is, at the vertical impact velocities experienced by the sled at the HHSTT, a gouging event is not predicted by the CTH simulations. This result indicates that a pure vertical impact scenario is insufficient to explain the observed gouging phenomenon at the HHSTT and consequently led to further investigation into the causal factors in the gouging event.

21.3.2 RAIL TOLERANCE IMPACT CASE

The second primary way to generate hypervelocity gouging is to have the sled shoe impact a rail discontinuity – sometimes referred to as "roughness" or an "asperity." The HHSTT has specific rail tolerances that it has determined to be necessary to prevent gouging from occurring [19].

The first rail tolerance is that the rail height may not vary more than 0.0635 cm over a length of 132 cm (0.025 in. over 52 in.). The specific angle of that discontinuity is not measured by the machine, which checks the rail – only the height is considered. Therefore, this rail height change can occur as a sharp discontinuity in height or as a gentle slope. Figure 21.4 illustrates this "rail height tolerance" as a sharp discontinuity.

The second rail tolerance is defined at the seams between rail sections. Where these sections join (by welds), the specified seam tolerance is a sloping discontinuity of 0.19 cm over 2.54 cm (0.075 in. over 1 in.). This "rail seam tolerance" defines a specific discontinuity height at a given angle. If the seam was smoothed further (over a longer distance), a smaller angle could be achieved. Figure 21.4 illustrates this tolerance at the prescribed limit defined by the HHSTT. Seams that are better than standards are the goal, but only those which do not meet this tolerance are re-accomplished.

These two rail tolerances can be considered to define "acceptable" rail discontinuities with respect to discontinuity height (not more than 0.0635 cm over the entire range of face angles) and face angle (not more than 4.29). Note that the rail seam tolerance actually violates the rail height tolerance, but this is considered acceptable at the rail seams. This concept is illustrated in Figure 21.5.

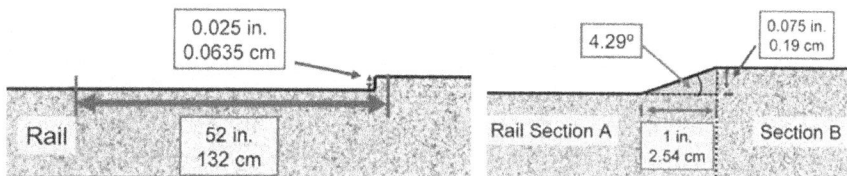

FIGURE 21.4 "Rail height tolerance" and "rail seam tolerance'".

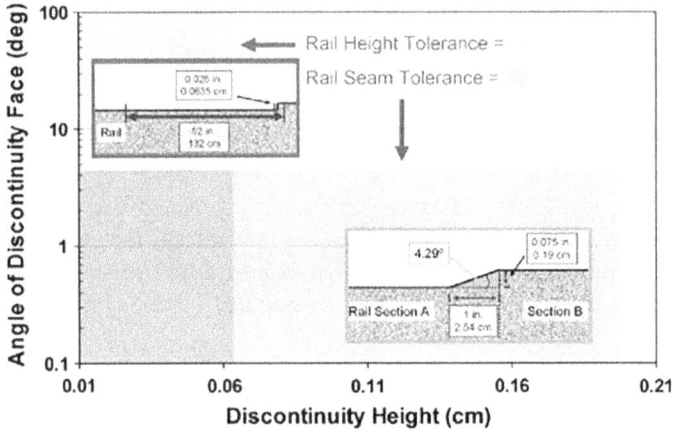

FIGURE 21.5   Rail alignment tolerance summary.

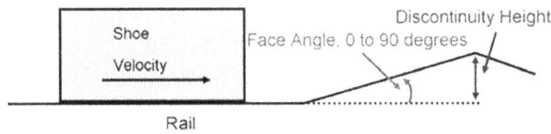

FIGURE 21.6   Model of rail tolerances.

The rail tolerances were simulated in CTH as a discontinuity of a specific height and a prescribed face angle. Figure 21.6 illustrates how these tolerances were modeled, and Figure 21.7 depicts how they appear within the CTH simulation. It should be noted that a "reverse discontinuity" in which the sled encounters a rail height that is lower than the current height did not lead to gouging in the simulations and was not therefore evaluated in detail.

Thirty combinations of discontinuity heights and face angles were simulated in CTH (over 720 computational hours on a massively parallel 64-bit computer cluster with 32 nodes) to determine the effect of these parameters on the initiation of hypervelocity gouging. It became clear almost immediately that there were threshold values for both face angle and discontinuity height that would cause either wear to the rail or a hypervelocity gouge. For instance, with a face angle of 1.85 or less, the discontinuity height did not matter; the sled would not gouge the rail. This established an angle limit for gouging. For a discontinuity height of 0.015 cm or less, the face angle could be 90, and gouging would still be avoided. This established a discontinuity height limit for gouging. The cases in between these extremes followed a discernable power–law relationship, as shown below.

The results of this modeling effort appear in Table 21.2. CTH simulations of rail tolerance impacts. The cases of gouging and wear also appear graphically in Figure 21.8 – which depicts the current HHSTT rail alignment tolerances. Of immediate note is that the current rail alignment tolerances allow combinations of

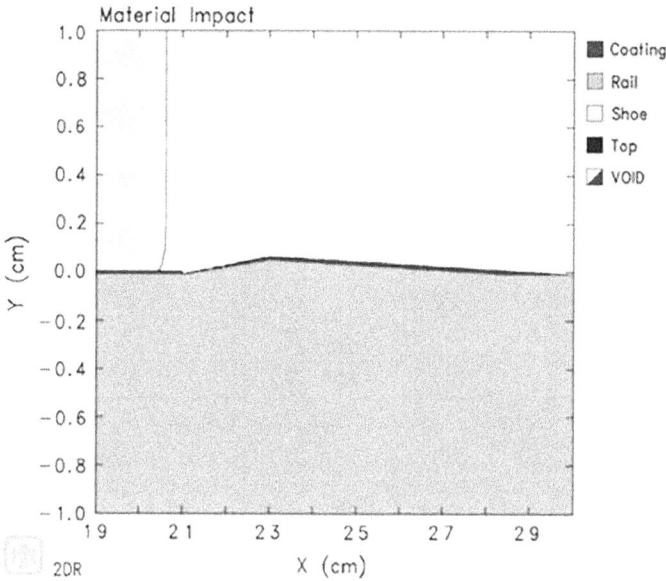

FIGURE 21.7 CTH model of rail tolerances.

face angle and discontinuity heights that lead to hypervelocity gouging in the CTH simulation.

The power–law fit to the line that corresponds to wearing impacts only, and not gouging, can be written as

$$\text{Face angle } (°) = \max\left[\left(9.174\times10^{-5} \times \text{Discontinuity (cm)}^{-3.3093}\right),1.85°\right] \quad (22.2)$$

21.3.3 CTH SIMULATION RESULTS COMPARED AGAINST EXPERIMENTAL DATA

As with the previous CTH simulation efforts, the authors sought to validate the results with experimentally observed phenomenon wherever possible. Previously examined gouged rail sections were reported to have unique characteristics, including easily identifiable microstructural alterations due to a high thermal load and rapid cooling after the gouging event had concluded [17–19]. Additional rail sections which had been in service at the HHSTT and had either experienced wear (but no visible damage) or had worn with visible damage (i.e., some material removal – a scrape) were also examined. Both of these rail sections also showed signs of "thermal damage" – i.e. microstructural changes due to high thermal load and rapid cooling [19]. These experimentally examined rail sections provide us with valuable validation points to which we can compare the CTH simulations.

For brevity, the three distinct cases of gouging, wear with damage, and wear without damage are examined here [19]. In each of these cases, a comparison will be made between the CTH simulation and the experimentally observed characteristics of the rail from the HHSTT.

TABLE 21.2
CTH Simulations of Rail Tolerance Impacts

Face Angle (°)	Discontinuity (cm)	Damage Type
0.014	0.07	Wear
1.6	0.055	Wear
1.6	0.06	Wear
1.65	0.07	Wear
1.85	0.0635	Wear
1.85	0.07	Wear
2	0.05	Wear
2	0.06	Gouge
2	0.0635	Gouge
2	0.07	Gouge
3	0.0635	Gouge
3	0.07	Gouge
4	0.07	Gouge
5	0.035	Wear
5	0.045	Gouge
10	0.03	Wear
10	0.035	Gouge
20	0.025	Wear
20	0.03	Gouge
20	0.05	Gouge
20	0.0635	Gouge
40	0.02	Wear
40	0.025	Gouge
60	0.018	Wear
60	0.022	Gouge
60	0.035	Gouge
90	0.01524	Wear
90	0.02	Gouge
90	0.03	Gouge
90	0.0635	Gouge

21.3.3.1 Case 1: Rail Wear without Visible Surface Damage

In order to understand the gouging phenomenon at the HHSTT, many sections of rail were examined to ascertain whether microstructural alterations were evident. The investigation initially focused on rail sections with significant gouges [17,18]. However, once it became clear that pronounced microstructural changes were present in those specimens, rail sections with wear-type damage were also examined [19]. Even some of the sections which were in-service at the HHSTT and had exhibited no visible material removal/damage had some level of microstructural change induced by high temperatures and rapid, post-impact cooling [19].

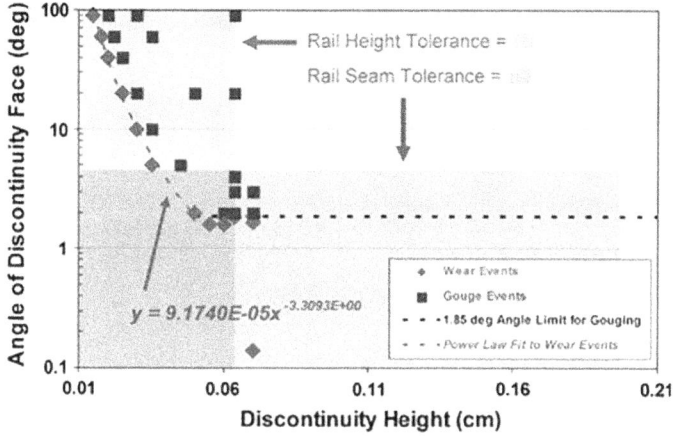

FIGURE 21.8   CTH gouge/wear predictions on rail alignment tolerance summary.

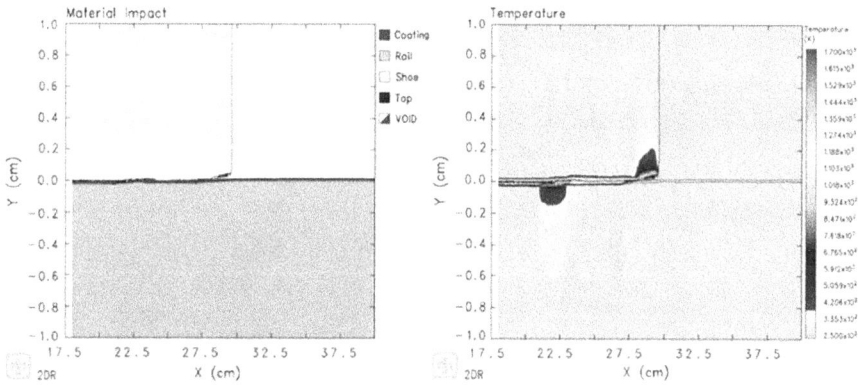

FIGURE 21.9   Wear impact case, no visible damage, at 20 μs.

Some of the impacts generated in Table 21.2 created a characteristic wear-type impact that generated this kind of microstructural change in the rail, yet did not measurably plastically deform the rail. In these cases, the post-impact rail would only exhibit erosion of the rail coatings (which is a constant occurrence at the HHSTT after a high-speed test) and would not have visible post-impact damage. Figure 21.9 illustrates a shoe impacting a 0.01524 cm discontinuity with a face angle of 90 (one of the cases summarized in Table 21.2). While very little permanent damage is done to the rail material (although the coating is affected), a thermal condition is induced into the depth of the rail that exceeds the austenitizing limit of 1080 steel (1,000 K) [17–19]. A comparison of the depth of austenitized material (and the subsequent alteration of the microstructure) is made with the HHSTT rail specimens in Figure 21.10 (a magnification of Figure 21.9). The CTH simulation matches the experimentally observed microstructural changes extremely well and offers an explanation to the rail's condition – despite the lack of visible plastic damage [19].

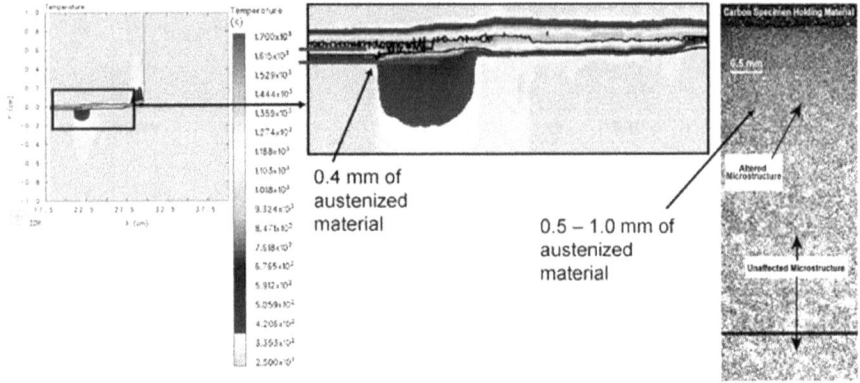

FIGURE 21.10 Wear impact case, no visible damage, at 20 µs, comparison to rail sample.

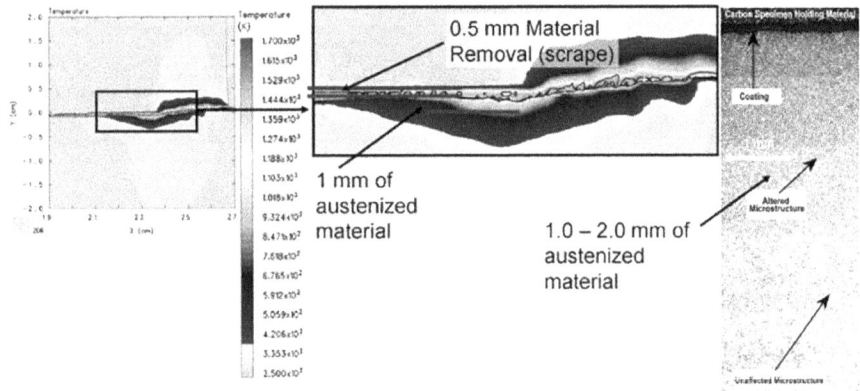

FIGURE 21.11 Wear impact case, visible damage, at 20 µs, comparison to rail sample.

21.3.3.2 Case 2: Rail Wear with Visible Surface Damage

In many cases, after a high-speed test at the HHSTT, the rails are inspected and minor scrapes (small amounts of damage to the coating and rail surface, typically less than 1.0 mm in depth) are discovered. In some cases (in which the damage depth is greater), the rail is replaced in response to these observations. In others, the scrape is sanded down and the depression is filled with the coating material. A rail section that experienced a scrape and was replaced was metallurgically examined. A more pronounced case of rail material microstructural alteration was observed [19].

The CTH simulations conducted also yielded cases in which material removal was predicted, but the characteristics of gouging were not attained. Figure 21.11 illustrates a shoe impact to a 1.65 face angle discontinuity. The CTH model resulted in a 0.5 mm scrape (which matched the rail specimen) and also predicted a very similar depth of microstructure change [19].

21.3.3.3 Case 3: Hypervelocity Gouging

In the first two cases, the result of the hypervelocity impact between the sled shoe and the rail was not gouging. Hypervelocity gouging is differentiated from material removal and gross mechanical damage by a material mixing between the shoe and the rail across the impact interface [3–6]. The primary goal of this study was to duplicate this phenomenon of hypervelocity gouging within CTH simulations, and then ascertain the parameters critical to its initiation. Prevention and/or mitigation of these hypervelocity gouging impacts will allow the HHSTT to continue to increase its test velocities and avoid catastrophic damage in the future.

By varying the impact parameters within the CTH simulation, gouging events were readily replicated. The hypervelocity gouging events yielded rail damage – both physical material removal and microstructural alteration – that matched closely the experimental examination [17–19] of a gouged rail specimen. Figure 21.12 illustrates the initiation of a gouging interaction, with the material of the shoe beginning to flow into the rail specimen and vice versa. This material mixing continues as the event develops. In this case, the discontinuity was 0.0635 cm at a face angle of 2 (one of the cases summarized in Table 21.2).

As the impact event progresses, the material mixing and pressure development increase. Figure 21.13 illustrates the impact at 40 μs. Figure 21.14 highlights the development of plastic strain-rate (validating the need for an accurate material strain-rate dependent flow model and EOS across the spectrum of 1–10^6/s strain rates) and temperature within the shoe and rail. Figure 21.15 compares the prediction of gouge depth and microstructural change to a post-gouge rail specimen previously examined

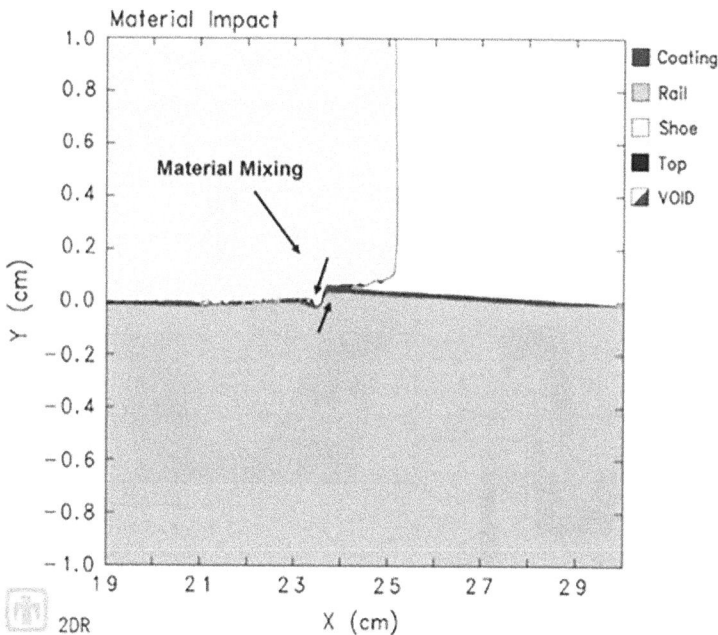

FIGURE 21.12 Gouge initiation, 15 μs.

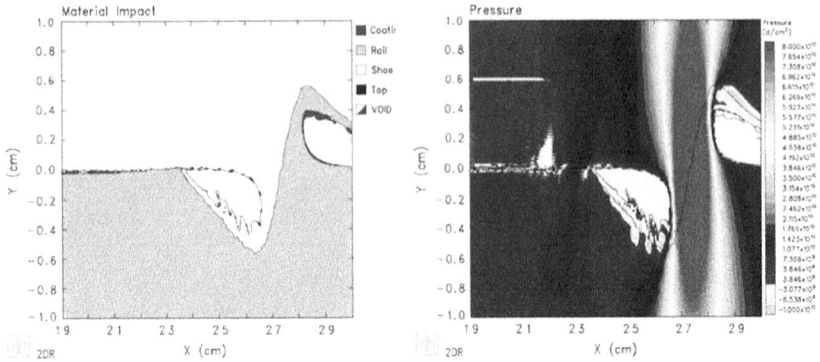

FIGURE 21.13   Gouge impact case, 40 μs, material mixing and pressure development.

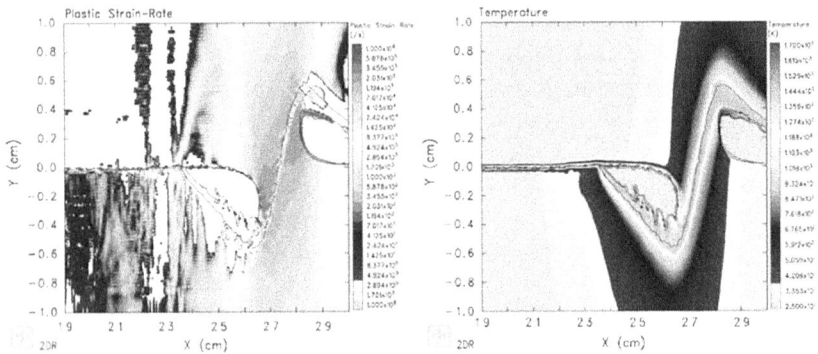

FIGURE 21.14   Gouge impact case, 40 μs, strain-rate and temperature development.

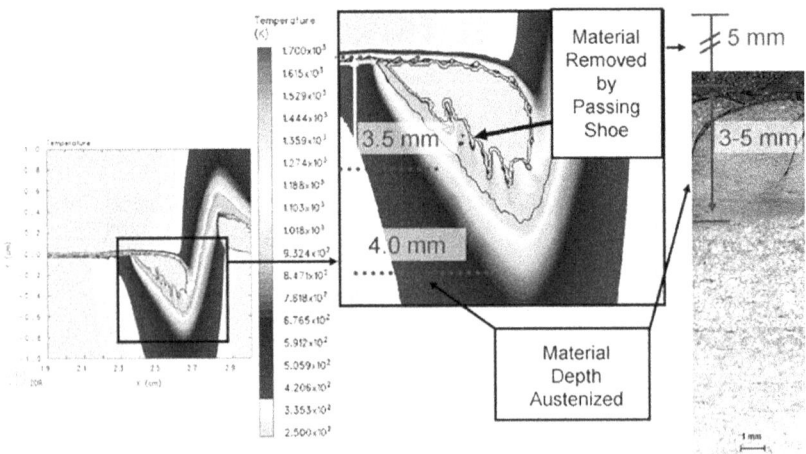

FIGURE 21.15   Gouge impact case, 40 μs, comparison to rail sample.

[17–19]. The gouges predicted by CTH were shown to match gouge characteristics, such as depth and length, from the field very well [19].

This representative case illustrates that the rail damage evident at the HHSTT can be accurately predicted by the simulation of the impact event with CTH. Of primary importance is that these hypervelocity gouges are predicted within the current HHSTT rail alignment tolerances and offer an explanation to the creation of these gouges during high-speed test runs. This, of course, makes it clear that the defined HHSTT rail alignment tolerances for safe operation are inadequate.

21.4   CONCLUSIONS

In this work, recently advanced and validated material flow model constants for both VascoMax 300 and 1080 steel are applied within CTH to model the HHSTT sled impact scenario. The goal of this effort was to delineate test parameters that directly contributed to the development of hypervelocity gouging impacts and to compare CTH simulation results to available post-impact rail specimens.

An extensive series of CTH simulations were conducted to attempt to determine the probable cause of the hypervelocity gouging during HHSTT test runs. During this analysis, the vertical impact case was determined to be highly unlikely due to the required vertical velocity to initiate gouging. The existing HHSTT rail alignment tolerances were examined within the CTH simulation, and it became clear that a large portion of the allowable operational space for rail tolerances led to hypervelocity gouging. Based on the simulation series, a new rail alignment tolerance was determined which promises to reduce or eliminate these occurrences of gouging. These tolerances are being applied to the HHSTT, and test runs of greater than 3.0 km/s are scheduled to resume in 2008.

Finally, the CTH simulations were compared against the experimental examination of HHSTT rail samples. All of the visibly damaged sections of rail exhibited microstructural changes due to a temperature development through the depth exceeding the austenitizing temperature of 1080 steel. In some of the rail samples, no visible damage was present, but a small amount of microstructural alteration was observed. The CTH simulations were able to accurately duplicate each of these cases, with variations of impact parameters. These impact parameters are well within the typical/ allowable range of rail alignment tolerances currently used by the HHSTT.

Therefore, a model within CTH has been developed and validated against experimentally observed phenomenon at the HHSTT. This simulation uses the latest material flow models and is able to model both hypervelocity gouging impacts and impacts characterized by wear. The result of this simulation is a modified rail alignment tolerance at the HHSTT to mitigate the occurrence of hypervelocity gouging impacts.

ACKNOWLEDGMENTS

The authors would like to acknowledge the invaluable support of the Air Force Office of Scientific Research (Dr. Neal Glassman and Dr. John Schmisseur, monitors). Also,

Dr. Michael Hooser from the HHSTT made critical contributions to this endeavor. Additionally, the technical advice of Dr. Theodore Nicholas of the Air Force Institute of Technology was instrumental to the research.

REFERENCES

1. Gerstle FP, Follansbee PS, Pearsall GW, Shepard ML. Failure of steel during high velocity sliding. *Wear* 1973;24:97–106.
2. Laird DJ, Palazotto AN. Effects of temperature on the process of hypervelocity gouging. *AIAA Journal* 2003;41(11):2251–260.
3. Laird DJ, Palazotto AN. Gouge development during hypervelocity sliding impact. *International Journal of Impact Engineering* 2004;30(2):205–223.
4. Szmerekovsky AG, Palazotto AN, Baker WP. Scaling numerical models for hypervelocity test sled slipper-rail impacts. *International Journal of Impact Engineering* 2006;32(6):928–946.
5. Szmerekovsky AG, Palazotto AN, Earnst MR. Numerical analysis for a study of the mitigation of hypervelocity gouging. In: *Proceedings of the 45th AIAA/ ASME/ASCE/ AHS/ASC Structures, Structural Dynamics, and Materials Conference*. Palm Springs, CA, 19–22 April. Reston, VA: AIAA; 2004. pp. 1–10.
6. Szmerekovsky AG, Palazotto AN. Structural dynamic considerations for a hydrocode analysis of hypervelocity test sled impacts. *AIAA Journal* 2006;44(6): 1–14.
7. Hertel ES. CTH: a software family for multidimensional shock physics analysis. In: *Proceedings of the 19th International Symposium on Shock Waves*, Marseille, France, 26–30 July, vol. 1. New York: Springer-Verlag; 1993. p. 377–382.
8. McGlaun JM, Thompson SL, Elrick MG. CTH: a three-dimensional shock wave physics code. *International Journal of Impact Engineering* 1990;10(1–4): 351–360.
9. Cinnamon JD, Palazotto AN, Brar NS, Kennan Z, Bajaj D. Johnson–Cook strength model constants for VascoMax 300 and 1080 steels. In: Furish MD, editor. *Proceedings of the 14th American Physical Society (APS) Topical Conference on Shock Compression of Condensed Matter*, Baltimore, MD, 31 July–3 August, vol. 845. College Park, MD: American Physical Society; July 2006. pp. 709–712.
10. Cinnamon JD, Palazotto AN, Kennan Z. Material characterization and development of a constitutive relationship for hypervelocity impact of 1080 steel and VascoMax 300. Proceedings of the 2005 hypervelocity impact symposium, Lake Tahoe, CA, 10–14 October, 2005. *International Journal of Impact Engineering* December 2006;33(1–12):180–189.
11. Cinnamon JD, Palazotto AN. Refinement of a hypervelocity model for the rocket sled test. In: *Proceedings of the 2005 American Society of Mechanical Engineers (ASME) International Mechanical Engineering Congress and Exposition*, ASME Paper IMECE2005-80004, Orlando, FL, 5–11 November. New York: ASME; 2005. pp. 1–10.
12. Cinnamon JD, Palazotto AN, Szmerekovsky AG, Pendleton RJ. Further investigation of a scaled hypervelocity gouging model and validation of material constitutive models. In: *47th AIAA/ASME/ASCE/AHS/ASC Structures, Structural Dynamics, and Materials Conference*, Newport, RI, 1–4 May. Reston, VA: AIAA; 2006. pp. 1–20.
13. Cinnamon John D, Palazotto Anthony N, Szmerekovsky AG, Pendleton RJ. Investigation of a scaled hypervelocity gouging model and validation of material constitutive models. *AIAA Journal* May 2007;45(5):1104–112.
14. Cinnamon John D, Palazotto Anthony N, Brar NS. Further refinement of material models for hypervelocity gouging impacts. In: *47th AIAA/ASME/ ASCE/AHS/ASC Structures, Structural Dynamics, and Materials Conference*, Newport, RI, 1–4 May. Reston, VA: AIAA; 2006. pp. 1–20.

15. Cinnamon John D, Palazotto Anthony N, Szmerekovsky AG. Further refinement and validation of material models for hypervelocity gouging impacts. *AIAA Journal* February 2008;46(2):317–327.
16. Cinnamon John D, Palazotto Anthony N. Further validation of a general approximation for impact penetration depth considering hypervelocity gouging data. *International Journal of Impact Engineering* August 2007;34(8):1307–1326.
17. Cinnamon JD, Palazotto AN. Metallographic examination of thermal effects in hypervelocity gouging. In: *Proceedings of the 2005 American Society of Mechanical Engineers (ASME) Pressure Vessel and Piping Conference*, ASME Paper PVP2005–71613, Denver, CO, 17–21 July. New York: ASME; 2005. pp. 1–9.
18. Cinnamon JD, Palazotto AN. Metallographic examination and validation of thermal effects in hypervelocity gouging. *American Society of Mechanical Engineers (ASME) Journal of Pressure Vessel Technology* February 2007; 129(1):133–141. ASME, New York.
19. Cinnamon JD. Analysis and simulation of hypervelocity gouging impacts, AFIT/ DS/ ENY/06-01. Ph.D. Dissertation. Air Force Institute of Technology; June 2006. pp. 1–312.
20. Zerilli FJ, Armstrong RW. Dislocation-mechanics-based constitutive relations for material dynamics calculations. *Journal of Applied Physics* March 1987; 61(5):1816–1825.
21. Johnson GR, Cook WH. A constitutive model and data for metals subjected to large strains, high strain rates, and high temperatures. In: *Proceedings of the Seventh International Symposium Ballistics*. The Hague, Netherlands: American Defense Preparation Organization; April 1983. pp. 541–547.

22 Nonlinear Dynamic Analysis of an Icosahedron Frame which Exhibits Chaotic Behavior[*]

*Lucas W. Just, Anthony M. DeLuca,
and Anthony N. Palazotto*
US Air Force Institute of Technology

CONTENTS

The research question addressed is whether a lighter than air vehicle (LTAV), which uses an internal vacuum to become positively buoyant, can be designed to provide extended loiter for U.S. Air Force applications. To achieve a vacuum, internal gases are evacuated from the vessel, which creates a dynamic response in the supporting structural frame. This chapter considers the frame of an icosahedron-shaped LTAV subject to external atmospheric pressure evacuated at varying rates. A static finite element analysis documented in previous research revealed a *snapback* phenomenon in the frame members under certain loading conditions. A nonlinear chaotic response was observed when a dynamic analysis was conducted with the same boundary conditions used in the static analysis. The chaotic response for a variety of boundary

[*] Reprinted with permission: Just, L. W., DeLuca, A. M., & Palazotto, A. N. (2017). Nonlinear dynamic analysis of an icosahedron frame which exhibits chaotic behavior. *Journal of Computational and Nonlinear Dynamics*, *12*(1).

DOI: 10.1201/9781003220978-28

conditions, generated by varying the rate of evacuation, similar to a ramp input, is determined. An analysis of the dynamic response is determined nonlinearly using a method that relies on a reference point distribution of external pressures to distribute the surface force across the frame. A novel method of combining the power spectral density with a Lyapunov exponent was used to determine the degree of nonlinearity and chaotic response for each boundary condition examined.

22.1   INTRODUCTION

The Air Force Institute of Technology has an interest in LTAVs in the shape of an icosahedron, which comprises a frame consisting of 12 vertices, 20 equilateral triangles, and 30 beams. The novel feature in this study is the lighter-than-air gases typically added to generate buoyancy in an LTAV are removed, and an evacuated vessel is used to achieve vertical lift. Thus, the frame of the vehicle is a critical supporting structure, requiring a dedicated structural and dynamic analysis of the frame. Metlen [1] and Rodriguez [2] conducted a static analysis on the (Dassault Systèmes, Velizy Villacoublay, France) frame utilizing ABAQUS finite element analysis (FEA) software, and discovered under certain conditions, the beam members demonstrated a nonlinear *snapback* behavior. The objective of the study was to analyze the nonlinear dynamics of the frame structure, which supports the loading related to the icosahedron under compressive stress due to the internal vacuum required to achieve buoyancy. This study expanded on Rodriguez's static research by conducting a full dynamic FEA, refining the finite element model (FEM) of the frame through matching the FEM's modal results from experimentally determined vibratory responses, and then further investigated the ramp loading conditions (evacuation rate), which resulted in nonlinear deformations. A novel adaptation of information chaos theory, through computing Lyapunov exponents and power spectral density analysis, was implemented to provide quantitative predictions of loading conditions, which foment the onset of structurally catastrophic system nonlinearities in the frame behavior. This chapter presents a robust nonlinear dynamic analysis of the same frame by evacuating the internal air as a function of time, resulting in a ramp load applied to the frame by the external ambient sea-level air.

22.2   BACKGROUND

The principle of using a vacuum to achieve positive buoyancy is centuries old; however, using an icosahedron frame with a membrane-like skin as a load-bearing structure is relatively new. This specific implementation was originally devised by Captain Trent Metlen of the Air Force Institute of Technology in 2013 [1]. An icosahedron is comprised of twenty equilateral triangles formed into a shape with 12 vertices, which lie within an encompassing sphere. A sphere is the ideal geometric shape for a structure to attain positive buoyancy using an inner vacuum because it displaces the greatest volume per surface area. The structure is also beneficially geometrically symmetric, so each frame member bears an equally distributed load, resulting in equivalent total stress carried within those members.

Following Metlen's research, Lieutenant Ruben Adorno-Rodriguez developed a FEM in ABAQUS and conducted a static structural analysis on the icosahedron-shaped LTAV design [2]. Figure 22.1 shows the FEM of the icosahedron structure with boundary conditions, and the external loading applied to the frame. The loads are applied through reference points at the center of gravity of each triangle, which are distributed to the beams using a coupling constraint in ABAQUS. The coupling constraint exerts the same force on the beans as would a load from a triangular skin membrane with a distributed surface pressure tied along its edges. Adorno-Rodriguez's research verified the applied force transferred to the beams was equal using both the reference point and coupling constraints to loads of the skin tied to the beams [2]. The resulting displacement data were collected at the midpoint node on the lower beams. The midpoint node was chosen because the icosahedron deforms symmetrically, and the midpoint nodes on all beams displace equally and thus have the highest displacement magnitude of any other node on the structure.

Figure 22.1 shows the top and bottom nodes are restricted in the x and y direction, while all other degrees-of-freedom are free to move. The boundary conditions applied to the structure in Figure 22.1 allow the frame to displace symmetrically, and move freely in the z-axis, representing a free floating structure. The absence of a z-axis boundary condition does not result in rigid body movement because the structure is symmetric, and the load applied to the bottom half is equivalent to the load applied to the top half of the structure, resulting in an equilibrium condition. However, the x (U1) and y (U2) degrees-of-freedom (DoF) at the top and bottom nodes are constrained to prevent rigid body movement in the x-y plane, which expedites FEA solution convergence. The boundary condition applied in Figure 22.1 was

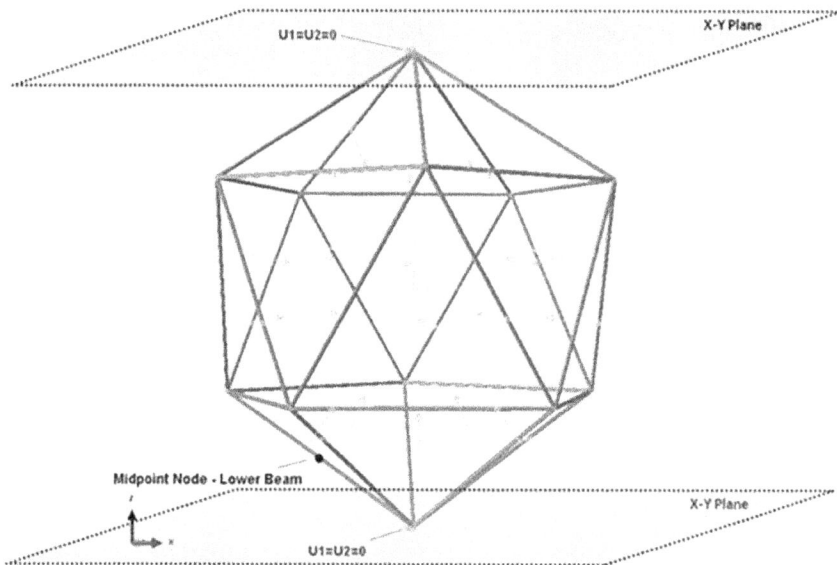

FIGURE 22.1 Icosahedron frame with boundary conditions (a and b vertex constrained in the x and y direction) and external applied loading.

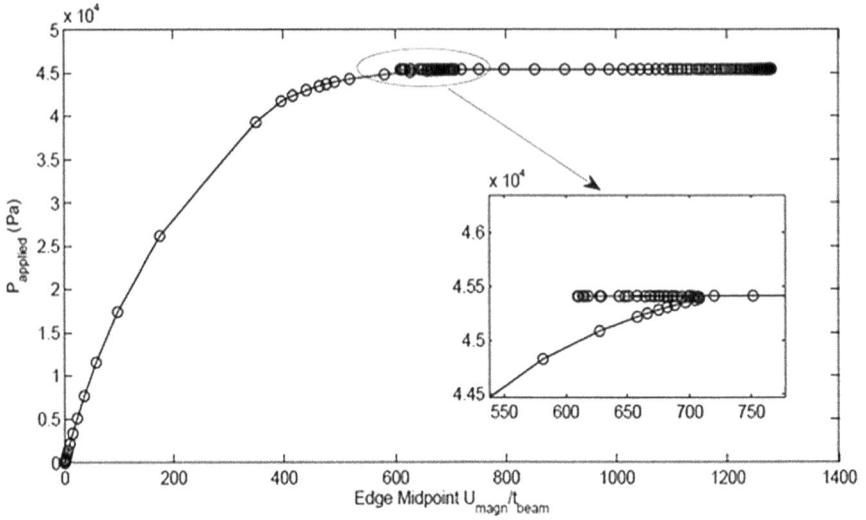

FIGURE 22.2   Boundary condition 2 (antisymmetric condition) load–displacement curve exhibiting *snapback* behavior.

TABLE 22.1
Dimensions and Material Properties for Icosahedron Frame

Property	Dimension	Units
Radius (center to vertex)	1.0 (0.3048)	ft (m)
Beam cross section radius	5.99510^{-2} (1.52310^{-3})	in (m)
Beam cross section thickness	2.99810^{-3} (7.61410^{-5})	in (m)
Beryllium density	115.12 (1844.0)	lb/ft^3 (kg/m^3)
Beryllium modulus of elasticity	6.33 (303.0)	lb/ft^2 (GPa)
Beryllium poisson's ratio	0.18	unit less

referred to as Boundary Condition 3 (BC3) in previous research, and that notation is preserved here. The frame produced symmetric responses to statically applied pressure loads, while the other boundary condition, (BC2), yielded nonsymmetrical behavior, which resulted in a sudden change in the slope of the applied pressure versus displacement curve, called a sudden *snapback*, and is shown in Figure 22.2. The load which results in an instantaneous change in the displacement direction is referred to as the *snapping* load. The dimensionality and material properties used in the FEM of the frame are listed in Table 22.1.

Figure 22.3 displays the magnitude of the displacement of the midpoint node, shown in Figure 22.1 above, as the load is applied from zero to one hundred percent of sea-level pressure (101,325 Pa). The applied load results from creating a vacuum by evacuating the inner air of the icosahedron. Since only frame loading is considered, the applied pressure is transferred to the frame through a reference point. The reference points are located at the centroid of the equilateral triangles between the

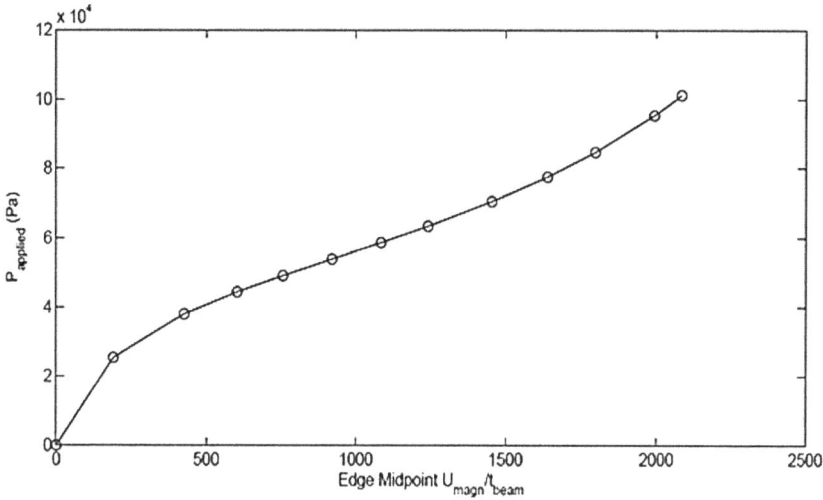

FIGURE 22.3 Boundary condition 3 (symmetric condition) load–displacement curve [2].

beams. Therefore, each beam has a specific pressure distribution determined by its area, defined by the triangular-shaped loading from the centroid to the vertices of each beam. Consequently, each beam supports a continuous line load. The displacement curve shown in Figure 22.3 is nonlinear; however, it is continuous, and the beam displacement increases steadily as the load increases.

Adorno-Rodriguez observed a *snapback* behavior present in the frame of the structure under certain boundary conditions where the symmetry condition was not preserved. An example of the *snapback* behavior is demonstrated in the pressure versus displacement curve shown in Figure 22.2, which resulted from changing the bottom node boundary condition in Figure 22.1 to a fully clamped constraint, and is referred to as BC2. The boundary conditions analyzed by Adorno-Rodriguez are critical because one of the boundary conditions examined exhibited *snapback* behavior. For BC3, the addition of the applied load created a symmetric displacement in all beam members and did not result in any erratic deformation behavior.

A dynamic analysis of the structure was conducted to characterize the *snapback* behavior presented in the frame under specific boundary conditions. The theory used to conduct the dynamic analysis is outlined in the Theory section.

22.3 THEORY

If a material is linearly elastic, the undamped dynamic response of a structure caused by the evacuation of the internal gas is obtained by recursively solving Eq. (22.1) [3].

$$[M]\{\ddot{D}\} + [K]\{D\} = \{R^{\text{ext}}\}$$
(22.1)

where D; $\ddot{D}$ = nodal position and acceleration, respectively, K = stiffness matrix, M = mass matrix, R^{ext} = externally applied loads

A solution to the forced, undamped Dynamic Equation of Motion (DEOM), shown in Eq. (22.1), can be solved with either an implicit forward differencing technique or through the use of an explicit backward or central differencing Euler integration method. Numerically solving the DEOM implicitly renders an unconditionally stable solution for the displacement, D, regardless of the size of the integration time-step. Numerical integration solves the DEOM for the displacement by assuming a constant acceleration, $\ddot{D}$, calculating a future value of $\dot{D}$; and then rearranging the equation to solve for the displacement, D, using the current node's displacement and the future acceleration. The use of implicit integration requires numerically calculating the inverse of the stiffness matrix, [K] and is therefore more computationally intensive, which does, however, benefit from being less sensitive to the size of the time-step, and hence, requires less iterations to achieve convergence. Implicit integration yields a conditionally stable solution; therefore, the stability of the solutions is inextricably dependent on choosing a sufficiently small time-step. Stiff problems, characterized by large deformations, which do not obey small angle assumptions, occur over a short time and thus require finer granularity to converge using a forward time marching scheme. As stated by Cook et al., "Implicit direct integration is suited to structural dynamics problems [and] nonlinearity can be accommodated without great trouble" [3]. To reduce the number of iterations and ensure stable solutions, an implicit integration solver was used in all of the subsequent ABAQUS FEM simulations.

Because the frame beam member force versus displacement behavior is nonlinear, the stiffness, [K], changes with time, and the dynamic response varies as a function of time. The system represented by Eq. (22.1) is continuous, but can be discretized to model the behavior of the icosahedron frame exposed to transient forces caused by evacuating the interior gas, given in Eq. (22.2) below, where n indicates each time increment, and R^{int} represents the internal force vector. The general form of the discrete solution is given in Eq. (22.3) below [3]

$$[M]\{\ddot{D}\}_n + \{R^{\text{int}}\}_n = \{R^{\text{ext}}\}_n \tag{22.2}$$

$$\{D\}_{n+1} = f\left(\{\dot{D}\}_{n+1}, \{\ddot{D}\}_{n+1}, \{D\}_n, \{\dot{D}\}_n, \{\ddot{D}\}_n, \dots\right) \tag{22.3}$$

ABAQUS uses a Newton-Raphson integration solver to find the roots of the DEOM when applying nonlinear analysis techniques to model frame members experiencing *snapback* behavior under specific BCs. An external time-varying load condition and a desired time-step are required to compute the nonlinear response. ABAQUS segments the simulation into time increments, calculates an equilibrium position at the end of each time increment, and computes the displacement, stress, and strain [4]. The user determines the time increment to be used, and whether it is a fixed or automatically applied constraint.

The dynamic time response of a system is dependent on the type of input forcing function applied. A ramp loading function is used to model the forces generated by the rate of internal gas evacuation, and the response of a single degree-of-freedom (SDOF) system to the ramp loading is the basis of the future dynamic analysis executed

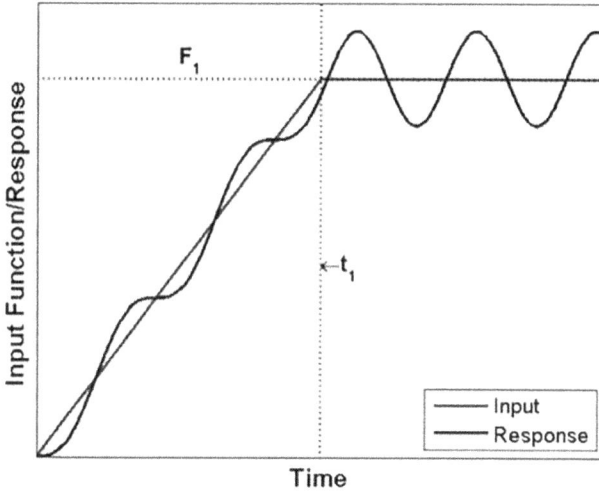

FIGURE 22.4 Ramp input and response for a 1D spring-mass system.

hereafter. Given the input forcing function provided by Eq. (22.4), with frame member structural characteristics identified, a solution to the DEOM can be computed, and the system response can be determined. The DEOM for a SDOF is given in Eq. (22.5) below, and the ramp response to the system is shown in Eq. (22.6) [5]. The ramp input and output response for a typical SDOF system are shown in Figure 22.4

$$r(t) = \int_{-\infty}^{t} u(\tau)d\tau = t * u(t) \tag{22.4}$$

$$\left(m\frac{d^2}{dt^2} + k \right) r(t) = r(t) \tag{22.5}$$

$$r(t) = \int_{-\infty}^{t} s(\tau)d\tau = \frac{1}{k\omega_n}(\omega_n t - \sin \omega_n t)u(t) \tag{22.6}$$

where k = spring stiffness, m = mass of system, r (t) = input ramp forcing function, r (t) = ramp response of system, $s(\tau)$ = step response of system, u (t) = {0 for $t_1 < 0$, and F_1 for $t_1 > 0$} step function

The SDOF system's response to a ramp input is easily calculated by integrating the unit step response, which is found by integrating the unit impulse function. This analysis methodology can be applied to compute the icosahedron's frame response to a time-varying evacuation rate as long as it remains a linear time invariant (LTI) system, thereby avoiding the *snapback* region. The midpoint nodes tracked in the frame follow the same oscillatory pattern shown in Figure 22.4. Further examination of the response in Figure 22.4 reveals the slope of the response curve produces a time-varying velocity of the mass. This movement indicates a wave of kinetic energy propagating through the structural members. When the maximum load is reached at maximum vacuum, the transient wave transitions to a standing wave, which indicates a stable response; however, under certain ramp loading conditions, the wave grows unstable, the system ceases to behave as an LTI system, and the dynamic response can no longer be calculated as a simple ramp response, and its behavior grows increasingly chaotic.

In addition to calculating the time response, ABAQUS was used to analyze the frequency response of the structure. The frequency response data were used to identify operating frequencies containing high energy content likely to cause rapid catastrophic failure of the structure, and to also predict the onset of chaotic motion. While the time domain response represents the system response of the beam to a single excitation frequency, the frequency response includes information about the energy content of all excitation frequencies excited by a periodic external force.

Given a random variable input, the frequencies which excite the system most are computed using the power spectral density (PSD) function. The power spectral density, $S_x(\omega)$, provides the amplitude of the energy content of a stochastic time series, as a function of its distribution across a frequency spectrum. The PSD is calculated by taking the Fourier transform of the autocorrelation function, $R_x(\tau)$; which is the cross-correlation of a time series signal with itself at one point in time compared to another point in time, shifted by the amount s. The mathematical definition of the autocorrelation function is given in Eq. (22.7), and the power spectral density function is thus defined as the Fourier transform of the autocorrelation function, shown in Eq. (22.8) below

$$R_x(\tau)= \lim_{T \to \infty} \frac{1}{T} \int_{-T/2}^{T/2} x(t)x(t+\tau)\,dt \tag{22.7}$$

$$S_x(\omega) = \int_{-\infty}^{\infty} R_x(\tau)e^{-i\omega\tau}\,d\tau \tag{22.8}$$

where $x(t)=$ time domain signal, R_x (τ) autocorrelation function of $x(t)$ as a function of time shift, τ, $S_x(\omega)=$ power spectral density function, $T=$ period of signal, $e^{-i\omega t} = \cos(\omega t) - i\sin(\omega t) =$ Euler representation of complex phase.

If the time domain signal is random, wide-sense stationary, ergodic, and varies from time-step to subsequent time-step, the PSD can be used to create a sequence of rapid amplitude changes along the frequency spectrum, which indicate the growth of instability, and hence chaos. When the change in amplitude is examined in this fashion, the PSD provides a tool to predict chaotic behavior. Previous research on the static loading of an icosahedron LTAV predicted a *snapback* behavior, which is assumed to be chaotic. Therefore, to develop a deeper understanding of the structural behavior and operating limits of an icosahedron-shaped LTAV for USAF use, a more complete study of this chaotic structural behavior is required.

Chaos is indicated by the irregular and unpredictable time evolution of nonlinear systems, whose behavior at a specific point in time cannot be predicted by information presently known about the system [6]. However, Baker and Gollub stipulate dynamical systems, which exhibit random, chaotic behavior, still must obey canonical

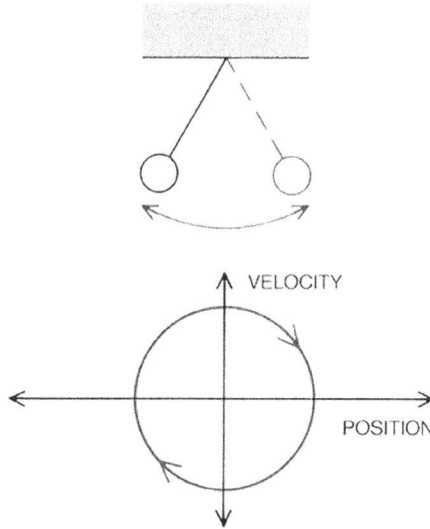

FIGURE 22.5   Single pendulum system (a) and phase-plane trajectory (b) [7].

laws of physics, and the governing equations of motion derived from those laws [6]. Chaotic behavior occurs when the governing DEOM of a system is nonlinear and the system demonstrates a time history sensitive to initial conditions [6]. Several mathematical indicators can show if a system possesses chaotic behavior. An analysis of the phase-plane trajectory, PSD plots, and the calculation of Lyapunov exponents all help to distinguish chaotic motion from nonchaotic motion and to quantify the chaos in a system's nonlinear response.

An explanation of two different, but related dynamical systems illustrates the difference between a system that displays chaos and the one that does not. A simple pendulum suspended by an inextensible string with known initial conditions and boundary conditions has a predictable periodic time domain response. Changing the initial conditions does not alter the nature of the response. The undamped pendulum motion is shown in Figure 22.5. A double pendulum is created by adding a second simple pendulum to the end of the first pendulum. This system exhibits a completely different response to small changes in the initial conditions than the single simple pendulum. For a certain range of initial conditions, the motion demonstrates chaotic behavior [7]. Figure 22.6 shows the trajectories of the double pendulum for two sets of different initial conditions. Slight changes in the initial conditions result in significant changes in the dynamic response of the system, which is indicative of chaotic motion.

A phase-plane history plot shows velocity versus position for some point on the structure over time. If the system is in static equilibrium, the phase-plane plot appears as a single point. If the system is dynamically stable, and has a periodic motion, the phase-plane plot has a trajectory appearing as a closed curve, known as an orbit. Considering the single pendulum with damping, a phase space diagram of the orbit is shown in Figure 22.7 [6]. The time domain response of a single pendulum periodically decays, which eventually converges to a single point, known as the attractor,

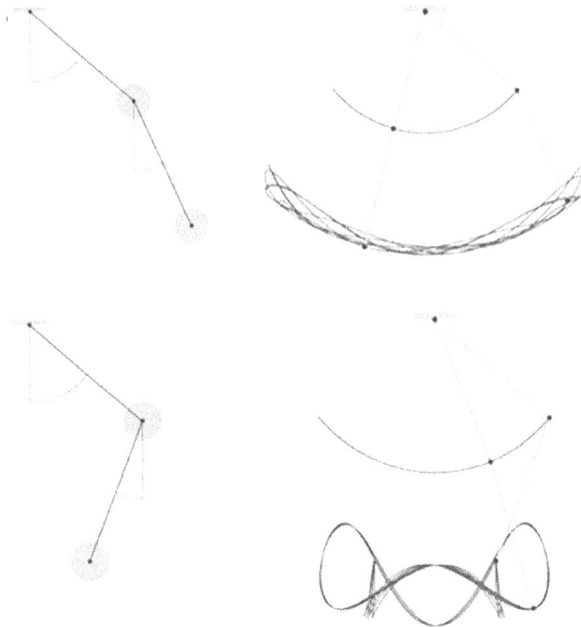

FIGURE 22.6 Double pendulum system with different initial conditions (a) and the trajectories of the two points corresponding to each pendulum (b) [8].

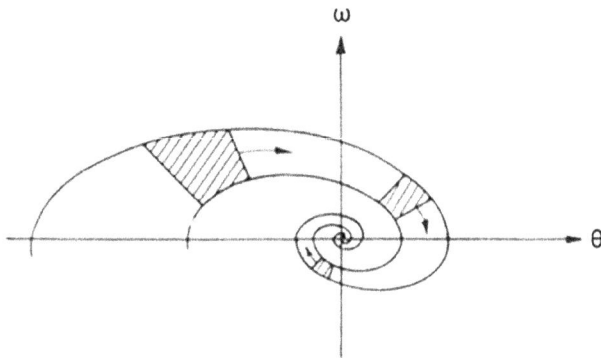

FIGURE 22.7 Phase space plot of single pendulum motion decaying to attractor [6].

regardless of the initial conditions. *Attractors* are geometric state-space representations, which characterize long-term behavior after transients subside and provide information about the settling behavior of a system [9]. *Attractors* can have various forms, with the simplest being a single point, shown at the origin of Figure 22.7. The phase-plane plot in Figure 22.7 shows two different trajectories with slightly different initial conditions. The trajectories eventually converge to a single point as they propagate through time. The next form of *attractor* is a closed loop orbit, shown in the phase-plane plot of Figure 22.5, and the final form of attractor is a multidimensional torus-shaped orbit. These three *attractors* are predictable and result from nonchaotic motion [9]. Chaotic *attractors*, however, have a more complicated geometric form, and the phase-plane plot consists of orbits with trajectories that fill the phase space as a function of the amount of chaos present in the system [10].

Plots of the PSD indicate the presence of chaotic behavior, depicted by the characteristics of the response function, shown in Eqs. (22.7) and (22.8). Nonchaotic PSD plots are smooth, with clear peaks at the frequencies of highest attenuation, while PSD plots of chaotic systems are more irregular and do not have discrete frequencies associated with the motion [10].

Another method used to measure the amount of chaos in a dynamic system is through the calculation of the Lyapunov exponents. As stated by Wolf et al., *Lyapunov exponents [have] proven to be the most useful dynamical diagnostic for chaotic systems. [They] are the average exponential rates of divergence or convergence of nearby orbits in phase space…Any system containing at least one positive Lyapunov exponent is defined to be chaotic* [11]. Equation 22.9 shows the time series method used to calculate the Lyapunov exponent from experimental displacement data [11]. An *attractor* is reconstructed using the time series data, and the trajectories of the reconstructed plot are analyzed to determine if convergence or divergence occurs from the current to subsequent orbits. The trajectory is traversed, and the distance between neighboring points of the trajectory is calculated, as well as the evolved length between points to determine convergence or divergence. If a neighboring point is crossing on a different trajectory, a replacement point is computed to correct the trajectory by incrementing the total steps, N, by one. A more thorough explanation of the details of the Lyapunov computation can be found in the paper by Wolf et al. [11]. The value of the Lyapunov exponent changes with each time-step and the final value is the sum of the previously calculated time increments. If the value of the summed Lyapunov exponent is negative, or equal to zero, nonchaotic, periodic motion is indicated. If the value is positive, chaotic motion is indicated, and two trajectories with nearly identical initial conditions will diverge. The magnitude of the Lyapunov exponent indicates the amount of chaos present in the system [10]. Equation 22.9 gives the solution procedure used to calculate the Lyapunov exponent for each set of ramped initial conditions [11].

$$\lambda_1 = \frac{1}{t_N - t_0} \sum_{k=1}^{N} \log_2 \frac{L_p'(t_k)}{L_p(t_{k-1})} \tag{22.9}$$

where $L_p(t_{k-1})$ = length between two points on the trajectory, $L_p'(t_k)$ = evolved length between two points at a later time, N = total number of replacement steps, t_N = time of current replacement step, t_0 = initial time, λ_1 = Lyapunov exponent.

22.4 METHODOLOGY

Certain aspects of a structural analysis problem have to be considered when investigating dynamic responses not necessary in the analysis of the static response. One consideration necessary to determine the dynamic response is the rate at which a load is applied. In a static analysis, the load is applied so the structure is held in equilibrium, and member velocity and acceleration are equal to zero. The dynamic response will appear similar to the static response if the load is applied at a sufficiently slow rate. Therefore, to evaluate the rate which a reasonable dynamic response could be produced, and to define the line between dynamic and static loading so that evacuation of internal gas can be carried out without incident, an analysis of ramped loads was considered. This rate determines how fast the vacuum is created inside the icosahedron structure.

The solution of the system's DEOM depends on the initial conditions applied to the structure at rest; therefore, the primary determinant of the presence of chaos is in the application of its initial conditions. The rates at which loads are applied change the initial velocity of the forces in the icosahedron frame, which change the initial conditions. The initial slope of the displacement curve in the ramp responses is the initial velocity experienced by the frame. Increasing the initial velocity will result in an increase in oscillations in the system, which will occur once the maximum load is reached, which increases the total system kinetic energy. When the amount of energy applied to the system exceeds the system's ability to absorb, a *snapback* behavior is seen in the frame members. However, if the kinetic energy applied to the system is below the ability of the system to absorb, the response is similar to the static response, and chaotic motions will not develop.

The study of chaotic behavior in a system requires a dynamic response dependent on initial conditions. As discussed previously, various ramp load conditions were applied to the frame. The displacement response to the ramp input was determined for the midpoint of each node examined. As the evacuation rate is increased, increasing the ramp slope, and thus, the resulting initial velocity experienced by each member, the difference between the static response and the dynamic response became apparent with larger oscillations occurring once the full load is applied. A rise time of 0.005 s, corresponding to a load rate of 4.053 MPa/s, sufficiently displayed the dynamic characteristics of interest in this research, which was the minimum load rate used for all remaining test cases.

The *snapback* behavior presented by Adorno-Rodriguez occurred in BC2 and was deemed unsymmetrical. However, *snapback* behavior developed for both boundary conditions, including BC3, when the load was applied dynamically within a certain percentage of sea-level pressure. The load which caused the *snapback* behavior to occur decreased when the load was applied dynamically. The initial conditions determined the amount the load could decrease and still produce a *snapback* response. These initial conditions drive the dynamics of the system, and for certain scenarios, lead to chaotic behavior. A clear difference between a slowly applied load and a quickly applied load was observed. The structure exhibits greater oscillatory behavior as the time over which the load is applied decreases. These results are utilized in studying the chaotic behavior associated with the frame when an unsymmetrical boundary condition is applied, or the applied load exceeded the system's ability to absorb kinetic energy. Table 22.3 shows the ramp force loading conditions used in the test case runs, and the specific frame boundary conditions applied at each vertex.

The load in which an instantaneous change in displacement direction occurs is referred to as the *snapping* load. In all cases, a follower force was applied at the reference point to replicate the pressure. This force is equal to the pressure distribution over the equilateral triangular area times the area. In addition, the force was directed toward the origin of the frame, making it a radial defined force.

In a static analysis, BC3 did not display the *snapback* behavior present in BC2. This research investigates the effect of dynamic loading on the structure using the same boundary conditions. Table 22.3 shows three dynamic loads when BC2 and BC3 are applied to determine if chaotic motion is present in the system response. The first two loads are below the *snapping* load, while the third load is above.

Four plots were generated for each load applied to the structure to determine if chaotic behavior exists. The first plot is the displacement versus time response for the given load. The second plot is the phase-plane trajectory, displaying velocity versus displacement. The third plot is the power spectral density plot for the given load, and the fourth plot shows the convergence of the Lyapunov exponent calculated from Eq. (22.9). The Lyapunov exponent convergence plots were created following the suggested method and algorithm developed by Wolf et al. [11].

A delay reconstruction of the attractor at each point is created. The delay reconstructed data is used to calculate an estimate for the Lyapunov exponent at each evolution of the data. Delay reconstructions of the *attractor* were computed using a delay parameter s, which was varied to avoid crossing and folding the trajectories within the *attractor*. The algorithm cycles through the trajectory based on input parameter values are used to calculate the Lyapunov exponent using Eq. (22.9) [11].

The input parameters, {*tau, evolve, dismin, dismax*}, are part of the computation algorithm outlined by Wolf et al. and vary based on the nature of the input time series data [11]. Variations of the parameters resulted in slightly different values for the Lyapunov exponent calculation, with the most significant differences arising from *tau* and *evolve*. The *dismax* parameter was selected based on the longest distance between points in the reconstructed attractor plot, and *dismin* was set to be smaller than the shortest distance between points. The *tau* and *evolve* parameters were chosen heuristically so the *attractor* does not appear to fold on itself, which results in a false-positive calculation of the Lyapunov exponent. Table 22.2 below shows the values of the algorithm parameters used in the exponent calculation for each load case.

TABLE 22.2
Lyapunov Exponent Input Parameters

Load #	Percentage of Sea-Level Pressure Applied	Ramp Duration (s)	Load Rate (MPa/s)	Boundary Condition
1	10	0.002	5.0663	BC3 (Frame)
2	20	0.005	4.053	BC3 (Frame)
3	40[a]	0.005	8.106	BC3 (Frame)
4	10	0.002	5.0663	BC2 (Frame)
5	20	0.005	4.053	BC2 (Frame)
6	40[a]	0.005	8.106	BC2 (Frame)

[a] Above dynamic snapping load for frame determined for the applied load rate.

TABLE 22.3
Loading Rates for Applied Pressure

Load #	Tau	Evolve	Dismin	Dismax
1	8	8	1×10^8	2×10^4
2	8	8	1×10^8	2×10^4
3	80	80	1×10^8	2×10^2
4	15	10	1×10^8	2×10^4
5	15	10	1×10^8	2×10^4
6	150	80	1×10^8	2×10^2

A time-step of 1×10^{-5}s was used to evaluate the varying loading conditions and was determined by evaluating a single beam of the icosahedron frame under a simply supported boundary condition, and an analytical value of the natural frequency for the simply supported beam was calculated. Next, ABAQUS was used to calculate the free decay response of the beam to an initial displacement by varying the time-step used for the solution. Each response was analyzed using its PSD function to compare the value of the natural frequency computed from the FEA solution, and then the FEA calculated natural frequencies were compared against the analytical values for the individual simply supported beam [5]. A time-step of 1×10^{-5}s resulted in less than 1% difference for the first five natural frequencies and was determined to be sufficient for the remaining analysis.

Finally, convergence studies were conducted in previous research to determine the optimal number and type of elements necessary to perform a static analysis of the structure [2]. The same input values were utilized from this previous research, and the individual beams comprising the structure were discretized into eight B32 beam elements. B32 beams in ABAQUS are Timoshenko beams that admit transverse shear deformation and use quadratic interpolation between nodes [4].

22.5 RESULTS AND DISCUSSION

The system response for the first load case is displayed in Figure 22.8. The applied load is below the static and dynamic *snapping* load. Inspection of the time displacement curve indicates that this input load does not cause a *snapback* response, resulting in stable, steady-state oscillation, which is purely periodic and not chaotic. The displacement curve for this input load has characteristics identical to an undamped single DoF system discussed in section III. A damping factor is not added to the beams in the icosahedrons FEM; therefore, the phase-plane trajectory maintains a single stable elliptical orbit, which does not decrease either in its semimajor axis or in its semiminor axis, indicative of a stable oscillating system. The PSD shows the system frequency response and reveals a resonant frequency at 1,500 Hz. Finally, the Lyapunov exponent convergences to a negative number in Figure 22.8, which indicates the response of the icosahedron frame for the first load number is predictable and not chaotic.

For the first load, the Lyapunov exponent was calculated using 4,500 data points, spaced at 1×10^{-5}s intervals. The initial 0.002s of data corresponding to the ramped

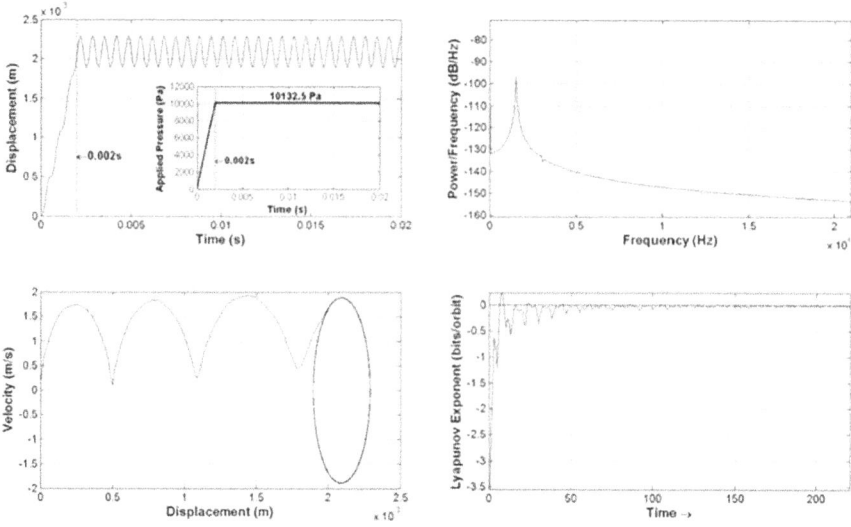

FIGURE 22.8 Load one displacement versus time response, PSD plot, phase-plane trajectory, and Lyapunov exponent convergence (clockwise from top-left) for nonchaotic, purely periodic motion.

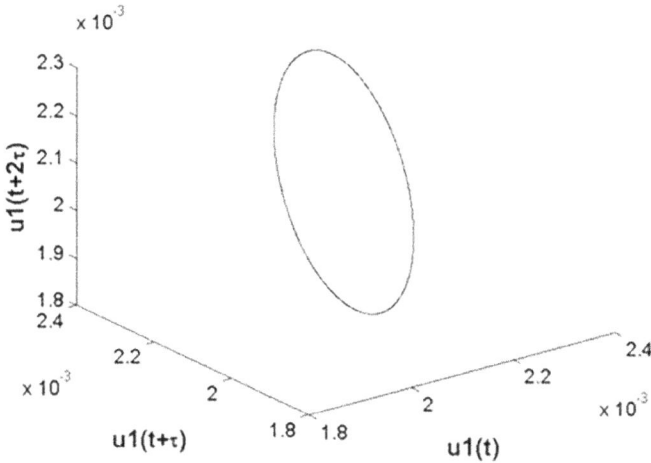

FIGURE 22.9 Load number 1 delay reconstructed attractor.

load is omitted from the calculation because the transient response is not considered in chaotic determinations. Figure 22.9 shows an example of the reconstructed attractor for the first load. As expected, for a purely periodic response, the *attractor* is simply a closed circuit orbit. The attractor is reconstructed in three dimensions because the system is three-dimensional and the plot is made of ordered triples comprising the displacement data separated by the delay parameter, τ. For example, one point has coordinates of $[u1(t), u1(t+\tau), u1(t+2\tau)]$.

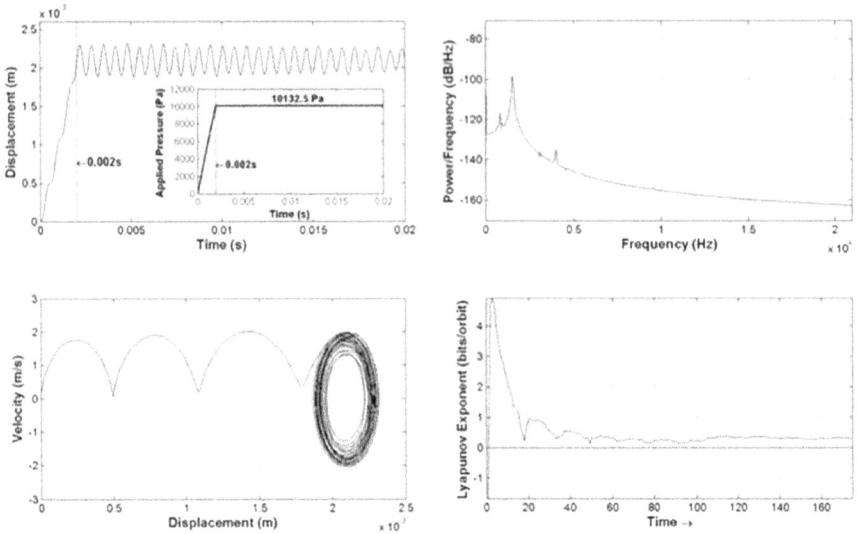

FIGURE 22.10 Load four displacement versus time response, PSD plot, phase-plane trajectory, and Lyapunov exponent convergence (clockwise from top-left) for slightly chaotic, perturbed periodic motion.

Load number two reveals nearly identical results as load number 1, and similarly, is not chaotic. The displacement curve, phase-plane trajectory, PSD plot, and Lyapunov convergence are nearly identical to those of the first load. There is a periodic steady-state oscillation present after the transient response subsides, resulting in a fixed orbit shown in the phase-plane trajectory. The PSD is smooth and has a clearly identifiable resonant frequency, while the Lyapunov exponent converges to a negative value. The input parameters for the Lyapunov exponent calculations were the same as those used in the first load case.

Load four and five have similar response curves; therefore, load case four results are shown here to represent both cases. Figure 22.10 reveals a mostly periodic displacement curve with small disturbances occurring throughout the response. The displacement curve of load five grows a bit more erratic, and the number of nonperiodic disturbances increases as the slope of the applied load increases. The phase-plane trajectories of the two responses are also similar, settling into an elliptical orbit of varying size. Variations in the orbit size correspond to the disturbances shown in the displacement plots. As the force increases in load five, the disturbances become larger, and more numerous, which create larger variations in the orbits of the phase-plane trajectory. The PSD plot of the two loads looks similar, with peak frequencies at 1,556 and 1,200 Hz, respectively. However, the increased pressure of load five creates more peaks than in the PSD plot of load four. The Lyapunov exponent convergence plot, shown in Figure 22.10, shares the same general characteristics, with the final steady-state convergence settling at a 0.303 bits/orbit. The number of bits/orbit is a measure of the conservation of the system's behavioral "memory". Wolf et al. referred to it as the amount of information retained in a system and stipulated

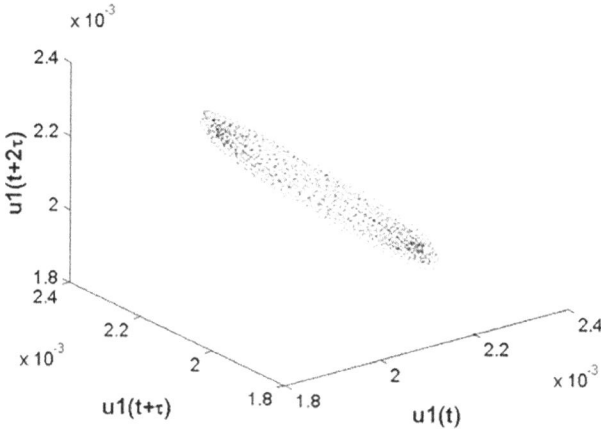

FIGURE 22.11 Load number four delay reconstructed attractor.

the Lyapunov exponent measures the rate of information conservation [11]. Given an estimate of the accuracy of a point in space, the accuracy of the icosahedron's beam displacement solution computed by ABAQUS, of at least one part per million (20 bits), the future location could not be predicted after 0.0424 s (20 bits ÷ 471 bits/second), which is approximately 65.9 orbits [11]. The accuracy claims are contingent upon making the assumption that after 0.0424 s, small levels of uncertainty will not allow the system of points to follow a controlled and predictable path, which will obfuscate the *attractor's* state-space rendering, thus 20 bits of new information can be gained from an additional measurement of the system [11]. If we start with Wolf's assumption of 20 bits of "good" displacement data, predictability of the midpoint node's position is lost after 65.9 and 53.95 orbits for loads four and five, respectively.

Figure 22.11 displays the attractor corresponding to load number four. The attractor for load number four does not form a closed loop associated with nonchaotic behavior. The response plots for loads four and five indicate boundary condition BC2 produces slightly chaotic behavior, decreasing in predictability as the load rate is increased. While the *snapback* behavior associated with the unsymmetrical boundary condition is not identified by applying either load four or five, the structure responds somewhat chaotically below the dynamically applied *snapping* load. This indicates small changes in the initial conditions cause significant changes in the response of the structure when BC2 is applied to the frame.

Load three is applied above the *snapping* load pressure and presents extremely different results, both quantitatively and qualitatively, which are shown in Figure 22.13. The displacement curve is not periodic, displays random vibration, and has amplitude approximately 100 times higher than load one. The *snapback* behavior can be seen as the displacement instantaneously changes direction. Additionally, the phase-plane trajectory does not follow a repeatable pattern, but it does generally remain within an elliptical envelope. The orbits of the trajectory fill a portion of the phase space, an indicator of chaotic behavior. Load three's PSD plot contains considerable noise and does not show a clearly defined resonant peak, a departure from the character

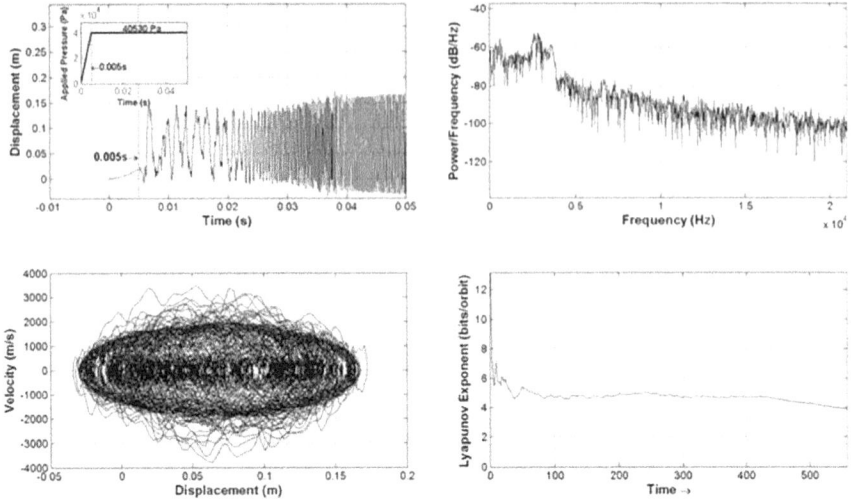

FIGURE 22.12   Load three displacement versus time response, PSD plot, phase-plane trajectory, and Lyapunov exponent convergence (clockwise from top-left) for highly chaotic motion.

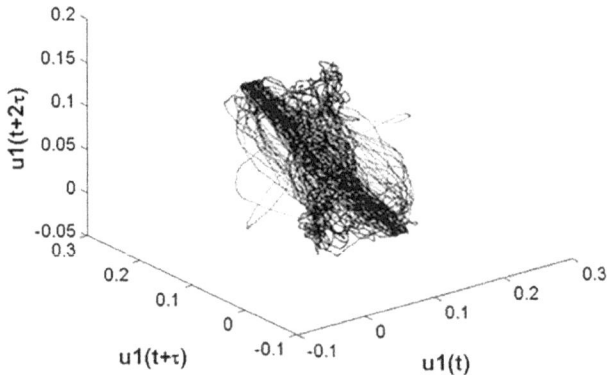

FIGURE 22.13   Load number 3 delay reconstructed attractor.

of the frequency spectrum shown in loads one and two. Finally, the convergence of the Lyapunov exponent is well above zero bits/orbit, indicating significantly chaotic behavior occurring above the dynamically applied *snapping* load. Assuming an initial 20 bits of good displacement data, load number three would lose predictability after only 0.0018 s, corresponding to a quarter of a single orbit. Figure 22.13 shows the delay reconstructed attractor for load three. The behavior of the attractor does comport with the previously examined periodic, nonchaotic conditions. The attractor fills nearly all of the phase space, consistent with chaotic behavior.

The large change in the Lyapunov exponent input parameters required to prevent folding in load case three was largely attributed to the change in amplitude in the displacement curve, as well as the decrease in the time-step necessary for

TABLE 22.4

Lyapunov Exponent for Different Applied Loads

Load Number	% of Sea-Level Pressure Applied	λ_1 (bits/s)	Dominant Orbital Period (s)	λ_1 (bits/orbit)
1	10[a]	−18.08	6.67×10^{-4}	−0.0121
2	20[a]	−16.39	8.33×10^{-4}	−0.0137
3	40[a]	10953.2	3.543×10^{-24}	3.8814
4	10	471.72	6.433×10^{-4}	0.303
5	20	435.91	8.333×10^{-4}	0.363
6	40[a]	10051.5	1.963×10^{-3}	19.67

[a] Above dynamic *snapping* load for frame determined for the applied load rate.

solution convergence. The time-step was reduced to 1×10^{-6}s for ABAQUS to converge on a solution.

The same loads applied to BC3 were applied to BC2 to confirm the presence of the *snapback* behavior originally identified in the static analysis. Load six was applied to BC2 like load three was applied to BC3. This dynamically applied load six was above the pressure required to create the *snapback* behavior for the symmetrical BC3 condition; therefore, it produced more chaotic results. The response is more chaotic than the response to load three applied to BC3.

Bold values indicate load conditions that are chaotic and nonlinear complete loss of center point displacement predictability after a single orbit.

The results of the Lyapunov exponent calculations are displayed in Table 22.4. Positive values of the Lyapunov exponent indicate chaotic behavior associated with the nonlinear dynamics which create a *snapback* behavior in the frame during load conditions above the *snapping* load, or as a result from applying the asymmetric boundary condition BC2. The higher the positive values of the Lyapunov exponent get, the more chaos is present in the system response.

The proof of chaotic behavior, and hence nonlinear dynamics, is a three-component verification. The nature of the PSD plots, the presence of chaotic delay-attractor orbits, and the magnitude of the Lyapunov exponents are the necessary and sufficient conditions to identify structural nonlinearities instigating chaotic system dynamics. All of these functions indicate an erratic transition from stability to dynamic instability based upon initial conditions. The use of the PSD to understand the system dynamical response is analogous to using the frequency response function (FRF) to identify system natural frequencies and harmonics. Locations of peaks in the PSD or FRF plots indicate a convergence of the driving frequency to a system's resonant frequency, which is dictated by the system characteristics. However, erratic peak geometry and rapid transition between several peaks over a short frequency span indicate a nonlinear response.

The development of rapidly changing PSD plots indicates instability, which is motivation to examine the phase-plane plot of velocity versus position at some fixed location as a function of time. Phase-plane plots for systems in static equilibrium are a point, while systems exhibiting a stable periodic response have a closed loop trajectory, and systems exhibiting nonlinear chaotic response have filled in trajectory orbits. Lyapunov exponents measure the amount of divergence or convergence of orbits in phase-plane space as a function of the distance between orbits. The exponents are summed across the entire time series signal, and a positive value indicates chaotic motion is present.

The ultimate objective of this entire research program at AFIT is to create an LTAV capable of executing long endurance loiter tasks fueled by an internal vacuum. The utility of this specific research effort was to verify the dynamics of how the vacuum can be safely developed. A safe rate of the internal pressure exhaust from the interior of the LTAV was required. A major motivating factor was if the speed of pressure exhaust was excessive, the effect on the thin membrane cover could create failure of the external membrane. The most conservative approach to this concern was to separate the frame from the membrane and evaluate the individual frame's response to the expected boundary and loading rate. Adorno-Rodriguez's static load analysis initially revealed the possibility of dynamic instability. Nonlinear displacement developed under a slow exhaust of the internal pressure leading to a snap-back phenomenon in certain frame members. To ameliorate this concern in normal thrust-producing vehicles, or even gas-fueled LTAVs, additional support members could be added to stiffen the external structure; however, since the object of investigation is a lighter than air vehicle, subject to an internal vacuum, which must be positively buoyant, the need to maintain minimum vehicle weight, with maximum strength is driving design considerations.

22.6   CONCLUSIONS

A nonlinear dynamic analysis of an icosahedron LTAV frame was conducted considering two different geometric boundary conditions. An interior beam center point was traced over time and its displacement evaluated. Several features of the frame displacement lead to certain conclusions:

- The FEA requires a fixed time-step of at least 1×10^{-5}s to obtain an accurate dynamic response of the icosahedron structure under ramp loading. Implicit direct integration yielded the best solution to the dynamic problems examined.
- A dynamic response of the icosahedron-shaped LTAV requires a dynamically applied load, and is highly dependent on the initial conditions. The characterization of the structural response is dependent on the magnitude of the applied load and the rate at which the load is applied.
- Previous research under static loading conditions identified the *snapping* load only occurred at BC2 at 45% of Sea-Level pressure. This research revealed pressures greater than 35% of Sea Level applied at a rate of 4.053 MPa/s resulted in a dynamic *snapping* load for *all boundary conditions.*
- At pressures greater than 4.053MPa/ s, the *snapback* displacement propagating through the frame was of a chaotic nature, which was confirmed by the Lyapunov exponent calculation, and the PSD and phase-plane orbit plots. The nonlinear chaotic response occurred in BC2 regardless of the magnitude of the load, or the rate at which it was applied, indicating

significant differences in the magnitude of the response with small changes in initial conditions. The frame members exhibited chaotic behavior under BC3 when the load applied was above the dynamic *snapping* load.

- The accuracy of the chaotic behavior prediction is dependent on the proximity of data displacement-time coordinates. Plots of the PSD, the phase-plane diagram, the delay attractor, and the Lyapunov exponent convergence are critical tools necessary to represent and quantify chaotic behavior.

- The presence of *snapback* behavior establishes a nonlinear chaotic beam response. Identifying the boundary conditions and the maximum vacuum evacuation rate are critical design parameters to avoid dynamic instabilities in the structure. Unmitigated chaotic instability causes structural damage to a single, or multiple beam members, resulting in displacement divergence from the remainder of the frame, and thus vertex separation or structural failure, which inexorably leads to subsequent membrane rupture and failure, ending with catastrophic vehicle failure, and ultimately mission failure.

ACKNOWLEDGMENT

The writers would like to thank Dr. Wolf, of Cooper Union in New York, for his time explaining the concept of information chaos theory, and providing insights into how it might be a useful tool in nonlinear system analysis. Also, thanks to Dr. David Stargel from the Air Force Office of Scientific Research (AFOSR) for giving an interest and the requisite intellectual and tactical motivation to pursue the subject area, and for providing the funding necessary to complete the research.

NOMENCLATURE

$D, \dot{D}, \ddot{D}$ = displacement, velocity, and acceleration vectors, respectively
K, k = stiffness matrix and individual stiffness
L_p = length between two points in reconstructed attractor
L_p' = evolved length between two points in reconstructed attractor
M, m = mass matrix and individual mass
N = number of replacement points
R^{ext} = externally applied load vector
R^{int} = internal force vector
R_x = autocorrelation function
S_x = power spectral density function
r = ramp input function
r = ramp response function
s = step response function
t = time
u = step input function
λ_1 = Lyapunov exponent
τ = time delay
Ω = frequency

REFERENCES

1. Metlen, T. T., 2013, "Design of a Lighter Than Air Vehicle That Achieves Positive Buoyancy in Air Using a Vacuum," M.S. Thesis, Air Force Institute of Technology, Wright-Patterson Air Force Base, OH.
2. Adorno-Rodriguez, R., 2014, "Nonlinear Structural Analysis of an Icosahedron and its Application to Lighter than Air Vehicles Under a Vacuum," MS Thesis, Air Force Institute of Technology, Wright-Patterson Air Force Base, OH.
3. Cook, R. D., Malkus, D. S., Plesha, M. E., and Witt, R. J., 2002, *Concepts and Applications of Finite Element Analysis*, 4th ed, Wiley, New York.
4. Dassault Syste'mes, 2011, "ABAQUS Analysis User's Manual," Dassault Syste'mes, Velizy Villacoublay, France.
5. Meirovitch, L., 2001, *Fundamentals of Vibrations*, McGraw-Hill, New York.
6. Baker, G. L., and Gollub, J. P., 1990, *Chaotic Dynamics: An Introduction*, Cambridge University Press, New York.
7. Shinbrot, T., Grebogi, C., Wisdom, J., and Yorke, J. A., 1992, "Chaos in a Double Pendulum," Am. *J. Phys.*, 60(6), p. 491.
8. The Wolfram|Alpha Team, 2015, "WolframAlpha Blog," Wolfram Alpha LLC, Champaign, IL, accessed Feb. 19, 2015 http://blog.wolframalpha.com/2011/03/03/from-simple-to-chaotic-pendulum-systems-in-wolframalpha/
9. Crutchfield, J. P., Farmer, J. D., Packard, N. H., and Shaw, R. S., 1986, "Chaos," *Sci. Am.*, 254(12), pp. 46–57.
10. Forral, A. E., 1993, "Comparison of the Dynamic Behavior of Composite Plates and Shells Incorporating Green's Strain Terms With the von Karman and Donnell Models," M.S. Thesis, Air Force Institute of Technology, Wright-Patterson AFB, OH.
11. Wolf, A., Swift, J. B., Swinney, H. L., and Vastano, J. A., 1985, "Determining Lyapunov Exponents From a Time Series," *Phys. D Nonlinear Phenom.*, 16(3), pp. 285–317.

Section 7

Optical Technologies

23 Quantitative Analysis of Cerium-Gallium Alloys Using a Hand-Held Laser-Induced Breakdown Spectroscopy Device[*]

Ashwin P. Rao and Michael B. Shattan
US Air Force Institute of Technology

Matthew T. Cook and Howard L. Hall
University of Tennessee

CONTENTS

23.1 INTRODUCTION

Laser-induced breakdown spectroscopy (LIBS) systems have shown increasing promise to bolster current capabilities for chemical analysis, particularly for cases involving the need for near real-time, standoff, and/or in situ sampling with little to no sample preparation. This is often desired in applications involving hazardous materials or the monitoring of industrial processes. Nuclear forensics is one such application where LIBS has found significant application [1]. Several studies have

---

[*] Reprinted with permission: Rao, A. P., Cook, M. T., Hall, H. L., & Shattan, M. B. (2019). Quantitative Analysis of Cerium-Gallium Alloys Using a Hand-Held Laser Induced Breakdown Spectroscopy Device. *Atoms*, 7(3), 84.

DOI: 10.1201/9781003220978-30

demonstrated the ability of LIBS to detect nuclear material in matrices relevant to the nuclear community, such as geological deposits [2,3], uranium ores [4,5], and surrogate nuclear debris [6]. Other studies validated the use of LIBS in nuclear safeguard applications, including analysis of IAEAswipe samples [7], nuclear reprocessing plant activities [8], and standoff detection of radiological threat materials [9]. Recently, Harilal et al. summarized the advancements in LIBS and other optical techniques to conduct isotopic analysis in laser-produced plasmas (LPPs) [10]. Portable hand-held LIBS (HH-LIBS) systems, such as the SciAps Z500-ER [11], have demonstrated the ability to detect uranyl fluoride contamination in metallic and sand substrates at a level of 250 ppm, as well as rare earths in uranium matrices [12,13]. This work demonstrates the capability of a HH-LIBS to conduct rapid chemical analysis of cerium alloys for the first time.

Cerium, a lanthanide series metal, is a common chemical surrogate for plutonium. Cerium metal has similar physical and chemical properties to plutonium [14], and previous studies have used cerium as a nonradioactive substitute for plutonium to gain insights into plutonium behavior [15–18]. This use of cerium makes it a substance of interest to the nuclear forensics community. A particularly interesting property of cerium is its behavior when alloyed with gallium.

Gallium is a common fuel stabilizer material, alloyed with plutonium for use in nuclear applications [19]. Upon extraction from reprocessing, plutonium metal exists in its α (monoclinic) phase; this phase is characterized by brittleness and large changes in atomic volume over small changes in temperature, making it a less than ideal candidate for machining into fuel rods or other nuclear components [20]. The δ (FCC) phase of Pu is malleable and much less sensitive to atomic volume changes; however, it only exists between 600 and 700 K [21]. A phase change could easily be achieved by heating plutonium metal; however, it cannot be stabilized at room temperature. In order to circumvent this problem, a stabilizer, such as gallium, is added in small amounts to the plutonium and alloyed through an annealing process. This stabilizes the δ phase at room temperature, allowing the metal to be machined [14]. Cerium can also be alloyed with gallium; Ce-Ga alloys have similar properties to Pu-Ga alloys, and studying them can provide useful insights into the behavior of Pu-Ga metals [15,17]. LIBS provides a promising avenue to analyze the Ce-Ga alloy production process. Being able to determine and map the Ga concentration rapidly in a metal alloy sample can provide an indication of the process used to create the sample, as the level of sample heterogeneity is sensitive to the temperatures, cooling rates, and equipment used in the alloy production process [22]. The use of LIBS to monitor a metallic alloy production process has been studied previously [23–25], but to our knowledge, this work represents the first study of a hand-held LIBS device to conduct such an analysis of a lanthanide metal alloy. The rest of the chapter is organized as follows: the metallurgical processes used to create the Ce-Ga samples are described in Section 23.2. The HH-LIBS settings and mathematical pre-processing routines used in the data analysis are described in Section 23.3. Strong neutral and ionic emission lines of both elements are identified and analyzed in Section 23.4. Section 23.5 discusses the univariate calibration curves generated from emission peak intensity ratios and the limits of detection calculated from the fit parameters. A multivariate regression model fit to the data is discussed in Section 23.6. This model was then used to conduct a surface mapping analysis of the Ga concentration variation across a sample; the results are presented in Section 23.7.

23.2   SAMPLE MANUFACTURING

The cerium and gallium alloy samples were made using a Thermo Scientific Thermolyne (Model Number FD1545M) resistive heating furnace. Cerium metal (99.9% purity) was obtained from Aldrich Chemistry, and gallium metal (99.99% purity) was obtained from Alfa Aesar. Preparation of the samples took place in an argon-filled glovebox with oxygen content nominally under 200 PPM. Between 10 and 20 g of cerium metal in chips of approximately 4 g, each was weighed using a mass balance (Mettler Toledo PR2003 DeltaRange). Gallium metal was then heated to its liquid state (approximately 60°C) and measured out using a glass pipette to the desired concentration within the Ce-Ga alloy. The combined Ce-Ga was placed in a magnesium oxide crucible obtained through Fisher Scientific and heated in the furnace to 850°C and held at that temperature for 8 h. The furnace temperature was then reduced to 480°C and held for 12 h to anneal the samples. After annealing, the furnace was turned off and allowed to cool via natural convection down to room temperature. The crucible containing the Ce-Ga alloy was removed and cracked with a hammer to release the sample. Samples were then exposed to ambient air and humidity to grow an oxide layer. For the scope of this work, the Ce-Ga samples were exposed to air for over 3 months.

23.3   SPECTRAL ACQUISITION AND PRE-PROCESSING

A commercially-available SciAps Z500-ER was used to collect spectral data from the alloy samples. The Z500-ER (Figure 23.1) is an industrial HH-LIBS device that uses a 5 mJ 1,064 nm Nd:YAG laser at a repetition rate of 10 Hz to ablate the surface of a sample. Spectral emissions are then collected by a group of four onboard spectrometers and recorded by the device computer. The gate delay of the device was varied, and an optimal delay of 450 ns was determined to give the best signal while minimizing noise. Ten cleaning shots were used to ablate through the surface of any oxide layer that had developed on the sample surface. Additionally, an argon gas purge was used before each data shot to minimize the presence of spectral lines of air in the data. The automated raster function of the device was used to sample eight different surface locations three times each, and an average spectrum was saved.

The extracted peak data was then processed in a method similar to the one used for UO_2F_2 detection with this device [12] and based on well-established spectral pre-processing techniques that aid in quantitative analysis [26,27]. First, a signal removal method algorithm was employed to subtract the baseline from the spectra. Next, a five-point Savitzky–Golay (SG) filter was employed after analysis determined that it removed continuum noise from the peak while maintaining the peaks and valleys in the spectra. Finally, a third-order noise median method (NMM) function was used to further remove noise from the peak wings. The applied filter is shown in Figure 23.2. The filter parameters were optimized to maximize noise reduction and

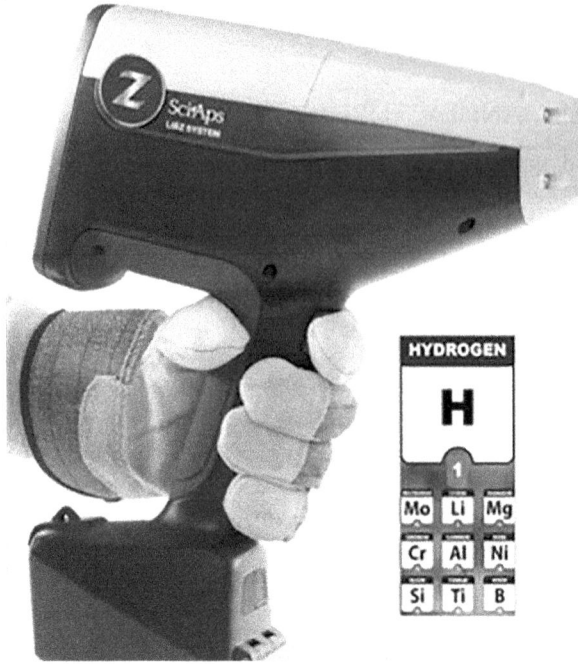

FIGURE 23.1　The SciAps Z500-ER hand-held (HH)-LIBS device used in this study.

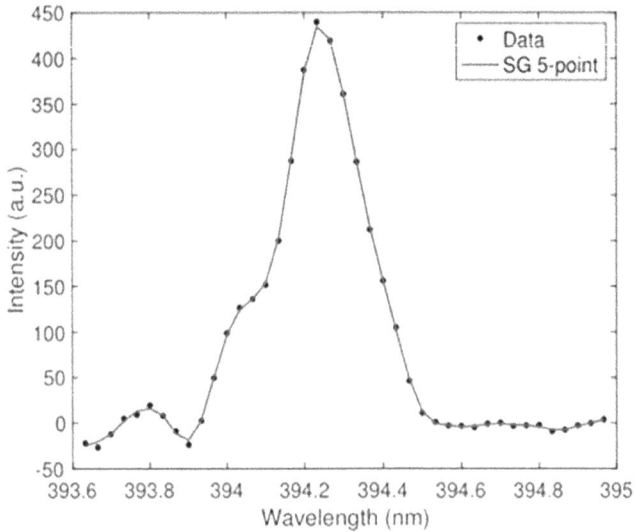

FIGURE 23.2　Ce II 394.3 nm emission peak after effective smoothing and noise reduction by a five-point Savitzky–Golay (SG) filter.

minimize peak information loss. Ensuring the filtering algorithms did not clip the peaks allowed for the highest accuracy in the peak ratio calculations for the calibration curves discussed further in Section 23.5.

23.4   ANALYTICAL LINE SELECTION

Four emission lines (Ga I 287.4 nm, Ga I 294.4 nm, Ce II 394.3 nm, and Ce II 413.8 nm) were identified for use in building calibration curves. These lines appeared across all gallium concentrations tested and were relatively free from other spectral interferences. These lines were extracted from each spectrum and processed according to the routine described in Section 23.3. Figure 23.3 displays the behavior of the two selected gallium emission lines with varying gallium concentration levels in the alloy sample. The peak intensity increased as the weight percent of gallium in the alloy increased, as expected. These lines showed good responsiveness to gallium concentrations, and their line shapes suggested that the plasma was optically thin and free from self-absorption for these emissions.

The Ce II peak behavior was similarly analyzed as a function of gallium concentration; the results are displayed in Figure 23.4. As expected, Ce II peak emission

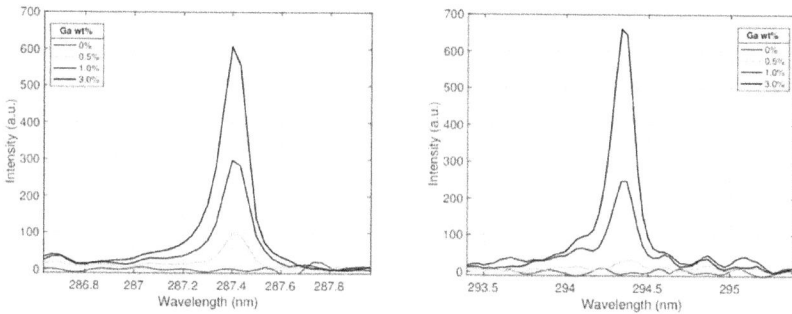

FIGURE 23.3   Ga I emission peaks centered at (a) 287.4 nm and (b) 294.4 nm. The peak intensities increase as the gallium concentration in the alloy increases.

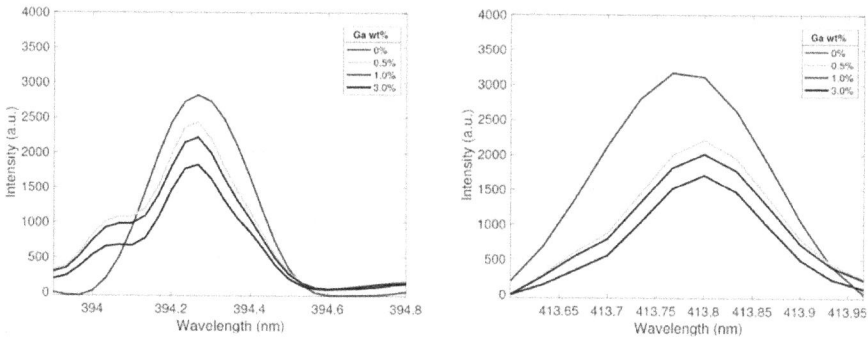

FIGURE 23.4   Ce II emission peaks centered at (a) 394.3 nm and (b) 413.8 nm. The peak intensities decrease as the Ga concentration in the alloy increases.

intensity decreased as the Ga concentration of the samples increased. The large drop in intensity between pure Ce and the 0.5% Ga sample was likely due to surface gallium crystallization diminishing the Ce peak intensity of some of the spectra taken from this lower concentration sample; this effect was then propagated to the averaged spectra. The recorded peak intensity ratios for the selected gallium and cerium emission lines were further analyzed to create calibration curves for the samples.

23.5   UNIVARIATE CALIBRATION CURVES AND LIMITS OF DETECTION

Four different calibration curves were fit for the purposes of conducting a univariate analysis of the spectral data. The ratio of each Ga I peak to each Ce II peak was taken across all concentrations. The first calibration curve sets using the Ga I 287.4 nm line are displayed in Figure 23.5. The data points from the intensity ratios are represented by the black dots; error bars were calculated from standard deviation measurements of the peak intensities between shots. The increasing magnitude of error corresponding to increasing concentrations is most likely due to the varying heterogeneity of the samples; the data indicated that as the Ga concentration of the sample increased, the more non-uniform the surface was. This would lead to higher variations in intensity between shots and yield a higher standard deviation for these samples. A weighted linear regression based on the error of each data point was employed to create the calibration curve for each dataset; this is represented by a solid line. Error terms of the fitting coefficients were calculated, and the confidence region is noted by the dashed lines. The largest error was seen in the intensity ratios from the 3% samples, as these peaks had the largest shot-to-shot variation. While emission line intensities were dependent on parameters such as laser energy and spot size, the intensity ratios remained relatively constant since individual line intensities scaled similarly with changing plasma temperature and plasma density. Therefore, line ratios from normalized spectra taken with other laser systems can be compared to these calibration curve sets.

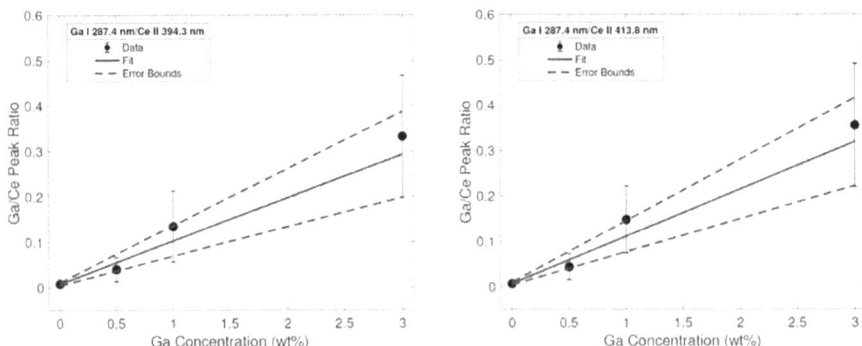

FIGURE 23.5   Calibration curve from line intensity ratios of the Ga I 287.4-nm emission to the (a) Ce II 394.3 nm (b) and the Ce II 413.8-nm emission lines. The regression fit is shown with a solid line, while errors in the regression coefficients are plotted with the dashed lines.

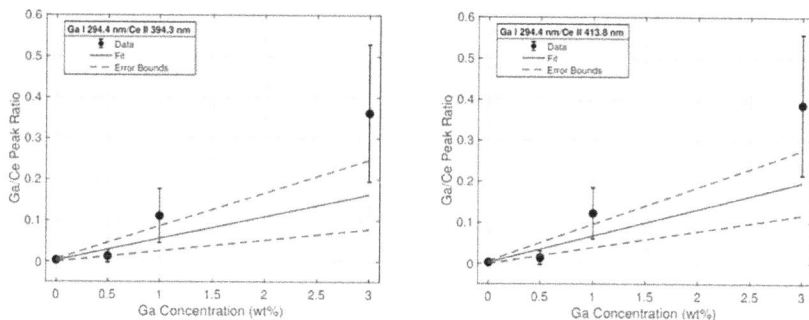

FIGURE 23.6 Calibration curve from line intensity ratios of the Ga I 294.4 nm emission to the (a) Ce II 394.3 nm (b) and the Ce II 413.8 nm emission lines.

TABLE 23.1

Fitting Parameters for Calibration Curves for the Equation $y = ax + b$ and limit of detection (LOD)

Line Ratio	a	δa	b	δb	R^2	LOD (wt%)
287.4 nm/394.3 nm	0.098	0.0294	0.0072	0.0036	0.9669	0.335
287.4 nm/413.8 nm	0.106	0.0320	0.0066	0.0032	0.9661	0.318
294.4 nm/394.3 nm	0.053	0.0251	0.0029	0.0033	0.4855	3.524
294.4 nm/413.8 nm	0.063	0.0243	0.003	0.0033	0.5637	3.429

The same analysis was conducted using intensities of the Ga I 294.4 nm peak, shown in Figure 23.6.

Upon initial visual inspection, it was clear that the fits for the calibration ratios generated using the 294.4 nm peak intensities were worse than those derived from the 287.4 nm peak in Figure 23.5. The low signal intensity of this line at low Ga concentrations made it difficult to achieve a good fit. These data clearly indicated the superiority of the 287.4 nm emission line for Ga detection purposes at low Ga weight percentages. LODs for each calibration curve were calculated according to the commonly used definition $3\sigma_d/s$. σ_d is referred to as the standard deviation of the blank and inferred from variations in spectra taken from cerium with no gallium. s refers to the slope of the calibration line. Table 23.1 lists all experimental fitting parameters along with the limits of detection for each calibration curve.

The tabulated R-squared values and LODs for each calibration set reflect the conclusions drawn from the observations in Figures 23.5 and 23.6. The coefficients of determination were much closer to unity for the calibration curves built using the Ga I 287.4 nm peak data. Additionally, the calculated LODs using these peaks were an order of magnitude less than those calculated from the 294.4 nm peak.

These calculations confirmed the superiority of the 287.4 nm line for quantitative analysis of gallium concentrations in Ce-Ga alloys, particularly at lower gallium concentrations.

23.6   MULTIVARIATE ANALYSIS

Multivariate analysis is commonly used in the analysis of complex spectra [28–30]. This technique numerically estimates how multiple independent input variables interact to produce the dependent response; this can result in a fit with less statistical uncertainty than a univariate calibration [31]. In this case, the intensity ratios of the Ga I 287.4 nm line to the Ce II 394.3 nm and 413.8 nm lines were analyzed using the multivariate regression function in MATLAB. More advanced multivariate techniques such as partial least squares (PLS) and principle component regression (PCR) are also commonly used in the LIBS community for quantitative elemental analysis. These techniques have the advantage of using the entire spectrum to inform their calibrations, at the cost of an increase in processing complexity. Given the limited computing power of the onboard systems of a hand-held instrument, these more advanced techniques were avoided. However, a simple MLS regression technique based on two well-defined, interference-free peak ratios appeared reasonable for this application. The generated coefficients from the MLS approach are displayed in Table 23.2.

These produced a regression fit of the following form:

$$y = b_0 + b_1 x_1 + b_2 x_2 \tag{23.1}$$

Here, y represents the Ga concentration in a sample, x_1 represents the peak intensity ratios I_{287}/I_{394}, and x_2 represents the ratio I_{287}/I_{413}. The regression equation is plotted with the experimental data in Figure 23.7 and includes the confidence interval based on the error bounds calculated for the tabulated fitting parameters.

The limit of detection using this multivariate method was calculated in the same manner as the univariate calibrations. The standard deviation σ_d was calculated by using the regression in Equation (23.1) with peak ratios taken from a blank shot series. The standard deviation in the calculated concentrations was then extracted. The slope s was calculated for the 3D line plot. Using the relation $3\sigma_d/s$ gave an LOD of 0.2435% Ga using the multivariate regression, a 23 percent improvement in LOD over the best-achieved univariate calibration, as well as a superior R-squared value.

TABLE 23.2

Fit Parameters for Multivariate Regression of Data Points Using Ga I 287.4 nm Line Intensities

Parameter	Value
b_0	0.006
b_1	2.934
b_2	5.886
δ_{b0}	0.001
δ_{b1}	0.341
δ_{b2}	0.333
R_2	0.994

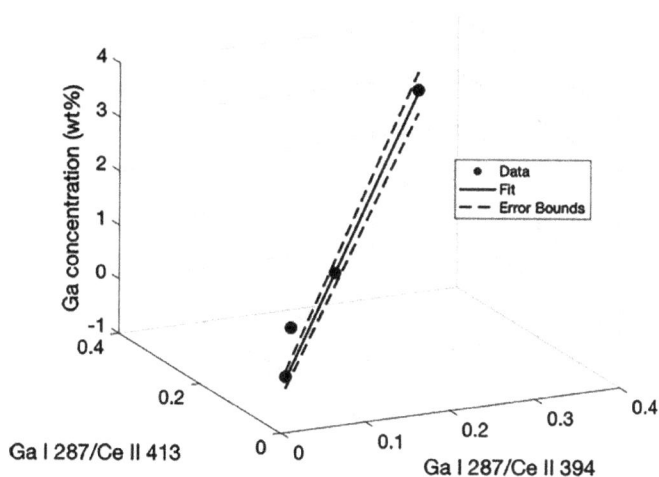

FIGURE 23.7 Multiple regression plot fitting ratios of the Ga I 287.4 nm peak intensity to a model determining Ga concentration based on the correlation between peak intensity ratios.

23.7 CONCENTRATION MAPPING

The data from the calibration curves were further used to analyze the gallium distribution across the surface of a Ce-Ga alloy sample. Mapping the gallium concentration can yield valuable insights into the alloy production process and help evaluate flaws in the alloying and annealing treatments. Determining how well distributed the gallium in an alloy can also help indicate its origin. This mapping process was conducted by mounting both the HH-LIBS and the alloy sample onto a laser table and securing them. An x–y translation stage was used to adjust the ablation location on the sample. Data points were taken at 1 mm intervals, and a 2D linear interpolation was used to determine the concentration between mapping points. Each point was ablated 10 times, and the average spectrum was recorded. The Ga/Ce peak ratios used in the multivariate calibration model were calculated at each data point; these ratios were then used to determine the gallium concentration from Eq. (23.1). The results of the mapping process on the three percent Ce-Ga alloy sample are shown in Figure 23.8.

The black line around the sample points represents the boundary of the sample piece. Each color reflects a gallium concentration between zero and six weight percent. Although the sample was created with amounts of each metal to reach an overall sample that was three percent gallium by weight, non-uniform cooling of the surface of the sample during the annealing process in the crucible led to the varying surface Ga distribution in Figure 23.8. The samples were created by pouring the molten metal into a room temperature crucible to mix them; upon contacting the crucible surface, the hot molten alloy can "flash freeze", causing phase changes and Ga crystallization along the surface boundary [32]. This rapid cooling prevents the gallium from migrating and distributing uniformly along the surface, leading to areas with higher amounts of gallium interspersed with zones of nearly no gallium [17]. This

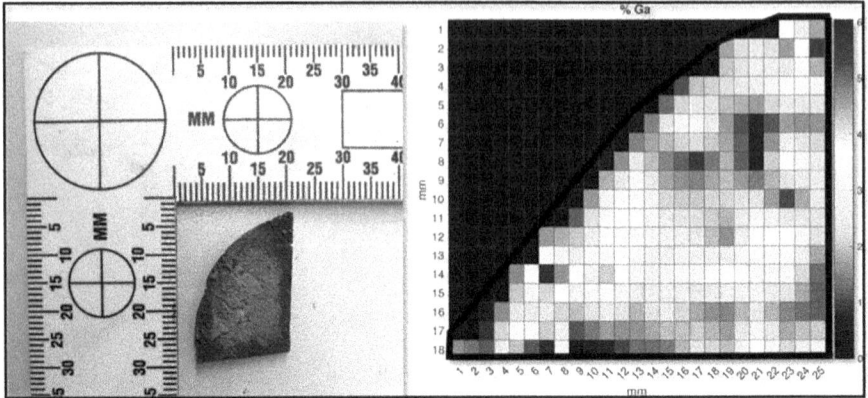

FIGURE 23.8   (a) Photograph of the 3% Ga cerium alloy sample used in the mapping analysis and (b) surface Ga concentration map of the sample. The color map represents the localized weight percent of gallium ranging from zero to six percent.

result is clearly reflected in Figure 23.8, where the yellow- to red-colored zones identify ablation points where gallium crystallized in the cerium lattice instead of diffusing due to the lower temperature. This experimental result proved the HH-LIBS device was capable of conducting surface concentration mapping analysis to evaluate the production quality of a cerium metal sample.

23.8   CONCLUSIONS

This study presented novel spectral line emission data of cerium-gallium alloys taken with a hand-held LIBS device. Univariate calibration curves were built from the ratios of selected atomic emission lines of gallium and cerium, and calibration fit parameters were tabulated. A limit of detection up to 0.318% was achieved from the univariate calibration curve approach, and it was determined that for the detection of gallium in lower quantities, the Ga I 287.4 nm line was the superior calibration standard due to its strong presence even at low weight percent concentrations of gallium. Multivariate analysis was used to improve the regression model and yielded an improved LOD of 0.2435%, making it superior to the univariate calibration fits. The multivariate regression model was then used to conduct a detailed surface mapping analysis of a cerium-gallium alloy, manufactured to have an overall gallium concentration of 3%. This allowed for a detailed map of the surface gallium distribution to be made, highlighting areas in which rapid cooling of the metal alloy in the crucible occurred during the annealing process, leading to an uneven diffusion of gallium atoms in the crystal lattice of the metal.

Author Contributions: Sample creation M.T.C.; data acquisition A.P.R. and M.T.C.; spectral data processing; A.P.R. and M.B.S.; sample mapping analysis: A.P.R. and M.T.C.; project supervisor: H.L.H.

Funding: This material is based on work supported by the U.S. Department of Homeland Security under Grant Award Number 2015-DN-077-ARI093, the Air Force

Office of Scientific Research (AFOSR) under Grant Award Number ENP19P906, and the Defense Threat Reduction Agency (DTRA) under Grant Award Number HDTRA 17-245-26.

Acknowledgments: The authors would like to acknowledge Duncan Brocklehurst of the Department of Nuclear Engineering at the University of Tennessee, Knoxville, for the preparation of the samples analyzed as part of this chapter.

Conflicts of Interest: The views and conclusions contained in this document are those of the authors and should not be interpreted as necessarily representing the official policies, either expressed or implied, of the United States Air Force, Department of Defense, U.S. Department of Homeland Security, or the United States Government.

REFERENCES

1. Bhatt, B.; Hudson Angeyo, K.; Dehayem-Kamadjeu, A. LIBS development methodology for forensic nuclear materials analysis. *Anal. Methods* **2018**, *10*, 791–798.
2. Barefield, J.E.; Judge, E.J.; Campbell, K.R.; Colgan, J.P.; Kilcrease, D.P.; Johns, H.M.; Wiens, R.C.; Mcinroy, R.E.; Martinez, R.K.; Clegg, S.M. Analysis of geological materials containing uranium using laser-induced breakdown spectroscopy. *Spectrochim. Acta B* **2016**, *120*, 1–8.
3. Klus, J.; Mikysek, P.; Prochazka, D.; Porizka, P.; Prochazková, P.; Novotny, J.; Trojek, T.; Novotny, K.; Slobodník, M.; Kaiser, J. Multivariate approach to the chemical mapping of uranium in sandstone-hosted uranium ores analyzed using double pulse laser-induced breakdown spectroscopy. *Spectrochim. Acta B* **2016**, *123*, 143–149.
4. Sirven, J.; Pailloux, A.; Baye, Y.; Coulon, N.; Alpettaz, T.; Gosse, S. Towards the determination of the geographical origin of yellow cake samples by laser-induced breakdown spectroscopy and chemometrics. *J. Anal. At. Spectrom.* **2009**, *24*, 451–459.
5. Kim, Y.; Han, B.; Shin, H.S.; Kim, H.D.; Jung, E.C.; Jung, J.H.; Na, S.H. Determination of uranium concentration in an ore sample using laser-induced breakdown spectroscopy. *Spectrochim. Acta B* **2012**, *75*, 190–193.
6. Shattan, M.B.; Gragston, M.; Zhang, Z.; John, D.; Auxier, I.; McIntosh, K.G.; Parigger, C.G. Mapping of uranium in surrogate nuclear debris using laser-induced breakdown spectroscopy (LIBS). *Appl. Spectrosc.* **2019**, *73*, 591–600.
7. Chinni, R.; Cremers, D.A.; Multari, R. Analysis of material collected on swipes using laser-induced breakdown spectroscopy. *Appl. Opt.* **2010**, *49*, C143–C152.
8. Sarkar, A.; Alamelu, D.; Aggarwal, S.K. Determination of thorium and uranium in solution by laser-induced breakdown spectrometry. *Appl. Opt.* **2008**, *4*, G58–G64.
9. Gaona, I.; Serrano, J.; Moros, J.; Laserna, J.J. Evaluation of laser-induced breakdown spectroscopy analysis potential for addressing radiological threats from a distance. *Spectrochim. Acta B* **2014**, *96*, 12–20.
10. Harilal, S.; Brumfield, B.; LaHaye, N.; Hartig, K.; Phillips, M. Optical spectroscopy of laser-produced plasmas for standoff isotopic analysis. *Appl. Phys. Rev.* **2018**, *5*, 021301.
11. SciAps. SciAps Z Brochure. 2014. Available online: https://www.sciaps.com/wp-content/uploads/2014/ 08/Z-brochureRGB.pdf (accessed on 21 July 2019).
12. Shattan, M.B.; Miller, D.J.; Cook, M.T.; Stowe, A.C.; Auxier, J.D.; Parigger, C.; Hall, H.L. Detection of uranyl fluoride and sand surface contamination on metal substrates by hand-held laser-induced breakdown spectroscopy. *Appl. Opt.* **2017**, *56*, 9868–9875.
13. Manard, B.T.; Wylie, E.M.; Willson, S.P. Analysis of rare earth elements in uranium using handheld laser-induced breakdown spectroscopy (HH LIBS). *Appl. Spectrosc.* **2018**, *72*, 1653–1660.

14. Marra, J. *Cerium as a Surrogate in the Plutonium Immobilized Form*; International Atomic Energy Agency (IAEA): Vienna, Austria, 2001.

15. Moore, M.; Tao, Y. *Aerosol Physics Considerations for Using Cerium Oxide CeO2 as a Surrogate for Plutonium Oxide PuO2 in Airborne Release Fraction Measurements for Storage Container Investigations*; Los Alamos National Lab. (LANL): Los Alamos, NM, USA, 2017.

16. Zheng, H.; Yueh, F.Y.; Miller, T.; Singh, J.; Zeigler, K.E.; Marra, J.C. Analysis of plutonium oxide surrogate residue using laser-induced breakdown spectroscopy. *Spectrochim. Acta B* **2008**, *63*, 968–974.

17. Gibbs, F.E.; Olson, D.L.; Hutchinson, W. Identification of a physical metallurgy surrogate for the plutonium—1 wt.% gallium alloy. *AIP Conf. Proc.* **2000**, *532*, 98–101.

18. *FY2015 Performance Evaluation Report*; Technical Report; NNSA: Washington, DC, 2015.

19. Steinzig, M.; Harlow, F.H. Characterization Of Cast Metals With Probability Distribution Functions. *MRS Proc.* **1999**, *538*.

20. Hecker, S.S. Plutonium: Coping with instability. *JOM* **2003**, *55*, 13–19.

21. Söderlind, P.; Zhou, F.; Landa, A.; Klepeis, J. Phonon and magnetic structure in δ-plutonium from density-functional theory. *Sci. Rep.* **2015**, *5*, 15958.

22. Johnson, C.G. *Metallurgy*, 4th ed.; American Technical Society: Chicago, IL, 1956.

23. Rai, A.K.; Yueh, F.Y.; Singh, J.P. Laser-induced breakdown spectroscopy of molten aluminum alloy. *Appl. Opt.* **2003**, *42*, 2078–2084.

24. Noll, R.; Sturm, V.; Aydin, Ü.; Eilers, D.; Gehlen, C.; Höhne, M.; Lamott, A.; Makowe, J.; Vrenegor, J. Laser-induced breakdown spectroscopy—From research to industry, new frontiers for process control. *Spectrochim. Acta Part B At. Spectrosc.* **2008**, *63*, 1159–1166.

25. Gruber, J.; Heitz, J.; Arnold, N.; Bäuerle, D.; Ramaseder, N.; Meyer, W.; Hochörtler, J.; Koch, F. In situ analysis of metal melts in metallurgic vacuum devices by laser-induced breakdown spectroscopy. *Appl. Spectrosc.* **2004**, *58*, 457–462.

26. Schulze, G.; Jirasek, A.; Yu, M.M.L.; Lim, A.; Turner, R.F.B.; Blades, M.W. Investigation of selected baseline removal techniques as candidates for automated implementation. *Appl. Spectrosc.* **2005**, *59*, 545–574.

27. Press, W.H.; Flannery, B.P.; Teukolsky, S.A.; Vetterling, W.T. *Numerical Recipes in C: The Art of Scientific Computing*, 2nd ed.; Cambridge University Press: Cambridge, 1992.

28. Guo, G.; Niu, G.; Shi, Q.; Lin, Q.; Tian, D.; Duan, Y. Multi-element quantitative analysis of soils by laser induced breakdown spectroscopy (LIBS) coupled with univariate and multivariate regression methods. *Anal. Methods* **2019**, *11*, 3006–3013.

29. Gottfried, J.L.; Harmon, R.S.; Lucia, F.C.D.; Miziolek, A.W. Multivariate analysis of laser-induced breakdown spectroscopy chemical signatures for geomaterial classification. *Spectrochim. Acta B* **2009**, *64*, 1009–1019.

30. Tiwari, P.K.; Awasthi, S.; Kumar, R.; Anand, R.K.; Rai, P.K.; Rai, A.K. Rapid analysis of pharmaceutical drugs using LIBS coupled with multivariate analysis. *Lasers Med. Sci.* **2018**, *33*, 263–270.

31. Larose, D.; Larose, C. *Data Mining and Predictive Analysis*; Wiley: Hoboken, NJ, 2015.

32. Predel, B. Ce-Ga (Cerium-Gallium). In *Ca-Cd–Co-Zr*; Springer: Berlin/Heidelberg, 1993; pp. 1–3.

24 Measurement of Electron Density and Temperature from Laser-Induced Nitrogen Plasma at Elevated Pressure (1–6 bar)[*]

Ashwin P. Rao
US Air Force Institute of Technology

Mark Gragston
University of Tennessee Space Institute

Anil K. Patnaik
US Air Force Institute of Technology

Paul S. Hsu
Spectral Energies LLC

Michael B. Shattan
US Air Force Institute of Technology

CONTENTS

---

[*] Reprinted with permission: Rao, A. P., Gragston, M., Patnaik, A. K., Hsu, P. S., & Shattan, M. B. (2019). Measurement of electron density and temperature from laser-induced nitrogen plasma at elevated pressure (1–6 bar). *Optics Express, 27*(23), 33779–33788.

DOI: 10.1201/9781003220978-31

24.1 INTRODUCTION

Laser-induced breakdown spectroscopy (LIBS) is a widely used technique for the spectroscopic analysis of gaseous [1], liquid [2], and solid [3] targets that utilize laser-induced plasma as a spectroscopic source. LIBS has been especially useful for combustion diagnostics, providing information on the local equivalence ratio [1,4,5], as well as determining the temperature of flames and heated gas flows [6–9]; this technique has also been used for simultaneously measuring and igniting combustible mixtures [5,10]. Interest in in-cylinder engines (up to 20 bar), gas turbine engines (up to 30 bar), detonations (up to a few hundred bars), thermal power stations (up to 42 bar), and planetary bodies (e.g., 90 bar at Venus) is driving spectroscopic investigations at high-pressure conditions. However, LIBS is very sensitive to conditions in which the plasma is generated, because effects such as spectral broadening or self-absorption are more prevalent at elevated pressure. Furthermore, recent work in this area has demonstrated that significant instability of the laser-induced plasma occurs at high-pressure conditions, impacting the measurement stability and precision [4]. To better understand how LIBS can be used for measurements under such high-pressure conditions, the plasma and spectral emission properties must be characterized to understand the physics of the laser-induced plasma [11]. Furthermore, short-gated LIBS is of particular interest for combustion analysis, where the measurement is taken approximately 100 ns after plasma creation but before more complex and highly-stochastic recombination processes dominate the signal [4,12].

For a typical LIBS measurement involving a gaseous sample, an intense pulsed laser is focused on the probe volume. The initial ionization at the target creates seed electrons that are accelerated by the remainder of pulse via inverse-Bremsstrahlung interactions [13,14], where the collisions between electrons accelerated by the laser field and, molecules and atoms significantly increase, resulting in avalanche ionization [14,15]. The resulting plasma then expands and cools, with various recombination and chemical reaction pathways dynamically changing the plasma emission characteristics. Early in the plasma lifetime (approximately hundreds of nanoseconds), the emission is largely broadband, owing to free–free and free–bound electron transitions [14,15]. As the plasma expands and cools, discrete emission bands emerge. During this process, ionic emissions are prevalent early (≤ 100 ns after the laser peak) but decay quickly, yielding emissions primarily of atomic neutrals followed by weak molecular lines as the gas returns to equilibrium [14,16].

These plasma emissions can be used to calculate many key plasma parameters essential for characterization. For example, the local electric field created by plasmas causes spectral broadening and line splitting. This phenomenon, known as the Stark effect, results in a broadening of the spectral line [17]. Thus, the Stark width of a spectral peak can be measured and used to calculate the electron density of the plasma. Additionally, the relative emission intensities of multiple spectral emissions from the same species can be analyzed to determine the corresponding excitation temperature. Under the partial local thermodynamic equilibrium conditions, this temperature will be equivalent to the electron temperature [18].

Nitrogen is abundant in combustion applications owing to its high concentration in the atmosphere. Hence, a large number of strong nitrogen spectral emission lines, such as the emission triplets at 566.6, 567.9, and 571.1 nm, are present in the LIBS signal of combustion mixtures; these emission lines are also free from spectral interference from other fuel elements, such as carbon and hydrogen. Thus, nitrogen is one of the ideal targets for diagnostic measurements in laser-induced plasma. Although Stark effect on hydrogen Balmer lines has been extensively investigated in the literature, only a few studies have been reported on the Stark broadening of the $N_{II}(568)$ emission triplet to measure electron density and temperature [19,20]. Furthermore, to the best of our knowledge, the influence of elevated pressure on Stark broadening measurements has never been reported for the determination of electron density. At elevated pressure, the Stark effect from the local electric field of plasmas is expected to become more pronounced, even for N_{II}. The enhanced Stark effect is caused by the increase in electron density with pressure. Only a few literature exists for plasma diagnostics at elevated pressures exceeding 1 bar [21]. In this work, we examine the Stark broadening of the N_{II} emission line (at around 568 nm) from a laser-induced nitrogen plasma at elevated pressure. Time-resolved spectra at various pressures are recorded, and electron densities are calculated from measured Stark width by using well-established relationships based on plasma parameters tabulated by Griem [22]. The deconvolved intensities of the N_{II} triplet emission lines at 566.6, 567.9, and 571.1 nm were determined; the Boltzmann distribution of the triplet is used to estimate the electron temperature under partial local thermodynamic equilibrium (pLTE) approximation. The Stark widths, electron densities, and electron temperatures are tabulated for pressures ranging from 1 to 6 bar.

The paper is organized as follows: in Section 24.2 the experimental setup for measurement of the N_{II} plasma emission at elevated pressure is presented; the spectral analysis methodology is described in Section 24.3; electron density and electron temperature measurements at elevated pressure are presented in Sections 24.4 and 24.5, respectively; the results are summarized in Section 24.6.

24.2   EXPERIMENT SETUP

A schematic of the experimental setup used is shown in Figure 24.1. An Nd:YAG laser (QuantaRay 290 Pro, Spectra-Physics) with an output wavelength of 532 nm and a pulse width of 10 ns provided pulses with an energy of 1,200 mJ/pulse and a repetition frequency of 10 Hz. The laser energy was controlled with a half-wave plate and polarizing beam splitter setup to reduce the beam energy to 80 mJ/pulse in the experiment. The beam was focused at the center of the high-pressure cell by using a plano-convex lens with $f = +150$ mm. LIBS signals were collected using a second spherical lens with $f = +150$ mm to direct light to a 0.25 m spectrometer (SpectraPro 2300i, Princeton Instruments) equipped with 1,200 grooves/mm grating blazed at 500 nm. An ICCD camera (PI-MAX4, Princeton Instruments) with a 1,024×1,024

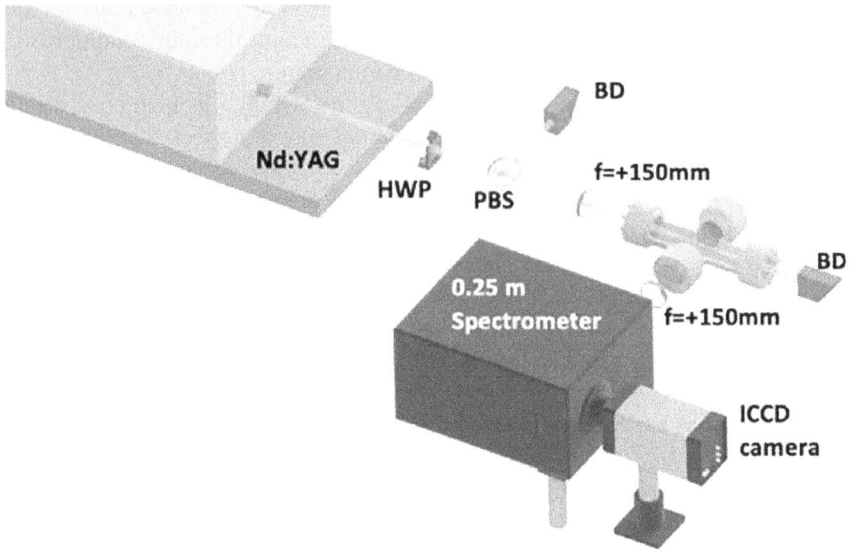

FIGURE 24.1 Schematic diagram of setup for time-gated spectrally resolved plasma emission measurements. PBS: polarizing beam splitter; HWP: half-wave plate; BD: beam dump.

pixel array was used for the collection of LIBS emission spectra with a spectral dispersion of 12.8 μm per pixel.

A tank of ultra-high purity nitrogen was coupled to the high-pressure cell. The pressure was regulated between 1 and 6 bars using a vacuum pump and manual gauges. The N_{II} spectra were recorded at camera gate delays from 60 to 260 ns (increasing by 40-ns steps) to monitor the temporal evolution of the electron density. All measurements were performed with a camera gate width of 20 ns for nitrogen. At each time delay, 150 frames of the spectra from the plasma were averaged for each pressure level.

24.3   SPECTRAL ANALYSIS: EXTRACTION OF THE STARK WIDTH AND DETERMINATION OF ELECTRON TEMPERATURE

The Stark effect has been reported in the literature to produce both broadening of spectral lines as well as shifts of the central emission wavelength; both of these parameters can be used for plasma diagnostic measurements [16,17,19,20,22,23]. In the current experiment, although Stark broadening of the N_{II} emission line was pronounced across all pressures and gate delay conditions in the recorded data, Stark shifts for this line were difficult to resolve and were often observable only at the earliest gate delay time, as reflected in Figure 24.2. For example, the measured Stark shift of the N_{II} (568) line in Figure 24.2 was approximately 0.03 nm, which was two orders of magnitude lower than the Stark broadening of this same line; similar behavior was reported by Konjevic et al. under atmospheric pressure conditions [20]. In contrast, the Stark shifts in the hydrogen emission lines [23] are very prominent

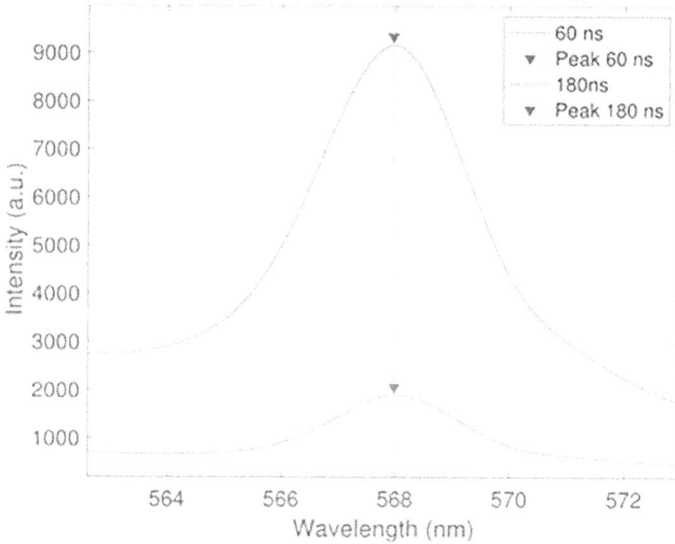

FIGURE 24.2 Recorded raw spectral emission at 2 bar for camera delays 60 and 180 ns. The dashed gray line represents the peak position of the spectrum at 60 ns delay. The observed Stark shift (w.r.t. the gray line) is visibly negligible. In contrast, the observed Stark broadening changed distinctly at two delay conditions.

(on the order of nanometers). Consequently, the Stark shift of the 568-nm N_{II} line would be a poor diagnostic metric with very high level of uncertainty in electron density estimations. Hence, only the Stark broadening of N_{II} was chosen for diagnostics of electron density.

To conduct a baseline subtraction on the emission peak data, a signal removal method was employed. The background subtraction is critical to ensuring the accuracy of measurements calculated from the recorded spectral broadening and intensity parameters [24]. A Voigt profile fitting (VPF) routine was used to extract the Stark width for each peak, which was estimated as the Lorentzian width of the Voigt profile [25]. The deconvolution process accounted for all peaks.

and their relative strengths of the nitrogen emission triplet centered around 568 nm, as shown in Figure 24.3.

Electron densities were calculated using the Lorentzian width of the 568 nm emission line according to the following well-known empirical formula [15,26–28]:

$$n_e = \left(\frac{\lambda_{1/2}}{2w} \right) \times 10^{16} \, \text{cm}^{-3} \tag{24.1}$$

In Eq. (24.1), $\lambda_{1/2}$ refers to the Stark full-width at half-maximum (FHWM), and w is the electron impact parameter, tabulated in Griem [22]. This method using the electron impact width parameter is commonly used to accurately calculate electron densities from the Stark width [15,26–28] and produces results that agree with other experimental measurements of Stark width for many different emission lines [20,22].

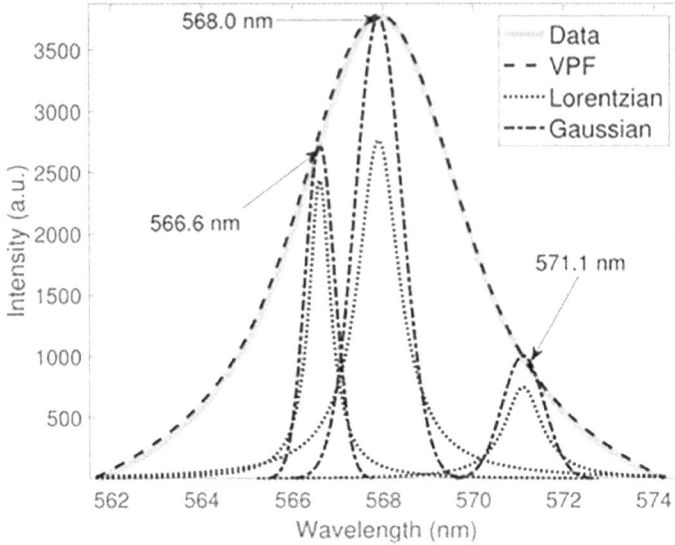

FIGURE 24.3 Example of peak deconvolution at 60 ns gate delay and 4 bar ambient pressure. The calculated Gaussian and Lorentzian components are displayed for the 566.6, 567.9, and 571.1 nm lines, plotted along with the experimental spectra and the Voigt profile fit (VPF).

Furthermore, the Boltzman distribution of the deconvolved N_{II} triplet is used to extract electron temperature employing the following relation [29]:

$$\ln\frac{I^{ki}\lambda}{g_k A_{ki}} = \frac{-E_k}{k_B T} + \ln\frac{hcN}{U(T)} \tag{24.2}$$

The left-hand side of Eq. (24.2) can be plotted against different upper-level energy transitions (E_k) to generate a Boltzmann plot, where the slope is proportional to the inverse of the electron temperature [26]. The terms I^{ki}, λ, g_k, and A_{ki} refer to line intensity of the excited ith degenerate k state, transition wavelength, excited energy level degeneracy, and transition probability for a given spectral emission line, respectively. These parameters were taken from the NIST Atomic Spectra Database [30]. The symbols E_k, k_B, T, h, c, N, and $U(T)$ refer to the upper-level energy, Boltzmann constant, plasma temperature, Planck's constant, speed of light, total species population, and partition function, respectively. By plotting the left side of Eq. (24.2) against the energy corresponding to the excited state for the emission line, the temperature can be determined from the experimental data using a linear fit to calculate the slope. The reciprocal of the slope determines the electron temperature in eV [29].

It may be noted that the validity of the Boltzmann temperature method relies on the assumption of the plasma being in a partial local thermodynamic equilibrium (pLTE). To determine whether the laser-induced plasma was in pLTE, the McWhirter criterion [31].

$$n_e > 1.6\times10^{12} T^{1/2}\Delta E_{nm}^3 \tag{24.3}$$

is employed. Here, T is the plasma temperature (K) and E_{nm} is the transition energy (eV). Note that Eq. (24.3) is a necessary but insufficient criterion for pLTE condition [18]. The existence of partial LTE is deduced from Eq. (24.3)–see Section 24.4 for details. The maximum and minimum electron densities estimated from Stark broadening are $2.8 \times 10^{17} \text{cm}^{-3}$ and $1.5 \times 10^{16} \text{cm}^{-3}$, respectively. Using the maximum (2.6 eV) and minimum (0.4 eV) electron temperatures, the calculated values of the right-hand side of Eq. (24.3) range from 1.2 to 2.9×10^{15}. For the whole range, the left-hand side is at least one order of magnitude higher than the right-hand side; hence, the McWhirter criterion and hence pLTE condition is always met in our experiment [18,32].

24.4 DERIVATION OF ELECTRON DENSITY FROM STARK BROADENING

The electron densities calculated using the extracted Stark widths using Eq. (24.1) at each pressure and gate delay are shown in Figure 24.4. The error represents one standard deviation of the Stark width measured from 150 laser shots; this error was propagated through the electron density calculations using basic rules of uncertainty analysis. The fits for each data set had adjusted R^2 values of 0.9998, 0.9899, 0.9981,

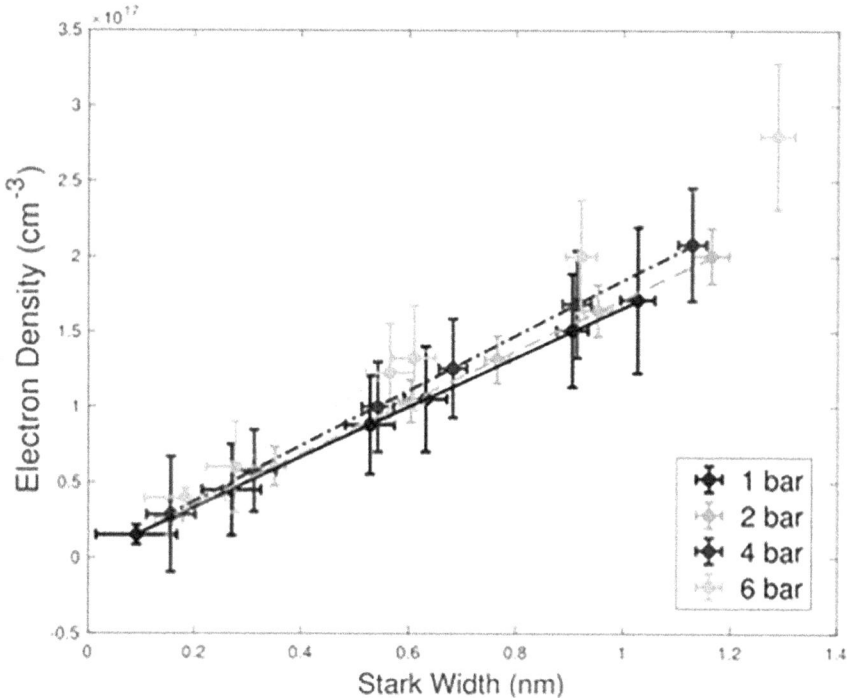

FIGURE 24.4 Electron density as a function of Stark width across different pressures. A linear fit was applied for each data set to generate the Eq. (24.4).

and 0.9972 for the atmospheric, 2, 4, and 6 bar conditions, respectively. The results clearly indicate that within 1–6 bar the electron density of the nitrogen plasma is linearly proportional to the Stark width of N_{II} (568) lines. This curve fitting, following the method used by Surmick and Parriger [23], was used to derive an empirical formula relating N_{II} Stark width and electron density, described in Eq. (24.4):

$$\Delta w = \beta(P)\left(\frac{n_e}{n_0}\right)^{1.01\pm0.05}. \tag{24.4}$$

Here, the first term $\beta(P)$ is an experimentally fitted function from Figure 24.4 representing the effects of elevated pressure on Stark broadening, given by the equation

$$\beta(P) = 0.6206e^{-0.034P}. \tag{24.5}$$

Here, P denotes the ambient pressure (bar), Δw is the Stark width (nm), n_e is the electron density (cm^{-3}), and n_0 is a normalization factor (10^{17} cm^{-3}). Equation (24.4) in conjunction with pressure-dependent exponential given in Eq. (24.5) represents a new empirical relation between the Stark width and electron density of a pure nitrogen plasma that includes the effects of ambient pressure from 1 to 6 bar. Equation (24.4) is accurate within a relative uncertainty of ±7% in measuring the electron density.

Next, the temporal evolution of electron density in N_2 plasma is presented in Figure 24.5 for 1–6 bar. Clearly, the increase in electron density with pressure is noticeable at earlier times and higher pressures (6 bar condition). The electron density is reduced at lower pressures and longer delays. Higher electron densities observed

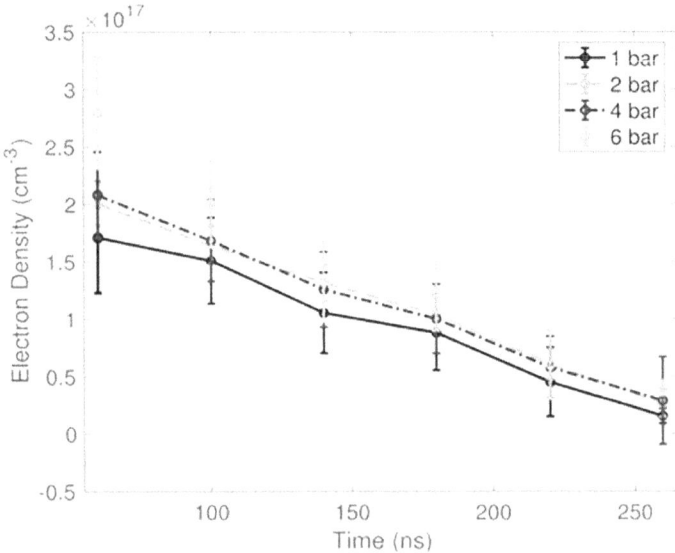

FIGURE 24.5 Temporal evolution of electron density in a nitrogen plasma calculated at pressures from 1 to 6 bar.

at earlier gate delays in the laser-induced plasma are dominated by avalanche ionization caused by inverse Brehmsstrahlung processes. These processes are typically accompanied by a significant increase in the Stark width of the spectral emission peaks as a result of the higher electron density [33,34]. As the plasma decays, the electron densities and the corresponding spectra resemble those of lower pressure cases, similar to the behavior seen in Figure 24.5. These results indicate that the N_{II} line could be useful in air plasma diagnostic measurements from LIBS at higher pressures. However, the usefulness of this line could be limited at pressures exceeding 6 bar and earlier gate delays (<100 ns) because of higher signal instability. With increasing pressure, the intermolecular distance decreases, the collisional scattering cross section increases adding high level of stochasticity to the LIBS signal generation process. It should be noted that the electron density decays quicker at higher pressures, indicating that the plasma recombines and cools faster at higher ambient pressure levels, similar to that reported in Ref. [11].

24.5 TIME-RESOLVED MEASUREMENT OF ELECTRON TEMPERATURE

The Boltzmann calculation method described in Section 24.3 was used to calculate the electron temperature from the time-resolved spectra obtained from the 150-shot-integrated peak intensity of the deconvolved 566.6, 567.9, and 571.1 nm emission lines starting at 60 ns gate delay and 40-ns intervals thereafter. The electron temperatures calculated from Eq. (24.2) are displayed in Figure 24.6.

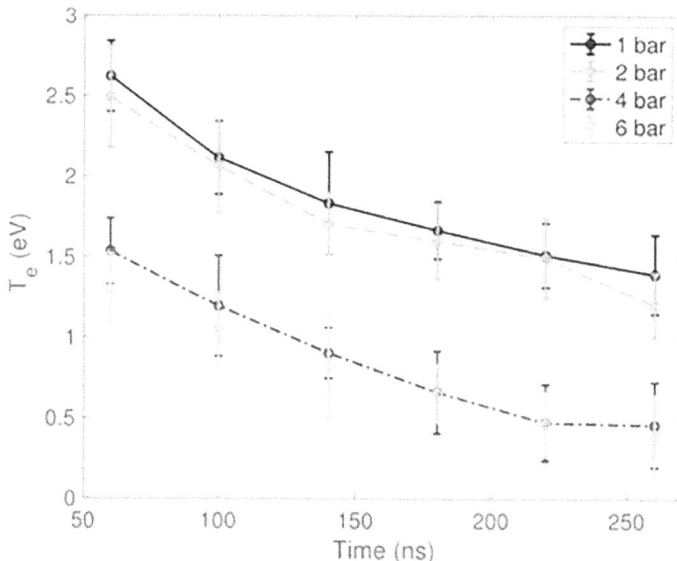

FIGURE 24.6 Temporal evolution of electron temperature at different ambient pressures. Each temperature was determined from the slope of the Boltzmann plot generated using the line intensity data at the corresponding gate delay.

TABLE 24.1

Exponential Fit Parameters of Equation $T_e = ae^{bt}$ for Different Ambient Pressures

Pressure (bar)	a	b
1	3.054	−0.0033
2	2.977	−0.0035
4	2.312	−0.0060
6	1.889	−0.0068

The results clearly indicate that the electron temperature is inversely proportional to the ambient gas pressure. The decay of plasma temperature T_e has been noted in several experimental studies [35–37]. At higher pressures, there is an increase in the collisional energy transfer between electrons and background gas, effectively reducing the temperature of the electrons. This pressure sensitivity is significant when the pressure is increased from 2 to 4 bar, but the trend is noticeably diminished at 6 bar. This diminished sensitivity could be because the electron densities generated at pressures above 4 bar are high enough to prevent an efficient recombination process, thus reducing the recombination rate that is important to lowering the overall plasma temperature. Using deconvolved emission line intensities, as opposed to resolved peak emission intensities, does affect the accuracy of the temperature measurements, as reflected in the larger statistical uncertainty of the temperature measurements mentioned previously. Additionally, the deviations increase as the pressure increases owing to this quenching process. The data in Figure 24.6 also indicates that ambient pressure affects the overall electron temperature as well as the rate of plasma cooling. An exponential decay function was fit to each data set; the fit parameters are recorded in Table 24.1.

The tabulated fit results show that increasing ambient pressure causes T_e to decay faster. The exponential decay of the measured time-resolved electron temperature values from this experimental series agrees with the data compiled by Zhang et al. [35] for the early gate delay times observed in this work. The overall temperature change in this time period can range from 0.9–1.3 eV (10,000–16,000 K); as evident from the data in Figure 24.6. All experimental data are summarized in Table 24.2 according to the gas pressure and gate delay. This table will help future high-pressure Stark effect investigations, where only limited literature exist that report any experimental investigation of nitrogen Stark broadening with respect to temperature and electron density.

24.6 SUMMARY AND CONCLUSIONS

We have reported the measurement of time-resolved Stark broadening of laser-induced plasma at 1–6 bar in a high-pressure pure nitrogen cell. The Stark width of the N_{II} (568) emission was employed to calculate the time-resolved electron density. We also derived an empirical formula for the nitrogen N_{II} line to determine the

TABLE 24.2
Summary of Experimental Data

P (bar)	Gate Delay (ns)	T_e (eV)	n_e (10^{17} cm^{-3})	Δw (nm)
1	60	2.621	1.712	1.027
	100	2.115	1.509	0.905
	140	1.832	1.053	0.632
	180	1.665	0.881	0.528
	220	1.511	0.450	0.270
	260	1.392	0.152	0.091
2	60	2.492	2.006	1.165
	100	2.062	1.641	0.953
	140	1.702	1.316	0.765
	180	1.596	1.039	0.604
	220	1.491	0.606	0.352
	260	1.201	0.263	0.153
4	60	1.534	2.081	1.129
	100	1.194	1.684	0.914
	140	0.902	1.257	0.682
	180	0.660	1.000	0.543
	220	0.472	0.577	0.313
	260	0.459	0.288	0.156
6	60	1.300	2.797	1.287
	100	1.049	2.002	0.922
	140	0.820	1.326	0.610
	180	0.651	1.228	0.565
	220	0.462	0.602	0.277
	260	0.399	0.398	0.183

electron density from the Stark width for the pressure range of 1–6 bar. Furthermore, we calculated the electron temperature from the Boltzmann plots of the experimental N_{II} triplet emission line intensities at elevated pressure. Our findings show that the electron temperature decreases with increasing ambient pressure, which could be due to the growing contributions of collisional energy transfer, whereas the rate of plasma cooling increased at higher pressures. Finally, the temperatures, densities, and Stark widths measured in this work were tabulated together, which may be helpful for future spectroscopic investigations of nitrogen in high-pressure reacting flows and plasmas [21].

One of the drawbacks of this investigation was the high uncertainty in Stark broadening, primarily due to the signal instability caused by avalanche ionization in the ns-laser-induced plasma. Future experiments of laser-induced plasma employing picosecond or femtosecond pulsed lasers may be a solution to high-pressure signal instability [38], which could increase the accuracy of the Stark broadening at high-pressure conditions.

FUNDING

Air Force Office of Scientific Research (ENP19P906).

ACKNOWLEDGMENTS

The authors sincerely thank Dr. Sukesh Roy for his thoughtful discussions on this project. Mr. Andy Kolonay of the Spectral Energies team was instrumental in assisting with experimental setup and hardware issues throughout the data collection campaign.

DISCLOSURES

The authors declare no conflicts of interest.

REFERENCES

1. Y. Wu, M. Gragston, Z. Zhang, P. S. Hsu, N. Jiang, A. K. Patnaik, S. Roy, and J. R. Gord, "High-pressure 1D fuel/air-ratio measurements with LIBS," *Combust. Flame* **198**, 120–129 (2018).
2. H. Suyanto, N. N. Rupiasih, W. T. B M. Manurung, and K. H. Kurniawan, "Qualitative analysis of Pb liquid sample using laser- induced breakdown spectroscopy (LIBS)," in *AIP Conference Proceedings*, vol. 1555, (2012), p. 14.
3. M. B. Shattan, M. Gragston, Z. Zhang, I. John D. Auxier, K. G. McIntosh, and C. G. Parigger, "Mapping of uranium in surrogate nuclear debris using laser-induced breakdown spectroscopy (LIBS)," *Appl. Spectrosc.* **73**(6), 591–600 (2019).
4. P. S. Hsu, M. Gragston, Y. Wu, Z. Zhang, A. K. Patnaik, J. Kiefer, S. Roy, and J. R. Gord, "Sensitivity, stability, and precision of quantitative Ns-LIBS-based fuel-air-ratio measurements for methane-air flames at 1–11 bar," *Appl. Opt.* **55**(28), 8042–8048 (2016).
5. T. X. Phuoc, "Laser-induced spark for simultaneous ignition and fuel-to-air ratio measurements," *Opt. Lasers Eng.* **44**(6), 520–534 (2006).
6. J. Lee, C. Bong, S. Lee, S. Im, and M. Bak, "Laser-induced breakdown thermometry via time-of-arrival measurements of associated acoustic waves," *Appl. Phys. Lett.* **113**(12), 123504 (2018).
7. T. W. Lee and N. Hegde, "Laser-induced breakdown spectroscopy for in situ diagnostics of combustion parameters including temperature," *Combust. Flame* **142**(3), 314–316 (2005).
8. J. Kiefer, J. Troeger, Z. Li, T. Seeger, M. Alden, and A. Leipertz, "Laser-induced breakdown flame thermometry," *Combust. Flame* **159**(12), 3576–3582 (2012).
9. A. P. Williamson, U. Thiele, and J. Kiefer, "Comparison of existing laser-induced breakdown thermometry techniques along with a time-resolved breakdown approach," *Appl. Opt.* **58**(14), 3950–3956 (2019).
10. H. Do, C. D. Carter, Q. Liu, T. M. Ombrello, S. Hammack, T. Lee, and K.-Y. Hsu, "Simultaneous gas density and fuel concentration measurements in a supersonic combustor using laser induced breakdown," *Proc. Combust. Inst.* **35**(2), 2155–2162 (2015).
11. A. K. Patnaik, Y. Wu, P. S. Hsu, M. Gragston, Z. Zhang, J. R. Gord, and S. Roy, "Simultaneous LIBS signal and plasma density measurement for quantitative insight into signal instability at elevated pressure," *Opt. Express* **26**(20), 25750–25760 (2018).

12. H. Do and C. Carter, "Hydrocarbon fuel concentration measurement in reacting flows using short-gated emission spectra of laser induced plasma," *Combust. Flame* **160**(3), 601–609 (2013).

13. Y. Wu, J. C. Sawyer, L. Su, and Z. Zhang, "Quantitative measurement of electron number in nanosecond and picosecond laser-induced air breakdown," *J. Appl. Phys.* **119**(17), 173303 (2016).

14. A. W. Miziolek, *Laser Induced Breakdown Spectroscopy* (Cambridge University, 2006).

15. D. Hahn and N. Omenetto, "Laser-induced breakdown spectroscopy (LIBS), Part I: Review of basic diagnostics and plasma-particle interactions: Still-challenging issues within the analytical plasma community," *Appl. Spectrosc.* **64**(12), 335A–336A (2010).

16. C. G. Parigger, A. C. Woods, and M. R. Rezaee, "Atomic hydrogen and molecular carbon emissions in laser-induced breakdown spectroscopy," *J. Phys.: Conf. Ser.* **397**, 012022 (2012).

17. H. Griem, *Spectral Line Broadening by Plasmas* (Academic Press, New York, 1974).

18. G. Cristoforetti, M. Aglio, and S. Legnaioli, "Local thermodynamic equilibrium in laser-induced breakdown spectroscopy: Beyond the McWhirter criterion," *Spectrochim. Acta, Part B* **65**(1), 86–95 (2010).

19. V. Milosavljevic and S. Djenize, "Measured Stark widths and shifts of NII, NIII and NIV spectral lines," *Astron. Astrophys., Suppl. Ser.* **128**(1), 197–201 (1998).

20. N. Konjevic and W. L. Wiese, "Experimental stark widths and shifts for spectral lines of neutral and ionized atoms," *J. Phys. Chem. Ref. Data* **19**(6), 1307–1385 (1990).

21. A. K. Patnaik, I. V. Adamovich, J. R. Gord, and S. Roy, "Recent advances in ultrafast-laser-based spectroscopy and imaging for reacting plasmas and flows," *Plasma Sources Sci. Technol.* **26**(10), 103001 (2017).

22. H. Griem, *Plasma Spectroscopy* (McGraw-Hill, New York, 1964).

23. D. M. Surmick and C. G. Parigger, "Empirical formulae for electron density diagnostics from Hα and Hβ Line profiles," *IRAMP* **5**, 73–81 (2014).

24. G. Schulze, A. Jirasek, M. M. L. Yu, A. Lim, R. F. B. Turner, and M. W. Blades, "Investigation of selected baseline removal techniques as candidates for automated implementation," *Appl. Spectrosc.* **59**(5), 545–574 (2005).

25. M. Zaghloul, "Computing the Faddyeeva and Voigt functions," *Software* **38**, 103001 (2011).

26. M. Suchonova, P. Veis, J. Karhunen, P. Paris, M. Pribula, K. Piip, M. Laan, C. Porosnicu, C. Lungu, and A. Hakola, "Determination of deuterium depth profiles in fusion-relevant wall materials by nanosecond LIBS," *Nucl. Mater. Energy* **12**, 611–616 (2017).

27. M. A. Gondal, Y. W. Maganda, M. A. Dastageer, F. F. Al-Adel, and A. Naqvi, "Study of temporal evolution of electron density and temperature for atmospheric plasma generated from fluid samples using laser induced breakdown spectroscopy," in *2013 Saudi International Electronics, Communications and Photonics Conference*, (2013), pp. 1–4.

28. P. W. Tawfik, "Calibration free laser-induced breakdown spectroscopy (LIBS) identification of seawater salinity," *Opt. Appl.* **38**, 103001 (2007).

29. V. K. Unnikrishnan, K. Alti, V. Kartha, C. Santhosh, G. P. Gupta, and B. M. Suri, "Measurements of plasma temperature and electron density in laser-induced copper plasma by time-resolved spectroscopy of neutral atom and ion emissions," *Pramana* **74**(6), 983–993 (2010).

30. A. Kramida, Y. Ralchenko, and J. Reader, "NIST Atomic Spectra Database,".

31. R. McWhirter, *Plasma Diagnostic Techniques* (Academic Press, New York, 1965).

32. D. Boker and D. Bruggemann, "Temperature measurements in a decaying laser-induced plasma in air at elevated pressures," *Spectrochim. Acta, Part B* **66**(1), 28–38 (2011).

33. C. Goueguel, D. L. McIntyre, J. P. Singh, J. Jain, and A. K. Karamalidis, "Laser-induced breakdown spectroscopy (LIBS) of a high-pressure CO_2-water mixture: Application to carbon sequestration," *Appl. Spectrosc.* **68**(9), 997–1003 (2014).
34. A. Effenberger and J. R Scott, "Effect of atmospheric conditions on LIBS spectra," *Sensors* **10**(5), 4907–4925 (2010).
35. S. Zhang, X. Wang, M. He, Y. Jiang, B. Zhang, W. Hang, and B. Huang, "Laser-induced plasma temperature," *Spectrochim. Acta, Part B* **97**, 13–33 (2014).
36. M. A. Hafez, M. A. Khedr, F. F. Elaksher, and Y. E. Gamal, "Characteristics of Cu plasma produced by a laser interaction with a solid target," *Plasma Sources Sci. Technol.* **12**(2), 185–198 (2003).
37. S. S. Harilal, R. C. Issac, C. V. Bindhu, V. P. N. Nampoori, and C. P. G. Vallabhan, "Optical emission studies of species in laser-produced plasma from carbon," *J. Phys. D: Appl. Phys.* **30**(12), 1703–1709 (1997).
38. P. S. Hsu, A. K. Patnaik, A. J. Stoll, J. Estevadeordal, S. Roy, and J. R. Gord, "Femtosecond-laser-induced plasma spectroscopy for high-pressure gas sensing: Enhanced stability of spectroscopic signal," *Appl. Phys. Lett.* **113**(21), 214103 (2018).

25 On-Chip Silicon Photonic Controllable 2×2 Four-Mode Waveguide Switch*

Cao Dung Truong and Duy Nguyen Thi Hang
Posts and Telecommunications Institute of Technology

Hengky Chandrahalim
US Air Force Institute of Technology

Minh Tuan Trinh
University of South Florida

CONTENTS

Multimode optical switch is a key component of mode division multiplexing in modern high-speed optical signal processing. In this chapter, we introduce for the first time a novel 2×2 multimode switch design and demonstrate in the proof-of-concept. The device composes of four Y-multijunctions and 2×2 multimode interference coupler using silicon-on-insulator material with four controllable phase shifters. The shifters operate using thermo-optic effects utilizing Ti heaters enabling simultaneous switching of the optical signal between the output ports on four quasi-transverse electric modes with the electric power consumption is in order of 22.5 mW and the

---

* Reprinted with permission: Truong, C. D., Nguyen Thi Hang, D., Chandrahalim, H., & Trinh, M. T. (2021). On-chip silicon photonic controllable 2 × 2 four-mode waveguide switch. *Scientific Reports,* *11*(1), 1–14.

switching time is 5.4 µs. The multimode switch exhibits a low insertion loss and a low crosstalk below −3 dB and −19 dB, respectively, in 50 nm bandwidth in the third telecom window from 1,525 to 1,575 nm. With a compact footprint of 10 µm×960 µm, this device exhibits a relatively large width tolerance of ± 20 nm and a height tolerance of ± 10 nm. Furthermore, the conceptual principle of the proposed multimode switch can be reconfigurable and scalable in multifunctional on-chip mode-division multiplexing optical interconnects.

Optical fiber communication systems and optical interconnects are now facing enormous demands on rapid bandwidth consumption of network traffics, especially in booming datacenters[1,2]. Besides, the growth of highly efficient computational systems leads to the increasing needs for high-bandwidth for information exchange centers[3]. To develop high-speed signal processing systems, some solutions have been proposed to scale the communication bandwidth aiming to overcome the barriers due to nonlinear limits[4,5]. Among them, the mode division multiplexing (MDM) technique has been considered as a promising solution for increasing communication bandwidth. In this method, the data was carried out using spatial orthogonality of guided modes. Since each mode is an independent channel thus potentially makes a myriad of single channel capacity for optical communications systems and optical interconnects when it combines with the wavelength division multiplexing (WDM) technique[6,7]. Various kinds of MDM systems have been proposed, for example, mode demultiplexer[8,9], mode converter[10,11], waveguide c rossing[12,13], mode selective s witch[14,15], mode add-drop m ultiplexer[16,17], and mode router[18].

In a multimode signal processing component, a multimode switch is one of the most important components to enable advanced processing functions for the MDM systems, especially, when combining MDM with WDM techniques, allowing an enhancement of data rates up to TBps in a multimode waveguide that has been demonstrated in recent w orks[19,20]. The functionalities of optical mode switches have been achieved in on-chip photonic devices with both input and output signals in the single-mode mechanism[21,22]. In multimode optical communication, multimode optically switching is a big challenge in interconnect systems. The difficulty in making switching for multimode waveguides comes from the contradictory design requirements: that is the confinement of light in an optical multimode waveguide with different modal distributions causing the highly dimensional complexity of the waveguide structures for realizing switching functionalities. In a communication optical fiber, the low contrast of refractive indices between the core and the cladding layers ($\Delta n \approx 5.10^{-3}$) and the weakly guided modes make it difficult to separate the guided modes for processing individually. In contrast, for guided modes in an optical waveguide with a high refractive index contrast-like silicon waveguides, ($\Delta n \approx 2$), the guided modes are strong. Moreover, the effective refractive index of silicon waveguides is highly independent on the modes so the interaction between guided modes is highly effective. Therefore, the modal transformation is also more flexible and convenient. An integrated multimode silicon waveguide could allow accessing specific modes for the reconfigurable switching system[23,24]. Currently, silicon waveguides widely use in the MDM system because of their advantages of wideband, high confinement of optical field, compact size, low power consumption, and especially CMOS-compatible devices[25].

Many approaches have been reported to construct silicon few-mode selective switches (silicon MSS) for the configuration of MDM networks since the fundamental mode and the higher-order modes can convert into single mode; therefore, these fundamental modes can be switched using spatial switching mechanisms. Stern et al.[26] have proposed a structure based on microrings to capable of multimode switching for the first time but that structure only supports two modes and selected two dedicated wavelengths. Recently, Zhang et al. have successively demonstrated a silicon 2×2 four-mode dual polarization optical switch[27] and a silicon 1×2 four-mode dual polarization optical switch[28]. However, both of those structures need asymmetric adiabatic couplers for realizing the (de)multiplexing functionalities in a multimode bus waveguide and Mach–Zehnder interferometers (MZI) for switching operation leading a large footprint and relatively complicated mechanism. Some other proposals for multimode switches either based on multiplexers/demultiplexers and waveguide crossing structures following Benes topologies[29,30] or use lots of relatively complicated microring resonators and waveguide crossing elements.

In this chapter, we propose a novel compact 2×2 four-mode optical switch enabling the switching operation of four modes simultaneously, which is based on Y-junction couplers and 2×2 multimode interference couplers. The proposed device could be constructed on silicon-on-insulator material and the switching operation of the device is performed via controllable phase shifters thanks to thermo-optic effect. The design, optimization, and characterization of the device are investigated using three-dimensional numerical simulations[31] based on commercial simulation tools from Rsoft's photonic device package.

25.1 WORKING PRINCIPLE AND OPTIMIZATION

The working principle diagram of the proposed 2×2 multimode switch is shown in Figure 25.1a and b. The proposed device consists of four identical and symmetric multi-branch 1×4 Y-junction couplers. The input sections have two 1×4 multi-branch Y-junctions that are symmetrically designed to guide the input multimode signals. The design uses three types of multimode interface (MMI) couplers composing of MMI-A, MMI-B, and MMI-C couplers. In this design, four MMI-A type couplers act as X-couplers between the two large symmetrical bridge of the device, eight MMI-B type couplers play the role of X-couplers between the internal branches, and eight MMI-C type couplers are 3-dB (50:50) couplers that divide and combine optical paths for the switching operations. Besides, the proposed device will enable us to simultaneously switch among four guided modes without blocking if we use four controllable phase shifters denoted PS_1 to PS_4 in order to combine suitable output optical signals.

The proposed device is constructed based on the structure of channel waveguides on silicon-on-insulator (SOI) material. The device is designed to support four transverse electric (TE) modes at the wavelength region of 1,550 nm. The refractive index of the silicon core layer and the corresponding silicon glass cover are $n_r = 3,465$ and $n_c = 1,445$ at the wavelength of 1,550 nm assume that these values do not change in the range of C-band.

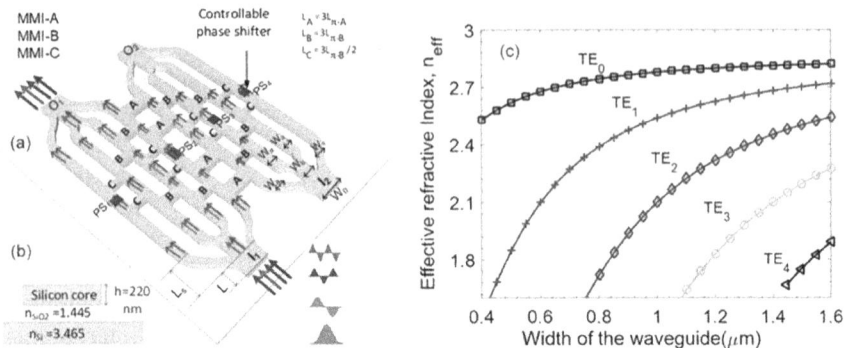

FIGURE 25.1   Device structure and TE mode characterization. Top view (a) and side view (b) of the conceptual diagram of the proposed four-mode switch. PS stands for phase shifter. Three different couplers, MMI-A, -B, and -C, with the lengths denoted as L_A, L_B, and L_C, respectively. (c) Effective indices for different guided modes of TE polarization as a function of input stem width via numerical simulation.

The 1×4 Y-junction coupler in this design composes of four sub-waveguides (as seen in Figure 25.1a), in which, the main bus waveguide at the input, four S-bent waveguides, and one straight waveguide connecting two sections of Y-junctions. We use a numerical simulation tool of mode solver to model the effective indices of the four lowest order quasi-TE modes from the input (including TE_0, TE_1, TE_2, and TE_3 modes) as a function of waveguide stem width W_0 of 1×4 Y-junction in the range from 0.3 to 1.6 µm, as presented in Figure 25.1c. The simulated results show that the stem waveguide only guides enough four quasi-TE modes in the range from 1.1 to 1.45 µm. In this design, we have initially chosen the width $W_0 = 1.4$ µm and the length of stem section $L = 20$ µm. The two S-bent waveguides on the two sides have the length in the propagation direction $L_s = 120$ µm and the width of two outer arms $W_a = 0.5$ µm. In the middle section, we use a straight waveguide whose width is $W_b = 0.7$ µm for supporting no more than two guided modes, which was utilized to connect to another Y-junction section with the width of outer arms as W_a. These parameters were optimized using a mode-sorting technique, similar to the phase-matching method suggested by D. Love et al.[32]. Using these parameters, the stem of input waveguides would couple to the first and the second-order modes on two outer wings converting to the fundamental modes and then couple to the fundamental and the third-order modes in the middle straight branch, which then continue to be divided to the fundamental modes at the outer wings of the second Y-junction substructure (O_1 in Figure 25.1a). Figure 25.2 presents the calculated mode field distributions at the input waveguide as well as the simulated electro-magnetic field patterns of four guided modes when propagating through the 1×4 Y-multijunction that were optimized by numerical simulation corresponding to the fundamental and three remaining high-order modes. In a 1×4 Y-junction, the fundamental mode and the third-order mode are distributed to two inner branches while the first- and the second-order modes are distributed to two outer branches at the output.

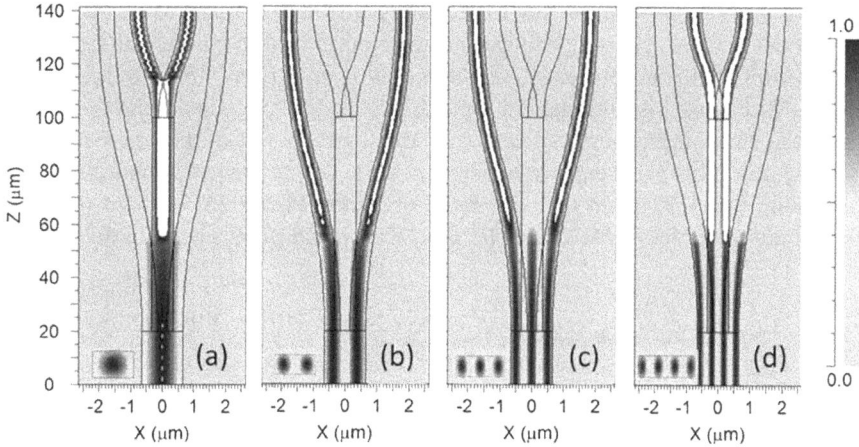

FIGURE 25.2 Electric field distribution. Numerical simulation of electric field distribution and mode solving calculation for the mode sorting in Y-multijunction waveguide for (a) TE_0, (b) TE_1, (c) TE_2, and (d) TE_3.

Note that the even modes will be divided into the pairs of in-phase fundamental modes and the odd modes will be divided into the pairs of counter-phase fundamental modes. The optical modes then continue to be guided through the waveguides and recombined at the outputs of the 1×4 Y-multijunctions with appropriate phase shifts, which are driven via controllable phase shifters (PSs) for switching to the desired output ports of the switches.

In this design, we used twenty 2×2 MMI couplers in three kinds named A, B, and C for switching operation (Figures 25.1a, 25.3). Three kinds of multimode interference couplers have the widths corresponding to W_A, W_B, and W_C and the lengths corresponding to L_A, L_B, and L_C, respectively. The working principle of a multimode interference coupler follows the Talbot effect[33]. In the general interference (GI) mechanism, the optical field is periodically reproduced along the propagation direction[34]. Self-imaging will be mirrored if the length of the multimode region $L_{MMI} = 3L_\pi$, in that case the MMI coupler will play the role of an X-coupler. However, self-imaging will be a mirrored pair of a photograph if the multimode region length $L_{MMI} = 3L_\pi/2$, and the MMI coupler will work as a 3 dB-coupler where L_π is the half-beat length which is defined as follows:

$$L_\pi = \frac{4n_{\text{eff}}W_e^2}{3\lambda}, \tag{25.1}$$

$$W_e = W_{MMI} + \frac{\lambda}{\pi}\left(n_{\text{eff}}^2 - n_c^2\right)^{-0.5} \tag{25.2}$$

here, W_e is the effective width calculated by the osmotic depth of the TE mode; is the operation wavelength; n_{eff} is the effective index; n_c is the refractive index of the cladding layer.

In this design, MMI-A and MMI-B couplers are the X-couplers while the MMI-C is the 3 dB-coupler with the lengths that were optimized by numerical simulation. The transmission for wavelength responses in a band from 1,525 to 1,575 nm of three MMI couplers is presented in Figure 25.3. For the X-couplers in Figure 25.3a and b, the switching is very efficient with the output power at the desired ports of about − 1 dB and is wavelength independent while output powers at undesired ports are smaller than − 22 dB in the wide band of 50 nm. Figure 25.3c shows the transmission curves of the MMI-C coupler with the optical field is equally divided into

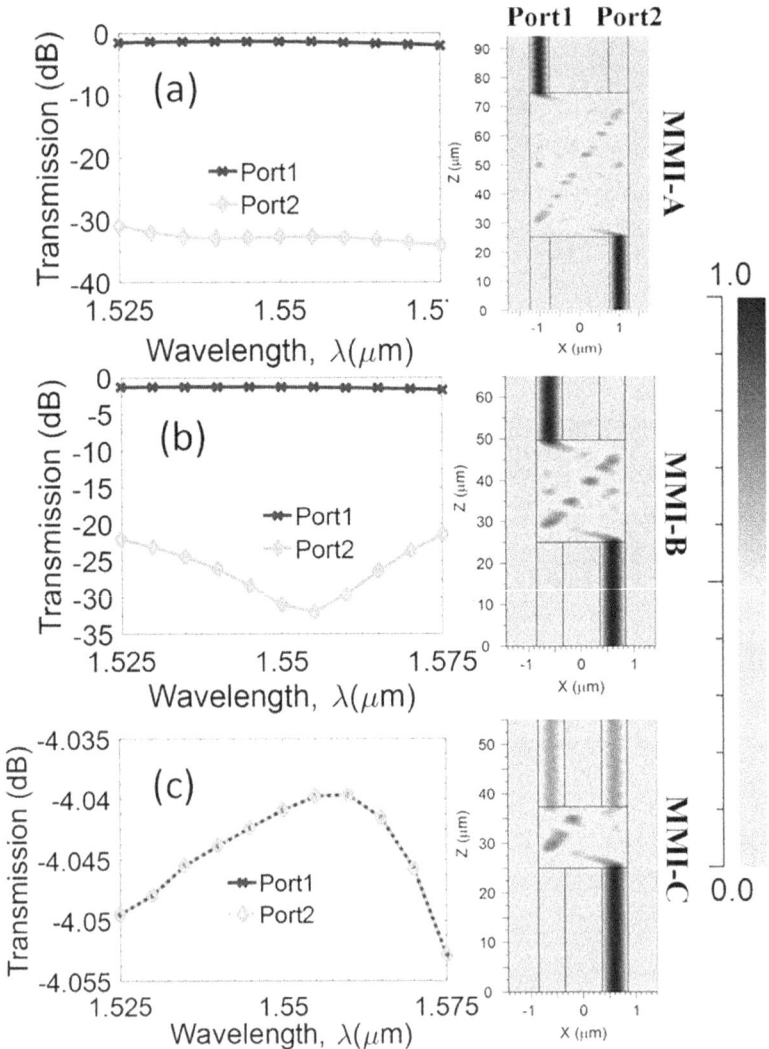

FIGURE 25.3 Transmission characteristics of three kinds of 2×2 MMI couplers on wavelength dependence. (a) MMI-A and (b) MMI-B are X-couplers with the lengths of $L_{MMI} = 3L_\pi$, (c) MMI-C is 3 dB-coupler with the length of $L_{MMI} = 3L_\pi/2$.

two output ports. The transmission outputs from two ports are identical with a small variation of 0.01 dB in the range of 50 nm; therefore, this coupler is a perfect wideband 3 dB coupler.

In this section, we will discuss the use of controllable phase shifters to combine or separate the optical fields. In order to achieve desirable simultaneously switching operation, four controllable phase shifters need to place at four branches as seen in Figure 25.1a. For each guided mode at the input port, after traveling through two different light paths in the circuit, will be recombined at the desired output port. Following the self-imaging principle when the multimode interference coupler in the role of X-coupler at the length of $L_{MMI} = 3L_\pi$, the image is a direct copy of the optical field at the input with the phase change of even or odd multiple of π. Hence, the interference mechanism occurring in this structure is similar to the interference mechanism in a Mach–Zehnder interferometer. Note that TE_0 and TE_3 modes are under influenced by the PS_1 and the PS_4, whereas TE_1 and TE_2 modes are under influenced by the PS_2 and the PS_3. The dependence of the output powers on the phase shift angles of controllable PSs for four modes can be written by explicit functions as follows:

$$P_{pn} = \frac{P_{in}}{2} \eta_{c,p} 10^{-\alpha L} \left(\sin^2 \left(\frac{\Phi_2}{2} \right) + \sin^2 \left(\frac{\Phi_3}{2} \right) \right), \tag{25.3}$$

$$P_{pn} = \frac{P_{in}}{2} \eta_{c,p} 10^{-\alpha L} \left(\cos^2 \left(\frac{\Phi_1}{2} \right) + \cos^2 \left(\frac{\Phi_4}{2} \right) \right), \tag{25.4}$$

$$P_{qn} = \frac{P_{in}}{2} \eta_{c,q} 10^{-\alpha L} \left(\sin^2 \left(\frac{\Phi_1}{2} \right) + \sin^2 \left(\frac{\Phi_4}{2} \right) \right), \tag{25.5}$$

$$P_{qn} = \frac{P_{in}}{2} \eta_{c,q} 10^{-\alpha L} \left(\cos^2 \left(\frac{\Phi_1}{2} \right) + \cos^2 \left(\frac{\Phi_4}{2} \right) \right), \tag{25.6}$$

where $p = \{0, 3\}$ and $q = \{1, 2\}$ are the orders of guided modes; $n = \{1, 2\}$ stands for the orders of the output ports. P_{in} is the input power of each mode; η_c, $\{p, q\}$ are the coupling efficiencies of the p, q-th order modes when coupled with the waveguide. α is the loss coefficient of silicon at the operation wavelength λ, typically $\alpha = 1$dB/cm for the wavelength of 1,550 nm[35]. L is the total device length; $\Phi_{1,2,3,4}$ are the phase shift angles under the control of the corresponding phase shifters $PS_{1,2,3,4}$, respectively.

In case all four controllable phase shifters are handled by only one external source such as a voltage-driven thermal source, the phase shift angles are the same: $\Phi_1 = \Phi_2 = \Phi_3 = \Phi_4 = \Phi$. Consequently, the combined optical fields at the output will be reformed to the mode shape like the input guided mode. Then, we can shorten the Eqs. (25.3–25.6) as:

$$P_{mn}(\Phi) = P_{in} \eta_{c,m} 10^{-\alpha L} \sin^2 \left(\frac{\Phi}{2} \right), \tag{25.7}$$

Figure 25.4 shows the output powers $P_{mn}(\Phi)$ obtained from a numerical simulation using Eq. (25.7) for four injected modes as a function of phase shift angle,

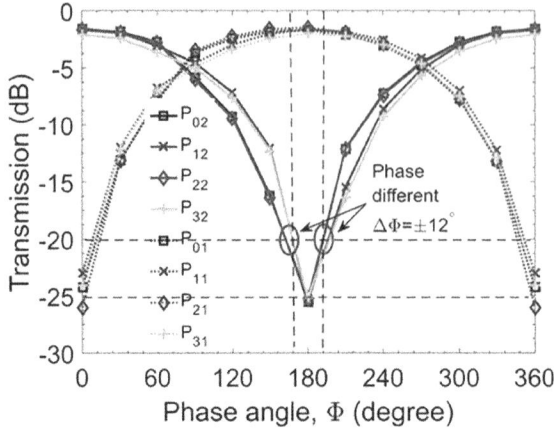

FIGURE 25.4 The transmission output powers for four guided modes at different ports as a function of the phase shift angle. Here P_{pn} and P_{qn} denoted the transmission powers of the mode orders p ($=0$ or 3) and q ($=1$ or 2) at the port n ($=1$ or 2). When the phase is of $0°$, $180°$, or $360°$ the outputs are switched. The circles indicate the operating phase range for crosstalked transmission lower than $-20\,dB$.

corresponding to bar (straight or ON state) and cross (OFF state) output ports. It is clear to see that the transmission curves are quite the same for all four characteristic modes in the harmonic shape curves. If the phase $\Phi = \pi$ radian, all four modes of TE_0, TE_1, TE_2, and TE_3 will be simultaneously switched to the bar output port, and if $\Phi = 0$ radian, all four modes will be simultaneously switched to the cross output port. In general, at some shifted angle different from an integer number of pi, the optical powers of each guided mode are divided into each output port, on the same side have the same convolution. The power splitting ratio between bar and cross output ports is given by $r_p = \tan^2\left(\frac{\Phi}{2}\right)$. In this case, the device plays the role of an arbitrary four-mode power ratio-splitting device. In addition, all cases of switching states at the desired output ports are highly efficient with the loss is not exceeded $1.25\,dB$ (the output transmission is larger than 75%) while the transmission at the undesired output ports is lower than $-23\,dB$.

To control the PSs, we employ the thermo-optic (TO) effect in which the refractive index of the core layer will be modified by thermal excitation resulting in a change of the phase of the propagating lightwave. The phase change can be calculated from the difference of index by[36]:

$$\Delta\Phi = k \cdot \Delta n \cdot L_h, \tag{25.8}$$

where L_h is the heater length to obtain the required phase shift of $\Delta\Phi$ and $k = 2\pi/\lambda$ is the wavenumber. The total index change, Δn, is determined by the thermal coefficient, dn/dT, of the material, in this study is silicon, and can be represented by linear relation as follows[37,38]:

$$\Delta n = K_c \cdot \Delta T \cdot \frac{dn}{dT}, \tag{25.9}$$

where K_c is the specific heat capacity; $\Delta T = T - T_0$ is the temperature change in the silicon waveguide core and $T_0 = 300$ K; is the thermo-optic coefficient for silicon material ($\frac{dn}{dT} = 1.84 \times 10^{-4}$ K^{-1} at the 1,550-nm wavelength).

In a TO switch, the minimum length to reach the phase shift of $\pm \pi$ radian, which is essential for switching operation, depends on the thermal crosstalk in the gap between the silicon core layer and heaters. A large heat transfer area or large heater length is necessary to make a large shift phase with a small temperature range. However, large geospatial factors will need a large power consumption, thus reducing performance and increasing operating costs. It also causes a large thermal impedance resulting in a high switching time. In addition, a large spatial size also prevents the device from integrating at subwavelength optics as well as on-chip integration.

To reduce the switching time for ultrafast applications, one can use a p–n junction structure as a controllable phase shifter by implanting p++ and n++ ions of group III/V in silicon ridge waveguide s tructures[39]. However, such carrier-depletion waveguides suffer a big loss from optical absorption by the ions. To overcome that drawback, in this study, we use driven heaters which are built by coating a metallic thin film of titanium (Ti) and electrical routing pads by coating a layer of aluminum (Al). The designed heaters have a reasonable length of about 200 μm. The driven heaters are placed at the position above the silicon core at a gap of $h_p = 0.7$ μm. The Ti film layers have a thickness of $\delta_{Ti} = 100$ nm and a width of $W_{Ti} = 1$ μm[40]. Figure 25.5 shows the structure and working principles of the numerically simulated thermo-optic phase shifters. In which, the distance h_p is determined by $h_p = h_{Si}/2 + h_{SiO_2} + \delta_{Ti}/2$, where h_{SiO_2} is the gap of the upper silica cladding layer sandwiched between the Ti-heater and the silicon core layer, as shown in Figure 25.5a. Figure 25.5b and c shows the distributions of the temperature rise (ΔT) and the index change (Δn) in the silicon core layer at the switching state "ON" by using multi-physics tools when the electric power is supplied to reach a required phase difference of π radian. As can be seen the heat distribution mainly focuses on the active region from the micro-heater to the silicon core at access arms and the maximal change is ~70 K on the Ti-heater surface. The distribution of the index change results in a nonuniform treatment that the highest variation is near the central region below the micro-heater. The phase shift angle increases linearly with the change of the temperature and it equals π radian when the maximum temperature difference increases a quantity of $\Delta T \sim 70$ K, as seen in Figure 25.5d.

The proposed four-mode switch could be patterned on a SOI standard wafer with 220-nm-thick silicon on 3-μm-thick BOX layer of silicon dioxide (SiO$_2$) using electron-beam lithography (EBL) and an inductively coupled plasma-induced reactive ion etching (ICP-RIE). A 1-μm-thick SiO$_2$ cladding layer will be deposited using a plasma-enhanced chemical vapor deposition (PECVD) process. After this SiO$_2$ layer, 100-nm-thick Ti heaters and 1-μm-thick Al contact pads will be deposited by the lift-off process[27]. Then, a 1.5 μm silica layer is deposited using the PECVD process to protect and isolate the silicon device layer. On top of this oxide layer, 100 nm thin film of Ti-metal is deposited as a high-resistance heater and 300 nm thin film of aluminum is deposited for the electrical routing using electron-beam evaporation. After the metallization process, a 300 nm SiO$_2$ layer is deposited as a protective layer for the heater that is etched away over the aluminum pad for an electrical probing signal.

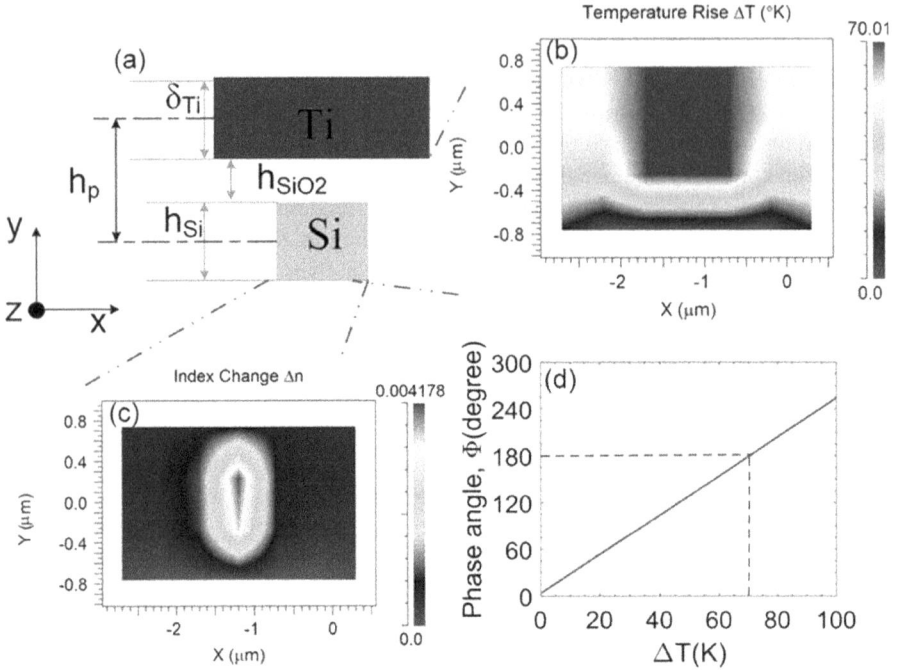

FIGURE 25.5 Structure of thermo-optic phase shifter. (a) The cross-view with the gap is SiO$_2$. Here $\delta_{Ti}=100$ nm, $h_{Si}=220$ nm, $W_{Ti}=1$ μm and $h_p=700$ nm. (b) The x–y plane shows distribution of temperature in the Si core when shifting phase is π radian. (c) Index change profile under the thermo-optic effect. (d) The simulated phase angle of the TO phase shifter as a function of temperature change, ΔT, from room temperature.

An electrical pulse may apply to drive the controllable phase shifters. This pulse generates heat and is transferred to silicon waveguide changing the refractive index of the silicon core making a phase shift. Currently, TO switches can work with an ultrafast switching time that is shorter than 10 μs[36,41].

25.2 CHARACTERISTICS EVALUATION

Figure 25.6 shows visual images of the electric field distribution that were carried out by numerical simulation for all switching states of four guided modes at the center wavelength of 1,550 nm in the bar and cross directions. Simulation results agree with the working principle of the proposed simultaneous 2×2 multimode switch as theoretically analyzed. The proposed switch operates at two states: each mode in four input modes entering port I$_1$ switched to equivalent order mode on the BAR side at port O$_1$ corresponding to "ON" state as seen in Figure 25.6a–d. These modes switched to the CROSS side corresponding to the "OFF" state as seen in Figure 25.6e–h.

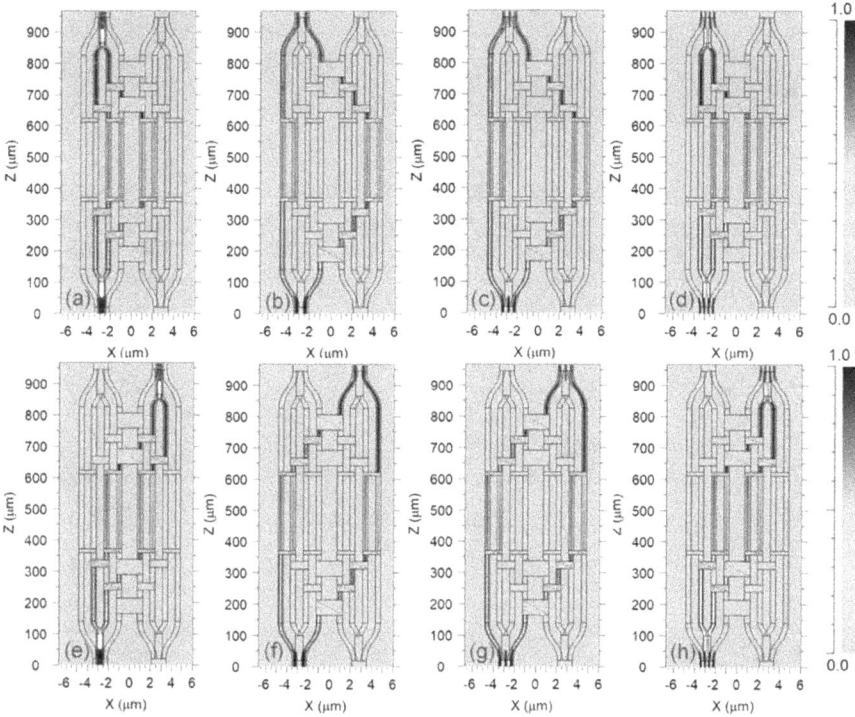

FIGURE 25.6 Contour plots of the electric field patterns for the working proposed 2×2 four-mode switch: (a), (b), (c) and (d) are for TE_0, TE_1, TE_2, TE_3 modes for ON state, respectively; (e), (f), (g) and (h) are for TE_0, TE_1, TE_2, TE_3 modes for OFF state, respectively.

To evaluate the optical performances of the proposed four-mode switch, the key parameters such as insertion loss (I.L) and crosstalk (Cr.T) are studied as a function of operation wavelength that can be calculated by following formulas:

$$I.L_{mn} = 10\log_{10}\left(\frac{P_{out}}{P_{in}}\right), \quad (25.10)$$

$$Cr.T_{mn} = 10\log_{10}\left(\frac{P'_{out}}{P_{in}}\right), \quad (25.11)$$

where P_{in} is the input power; P_{out} is the output power at the desired output port; P'_{out} is the output power at the undesired output port; $n = \{1, 2\}$ stands for the order of output ports O_1 and O_2; $m = \{0, 1, 2, 3\}$ stands for the order of the guided modes when propagating through the device.

Because of the chromatic dispersion, the mode coupling and mode confinement efficiencies of the optical field are dependent on the operation wavelength. We consider wavelength responses of the insertion loss and crosstalk in the window of 1.525–1.575 µm for the propagation states from the input I_1 to the output ports O_1

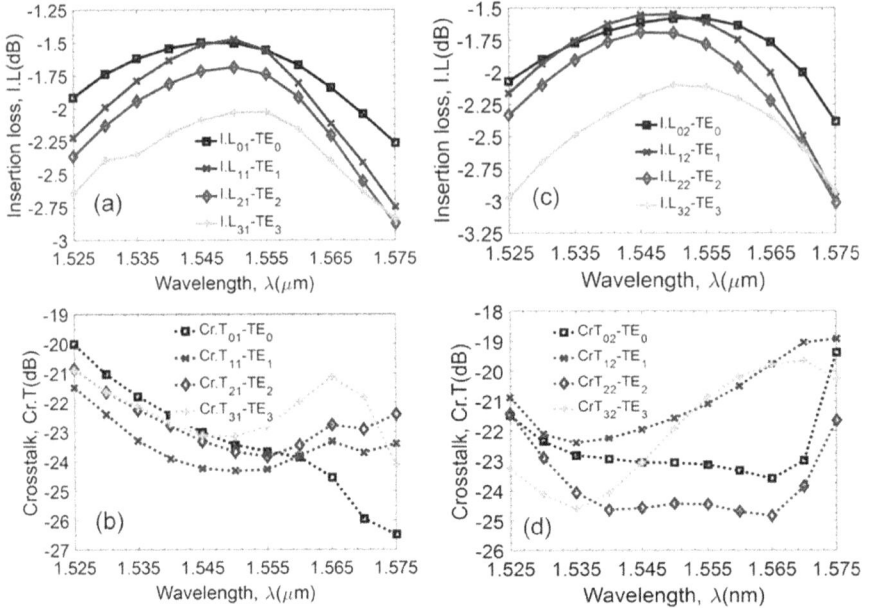

FIGURE 25.7 The wavelength-dependent characteristics curves of the optical performance of the device. (a) Insertion loss (I.L) and (b) crosstalk (Cr.T) for four modes according to the ON switch mechanism from the input I_1 to the output O_1. (c) Insertion loss and (d) crosstalk according to the OFF switch mechanism from the input I_1 to the output O_2.

(Figure 25.7a and b) and O_2 (Figure 25.7c and d). As can be seen from these figures, I.L varies from −1.5 to −3 dB and Cr.T varies from −19 to −27 dB for both cases of ON and OFF states. This low crosstalk demonstrates the excellent performance of the proposed multimode switch in a relatively wide bandwidth of 50 nm indicating advantages of the device in terms of low loss, low crosstalk, and wideband. The MMI couplers constructed for the crossover couplers and the 3-dB couplers have relatively flat responses as a function of the wavelength-spectral response, which the highest transmission attains at the central wavelength of 1,550 nm, as shown in Figure 25.3a and b. Besides, the suggested multimode switch utilized Y-multijunctions resulting in different wavelength responses for different guided modes. In addition, the operation of the multimode switch depends on the phase shifters. However, the phase difference in the phase shifters is a function of dn/dT, and therefore, the required phase difference (π radians) is temperature-dependent and the wavelength-dependent. For each different guided mode at each different wavelength, the temperature needed to transit the phase angle of π radian is different. The proposed device is optimized for working at a central wavelength of 1,550 nm. As a result, accumulative transmissions are vigorously rolled off following the wavelength response around the central wavelength of 1,550 nm, thus limiting the wavelength bandwidth, as shown in Figure 25.7. Noted that, controllable phase shifters are only optimized for the central wavelength 1,550 nm, and the 50-nm bandwidth is suitable for an optical switch.

Besides I.L and Cr.T, the figure-of-merit (FOM) criterion is also an important parameter to evaluate device performances. It is well known that a single-mode system can achieve a good performance of transmission in a wideband; however, there is a limit in spectral efficiency. In contrast, a multichannel system like the MDM–WDM hybridization system has to sacrifice the optical performances for spectral efficiency because of issues on nonlinear effects, high modal crosstalk, and multimode dispersion. Therefore, a photonic integrated circuit (PIC) applying for the MDM–WDM systems needs to consider the optimized performance in order to get a high FOM value for two aspects of wavelength range and number of guided modes. There are some different definitions of FOM used in designing the PICs, such as FOM standards for the TO switches versus material[42], consumption power[43], geometrical size[44], etc. However, in this design, our FOM standard is applied for optimizing optical performances on the number of guided mode and wavelength, which is defined as f ollows[45]:

$$FOM = 1 - (1-\alpha) \cdot \frac{1}{2M} \cdot \sum_{n=0}^{3} \left| x_n - \frac{1}{2} \right| - \alpha \cdot \frac{1}{2M} \cdot \sum_{n=0}^{3} |y_n|, \qquad (25.12)$$

where M is the wavelength resolution scanned in the range from 1.525 to 1.575 μm. α is the factor with the value is within the range ($0 \leq \alpha \leq 1$) as a trade-off between I.L and Cr.T; x, y are corresponding to I.L and Cr.T, respectively.

Equation (25.12) satisfies the following inequality: $0 < FOM \leq 1$. It will be ideal if FOM reaches unity. Figure 25.8 shows that FOM is ideal at the operation wavelength of 1,550 nm. In the wavelength range of 1,525–1,575 nm, FOM can get the value from 0.92 to 1 for the switching state I_1O_1 (ON state) and from 0.88 to 1 for the switching state I_1O_2 (OFF state). This demonstrates the high performance of the device in the wideband of 50 nm of the third telecom windows, especially at the central wavelength of 1,550 nm.

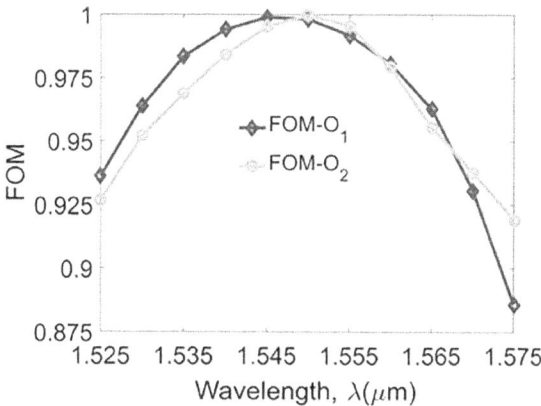

FIGURE 25.8 Figure-of-merit. Simulated FOM as a function of the operating wavelength for two states: ON, output O_1 and OFF, output O_2 in the wavelength window of 50 nm.

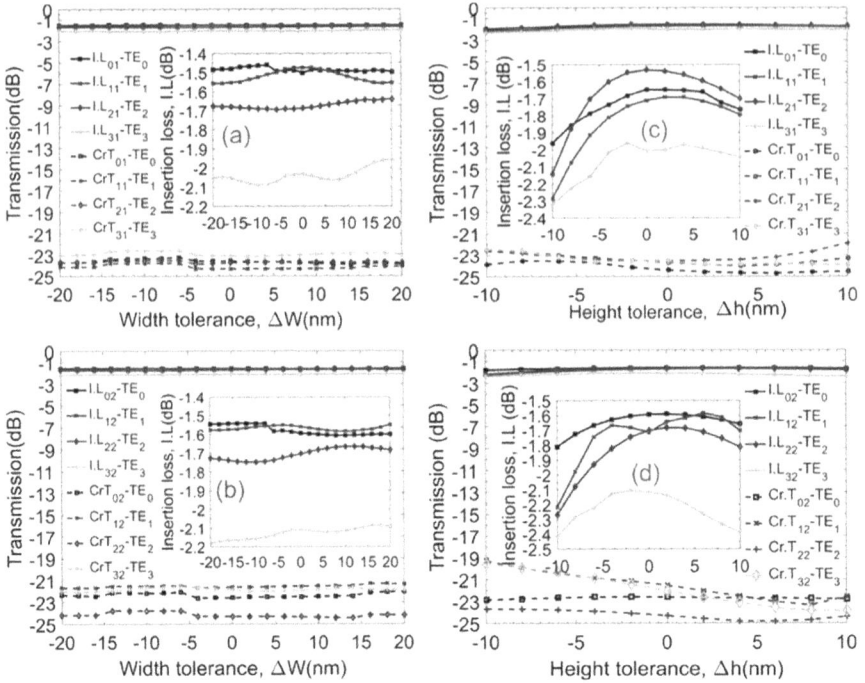

FIGURE 25.9 Insertion loss and crosstalk of the proposed device for four operation modes. For the ON state (a), from the input I_1 to the output O_1 and OFF state (b), from the input I_1 to the output O_2, as a function of a width tolerance ΔW. For the ON (c) and OFF (d) states as a function of height tolerance Δh.

Fabrication tolerances are also important for designing photonic devices, especially for simulation-based designs. Because errors of material and geometrical parameters strongly affect the working performances. For example, the quality of SOI wafer depends on suppliers in terms of geospatial tolerances, high wall roughness, the purity of silicon, crystallinities, etc. Besides, the accuracy of fabricated patterns strongly depends on electron-beam writing or deep ultraviolet (DUV) photolithography technologies. Also, the simulation tolerances are depending on the accuracy of the simulation models as well as the simulation algorithms. Therefore, we investigated the tolerances for the proposed switch in terms of the waveguide width and waveguide height. The width tolerances ΔW are shown in Figure 25.9a and b for two outputs corresponding to the ON and OFF states when the guided modes are injected into the input I_1 with a change of the width within ± 20 nm. Simulated data shows that I.L fluctuates very little around -1.4 dB to -1.7 dB while Cr.T keeps stable with the value is smaller than -21 dB for four guided modes. Also, the high tolerances, Δh, presented in Figure 25.9c and d show I.L fluctuates around -2 dB and Cr.T is less than -19 dB for both switching states ON and OFF. These tolerances are investigated in the variation of the height tolerances equally as ± 10 nm. Such relatively high tolerances on the aspects of geometrical dimensions are acceptable thanks to the current advancement of fabrication technology in the state-of-the-art electron-beam writing.

The power consumption of the TO phase shifters is often evaluated by the essential power to achieve a phase shift of π radians (P_π). This is a crucial measure parameter in the operation of a thermo-optic optical switch. It is desirable to obtain the product of $P_\pi.\tau = H.\Delta T_\pi$ as an optimal value during the operation process of the optical switch[36]. Here, H stands for the heat capacity, ΔT_π is the temperature change from a cold state to a hot state to attain the expected phase shift, and τ is the switching time related to the fall time or the rise time in the phase shifter temporal response. The switching electric consumption power is determined by the following equation utilizing a modified two-dimensional treatment of the heat flow on the lateral spreading[46]:

$$P_\pi = \frac{\lambda \kappa_{SiO_2}\left(\frac{W_{PS}}{h_{SiO_2}} + 0.88\right)}{\left|\frac{dn}{dT}\right|_{Si}} \tag{25.13}$$

where $\kappa_{SiO_2} = 1.4$ W/(m K) is the thermal conductivity of SiO_2, λ is the operation wavelength, and W_{PS} is the width of the Ti-metal film on the lateral direction. The switching time can be considered as a consequence of the required transport time of the heat flow propagation from the micro-heater to the silicon core layer along the active length of the TO phase shifter relating to the consumption power by the relation[47,48]:

$$\tau = \frac{\pi \lambda \rho_{SiO_2} C_{SiO_2} A}{e P_\pi \left|\frac{dn}{dT}\right|_{Si}}, \tag{25.14}$$

where $\rho_{SiO_2} = 2.203$ g/cm³ is density of silica, $C_{SiO_2} = 0.703$ J/(g K) is specific heat capacity of silica; $A = (2L_{th} + W_{PS})(h_{Si} + h_{SiO_2})$ is the effectively heated cross-section area relating to geometry parameters of the TO phase shifter; $h_{Si} = 220$ nm is the height of the silicon core; L_{th} is the thermal diffusion length measured by taking the distance where the maximum temperature laterally decreased at $1/e^2$ away from the silicon waveguide.

From Eqs. (25.13), (25.14) the switching time can be deduced as:

$$\tau = \frac{\pi \lambda \rho_{SiO_2} C_{SiO_2} (2L_{th} + W_{PS})(h_{Si} + h_{SiO_2})}{e \kappa_{SiO_2}\left(\frac{W_{PS}}{h_{SiO_2}} + 0.88\right)}, \tag{25.15}$$

Besides, the trade-off between power consumption and switching speed is unavoidable as well as integrated size and long range propagation. For the TO optical switch, switching time is designed for several microseconds, and therefore the electric power consumption should be kept in a few tens of m W[40]. Figure 25.10 exhibits the simulated electric power consumption and the switching time as a function of the distance h_p. As can be seen, the electric power needed to achieve a phase shift of π radians increase when the distance h_p increases. However, the switching time increases to the

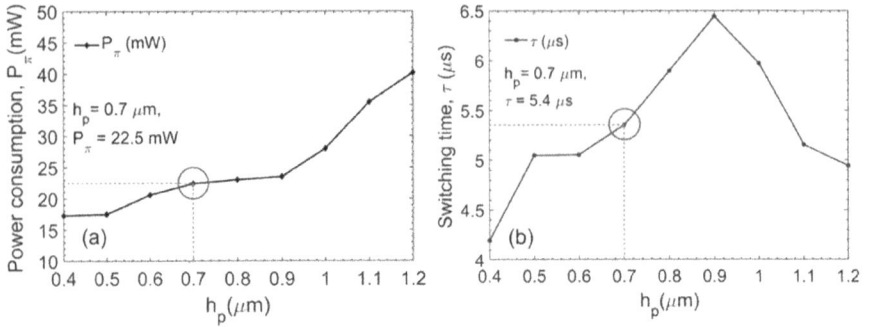

FIGURE 25.10   Electric power consumption and switching time. The simulated power consumption (a) at the switching level and the switch time (b) of the phase shifter as a function of the distance h_p between the Ti-heater and the silicon core layer. The circles indicate the values at $h_p = 0.7$ µm corresponding to $h_{SiO_2} = 540$ nm.

maximum value of 6.4 µs at h_p to 0.9 µm and decreases when h_p increases further. At the selected physical distance of $h_p = 0.7$ µm (or $h_{SiO_2} = 0.54$ µm), P_π obtained a relatively small value of 22.5 mW with a switching time of 5.4 µs.

To reduce the power consumption P_π as well as the desired temperature difference (ΔT_π) in the switching operation, the length of the thermo-optic phase shifters must be increased, thus making the PS length larger and also switching time slower. Furthermore, the power consumption can be reduced when the gap between the metallic heater and the silicon core decreases but should be properly chosen minimizing the impact of the plasmonic effect on the propagation of lightwave in term of phase shift and optical loss. In the plasmonic regime, the p-polarized optical modes (TM modes) is confined along the metal–dielectric interface. Hence, the Ti-heater strongly affects the TM modes than the TE modes in term of hybrid plasmonic effect when the gap between the Ti-heater and the silicon core is smaller than a limit. This is depicted in Figure 25.11 where mode profiles of both TE and TM polarization states are simulated by the finite element method (FEM) for several gaps of h_{SiO_2} corresponding to 540, 30, and 10 nm at the wavelength $\lambda = 1,550$ nm. Simulation results show that the plasmonic mode is significant when $h_{SiO_2} \leq 30$ nm for TM modes and only becomes significant when $h_{SiO_2} \leq 10$ nm for TE modes. When optical fields are in the plasmonic modes, the imaginary part of the dielectric constant becomes larger leading to a larger conductive absorption. This effect restricts the propagation length of the optical fields in the photonic device. In our proposed optical switch, which was designed for the operation of TE modes with the distance $h_p = 700$ nm ($h_{SiO_2} = 540$ nm), the impact of plasmonic effect can be neglected because the optical fields are always preserved in photonic modes.

25.3   DISCUSSION

The proposed device is a 2×2 multimode switch enabling the simultaneously four-mode switching operation in a multimode waveguide for the MDM and hybrid

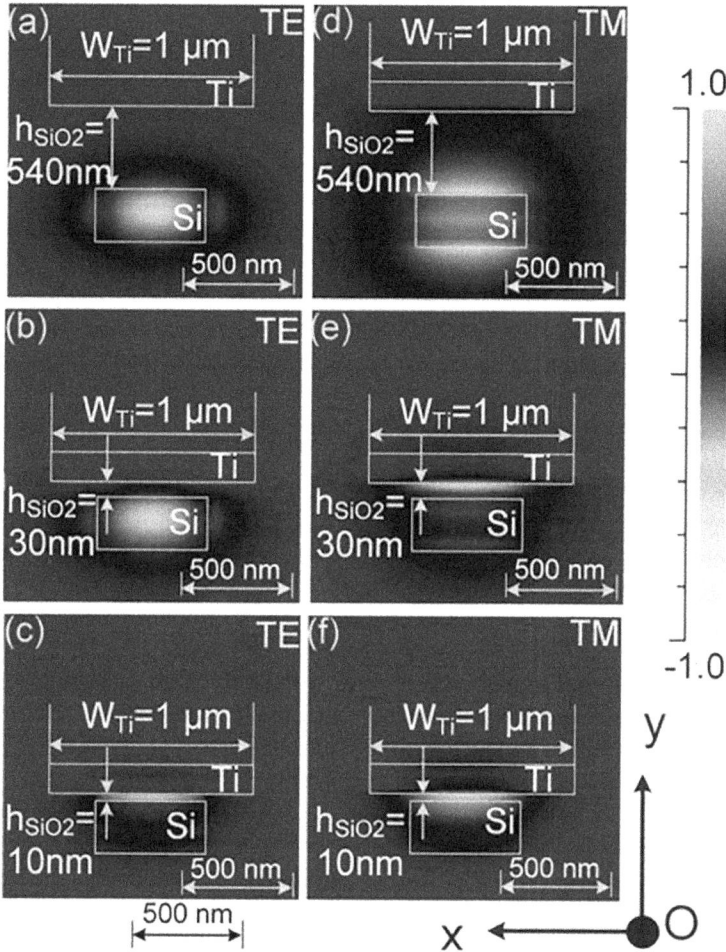

FIGURE 25.11 Mode profile ($|E_y|$) in Ti-heater-based phase shifter. Modes simulated by the FEM method for both TE and TM polarization states at some gaps, h_{SiO_2}, of 540, 30, and 10 nm at $\lambda = 1,550$ nm.

WDM–MDM applications. In order to couple the switch with the WDM–MDM system, grating couplers[49,50] or edge-couplers[51,52] are required to couple high-order modes from the few-mode fibers to the silicon waveguides for realizing the functionality of a wavelength division multiplexed few-mode fiber switch. In another way, sub-wavelength gratings (SWG) consisting of periodically arranged dielectric particles with dimensions much smaller than the wavelength are utilized for highly efficient coupling from single-mode fibers to a silicon single-mode waveguide-based chip in a compact footprint over a broad wavelength bandwidth[53,54]. Then, fundamental modes in single-mode silicon waveguides are multiplexed into a multimode waveguide and vice versa by mean of MMI couplers[41,55], ring resonators[3,56], or adiabatic couplers[57] before connecting to the multimode switch. Finally, single-mode optical fibers are coupled to the WDM system via the WDM multiplexer following the ITU-T G.694.1 recommendation to complete a hybrid WDM–MDM system. In another application scenario, single-mode optical fibers carrying each dedicated wavelength are coupled to each dedicated silicon single-mode waveguide via curved Bragg gratings allowing the propagation of light in the third telecom windows of 1,550-nm at first. Then, a set of individual wavelengths are multiplexed into a single-mode silicon waveguide carrying the total traffic of a WDM-channels group via an arrayed waveguide grating (AWG)[58]. Next, each group of the single-mode silicon waveguide is multiplexed to a multimode bus waveguide employing the phase-matched technique as adiabatic couplers[57]. Finally, four groups into a four-mode waveguide are connected to an input/output port of the designed 2×2 four-mode switch. The constructed structure can permit the switching operation of four WDM groups in the 2×2 configuration. Recently, adiabatic couplers are popularly used for multiplexing low-order modes into a multimode waveguide that such structures can carry up to dozens of optical modes in a multimode waveguide[9,59,60].

Compared to some related recent studies on multimode switches, our proposed switching device either supports the operation of four guided modes instead of two guided modes[27] or has much simpler and more compact than those reported elsewhere[61,62]. To the best of our knowledge, this novel design and demonstration by numerical simulation in the proof-of-concept of the 2×2 four-mode switch has not been proposed before. The device enables the switching operation for simultaneous four guided modes of TE polarization states without the need of high-order modes exchange and complicated spatial switching structures. Moreover, the proposed optical switch can operate as a multifunctional-multimode processing device, such as an arbitrary ratio multimode power splitter that can be realized with only one control bit for diving the multimode power simultaneously. In addition, the proposed structure potentially be used as a building block to scale up the higher number of guided modes.

25.4   CONCLUSIONS

In conclusion, we have designed and optimized a novel 2×2 four-mode optical switch using numerical simulations for the first time. The switch can operate in a 50 nm wide wavelength band from 1,525 to 1,575 nm with the low insertion loss and cross-talk that are lower than −3 and −19 dB, respectively. We also pointed out that the hybrid plasmonic effect between the metallic heater and the Si waveguide of the phase shifter does not effect to the device performance in the chosen parameters. In addition, the device has relatively large geometry tolerances correspondingly to ±20 and ±10 nm of width and height tolerances, respectively. Furthermore, the device is very compact in a small area of 10 μm×960 μm offering a huge potential for applications in very large scale integrated photonic circuits as well as in mode and WDM switching systems.

ACKNOWLEDGMENTS

Research is funded by Vingroup Innovation Foundation (VINIF) annual research grant program under Project Code VINIF.2019.DA12.

AUTHOR CONTRIBUTIONS

C.D.T. sketched the scientific idea for the project. C.D.T. and M.T.T. conceived the project. C.D.T. and D.N.T.H. carried out the simulation, characterizations, and analyzed the data. H.C. pointed out significant and relevant features on the technical aspect. C.D.T., D.N.T.H., H.C., and M.T.T. wrote the manuscript. All authors discussed the results and commented on the manuscript.

COMPETING INTERESTS

The authors declare no competing interests.

REFERENCES

1. Gu, M., Li, X. & Cao, Y. Optical storage arrays: a perspective for future big data storage. *Light Sci. Appl.* 3, e177 (2014).
2. Wu, X., Zhu, X., Wu, G.-Q. & Ding, W. Data mining with big data. *IEEE Trans. Knowl. Data Eng.* 26, 97–107 (2014).
3. Wu, X., Huang, C., Xu, K., Shu, C. & Tsang, H. K. Mode-division multiplexing for silicon photonic network-on-chip. *IEEE J. Light. Technol.* 35, 3223–3228 (2017).
4. Essiambre, R. J. & Tkach, R. W. Capacity trends and limits of optical communication networks. *Proc. IEEE* 100, 1035–1055 (2012).
5. Ellis, A. D., Zhao, J. & Cotter, D. Approaching the non-linear Shannon limit. *J. Light. Technol.* 28, 423–433 (2010).
6. Li, G., Bai, N., Zhao, N. & Xia, C. Space-division multiplexing: the next frontier in optical communication. *Adv. Opt. Photonics* 6, 413 (2014).
7. Yang, Q., Bergman, K., Hughes, G. D. & Johnson, F. G. WDM packet routing for high-capacity data networks. *J. Light. Technol.* 19, 1420–1426 (2001).
8. Tang, R. et al. Reconfigurable all-optical on-chip MIMO three-mode demultiplexing based on multi-plane light conversion. *Opt. Lett.* 43, 1798–1801 (2018).
9. Jiang, W., Miao, J. & Li, T. Compact silicon 10-mode multi/demultiplexer for hybrid mode- and polarisation-division multiplexing system. *Nat. Sci. Rep.* 9, 1–15 (2019).
10. Perez-Galacho, D. et al. Integrated mode converter for mode division multiplexing. In Proc. SPIE 9891, Silicon Photonics and Photonic Integrated Circuits V 98910B (2016).
11. Sun, C., Yu, Y., Chen, G. & Zhang, X. Integrated switchable mode exchange for reconfigurable mode-multiplexing optical networks. *Opt. Lett.* 41, 3257 (2016).
12. Chang, W. et al. Ultracompact dual-mode waveguide crossing based on subwavelength multimode-interference couplers. *Photonics Res.* 6, 660–665 (2018).
13. Han, H.-L. et al. High performance ultra-compact SOI waveguide crossing. *Opt. Express* 26, 25602 (2018).
14. Jia, H. et al. Mode-selective modulation by silicon microring resonators and mode multiplexers for on-chip optical interconnect. *Opt. Express* 27, 2915 (2019).
15. Priti, R. B. & Liboiron-Ladouceur, O. Reconfigurable and scalable multimode silicon photonics switch for energy efficient modedivision-multiplexing systems. *J. Light. Technol.* 37, 3851–3860 (2019).
16. Miller, D. A. B. Reconfigurable add-drop multiplexer for spatial modes. *Opt. Express* 21, 20220 (2013).
17. Wang, S. et al. On-chip reconfigurable optical add-drop multiplexer for hybrid wavelength/mode-division-multiplexing systems. *Opt. Lett.* 42, 2802 (2017).

18. Sherwood-Droz, N. et al. Optical 4×4 hitless Silicon router for optical networks-on-chip (NoC). *Opt. Express* 16, 15915–15922 (2008).

19. Mogami, T. et al. 1.2 Tbps/cm² enabling silicon photonics IC technology based on 40-nm generation platform. *J. Light. Technol.* 36, 4701–4712 (2018).

20. Soma, D. et al. 257-Tbit/s weakly coupled 10-mode C+L-band WDM transmission. *J. Light. Technol.* 36, 1375–1381 (2018).

21. Zhang, Y., Zhang, R., Zhu, Q., Yuan, Y. & Su, Y. Architecture and devices for silicon photonic switching in wavelength, polarization and mode. *J. Light. Technol.* PP, 1 (2019).

22. Qiao, L., Tang, W. & Chu, T. 32×32 Silicon electro-optic switch with built-in monitors and balanced-status units. *Sci. Rep.* 7, 42306 (2017).

23. Dai, D., Wang, J. & He, S. Silicon multimode photonic integrated devices for on-chip mode-division-multiplexed optical interconnects. *Prog. Electromagn. Res.* 143, 773–819 (2013).

24. Xiong, Y., Priti, R. B. & Liboiron-Ladouceur, O. High-speed two-mode switch for mode-division multiplexing optical networks. *Optica* 4, 1098 (2017).

25. Williams, C., Banan, B., Cowan, G. & Liboiron-Ladouceur, O. A source-synchronous architecture using mode-division multiplexing for on-chip silicon photonic interconnects. *IEEE J. Sel. Top. Quantum Electron.* 22, 473–481 (2016).

26. Stern, B. et al. On-chip mode-division multiplexing switch. *Optica* 2, 530 (2015).

27. Zhang, Y., He, Y., Zhu, Q., Qiu, C. & Su, Y. On-chip silicon photonic 2×2 mode- and polarization-selective switch with low intermodal crosstalk. *Photonics Res.* 5, 521–526 (2017).

28. Zhang, Y. et al. Silicon 1×2 mode- and polarization-selective switch. In *2017 Opt. Fiber Commun. Conf. Exhib. OFC 2017—Proc.* 1–3 (2017). https ://doi.org/10.1364/OFC.2017.W4E.2.

29. Benes, V. E. Algebraic and topological properties. *Bell Syst. Tech. J.* 41, 1249–1274 (1962).

30. Jia, H. et al. WDM-compatible multimode optical switching system-on-chip. *Nanophotonics* 8, 889–898 (2019).

31. Le, K. Q., Godoy-Rubio, R., Bienstman, P. & Hadley, G. R. The complex Jacobi iterative method for three-dimensional wide-angle beam propagation: erratum. *Opt. Express* 16, 21942 (2008).

32. Riesen, N. & Love, J. D. Design of mode-sorting asymmetric Y-junctions. *Appl. Opt.* 51, 2778–2783 (2012).

33. Soldano, L. B. & Pennings, E. C. M. Optical multi-mode interference devices based on self-imaging: principles and applications. *J. Light. Technol.* 13, 615–627 (1995).

34. Bachmann, M., Besse, P. A. & Melchior, H. Overlapping-image multimode interference couplers with a reduced number of self-images for uniform and nonuniform power splitting. *Appl. Opt.* 34, 6998–6910 (1995).

35. Rickrnan, A. G. & Reed, G. T. Silicon-on-insulator optical rib waveguides: loss, mode characteristics, bends and y-junctions. *IEE Proc. Optoelectron.* 141, 391–393 (1994).

36. Rosa, Á., Gutiérrez, A., Brimont, A., Griol, A. & Sanchis, P. High performance silicon 2×2 optical switch based on a thermo-optically tunable multimode interference coupler and efficient electrodes. *Opt. Express* 24, 191 (2016).

37. Komma, J., Schwarz, C., Hofmann, G., Heinert, D. & Nawrodt, R. Thermo-optic coefficient of silicon at 1550 nm and cryogenic temperatures. *Appl. Phys. Lett.* 101, 4–8 (2012).

38. Harris, N. C. et al. Efficient, compact and low loss thermo-optic phase shifter in silicon. *Opt. Express* 22, 10487 (2014).

39. Mishra, D. & Sonkar, R. K. Design and analysis of a graded-index strained $Si_{1-x}Ge_x$ optical PN phase shifter design and analysis of a graded-index strained $Si_{1-x}Ge_x$ optical PN. *IEEE Photonics J.* 10, 1–14 (2018).

40. Yang, L. et al. General architectures for on-chip optical space and mode switching. *Optica* 5, 180 (2018).

41. Priti, R. B. & Liboiron-ladouceur, O. A reconfigurable multimode demultiplexer/switch for mode-multiplexed silicon photonics interconnects. *IEEE J. Sel. Top. Quantum Electron.* 24, 1–10 (2018).

42. Vo, W. & Bragg, F. Modeling of a vertical hybrid plasmonic switch. *IEEE Photonics Technol. Lett.* 30, 997–1000 (2018).

43. Watts, M. R. et al. Adiabatic thermo-optic Mach–Zehnder switch. *Opt. Lett.* 38, 733 (2013).

44. Jacques, M. et al. Optimization of thermo-optic phase-shifter design and mitigation of thermal crosstalk on the SOI platform. *Opt. Express* 27, 10456–10471 (2019).

45. Chang, W. et al. Inverse design and demonstration of an ultracompact broadband dual-mode 3 dB power splitter. *Opt. Express* 26, 24135 (2018).

46. Liu, K., Zhang, C., Mu, S., Wang, S. & Sorger, V. J. Two-dimensional design and analysis of trench-coupler based Silicon Mach–Zehnder thermo-optic switch. *Opt. Express* 24, 15845 (2016).

47. Passaro, V. M. N., Magno, F. & Tsarev, A. V. Investigation of thermo-optic effect and multi-reflector tunable filter/multiplexer in SOI waveguides. *Opt. Express* 13, 3429 (2005).

48. Espinola, R. L., Tsai, M. C., Yardley, J. T. & Osgood, R. M. Fast and low-power thermooptic switch on thin silicon-on-insulator. *IEEE Photonics Technol. Lett.* 15, 1366–1368 (2003).

49. Wohlfeil, B., Stamatiadis, C., Zimmermann, L. & Petermann, K. Compact fiber grating coupler on SOI for coupling of higher order fiber modes. In *2013 Optical Fiber Communication Conference and Exposition and the National Fiber Optic Engineers Conference (OFC/NFOEC), Anaheim, CA* 1–3 (2013). https ://doi.org/10.1364/ofc.2013.oth1b.2.

50. Watanabe, T. et al. Coherent few mode demultiplexer realized as a 2D grating coupler array in silicon. *Opt. Express* 28, 36009–36019 (2020).

51. Melati, D., Alippi, A. & Melloni, A. Reconfigurable photonic integrated mode (de)multiplexer for SDM fiber transmission. *Opt. Express* 24, 12625 (2016).

52. He, A., Guo, X., Wang, K., Zhang, Y. & Su, Y. Low loss, large bandwidth fiber-chip edge couplers based on silicon-on-insulator platform. *J. Light. Technol.* 38, 4780–4786 (2020).

53. Cheben, P., Halir, R., Schmid, J. H., Atwater, H. A. & Smith, D. R. Subwavelength integrated photonics. *Nature* 560, 565–572 (2018).

54. Halir, R. et al. Subwavelength-grating metamaterial structures for silicon photonic devices. *Proc. IEEE* 106, 2144–2157 (2018).

55. Uematsu, T., Ishizaka, Y., Kawaguchi, Y., Saitoh, K. & Koshiba, M. Design of a compact two-mode multi/demultiplexer consisting of multimode interference waveguides and a wavelength-insensitive phase shifter for mode-division multiplexing transmission. *J. Light. Technol.* 30, 2421–2426 (2012).

56. Luo, L.-W. et al. WDM-compatible mode-division multiplexing on a silicon chip. *Nat. Commun.* 5, 1–7 (2014).

57. Chenlei Li, D. D. Low-loss and low-crosstalk multi-channel mode (de)multiplexer with ultrathin silicon waveguides. *Opt. Lett.* 42, 2370–2373 (2017).

58. Proietti, R., Cao, Z., Nitta, C. J., Li, Y. & Yoo, S. J. B. A scalable, low-latency, high-throughput, optical interconnect architecture based on arrayed waveguide grating routers. *J. Light. Technol.* 33, 911–920 (2015).

59. Beppu, S. et al. Weakly coupled 10-mode-division multiplexed transmission over 48-km few-mode fibers with real-time coherent MIMO receivers. *Opt. Express* 28, 19655–19668 (2020).

60. Dai, D. et al. 10-Channel mode (de)multiplexer with dual polarizations. *Laser Photonics Rev.* 1700109, 1–9 (2017).

61. Priti, R. B., Bazargani, H. P., Xiong, Y. & Liboiron-Ladouceur, O. Mode selecting switch using multimode interference for on-chip optical interconnects. *Opt. Lett.* 42, 4131–4134 (2017).

62. Jia, H. et al. Four-port mode-selective silicon optical router for on-chip optical interconnect. *Opt. Express* 26, 9740–9748 (2018).

26 Three-Dimensional Fabry–Pérot Cavities Sculpted on Fiber Tips Using a Multiphoton Polymerization Process[*]

Jonathan W. Smith and Jeremiah C. Williams
US Air Force Institute of Technology

Joseph S. Suelzer and Nicholas G. Usechak
US Air Force Research Laboratory

Hengky Chandrahalim
US Air Force Institute of Technology

CONTENTS

26.1 INTRODUCTION

The Fabry–Pérot (FP) cavity is an important optical component with many applications. A basic FP cavity consists of two parallel reflective surfaces separated by a chosen distance and encapsulating air, vacuum, or another media with refractive

[*] Reprinted with permission: Smith, J. W., Williams, J. C., Suelzer, J. S., Usechak, N. G., & Chandrahalim, H. (2020). Three-dimensional Fabry–Pérot cavities sculpted on fiber tips using a multiphoton polymerization process. *Journal of Micromechanics and Microengineering*, *30*(12), 125007.

DOI: 10.1201/9781003220978-33

index (RI) n. Multiple beam interference between the two surfaces causes transmission through the cavity to peak at specific wavelengths of maximum coherent interference, while others are reflected. At the micron-scale, this enables the FP cavity to propagate a small number of optical modes compared to other optical cavities such as ring resonators, photonic crystals, and distributed feedback gratings [1]. The FP cavity can also achieve large quality factors, with values as high as 10^5 reported [2]. It is easily accessible to the environment and, unlike devices such as the ring resonator, the FP cavity does not require the substance inside the cavity to have a different RI than the substance outside the cavity [3]. While often beneficial, the open nature of the FP cavity means it lacks lateral confinement and loses some resonant light off the edges of the mirrors. Flat FP cavities are highly sensitive to misalignment, and any misalignment, even one of several degrees, between the mirrors will significantly lower a cavity's quality factor [1]. One popular way to overcome this sensitivity is by using one or more curved mirrors [3–5], although this often increases the complexity of fabrication. This work introduces an innovative fabrication process that greatly simplifies the realization of complex geometries on virtually any substrate.

Many advantages of the FP cavity have made it a key component of a myriad of applications. When used to form a laser cavity, a variety of exotic gain media have recently been explored including biological tissues [6], silicon nanowires [7], and optical fluids [1,3]. Miniaturized tunable lasers [8,9] and tunable optical filters [10] have also been realized by integrating an FP cavity with microelectromechanical systems. The accessibility of the cavity has also made it a powerful tool for spectroscopy. It has been used in on-chip microfluidics [11], human breath analysis [12], interrogation of living cells [13], and compact imaging spectrometers [14]. The FP cavity is also set to play a key role in the emerging field of quantum computing, with cavity quantum electrodynamics at the forefront of many advances. It has been demonstrated in a photon emission source [4,15], in strong coupling to a trapped atom [16], and in frequency splitting of polarization Eigen modes [17,18].

The difference between two resonant wavelengths in a FP cavity, the cavity's free spectral range (FSR), is determined by the distance between the mirrors and the refractive index of the medium inside the cavity. Sensors can detect phenomena that affect these factors, and have found many applications to include sensing gravitational waves [19], acceleration [20], pressure and refractive indices of liquids [21], temperature [22], force [23], and even gas composition [24].

Optical fibers present a powerful platform to both form and interrogate FP cavities due to their small form factor, low-loss, and well-behaved transverse optical mode structure. Promising applications for fiber-integrated FP cavities include optofluidic in-fiber lasers [25,26] and miniaturized high-sensitivity sensors [21–24]. Poor lateral confinement and misalignment sensitivity continue to plague fiber-based FP cavity devices and represent significant design challenges. The fiber itself is also an exotic substrate due to its geometry, which renders it incompatible with many planar microfabrication processes. A variety of techniques have been explored to overcome these challenges and create FP cavities on optical fibers. One device was fabricated by splicing a segment of hollow-core optical fiber (HOF) to a single-mode fiber (SMF) and capping the HOF with a segment of photonic crystal fiber (PFC) [24]. While this design can interrogate gasses, liquids would have difficulty reaching the cavity through the small openings in the PFC. Splicing various types of optical fibers also requires precise alignment and may be difficult to repeat reliably. Another successful on-fiber FP resonator was made by ion milling a cavity into a tapered SMF probe [22]. While the environment is easily accessed by this cavity, the fabrication process is complex and laborious, involving CO_2 laser pulling and metal deposition before the ion milling. Another group used the photoactive polymer SU-8 to construct a suspended polymer cavity on a fiber tip [21]. The resulting device can interrogate liquid or gas, and the fabrication process enables 2D freedom with a digital mirror. But this process is also relatively complicated, requiring a spray coat, bake, and UV exposure for each individual layer. This also limits the 3D structures that can be realistically built.

The devices presented in this work were fabricated using a simple process that requires only mounting the fiber into a two-photon polymerization (2PP) system from Nanoscribe GmbH and chemical development [27]. This technique enabled us to realize 3D free-form geometries—a feat that cannot be accomplished using other methods on this spatial scale. This enables the use of nonplanar components to improve device performance, such as in our use of curved mirrors to create a hemispherical FP cavity to significantly reduce misalignment susceptibility. Our method can create these 3D components with submicron precision. The three on-fiber FP cavity designs that were fabricated and tested are depicted in Figure 26.1. The hemispherical device achieved the greatest extinction ratio of the three designs tested and highlights the power of the design freedom afforded by this process. The unreleased device demonstrated an extinction ratio of around 1.90, the released device achieved 61, and the hemispherical device achieved 253, providing a strong signal to observe changes in the FSR of the device. We were also able to extract the reflectance of the photopolymer by fitting an Airy distribution to the reflection spectrum. This yielded a reflectance between 0.2 and 0.3 for the polymerized resin. The dual-cavity devices allow for interrogation of an interstitial medium in the

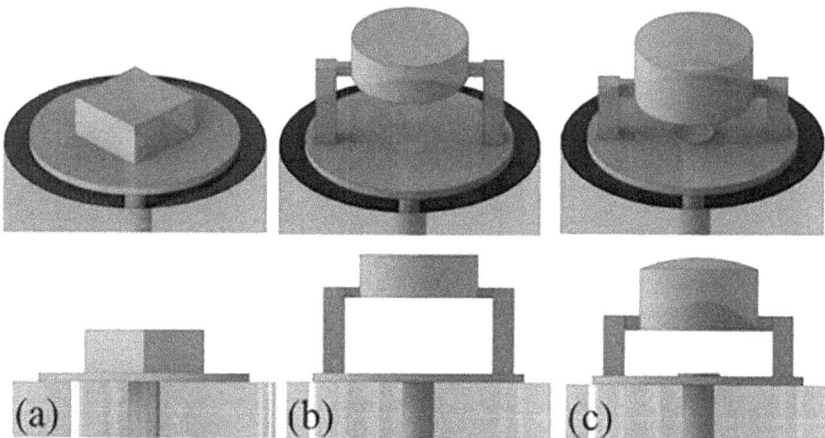

FIGURE 26.1 Schematic views of the FP cavity devices. (a) The single-cavity, (b) the released dual-cavity, and (c) the released hemispherical dual-cavity devices on cleaved ends of optical fibers.

first, open cavity while simultaneously referencing the static reflection spectrum of the second, solid polymer cavity. These advanced features, which are very difficult or impossible to achieve with traditional planar microfabrication techniques, were fabricated in a single patterning step. The speed and simplicity of fabrication enable rapid prototyping and iterative design processes to realize complicated devices and advanced features.

26.2   FABRICATION PROCESS

The maskless 2PP fabrication process used to fabricate all devices demonstrated in this work is outlined in Figure 26.2. First, the optical fiber was stripped, cleaned, and cleaved to create a flat platform with access to the core of the fiber, as illustrated in Figure 26.2a. The optical fiber used in this work was F-SM1500-9/125-P fiber from Newport Corporation. The cleaved fiber was secured into a Newport FPH-S fiber chuck and mounted into a custom jig that aligned the cleaved fiber face orthogonally to the laser aperture of the Nanoscribe GmbH system. The

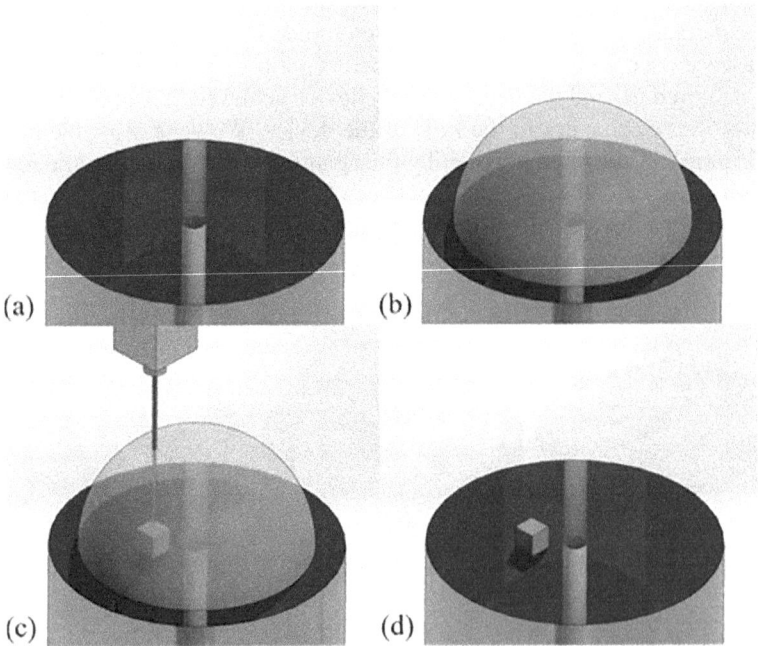

FIGURE 26.2   An example of the two-photon polymerization microfabrication process flow (devices fabricated in this work were centered on the fiber core). (a) The fiber tip was properly cleaned, cleaved, and mounted on a laser machining station. (b) Photosensitive IP-DIP resin (marketed by Nanoscribe GmbH) was deposited on the fiber tip. (c) A femtosecond laser was then focused in the resin to polymerize portions of resin layer by layer. (d) A chemical developer was used to remove non-polymerized resin, releasing the solidified structure.

uncured, liquid photoactive polymer resin (Nanoscribe's IP-DIP) was deposited onto the cleaved fiber face, as shown in Figure 26.2b. The resin can be deposited in any thickness or shape that encloses the desired build volume. Thus, several traditional photoresist deposition steps, such as spin-coating and pre-baking, were eliminated.

Once mounted, the resin was selectively exposed to ultrashort laser pulses with a wavelength of 780 nm, a repetition rate of 80 MHz, and a pulse duration of 120 fs by the Nanoscribe GmbH laser writing system. The resin only solidified when subjected to the nonlinear optical process of 2PP. Simultaneous absorption of two photons was necessary to polymerize the resin, which only occurred in a small portion of the focused laser beam [28]. The volume of the beam initiating 2PP can be scaled to offer a balance of resolution and speed. The minimum volume for this system was 150 nm wide by 150 nm long by 200 nm tall. This focal point was scanned through the resin according to a computer-aided design file to solidify the desired structure. This is pictured in Figure 26.2c. This system has a maximum scan speed of 2 mm/s, but small features and optical quality curvatures require significantly slower scan speeds. The FP cavity devices in this work were fully polymerized in less than 15 min. The x–y aspects of each layer were controlled by a galvanometer, while the z direction was controlled with a piezoelectric actuator.

Once the desired volume had been polymerized, the fiber tip was submerged in propylene glycol methyl ether acetate (PGMEA) for 20 min. This common solvent removed the unexposed resin, releasing the polymerized structure. Finally, the fiber tip was submerged in isopropanol for another 20 min to clean off the PGMEA. The result was the desired 3D structure of polymerized resin as illustrated in Figure 26.2d.

While significantly faster than other fabrication methods, the stepwise nature of the laser scanning process introduced striations into the surface finish of the devices. Planar FP cavities require a flat reflective surface, and hemispherical FP cavities require a smooth spherical mirror, and it was not known if the devices created here had an optical quality surface finish. Also of concern, features with a height equal to one-half or one-quarter of the wavelength of interest could introduce destructive interference and create an anti-reflective surface. To analyze the surface finish, we fabricated a sample structure onto an indium-tin-oxide-coated glass slide to mount into an atomic force microscope (AFM).

The resultant AFM scans are included in Figure 26.3. The expected striations from the scanning process were present at regular intervals. The surface finish, including these features, was found to have a roughness of approximately 60 nm, with the peak-to-peak difference averaging 120 nm. This work focused on using wavelengths in the 1,460–1,640 nm range to probe the FP structures fabricated on the fiber ends. Therefore, the fabrication variations in surface roughness are significantly smaller than the wavelengths of interest, and far less than one-half or one-quarter wavelength interval which would lead to their own interference effects. In fact, the structures fabricated by this process were confirmed to have a roughness below $\lambda/10$ which is consistent with an optical quality surface finish in the wavelengths of interest.

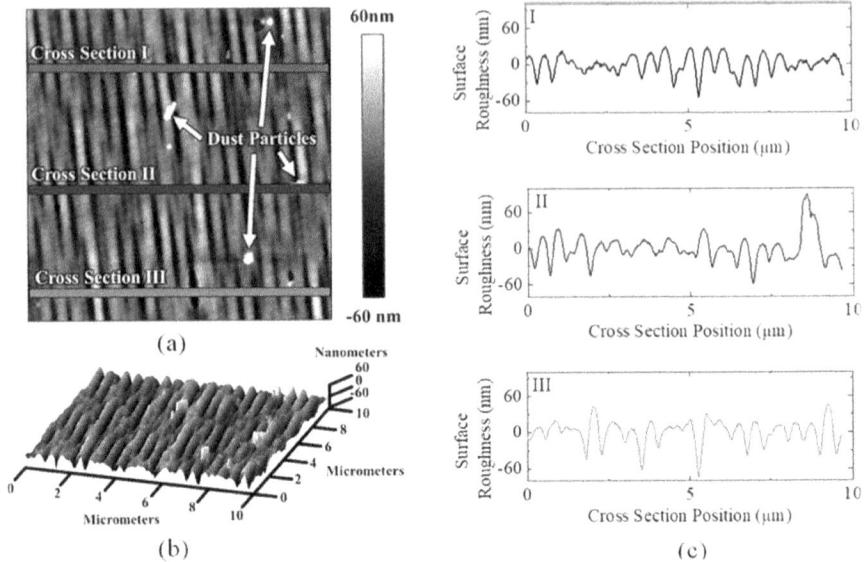

(a)

(b)

(c)

FIGURE 26.3   Surface analysis of an optical flat fabricated with two-photon polymerization by ultrashort laser pulses. (a) AFM image of the surface interrogated. (b) A 3D rendering of the AFM scan showing surface topography. (c) Three cross sections throughout the surface to quantify roughness.

26.3   DEVICE CHARACTERIZATION

26.3.1   MEASUREMENT SETUP

The FP cavities were characterized according to the measurement setup pictured in Figure 26.4. All experiments were performed in a temperature-controlled laboratory that has a temperature of 20°C with fluctuations of less than 2°C. A tunable laser source (Agilent 81600B) was connected to the first port of an optical circulator. This was swept from 1,463 to 1,634 nm during each measurement. The SMF with an FP cavity device was fusion spliced to another SMF terminating in an FC/APC connector using a Fujikura FSM-100P ARC Master fiber splicer. This was connected to the second port of the optical circulator. The third port of the optical circulator was connected to a Newport universal fiber optic detector. This photodetector interfaced with a Newport 1830-C optical power meter, whose output was visualized and stored using a Keysight D509254A digital storage oscilloscope.

The variable laser source was then swept from 1,463 nm to 1,634 nm and the reflection from the FP cavity device was isolated by the optical circulator. The photodetector and power meter transduced the optical power into a voltage that was recorded by the oscilloscope. Optical resonances within the FP cavity caused a peak in transmission through the cavity which was observed as a dip in reflection intensity. This technique allowed the devices to be used remotely, with the bulky input and measurement components geographically separated from the device. In this work, the optical circulator and input fiber were polarization maintaining, while the device fiber (and

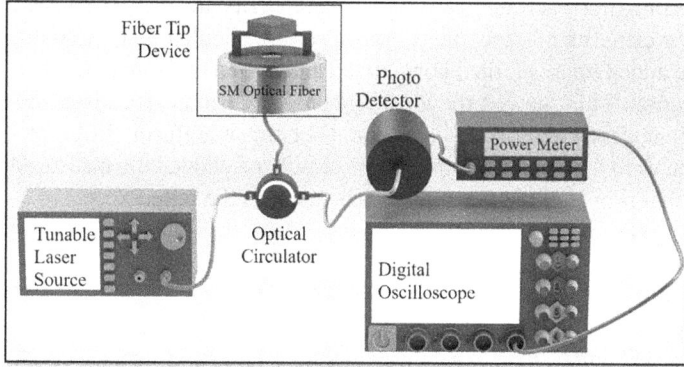

FIGURE 26.4   Diagram of the experimental setup used to characterize the reflection spectrum of each device in air at room temperature.

the fabricated FP cavities) were not. Polarization maintaining fibers can be used to insure operation using a single polarization which could prove important for very long standoff detectors where polarization mode dispersion could reduce fringe visibility.

The reflection spectrum of each device was measured in volts read by the oscilloscope at a given wavelength. The extinction ratio reported for each device was calculated using the ratio between the mean of the four lowest reflection dips and the mean of the four highest reflection peaks. The mean of the four highest peaks is referred to as the high reflection intensity, and the mean of the four lowest dips is referred to as the low reflection intensity. Assuming incident light normal to each cavity, the theoretical FSR of an FP cavity is calculated according to $\Delta\lambda_{FSR} = \lambda_0^2/2nl$, where λ_0 is the central wavelength of the transmission peak (and reflection dip), n is the RI of the cavity medium, and l is the length of the cavity. For our devices, we considered a hypothetical transmission peak at 1,550 nm, an IP-DIP refractive index of 1.504, and a refractive index of 1 in air. The RI was calculated by interpolating data provided by Nanoscribe, and 1550 nm is a common telecom wavelength in the middle of our laser's bandwidth. All calculations assume room temperature.

The transmission through a FP resonator can be modeled by the Airy distribution, which calculates the internal resonance enhancement factor for light of a given wavelength based on the physical properties of the cavity [29]. The generic Airy distribution for two mirrors of equal reflectance is calculated with, $A = \left[(1-R)^2 + 4R\sin^2(\phi)\right]^{-1}$, where R is the reflectance of the mirrors, and 2ϕ is the single-pass phase shift between the mirrors [29]. This is calculated with, $2\phi = 2\pi v/\Delta v_{FSR} \approx 2\pi\lambda / \Delta\lambda_{FSR}$. The intensity of light reflected back from the cavity, as was measured in this work, is inversely proportional to the transmission intensity.

We extracted the reflectance of the mirrors in our devices by fitting an Airy distribution to the measured reflection spectrum. To create a comparable waveform, we selected the FSR and first λ_0 from our measurements, and centered the phase shift at the initial resonant wavelength by subtracting it from λ to determine the single-pass phase shift in relation to the resonant wavelength, $2\phi' = 2\pi(\lambda - \lambda_0)/\Delta\lambda_{FSR}$. The distribution was also normalized and scaled to the maximum and minimum voltage readings for each device. For the dual-cavity devices, the FSR and initial resonant wavelengths of each cavity were used to calculate two Airy distributions, which were added together, then normalized and scaled to the magnitude of the measured reflection. This showed the ideal response of each device given the measured FSR, resonant wavelength, and magnitude. With this waveform, different values of R were chosen until the magnitude and shape closely resembled the measured response. The value that provided the best match was taken as the reflectance.

26.3.2 Measurement Results

A scanning electron microscope (SEM) image of the fabricated single-cavity device is presented in Figure 26.5a, while the measured reflection intensity is presented in Figure 26.5b. The cavity was formed by a 17.58 μm long, 40 μm by 40 μm rectangle, resulting in a theoretical FSR of 45.43 nm. The measured average $\Delta\lambda_{FSR}$ was 42.09 nm, showing a variation of only 3.34 nm. The device's high reflection intensity was 19.29 μW corresponding to a voltage of 132.58 mV. The low reflection intensity was 10.17 μW with a voltage of 69.89 mV, yielding an extinction ratio of 1.90. Fitting the Airy distribution to these results gave a reflectance of 0.01. This low value was caused by the thicker fiber-polymer interface, as the released devices show significantly higher reflectance. The corresponding Airy distribution is included in Figure 26.5c.

The single-cavity device confirmed that the 2PP method successfully produced optical elements for planar FP resonators. The released dual-cavity device represents a significant improvement in functionality over the single-cavity device because its first cavity is open to the environment. The dual-cavity device also improved the extinction ratio of the reflection spectrum, which is demonstrated in Figure 26.6b. The first cavity was 35 μm tall and filled with air, leading to a theoretical FSR of 34.32 nm. The polymer cavity was formed by a 56 μm diameter, 20 μm tall disk with a theoretical FSR of 39.94 nm. The device is pictured in Figure 26.6a. When measured, the air cavity had an average FSR of 36.24 nm and the polymer cavity had an average FSR of 36.07, agreeing within five nanometers of the theoretical values. The high reflected intensity of this device was 68.24 μW, corresponding to a voltage of 468.98 mV, and the low reflected intensity was 1.12 μW reading a voltage of 7.67 mV. This gives the device an extinction ratio of 61. Fitting the Airy distribution, which is included in Figure 26.6c, yielded a reflectance of 0.3.

Suspending the polymer structure over an air gap allows various interstitial media to be introduced into the first cavity. Optofluidic dies, quantum dot suspensions or other gain media could be inserted to create fiber-tip lasers. The RI of an unknown gas or liquid can also be determined by immersing the dual-cavity device and comparing the shifted FSR to a reference.

Furthermore, by including both a solid polymer cavity and an open cavity, an RI sensor with this device would be self-referencing and temperature immune. If a single open cavity sensor was exposed to both a change in temperature and interstitial medium, the FSR of the device would shift due to the new RI of the cavity, and the new cavity length introduced by thermal expansion of the polymer. It would be very difficult to decouple each effect from the observed FSR shift. The released dual-cavity device would be able to isolate a change in RI from the

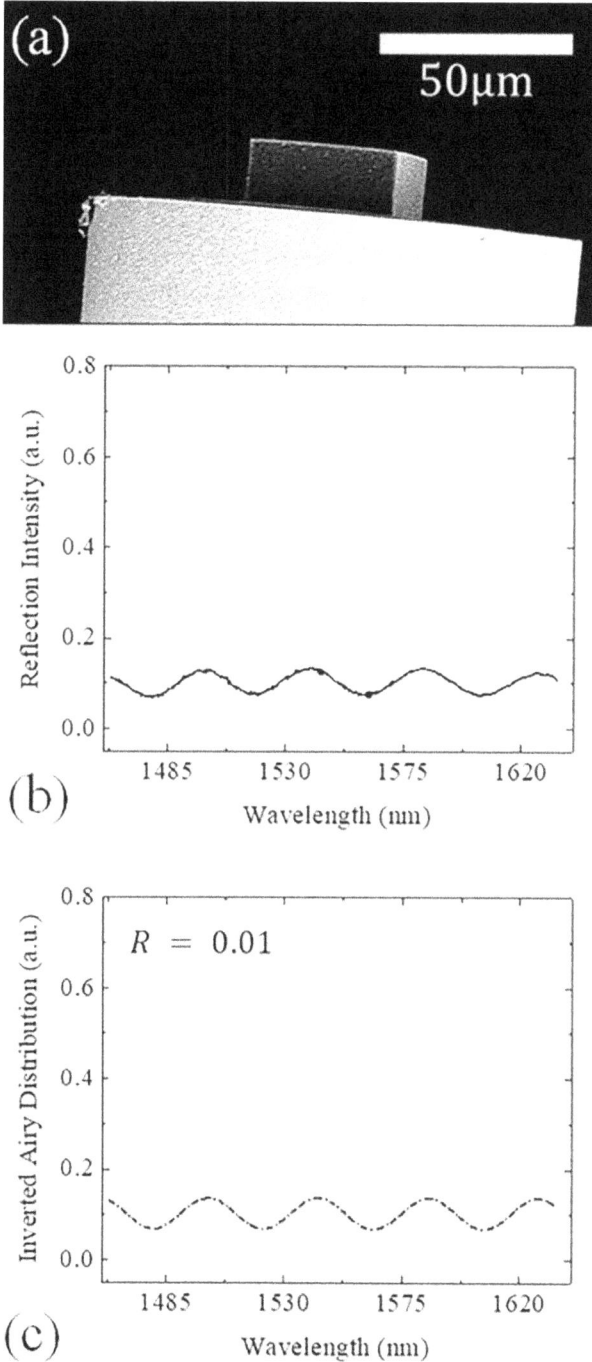

FIGURE 26.5 Measurement results from the single-cavity device. (a) SEM image of the device, a 40 µm by 40 µm and 17.58 µm tall rectangle. (b) Measured reflection intensity as a function of wavelength. (c) Airy distribution with $R = 0.01$.

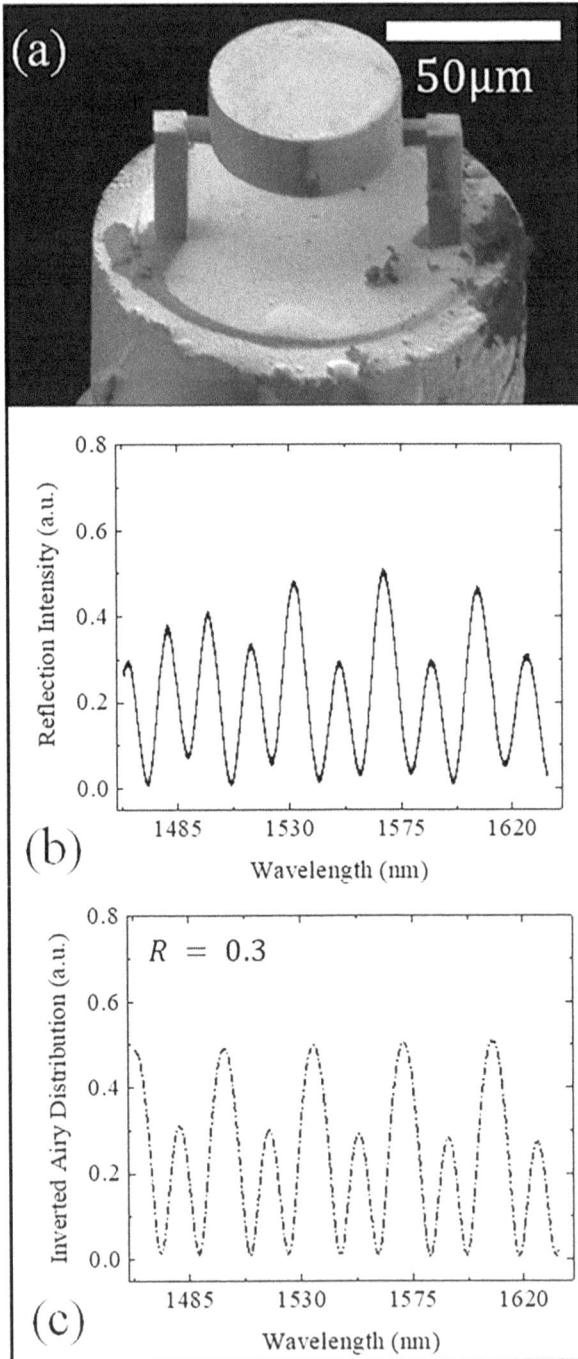

FIGURE 26.6 Measurement results from the flat dual-cavity device. (a) SEM image of the dual-cavity device, a 56 μm diameter, 20 μm tall disk suspended above a 35 μm air cavity. These cavity lengths match the hemispherical device. (b) Measured reflection intensity as a function of wavelength. (c) Airy distribution with $R = 0.3$.

effect of thermal expansion because the solid cavity would only experience the thermal effects. One could determine the thermal effects from the FSR shift in the polymer cavity, calculate the corresponding effects on the open cavity, and subtract them from the open cavity FSR shift to isolate the changes in the interstitial medium.

While the dual-cavity device enables many applications, there was a risk that the response from one cavity would interfere with the response from another. If the resonant wavelengths are too close and the width of the resonance is too large, different peaks could not be resolved. Furthermore, light reflected from one cavity could destructively interfere with light resonating in another, removing part of the signal. Fortunately, this kind of interference can be avoided by properly designing the constituent optical cavities. The thickness of all FP cavities and interstitial medium gaps were specifically designed to produce a good number of distinct optical resonances around 1,550 nm wavelength. The aforementioned equation of $\Delta\lambda_{FSR} = \lambda_0^2 2/2nl$ was used as a theoretical guideline to define the thickness of the polymer cavity and interstitial medium. Peaks from both cavities are clearly resolvable, as seen in Figure 26.6b, and the extinction ratio improved markedly over the single-cavity device. Since the signal quality improved with the addition of the second cavity, interference between the cavities does not seem to degrade the response.

The hemispherical device enjoys all the utility of the flat dual-cavity device while adding the many benefits of curved mirrors. The hemispherical mirrors reduced diffraction losses and improved lateral confinement to produce a more consistent peak transmission. Within the reflection dips, the flat dual-cavity device showed a variance of 0.53 mV, while the hemispherical counterpart achieved a variance of only 0.026 mV. The hemispherical FP cavity also had the largest extinction ratio observed, with a high reflection intensity of 78.162 µW reading 537.17 mV, and a low reflection intensity of 0.31 µW reading 2.12 mV. This gave the hemispherical device an extinction ratio of 253. An SEM image and the reflection spectrum of the hemispherical FP cavity are shown in Figure 26.7a–b, respectively. The distance between the center of the inner mirror and the face of the fiber was 35 µm, with a theoretical FSR of 34.32 nm. Within the polymer gap, the concave-convex resonator was 20 µm long between centers for a theoretical FSR of 39.94 nm. The FSR of the air cavity was measured to be 36.31 nm, and the FSR of the polymer cavity was measured to be 36.19 nm. The Airy distribution fit the measured response with a reflectance of 0.2. This value is lower than the flat cavity, although the curved mirror reduces losses. The drop is likely caused by the thin polymer feature fabricated over the surface of the fiber. While the feature does not function as intended, as a third curved mirror, the fiber-polymer interface it creates explains the loss in reflectance.

Like all curved-mirror FP cavities, hemispherical resonators are significantly less sensitive to misalignment, making the device more robust in the face of vibrations or impacts to the fiber. In addition, they can be used at higher incident intensities without the loss of resolution that occurs in planar FP cavities. The hemispherical device represents the power of our fabrication technique to utilize 3D freedom to create advantageous geometries that cannot otherwise be realized.

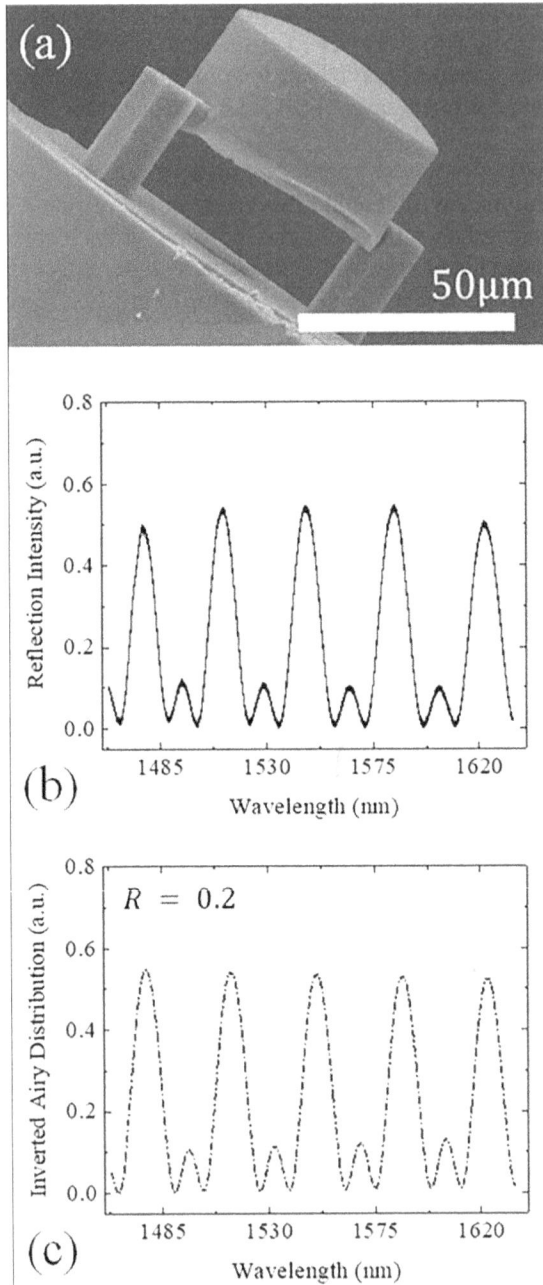

FIGURE 26.7 Measurement results from the hemispherical dual-cavity device. (a) SEM image of the device, a 56 μm diameter, 20 μm tall cylinder measured from each center of curvature. The top mirror has a 75 μm radius and the inner mirror has a 35 μm radius. The center of the inner mirror is suspended 22.5 μm above the face of the fiber. (b) Measured reflection intensity as a function of wavelength. (c) Airy distribution with $R = 0.2$.

26.4   FIBER TIP 3D REMOTE SENSORS

The successful outcome of this work will bring a new generation of 3D monolithically integrated multifunctional fiber tip remote sensors that have a plethora of applications. As a proof of concept, we characterized the FP resonant wavelength shifts of the single-cavity device (the device shown in Figure 26.5a) and the flat dual-cavity device (the device shown in Figure 26.6a) due to fluctuations in the surrounding temperature.

As the background temperature changes, the FP resonant wavelength shifts because the cavity undergoes a spectral shift due to the changes in refractive index (thermo-optic effect) and thermal expansion (thermo-elastic effect) of the sensing material. The changes in resonant wavelength at mode order m due to thermo-optic and thermo-elastic effects can be expressed as $\Delta \lambda_0 / \lambda_0 = \Delta n / n + \Delta l / l$.

The unreleased or single-cavity heat sensor was first characterized according to the measurement setup described in Figure 26.8a. The unreleased sensor exhibited a linear wavelength shift as a function of temperature with a sensitivity of ~46 pm/°C from 20°C to 60°C and ~5.5 pm/°C from 60°C to 120°C as shown in Figure 26.8b. The released or dual-cavity heat sensor was then characterized according to the same measurement setup. The released sensor demonstrated a linear wavelength shift as a function of temperature with a sensitivity of ~50 pm/°C from 20°C to 120°C as presented in Figure 26.8c.

The solidly-mounted sensor showed a linear response up to around 60°C as shown in Figure 26.8b. We have performed multiple measurements on the solidly-mounted sensor and obtained very similar results. Our measurements of the solidly-mounted sensors consistently indicated linear responses with slopes ranging from 45 to 50 pm/°C and from 5 to 6 pm/°C for temperature variations from 20°C to 60°C and 60°C to 120°C, respectively. We believe this is due to the anchoring scheme of the device. The solidly-mounted sensor was anchored to the fiber tip by the entire bottom of the 3D fabricated structure. This anchoring scheme has limited the linear expansion of the device due to an increase in temperature.

According to the data provided by Nanoscribe GmbH, the thermo-optic coefficient of the IP-DIP is approximately -2.6×10^{-4}/K at 1,550 nm. Using the experimental data and the thermo-optic coefficient of the IP-DIP given by Nanoscribe GmbH, the coefficient of thermal expansion of the IPDIP was calculated to be 2.9×10^{-4}/K.

26.5   CONCLUSION

In conclusion, we have demonstrated three FP resonator designs fabricated directly onto the cleaved ends of low-loss optical fibers. Our fabrication technique is simple, fast, and enables true 3D freedom to realize complex features, such as optical elements, which are difficult or impossible to create with traditional microfabrication methods. Two-photon polymerization with ultra-fast laser pulses creates devices on fiber tips in less than 15 min with a single writing step. Each device improved the quality of the FP cavity's reflection response. The single-cavity featured an extinction ratio of 1.90, the released dual planar cavity device obtained an extinction ratio of 61, and the hemispherical cavity device obtained an extinction ratio of 253.

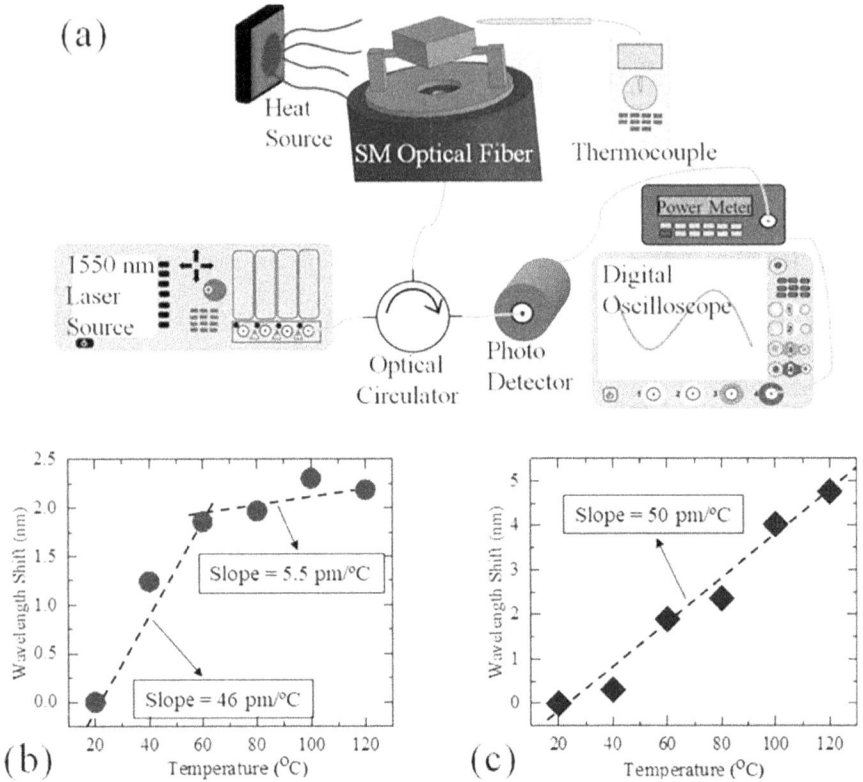

FIGURE 26.8 (a) Schematic of the measurement setup to characterize the thermal sensor on an optical fiber tip. (b) Wavelength shifts as responses to temperature variations demonstrated by the single-cavity sensor. (c) Wavelength shifts as responses to temperature variations demonstrated by the flat dual-cavity sensor.

The reflectance of the direct fiber-polymer interface was estimated to be 0.01, while the reflectance of IP-DIP was estimated to be between 0.2 and 0.3, both over 1,463 to 1,634 nm. The dual-cavity device promises increased utility as the open cavity is able to interact with its environment and reference changes in RI to the solid cavity. The hemispherical device brings the benefits of a curved-mirror FP resonator such as improved alignment insensitivity and constant resolution at increased intensity, while also providing a consistent resonant intensity across the spectrum of interest.

We have demonstrated the usability of the single-cavity and dual-cavity FP resonators as thermal radiation sensors. Although the single-cavity sensor has better structural robustness, the dual-cavity sensor has demonstrated a better extinction ratio, thermal sensitivity, and linearity compared to the single-cavity sensor.

These FP cavity devices invite numerous applications, such as on-fiber lasers and various sensors. Our future work specifically hopes to explore reflective coatings to improve the reflectance and quality factor of the reflection spectrum. Micron-scale plano and hemispherical cavities with have demonstrated quality factors as high as 10^5 by adding a reflective coating [2], and the next devices we are producing aim to achieve a reflectance of 0.9 or higher. The speed of our fabrication process enables us to take an iterative design approach and explore several reflective coating options. A more reflective FP cavity will have much higher resolution and produce measurable responses for small changes in the RI or length of the cavity to detect different phenomena such as electromagnetic radiations, acoustic waves, temperature, and pressure changes, displacements, and hazardous material concentrations in both gas and liquid form. Future work includes determining the suitability of the next generation devices to sense some of these phenomena.

ACKNOWLEDGMENTS

The views expressed in this paper are those of the authors and do not reflect the official policy or position of the United States Air Force, Department of Defense, or the US Government. We thank Abigail Juhl and Ecklin Crenshaw for training and assistance with the Nanoscribe, and Richard Johnston and Adam Fritzshe for assistance in the AFIT Nanofabrication and Characterization Facility. This work was funded by the Air Force Office of Scientific Research (AFOSR), Grant F4FGA08338J001.

REFERENCES

1. Wu X, Wang Y, Chen Q, Chen Y-C, Li X, Tong L and Fan X 2019 High-Q, low-mode-volume microsphere-integrated Fabry–Perot cavity for optofluidic lasing applications *Photonics Res.* 7 50–60.
2. Wu X, Chen Q, Wang Y, Tan X and Fan X 2019 Stable high-Q bouncing ball modes inside a Fabry-Pérot cavity *ACS Photon.* 6 2470–8.
3. Wang W, Zhou C, Zhang T, Chen J, Liu S and Fan X 2015 Optofluidic laser array based on stable high-Q Fabry–Pérot microcavities *Lab Chip* 15 3862–9.
4. Toninelli C, Delley Y, Stoferle T, Renn A, Gotzinger S and Sandaghdar V 2010 A scanning microcavity for in situ control of single-molecule emission *Appl. Phys. Lett.* 97 021107.
5. Hunger D, Deutsch C, Barbour R J, Warburton R J and Reichel J 2012 Laser microfabrication of concave, low-roughness features in silica *AIP Adv.* 2 012119.
6. Chen Y C, Chen Q, Zhang T, Wang W and Fan X 2017 Versatile tissue lasers based on high-Q Fabry–Pérot microcavities *Lab Chip* 17 538–48.
7. Duan X, Huang Y, Agarwal R and Lieber C M 2003 Single-nanowire electrically driven lasers *Nature* 421 241–4.
8. Cai H, Liu B, Zhang X M, Liu A Q, Tamil J, Bourouina T and Zhang Q X 2008 A micromachined tunable coupled-cavity laser for wide tuning range and high spectral purity *Opt. Express* 16 16670–9.
9. Masson J, St-Gelais R, Poulin A and Peter Y A 2010 Tunable fiber laser using a MEMS-based in plane Fabry-Pérot filter *IEEE J. Quantum Electron.* 46 1313–9.
10. Saadany B, Malak M, Kubota M, Marty F, Mita Y, Khalil D and Bourauina T 2006 Free-space tunable and drop optical filters using vertical Bragg mirrors on silicon *IEEE J. Quantum Electron.* 12 1480–8.
11. St-Gelais R, Masson J and Peter Y A 2009 All-silicon integrated Fabry–Pérot cavity for volume refractive index measurement in microfluidic systems *Appl. Phys. Lett.* 94 243905.
12. Thorpe M J, Balslev-Clausen D, Kirchner M S and Ye J 2008 Cavity-enhanced optical frequency comb spectroscopy: application to human breath analysis *Opt. Express* 16 2387–97.

13. Song W Z, Zhang X M, Liu A Q, Lim C S, Yap P H and Hosseini H M M 2006 Refractive index measurement of single living cells using on-chip Fabry-Pérot cavity *Appl. Phys. Lett.* 89 203901.

14. Pisani M and Zucco M 2009 Compact imaging spectrometer combining Fourier transform spectroscopy with a Fabry-Perot interferometer *Opt. Express* **17** 8319–31.

15. Snijders E, Frey J A, Norman J, Post V P, Gossard A C, Bowers J E, van Exter M P, Loffler W and Bouwmeester D 2018 Fiber-coupled cavity-QED source of identical single photons *Phys. Rev. Appl.* 9 031002.

16. Colombe Y, Steinmetz T, Dubois G, Linke F, Hunger D and Reichel J 2007 Strong atom–field coupling for Bose–Einstein condensates in an optical cavity on a chip *Nature* 450 272–6.

17. Garcia S, Ferri F, Ott K, Reichel J and Long R 2018 Dual-wavelength fiber Fabry-Perot cavities with engineered birefringence *Opt. Express* 26 22249–63.

18. Uphoff M, Brekenfeld M, Rempe G and Ritter S 2015 Frequency splitting of polarization eigenmodes in microscopic Fabry–Perot cavities *New J. Phys.* **17** 013053.

19. Wise S, Mueller G, Reitze D, Tanner D B and Whiting B F 2004 Linewidth-broadened Fabry–Perot cavities within future gravitational wave detectors *Class. Quantum Grav.* 21 S1031–S1036.

20. Zandi K, B´elanger J A and Peter Y A 2012 Design and demonstration of an in-plane silicon-on-insulator optical MEMS Fabry–Pérot-based accelerometer integrated with channel waveguides *J. Microelectromech. Syst.* 21 1464–70.

21. Yao M, Ouyang X, Wu J, Zhang A P, Tam H Y and Wai P K A 2018 Optical fiber-tip sensors based on in-situ μ-printed polymer suspended-microbeams *Sensors* 18 1825.

22. Kou J L, Feng J, Ye L, Xu F and Lu Y Q 2010 Miniaturized fiber taper reflective interferometer for high temperature measurement *Opt. Express* 18 14245–50.

23. Liu X, Iordachita I I, He X, Taylor R H and Kang J U 2012 Miniature fiber-optic force sensor based on low-coherence Fabry-Pérot interferometry for vitreoretinal microsurgery *Biomed. Opt. Express* 3 1062–76.

24. Quan M, Tian J and Yao Y 2015 Ultra-high sensitivity Fabry–Perot interferometer gas refractive index fiber sensor based on photonic crystal fiber and Vernier effect *Opt. Lett.* 40 4891–4.

25. Aubry G, Kou Q, Soto-Velasco J, Wang C, Meance S, He J J and Haghiri-Gosnet A M 2011 A multicolor microfluidic droplet dye laser with single mode emission *Appl. Phys. Lett.* 98 111111.

26. Gong C-Y, Gong Y, Zhang W-L, Wu Y, Rao Y-J, Peng G-D and Fan X 2008 Fiber optofluidic microlaser with lateral single mode emission *J. Sel. Topics Quantum Electron.* 24 0900206.

27. Smith J W, Suelzer J S, Usechak N G, Tondiglia V P and Chandrahalim H 2019 3-D thermal radiation sensors on optical fiber tips fabricated using ultrashort laser pulses (IEEE) pp 649–52.

28. Lee K-S, Kim R H, Yang D-Y and Park S H 2008 Advances in 3D nano/microfabrication using two-photon initiated polymerization *Prog. Polym. Sci.* 33 631.

29. Ismail N, Kores C C, Geskus D and Pollnau M 2016 Fabry-Pérot resonator: spectral line shapes, generic and related Airy distributions, linewidths, finesses, and performance at low or frequency-dependent reflectivity *Opt. Express* 24 16366.

Index

For Product Safety Concerns and Information please contact our EU representative GPSR@taylorandfrancis.com
Taylor & Francis Verlag GmbH, Kaufingerstraße 24, 80331 München, Germany

www.ingramcontent.com/pod-product-compliance
Lightning Source LLC
Chambersburg PA
CBHW060420220326
41598CB00021BA/2230